PLEASE STAMP DATE DUE, BOTH BELOW AND ON CARD

DATE DUE	DATE DUE	DATE DUE	DATE DUE
JUN 1 9 1987			
MAR 2 2 1993			

WITHDRAWN
CALTECH LIBRARY SERVICES

The Deep Proterozoic Crust in the North Atlantic Provinces

NATO ASI Series

Advanced Science Institutes Series

A series presenting the results of activities sponsored by the NATO Science Committee, which aims at the dissemination of advanced scientific and technological knowledge, with a view to strengthening links between scientific communities.

The series is published by an international board of publishers in conjunction with the NATO Scientific Affairs Division

A	Life Sciences	Plenum Publishing Corporation
B	Physics	London and New York
C	Mathematical and Physical Sciences	D. Reidel Publishing Company Dordrecht, Boston and Lancaster
D	Behavioural and Social Sciences	Martinus Nijhoff Publishers
E	Engineering and Materials Sciences	The Hague, Boston and Lancaster
F	Computer and Systems Sciences	Springer-Verlag
G	Ecological Sciences	Berlin, Heidelberg, New York and Tokyo

Series C: Mathematical and Physical Sciences Vol. 158

The Deep Proterozoic Crust in the North Atlantic Provinces

edited by

Alex C. Tobi

Earth Science Institute, State University,
Utrecht, The Netherlands

and

Jacques L. R. Touret

Earth Science Institute, The Free University,
Amsterdam, The Netherlands

D. Reidel Publishing Company

Dordrecht / Boston / Lancaster

Published in cooperation with NATO Scientific Affairs Division

Proceedings of the NATO Advanced Study Institute on
The Deep Proterozoic Crust in the North Atlantic Provinces
Moi, Norway
16-30 July, 1984

Library of Congress Cataloging in Publication Data
The Deep Proterozoic crust in the North Atlantic provinces.

(NATO ASI series. Series C, Mathematical and physical sciences; vol. 158)
"Proceedings of the NATO Advanced Study Institute on the Deep Proterozoic Crust in the North Atlantic Provinces, Moi, Norway, 16-30 July, 1984"—T.p. verso.
Includes index.
1. Geology, Stratigraphic—Pre-Cambrian—Congresses. 2. Geology—North Atlantic Ocean Region—Congresses. 3. Earth—Crust—Congresses. I. Tobi, Alex C., 1924- II. Touret, Jacques L. R., 1936- III. NATO Advanced Study Institute on the Deep Proterozoic Crust in the North Atlantic Provinces (1984: Moi, Norway) IV. Series.
QE653.D44   1985   551.7'15   85-14484
ISBN 90-277-2101-7

Published by D. Reidel Publishing Company
P.O. Box 17, 3300 AA Dordrecht, Holland

Sold and distributed in the U.S.A. and Canada
by Kluwer Academic Publishers,
190 Old Derby Street, Hingham, MA 02043, U.S.A.

In all other countries, sold and distributed
by Kluwer Academic Publishers Group,
P.O. Box 322, 3300 AH Dordrecht, Holland

D. Reidel Publishing Company is a member of the Kluwer Academic Publishers Group

All Rights Reserved
© 1985 by D. Reidel Publishing Company, Dordrecht, Holland.
No part of the material protected by this copyright notice may be reproduced or utilized in any form or by any means, electronic or mechanical, including photocopying, recording or by any information storage and retrieval system, without written permission from the copyright owner.

Printed in The Netherlands.

TABLE OF CONTENTS

Preface ix
Acknowledgements xii

THE DEEP PROTEROZOIC CRUST IN THE NORTH-ATLANTIC PROVINCES

1. Opening address 1
    Knut S. Heier (*)

2. Precambrian geodynamical constraints 3
    Nico J. Vlaar (*)

3. Seismic reflection results from Precambrian Crust 21
    Scott B. Smithson, William R. Pierson, Sharon L. Wilson and Roy A. Johnson

4. Proterozoic anorthosite massifs 39
    Ron F. Emslie (*)

5. Sm-Nd isotopic studies of Proterozoic anorthosites: systematics and implications 61
    Lewis D. Ashwal and Joseph L. Wooden

6. Temperature, pressure and metamorphic fluid regimes in the amphibolite facies to granulite facies transition zones 75
    Robert C. Newton (*)

7. Fluid enhanced mass transport in deep crust and its influence on element abundances and isotope systems 105
    William E. Glassley and David Bridgwater

8. C-O-H fluid calculations and granulite genesis 119
    William M. Lamb and John W. Valley

THE PROTEROZOIC PROVINCES IN NORTH AMERICA AND GREENLAND

9. Tectonic framework of the Grenville Province in Ontario and western Quebec, Canada 133
    Antony Davidson (*)

10. A 1650 Ma orogenic belt within the Grenville Province
    of northeastern Canada   151
       Andre Thomas, Gerald A.G. Nunn
       and Richard J. Wardle

11. A reassessment of the Grenvillian Orogeny in western
    Labrador   163
       Toby Rivers and Gerald A.G. Nunn

12. Geological evolution of the Adirondack mountains:
    A review   175
       John M. McLelland and Yngwar W. Isachsen (*)

13. Polymetamorphism in the Adirondacks: wollastonite at
    contacts of shallowly intruded anorthosite   217
       John W. Valley

14. Pb-isotopic studies of Proterozoic igneous rocks,
    West Greenland, with implications on the evolution
    of the Greenland shield   237
       Feiko Kalsbeek and Paul N. Taylor

15. Correlations between the Grenville Province and
    Sveconorwegian orogenic belt - implications for Pro-
    terozoic evolution of the southern margins of the
    Canadian and Baltic Shields   247
       Charles F. Gower

         THE PROTEROZOIC PROVINCES IN SOUTH SCANDINAVIA:
               GEOLOGICAL EVOLUTION AND STRUCTURE

16. The evolution of the South Norwegian Proterozoic as
    revealed by the major and mega-tectonics of the
    Kongsberg and Bamble sectors   259
       Ian C. Starmer

17. Tectonic environment and age relationships of the
    Telemark Supracrustals, southern Norway   291
       Tom S. Brewer and Dennis Field

18. Geotectonic evolution of southern Scandinavia in
    light of a late-Proterozoic plate-collision   309
       Torgeir Falkum (*)

19. The Mandal-Ustaoset line, a newly discovered major
    fault zone in South Norway   323
       Ellen M.O. Sigmond (*)

20. Terrane displacement and Sveconorvegian rotation of the Baltic shield: A working hypothesis ... 333
    Tore Torske

21. Proterozoic development of Bohuslän, south-western Sweden ... 345
    Lennart Samuelsson and Karl Inge Åhäll

22. Late Presveconorwegian magmatism in the Östfold-Marstrand Belt, Bohuslän, SW Sweden ... 359
    Karl Inge Åhäll and John S. Daly

23. The West Uusimaa complex, Finland: an early Proterozoic thermal dome ... 369
    Laszlo Westra and Jan Schreurs

THE PROTEROZOIC PROVINCES IN SOUTH SCANDINAVIA: GEOCHRONOLOGY AND GEOCHEMISTRY

24. Geochronological framework for the Late-Proterozoic evolution of the Baltic Shield in South Scandinavia ... 381
    Rob H. Verschure (*)

25. Isotope geochronology of the Proterozoic crustal segment of southern Norway: A review ... 411
    Daniel Demaiffe (*) and Jean Michot (*)

26. Neodymium isotope evidence for the age and origin of the Proterozoic of Telemark, South Norway ... 435
    Julian F. Menuge

27. The Rogaland anorthosites: facts and speculations ... 449
    Jean-Claude Duchesne (*), Robert Maquil and Daniel Demaiffe (*)

28. Metamorphic zoning in the high-grade Proterozoic of Rogaland-Vest Agder, SW Norway ... 477
    Alex C. Tobi (*), Gé A.E.M. Hermans, Cornelis Maijer and J. Ben H. Jansen (*)

29. Geothermometry and geobarometry in Rogaland and preliminary results from the Bamble area, South Norway ... 499
    J. Ben H. Jansen (*), Rob J.P. Blok, Ariejan Bos and Mies Scheelings

30. Fluid regime in southern Norway: The record of fluid inclusions ... 517
    Jacques L.R. Touret (*)

31. Geochemical constraints on the evolution of the
    Proterozoic continental crust in southern Norway
    (Telemark Sector)   551
    P. Craig Smalley and Dennis Field

32. Geochemical evolution of the 1.6 - 1.5 Ga-old
    amphibolite-granulite facies terrain, Bamble Sector,
    Norway: dispelling the myth of Grenvillian high-
    grade reworking   567
    *Dennis Field, P. Craig Smalley, R.C. Lamb
    and Arne Råheim*

33. A preliminary study of REE elements and fluid in-
    clusions in the Homme granite, Flekkefjord,
    South Norway   579
    *Torgeir Falkum (*), Jens Konnerup-Madsen and
    John Rose-Hansen*

SUBJECT INDEX   585

(*) Invited Speaker, NATO ASI.

PREFACE

The Proterozoic terrains of South Scandinavia and the Grenville Province in North-America have many common features : Regional high-grade metamorphism (Granulite-facies), anorthosites, etc. They are separated by the Caledonian orogeny and, above all, by the Atlantic Ocean. During the time of the great continental drift controversy, few people were ready to admit that both sides on the Atlantic were once an unique province. Now everybody agrees on the Post-paleozoic age of the Atlantic and, consequently, on the intrinsic homogeneity of the much older rocks which occur around it. But a detailed comparison is not easy. The Grenville Province is much larger than South Scandinavia, both regions have been investigated by a great number of different schools, using various methods, approaches and concepts. After several attempts, and long discussion, it was felt by a small group of individuals, that literature study would not be enough and that nothing could replace the direct contact, in the field, of specialists who had a first-hand knowledge of all involved regions. The formula of a NATO Advanced Study Institute, which gives a unique opportunity to meet and mix people of various origin and levels, came almost by itself.

Much work was needed, much help has been obtained, as detailed in the "Acknowledgements". We present here the results of the lectures given during the Institute, not only by invited speakers, but also by participants who wanted to contribute and whose presentation was retained by the organizing committee. We were extremely surprised by the number and quality of these contributions and by the desire of all authors to cope with the rather complicated rules of manuscript presentation.

This enforces us in the idea that the scope of the Institute was well founded and that it did correspond to a real demand from the geological community.

For the beauty and exceptional character of many minerals and rocks,

South Scandinavia - and especially Southern Norway - has played a key role in the development of earth sciences : Kongsberg native silvers were king presents at the time of Enlightment and names like T. Scheerer, W.C. Brögger, A. Lacroix (to cite a few among a long list) have given an international reputation to the region. Before 1960, many minerals and rocks were first identified and named from localities in S. Norway, e.g., norite by Esmark (1838). Famous were, among others, C.F. Kolderup, later followed by P. Michot, for their study of the anorthosites of SW Norway, and J.A.W. Bugge for his studies on the Arendal region.

In 1960, the Baltic countries presented a synthesis of their knowledge in reports, geological maps and excursion guides on the occasion of the 21st International Congress in Copenhagen. Among the scientists involved T.F.W. Barth, one of the first to write a book on "Theoretical Petrology", deserves special mention. He contributed in many fields and, as far as we know, was the first to organize a Nato Advanced Study Institute in the earth sciences (Oslo 1962, first meeting on feldspars).

Under his guidance, the Mineralogical-Geological Museum in Oslo became during the sixties an unique meeting place for young geologists from all over the world; several participants to this meeting, notably two organizers (T. Falkum and J. Touret), were his direct students and he suggested several of the research topics reported in the present volume.

Soon after 1960 tools became available to start four-dimensional modelling of the evolution during the Precambrian and to interpret the significant resemblances which had been known for a long time between the Grenville and South Scandinavia : refinement of geochronology, geothermometry and geobarometry, fractionation sequences of magmatic rocks, distribution of trace elements, and last but not least plate tectonics. These then are among the topics extensively treated in this volume. The emphasis is on South Scandinavia : the assembled reports give a fair impression of the work now being done. During the Institute the lectures were supplemented by a 10-day excursion, during which much of the discussions took place. After its publication in the series of the Norwegian Geological Survey (NGU) the excursion guide may serve as a companion to this volume; in a preliminary form it was presented to the participants of the Institute.

Papers with a more general bearing, and from other regions, notably from the Grenville Province, contributed significantly to the aim of comparing the various "North-Atlantic" terrains.

The editors modestly state that the moment to organize this Institute appears to be well-chosen. The subject fits in the Lithosphere Program which has just started, and the excursion section almost coincides with one of the planned E-W traverses of the European Geotraverse. The new edition of the geological map 1:1.000.000 of Norway and the first edition of the Mandal sheet 1:250.000 have just been published, and the Arendal sheet will follow soon.

We think most participants enjoyed the contacts the Institute provided. Of course the comparison could not be completed. Only few authors adapted their text after the discussions, so many conclusions remain for the readers to draw.

There are a few obvious points of disagreement (e.g. the age of metamorphic events in South-Eastern Norway). Each author has been completely free to express his own opinion, under his own responsibility, and we feel that the variety of interpretation reflects the never ending interest of the region : much remains to be done and the present volume is less a definite conclusion than an introduction for further work. Some participants have commented that the much greater size of the Grenville Province (including Labrador) often called for a different approach. They will have much to show us! It was tentatively suggested during the Institute that a second meeting should be planned in North America in about four years from now. Our understanding of Proterozoic evolution rapidly increases : no doubt a more coherent picture of the North-Atlantic provinces will result next time !

The Editors

ACKNOWLEDGEMENTS

First, the editors wish to express their gratitude to the other members of the organizing committee, whitout whose help this Institute would never have taken place. Members were:

        Jacques Touret, Amsterdam (director, editor)

        Alex C. Tobi, Utrecht (treasurer, editor)

        Cornelis (Cees) Maijer, Utrecht (editor excursion guide)

        Rob H. Verschure, Amsterdam (secretary)

        Ellen Sigmond, Trondheim (representative NGU)

        R. Keith O'Nions, Cambridge

        Jean-Clair Duchesne, Liège

        Torgeir Falkum, Aarhus

        Jacques Martignole, Montréal

The acknowledgements to follow are also made on their behalf. With the aim of the Institute in mind, we are grateful for the response the aim of the Institute in mind, we are grateful for the response obtained from the USA and Canada. In general, we thank all participants for their vived interest and their significant contributions.

Thanks are due to the NATO Scientific Affairs Division for allotting two grants. The first was obtained by H.R. Wynne-Edwards to explore the comparison of the North-Atlantic provinces. J. Touret and T. Torske benefited from this grant by joining him in the field. The second is that which led to this Institute. We have greatly appreciated the smooth and efficient engineering of the granting procedure.

Much additional research was needed to complete the excursion program and to fill gaps in the lecture and poster data.

Besides constant help and support from Dutch Universities (Rijksuniversiteit Utrecht and Vrije Universiteit Amsterdam) and from A.W.O.N. (Earth Sciences Research in the Netherlands, a subdivision of the Netherlands Organization for the Advancement of Pure

ACKNOWLEDGEMENTS

Research (ZWO)), these activities were made possible by substantial grants from the Dr. H.M.E. Schürmann Foundation, which are here gratefully acknowledged.

N.G.U. kindly paid for one of the touring cars, was helpful in getting permission to enter private grounds, and provided geological maps for each of the participants. Torgeir Falkum and Ellen Sigmond kept the cost of lodging at a reasonable level, so that all participants can receive a free copy of this volume. We thank the municipality of Arendal for a memorable reception and the Lundheim folkehøgskole in Moi, the Phønix Hotell in Arendal and the Farsund Fjell Hotell for their splendid hospitality. We thank Thor for remaining at sufficient distance to allow for a sun-bathed excursion most of the time, and the bus drivers of Lillesand & Topdalens Bilruter A.S. Kristiansand for their punctual and friendly service. The student-assistants are thanked for their indispensable services (including the 'dispensary' of beer!). Our secretary, Mrs. D. Huisman-Erkens is thanked for many typing and administrative jobs, and the drawing and photography departments of the Rijksuniversiteit, Utrecht for other contributions to prepare the manuscripts of excursion guide and proceedings. All these required occasionally high tension to meet the deadlines.

Finally we thank publisher Reidel for his finishing touch: the last deadline is his!

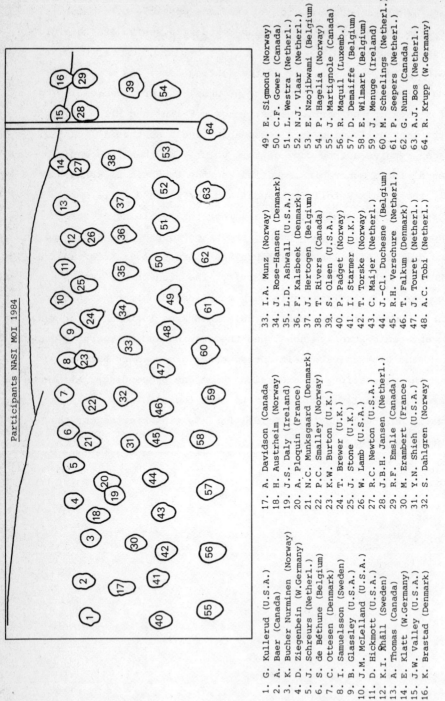

Participants NASI MOI 1984

1. G. Kullerud (U.S.A.)
2. A. Baer (Canada)
3. K. Bucher Nurminen (Norway)
4. D. Ziegenbein (W.Germany)
5. J. Schreurs (Netherl.)
6. S. de Béthune (Belgium)
7. C. Ottesen (Denmark)
8. I. Samuelsson (Sweden)
9. B. Glassley (U.S.A.)
10. J.M. McLelland (U.S.A.)
11. D. Hickmott (U.S.A.)
12. K.I. Åhäll (Sweden)
13. A. Thomas (Canada)
14. E. Klatt (W.Germany)
15. J.W. Valley (U.S.A.)
16. K. Brastad (Denmark)

17. A. Davidson (Canada)
18. H. Austrheim (Norway)
19. J.S. Daly (Ireland)
20. A. Ploquin (France)
21. N.C. Munksgaard (Denmark)
22. P.C. Smalley (Norway)
23. K.W. Burton (U.K.)
24. T. Brewer (U.K.)
25. J. Stone (U.K.)
26. W. Lamb (U.S.A.)
27. R.C. Newton (U.S.A.)
28. J.B.H. Jansen (Netherl.)
29. R.F. Emslie (Canada)
30. M. Erambert (France)
31. Y.N. Shieh (U.S.A.)
32. S. Dahlgren (Norway)

33. I.A. Munz (Norway)
34. J. Rose-Hansen (Denmark)
35. L.D. Ashwall (U.S.A.)
36. F. Kalsbeek (Denmark)
37. J. Hertogen (Belgium)
38. T. Rivers (Canada)
39. S. Olsen (U.S.A.)
40. P. Padget (Norway)
41. I. Starmer (U.K.)
42. T. Torske (Norway)
43. C. Maijer (Netherl.)
44. J.-Cl. Duchesne (Belgium)
45. R.H. Verschure (Netherl.)
46. T. Falkum (Denmark)
47. J. Touret (Netherl.)
48. A.C. Tobi (Netherl.)

49. E. Sigmond (Norway)
50. C.F. Gower (Canada)
51. L. Westra (Netherl.)
52. N.J. Vlaar (Netherl.)
53. E. Nzojibwami (Belgium)
54. P. Hagelia (Norway)
55. J. Martignole (Canada)
56. P. Maquil (Luxemb.)
57. D. Demaiffe (Belgium)
58. E. Wilmart (Belgium)
59. J. Menuge (Ireland)
60. M. Scheelings (Netherl.)
61. P. Seepers (Netherl.)
62. G. Nunn (Canada)
63. A.J. Bos (Netherl.)
64. R. Krupp (W.Germany)

not on photograph: T. Clifford (S.Africa), G. Godard (France), M. Ryan (U.K.) and R. Wilson (Denmark)

THE DEEP PROTEROZOIC CRUST IN THE NORTH ATLANTIC PROVINCES: OPENING ADDRESS

K.S. Heier
Director, Geological Survey of Norway
P.O. Box 3006
7001 Trondheim
Norway

The Proterozoic period dates from the end of the Archean at about 2500 Ma to the Cambrian at 570 Ma. Thus it covers about 2 Ga of Earth history and is the longest geological period with a record on Earth. Only the Archean covers a similar timespan but as yet we have no record of surface rocks older than 4.2 Ga.

In a recent article in Terra Cognita (1984, 4, 149-150) on Major divisions of Earth history, Preston Cloud argues that Earth history is punctuated by four main events, separating four major historical intervals. The events are (1) the Earth's accretion, accompanied or followed by segregation of the core and mantle, (2) the initial growth of continental crust, (3) the general onset of epicratonal sedimentation, and (4) the wide appearance of animal life with feedback to the sedimentary record. The third of these events marks the beginning of the Proterozoic, an interval of prevailingly epicratonal sedimentation and potassic granitization leading up to the Phanerozoic, characterized by the presence of manifest animal life.

The Proterozoic marks the time when earth evolution and geological processes affecting the crust became similar to those at present and the Huttonian theorem "the present is the key to the past" may be applied. On a global scale the Archean - Proterozoic boundary is a very productive period as regards ore formation as is also the period from the beginning of the Paleozoic to the present. In contrast the whole of the Proterozoic appears to have been singularly unproductive though notable exceptions occur. In Norway for instance, the large ilmenite deposit at Tellnes dates from the upper Proterozoic.

Norway offers an almost ideal opportunity for a study of the Proterozoic period. The Archean - Proterozoic boundary can be studied in Lofoten and in Finnmark, and gradual transitions into the Cambrian are exhibited by the Eocambrian sediments in Finnmark and central Norway.

The Archean basement is not present in the Precambrian of South Norway. The oldest basement rocks in the areas covered by our excursions are about 1.7 Ga old. Most of the crust forming processes have taken place in the 1250 to 1000 Ma interval. In Fennoscandia this has been termed the Sveconorwegian period. Internationally, it is better known as the Grenville period after the Grenville province in Canada. The recognition of the parallelism of the Sveconorwegian and Grenville periods results in the hypothesis that they were also closely linked in space at that time. This again has led to interesting models for continental splitting and movements after Grenville time. The idea that the Baltic shield and the Canadian shield formed one continent in the pre-Grenville period makes geological comparisons between the two continents particularly relevant.

The Precambrian rocks of southern Norway can be divided into four provinces:

(1) the volcanic and sedimentary rocks of the Telemark suite;
(2) the gneiss-migmatite-granite complex;
(3) the Kongsberg-Bamble area which is separated from the others by a major fault; and
(4) the Egersund anorthosite - mangerite complex.

In a paper to be presented at this meeting Ellen Sigmond demonstrates that the gneiss-migmatite-granite complex is divided into two blocks by a large tectonic line which she terms the Mandal-Ustaoset Fault Zone. This tectonic line may represent a continent-continent collision suture, a collision which in that case must have taken place more than 1200 Ma ago.

The rocks of this area have been the subject of study by a truly international group for several decades. The Belgians and Dutch have studied the Egersund complex and surrounding migmatites in particular; the Danes have completed the 1:250 000 scale Mandal sheet; the British and French have concentrated on the Bamble region between Arendal and Kragerø. The Norwegians have been scattered all over the region with major efforts within the volcanic and sedimentary rocks of the Telemark suite and the Iveland-Evje pegmatite province.

This is the first time students from these groups have been brought together for a meeting specially devoted to the geological problems of the area. I wish you all have an interesting meeting which will not, I believe, mark the end of this period of international participation in scientific endeavours in this region. Rather I expect that the opinions which will be brought forward and the discussions they will provoke will stimulate you to attack the problems with renewed zeal and insight. Only in this way will the meeting prove truly successful.

PRECAMBRIAN GEODYNAMICAL CONSTRAINTS

N.J. Vlaar
University of Utrecht
Dept. of Theoretical Geophysics
P.O. Box 80.021
3408 TA Utrecht
The Netherlands

ABSTRACT. Geotectonics during the earth's history is governed by the temperature of the upper mantle. This temperature determines melting and segregation of the basaltic component from solid undepleted peridotite. Temperature also determines the basalt-eclogite transition. Assuming that Archaean mantle temperature was above the peridotite sosidus at some time, the basalt fraction of the peridotitic upper mantle, due to its low density, had to accumulate in the shallow mantle. This situation governs Archaean geology and can explain the existence of both the greenstone belts and the high-grade terrains of that era. Plate tectonics became operative in the early Proterozoic. The plate tectonic style differed from the present one because of more buoyant oceanic lithosphere. This had a dominant influence on Proterozoic magmatism and tectonics.

INTRODUCTION

Though much of Precambrian geology may be explained by a uniformitarian approach, many problems remain to be solved as no modern equivalents, in terms of which the Archaean and the Proterozoic can be tackled, appear to be at hand.

An important constraint is that no evidence has been preserved for the existence of Precambrian oceans. Modern plate tectonics is based on the study of the oceanic crust and upper mantle. The consequences of plate tectonics for Phanerozoic intracontinental tectonics and magmatism and also for mountain building, are poorly understood. Moreover, the extrapolation of present-day knowledge to the Precambrian has not resulted in a thorough understanding. Studies of the Precambrian are mostly restricted to geological and petrological studies of the continental masses, lacking knowledge of concurrent tectonic involvement of the oceanic realm. Moreover, it appears to lead to conflicting ideas when the subject matter is approached departing from hypotheses of global tectonic mechanisms during the earth's history. Usually, this matter is treated in the very general and abstract setting of global convection theories. It should be noted that these

theories have not yet resulted even in an understanding of the
Phanerozoic build up of the continents. The main questions to be
answered require an adaptation of current knowledge to circumstances
which appear to be rather different from modern ones.

Some of the current problems related to Precambrian geology
concern:
(1) Archaean high-grade terrains and greenstone belts. This problem
leads to an apparent paradox: The pressure-temperature regime of high-
grade terrains as studied by their mineral assemblages give
evidence of a relatively cool geothermal regime when taken into account
the high radio-active heat production at that time. The low-grade
greenstone belts, by the presence of high melting temperature komatiites,
on the contrary, indicate higher mantle temperature.
(2) The growth of the continental volume by subcrustal accretion.
By these processes a very thick continental crust must have been
generated during the Archaean and Proterozoic. In the Archaean the
character of the material added from below is not well understood,
though remobilisation and granitisation of the lower crust appear
to be indicated. In the Proterozoic the massive addition of anorthosite
leading to considerable crustal thickening is subject to debate.

Most studies of the Precambrian are, due to its very nature,
confined to the continental crust. Relatively few efforts have been
made to put the continental data in a global perspective. The latter
mostly boils down to the question whether plate tectonics was operative
or not. Usually, conclusions are drawn from the geothermal state as
deduced from Archaean high-grade terrains. Bickle (1978), departing
from the assumption of a stationary geothermal state of Archaean
continental crust, gives arguments for a high heat flow from the mantle
underneath. He concludes that such higher heatflow requires plate
tectonics as a mechanism for the heat loss of the earth. England (1979),
favouring a lower continental geothermal gradient, comparable to a
modern one, concludes that the larger heat loss from the earth should
have taken place by a faster plate tectonic process in the oceanic
realm. This author, however, also considers non-stationary geothermal
states. Wells (1980) presents transient geotherms due to magmatic
under- and over-accretion and subsequent crustal thickening. However,
in order to comply with temperatures from mineral assemblages in the
high-grade terrains, he has to invoke magma temperatures which are very
low compared to those derived from komatiites. Jarvis and Campbell
(1983) dwell on the paradox of the high komatiite temperatures and
the low geothermal gradients in Archaean high-grade crust. On the basis
of convection-theoretical arguments they arrive at a model of mantle
convection in which low mantle temperatures are compatible with a lower
boundary layer from which komatiites could be derived.

In general, not much attention has been given to processes taking
place in the oceanic realm. This may be partly due to the circumstance
that no ocean lithosphere older than 200 Ma has survived and partly
because present-day plate tectonic theory does not give clear answers
for Precambrian and modern geology. Baer (1981) gives arguments that
subduction could not start before some 1 Ga ago. However, he based his
arguments on stationary geotherms, and did not take into account the

implications of subduction of young and stably stratified oceanic lithosphere, as is done in the present paper. Moreover, he is only concerned with the late Proterozoic tectonic style.

In some more recent papers, Arndt (1983), and Nisbet and Fowler (1983), address the oceanic buoyancy problem in the Archaean. They require the oceanic crust to have a dominantly komatiitic composition which would enhance plate tectonic motions. However, these authors did not take into account that plate tectonics requires stable stratification of the oceanic lithosphere. Vlaar (1975), and Oxburgh and Parmentier (1977), addressed the problem of stability of the oceanic lithosphere. Vlaar and Wortel (1976) demonstrated that subduction behaviour is dependent on the stability of the oceanic lithosphere and hence on its compositional stratification and age. Young and buoyant oceanic lithosphere, after subduction will reside in the upper part of the upper mantle. Very young lithosphere and oceanic ridge systems will be subducted subhorizontally and introduce thermal and density anomalies below the upper lithospheric plate. Vlaar (1983) presented some magmatic and tectonic consequences of this mechanism, which was termed "lithospheric doubling".

Cloetingh et al. (1983) gave evidence that young oceanic lithosphere is more liable to subduction than old lithosphere, because of its small strength. The effect of increasing age on strength surpasses the effect of negative buoyancy. Older lithosphere, therefore, is more resistant to subduction than younger one, particularly when it is loaded with sediments and is subject to a compressive regime. Therefore, the generation of fold belts is more likely to take place upon the closing of small and young sedimentary oceanic basins than mature oceans. Vlaar and Cloetingh (1984) presented a scenario for the Alpine orogeny, taking into account the consequences of age dependent subduction.

Above concepts have been further developed in the present paper and applied to Precambrian magmatism and tectonics.

KOMATIITES

Komatiites are ultramafic lavas found in the lower part of a sequence of sediments, volcanoclastics, and mafic and ultramafic rocks, constituting the low-grade greenstone belts of the Archaean (Nisbet, 1982). The mafics and ultramafics are extrusions of basalt, basaltic komatiites and komatiites. The komatiites have an extrusion temperature of at least 1650°C (Green et al., 1975), and are derived from a dry MgO-rich and hot magma. The extrusion temperature is indicative of high upper mantle temperatures in the Archaean. Green (1975) considers komatiite to be derived from single stage high degree partial melting of a dry peridotite (pyrolite) source at depths greater than 200 km. The komatiite magma is assumed to result from melting of a rising peridotite diapir along an adiabatic ascent path, which sets the temperature at 200 km depth at 1850°C. However, Green (1975) did not take into account that the latent heat of melting above the solidus should add another 100°C to the temperature at 200 km depth. Nisbet et al. (1979) require an even deeper mantle source for komatiite liquid to be formed from a dry peridotite. If latent heat of melting is taken into account, the intersection of the adiabatic P-T path of the rising peridotite solid with the dry peridotite solidus shifts to depths of some 400 km. However, in above schemes the amount of komatiite generated would be far in excess of the amounts actually found in greenstone belts.

Sleep and Windley (1982) demonstrate that a small volume of komatiite liquid could be generated at shallower depths from a rising dry peridotite diapir by fractional melting of the depleted peridotite residue after extraction of the basaltic component. They consider a shift of $100°C$ in temperature due to the latent heat of melting at approximately 30% of fractional melting of the original dry peridotite. Taking an extrusion temperature of $1650°C$, their scheme gives a dry peridotite source temperature of $1800°C$ at 150 km depth. This situation would give rise to a small fraction of komatiite melt generated from the depleted peridotite above the eutectic. In view of the high density of komatiite liquid (> 3 g cm^{-3}, (Nisbet, 1982)), it does not appear to be probable that hereby komatiite magma would extrude at the earth's surface at the expense of basaltic magma. Therefore, a model in which basalts and komatiites in greenstone belts originate from separate sources, which in turn may be derived from earlier melting of dry peridotite, might be preferable. A prime candidate suitable for this mechanism is a subducted oceanic lithosphere, which, when created at an oceanic spreading center, is differentiated into a basaltic and a depleted peridotite component.

Weaver and Tarney (1979) put forward a number of arguments in favour of a depleted peridotite source for komatiite, a.o. the low trace element abundances for the latter, indicating a depleted source. These authors place this source at greater depths in a steeply dipping, subducted oceanic lithosphere, which, as we assume, must be inherently cool. Therefore, basalt being transformed into eclogite under those P-T conditions, it is difficult to envisage simultaneous melting of komatiite and more abundant melting of eclogite. We shall require a hotter and differentiated source.

In the following, we shall depart from an average Archaean upper mantle temperature well above the basalt-eclogite transition. Hereby above arguments do not appear to be valid.

THE ARCHAEAN OCEANIC UPPER MANTLE AND CRUST

For modern plate tectonics to be operative, at least two requirements have to be satisfied. Firstly, the upper lithospheric layers, for part of their life time, should be gravitationally stably stratified with respect to underlying mantle. Secondly, the upper layers should have sufficient strength to guide the stress field caused by driving forces, such as ridge push and slab pull.

We can represent the cooling history of the oceanic lithosphere after spreading from an oceanic spreading ridge by a simple cooling half-space model (e.g. Turcotte and Schubert, 1982). Cooling from the surface of a half-space with uniform initial temperature T_m is given

by
$$T(z, t) = T_m \, \text{erf}(z/(2kt)^{\frac{1}{2}}) \qquad (1)$$

where k is thermal diffusivity, z the depth, and t lithospheric age. The parameters determining the thickness of the strong upper layer are the initial temperature T_m and the age t of the spreading lithosphere. The lower boundary of the mechanically strong part is determined by the depth of a specific isotherm. For a modern spreading ridge an initial temperature T_m = 1200°C often is taken. Komatiite melting temperatures indicate that T_m = 1650°C in the Archaean. The isotherm determining the lower boundary of the strong part of the lithosphere thus is at a depth of about 3/4 of modern values. This implies, all other factors being assumed to be equal, that 30 Ma old lithosphere in the Archaean could be assigned a thickness of some 40 km. This thickness would be sufficient for plate tectonic behaviour of Archaean oceanic lithosphere, permitting a large compressive or tensional state, not unlike that of modern plates.

The cooling near the surface favours increasing negative buoyancy which for modern lithosphere is counteracted by compositional buoyancy of the near surface layering.

Oxburgh and Parmentier (1977) express the stability of the system by means of a 'density defect thickness':

$$\delta = \int \frac{\rho_m - \rho}{\rho_m} \, dz$$

where ρ_m is the undepleted mantle density, and ρ the density of the segregated upper layers. Integration extends over the range where $\rho_m - \rho \neq 0$. δ can be split into a thermal and compositional part: $\delta_{th} + \delta_{comp}$. For the cooling halfspace, represented by (1), the negative buoyancy due to cooling of the upper layers is given by:

$$\delta_{th} = -\frac{2}{\sqrt{\pi}} \alpha \, T_m \sqrt{kt} \qquad (2)$$

where α is the coefficient of thermal expansion.

The transition from stable to instable stratification takes place when

$$\delta_{comp} + \delta_{th} = 0$$

this is, when the age of the lithosphere is

$$t = \frac{(\delta_{comp})^2 \pi}{4 k \, \alpha^2 T_m^2} \qquad (3)$$

For modern oceanic lithosphere Oxburgh and Parmentier (1977) derive t = 30 Ma with an initial T_m = 1200°C. Whether plate tectonic mechanisms comparable to the present were operative in the Archaean thus depends crucially on the stability of the crust-upper mantle layering.

At present, oceanic crust and lithosphere are generated at a spreading ridge. They are assumed to be derived from a dry-peridotite upper mantle. Underneath the ridge, a rising diapir of undepleted peridotite, fractionates into a melted basaltic fraction and a solid depleted peridotite residue. The melt accumulates in a shallow (< 10 km?) magma chamber, where the future basaltic crust is generated, leaving olivine cumulates at the bottom. By continuing spreading at the ridge, this process of fractionation is maintained and the oceanic crust is created. The generation of a 6 km basaltic crust by some 30% melting of lherzolite leaves a depleted peridotite layer underneath of some 14 km thick, which in turn rests on the undepleted mantle. The initial stable layering of this system becomes unstable, due to cooling from the surface, at about 30 Ma of lithospheric age. At the central graben of the ridge, because of hydrothermal action, the upper basaltic layer is cooled rapidly. The system thus appears to be stably stratified immediately following its creation. Sleep and Windley (1982) assume large scale (30%) melting to have taken place at shallow depth (< 40 km), though small scale melting below this depth might have been possible. Applying analogous scenarios to the Archaean mantle turns out not to be feasible.

The assumption that Archaean upper mantle with average undepleted peridotite composition was differentiated from the lower mantle, leads to large scale internal layering of the upper mantle. The basalt, being produced by partial melting of upper mantle peridotite, due to high temperatures in excess of some $1800°C$ is bound to accumulate near the earth's surface. The high temperature prohibits basalt to be transformed into eclogite and to be recycled in the upper mantle again. Following the single requirement of an upper mantle of average lherzolite constitution, we infer an Archaean stratified upper mantle of a thick (50 - 100 km?) predominantly basaltic upper layer on top of a three times thicker depleted peridote layer. The basalt could only be recycled again into the upper mantle, after the earth had cooled down sufficiently for basalt to be converted to eclogite at shallower depth. The sinking eclogite, reacting with the depleted peridotite could then give undepleted peridotite again, the latter to become the source of new modern lithosphere.

A form of plate tectonics involving a thin basaltic crust would be self-prohibitive in the Archaean because of accumulation of a thick layer of melted basalt growing from the earth's surface downward. Hence, departing from a roughly lherzolitic composition of the already segregated upper mantle, high temperatures in the Archaean would induce basalt to concentrate in the upper part of the upper mantle, leaving an even thicker depleted peridotite layer underneath. The thick upper basaltic layer, due to cooling at the earth's surface, must suffer internal instability and, therefore, prohibits a plate tectonic situation. Though some internal gradual layering is feasible, the discrete layering of the modern oceanic lithosphere, involving a thin buoyant layer of basalt resting on top of a likewise thin depleted peridotite layer, must be ruled out.

When assuming a temperature of 1800°C at a depth of some 100 km, the rising of lherzolite above the dry peridotite solidus, according to Sleep and Windley (1982) should have resulted in small scale fractional melting of depleted peridotite into komatiite. High density komatiite magma then should accumulate at the base of the thick basaltic layer. The internal layering of the upper basaltic layer would grade downward from basaltic to basalt komatiitic to komatiitic with increasing density.

The thermal and dynamical state of the Archaean upper mantle can be envisaged as follows:

(1) At some stage of its evolution, the upper layer, though being slightly and continuously layered with respect to density and composition, could be internally instable due to the large temperature gradient between bottom and top. The solid layers below could be expected to be either stably layered or to be subject to slow convection. In both cases, it may be assumed that the temperature at the top of the lower layers is kept at a constant temperature of some 1800°C, the peridotite solidus temperature at about 100 km depth.

(2) At this hypothetical stage, the upper layer is to be considered to be melted, except for a thin surface layer which solidifies by cooling from the surface downwards. However, the near surface compositional stratification being nearly absent, the 'density defect thickness' being negative, this surface layer will be unstable shortly after its creation and will sink into the mantle. At the initial stage of sinking, by partial melting of the basalt at shallow depth, a light tonalite fraction may be segregated and remain floating on top to form 'tonalite islands' on the dominantly fluid basaltic upper part of the upper mantle (Condie, 1984). Accretion of these islands will result in the accretion of an Archaean 'proto-continent'. The solidified basalt on its way down, at this stage, will cross the basalt liquidus at shallow depth and, therefore, remelt again.

(3) The dominantly fluid state of the upper basaltic layer promotes rapid cooling due to internal convective motion. This results in 'tonalite islands' and progresses toward a more stable stratification until fluid motions stagnate. The solid upper mantle below, however, being also cooled from the top downward, at this stage, will start to be heated again. The bottom of the upper layer will be heated to 1800°C, whereby instable stratification sets in again.

(4) By the foregoing scenario, the first material to be involved in acquiring a high temperature is the komatiite residue at the bottom and next the more shallow komatiite basalt layers and last the material of basaltic composition.

We envisage a cyclic process of phases of stagnation and episodic upwelling of the lower strata involving komatiite, basalt komatiite and basalt. The beginning of such a cycle is marked by the extrusion of high temperature komatiite at the surface, to be followed by less dense ultramafic and mafic liquids. The rising magma blobs may surface in the oceanic realm and be recycled back into the mantle, or impinge at the base of the crust of a 'tonalitic' protocontinent.

(5) The latter then should give rise to anatexis of the lower crust leading to granite diapirism, crustal thinning and stretching, and possibly incipient rifting of the protocontinent. Subsidence of the surface due to stretching (McKenzie et al., 1980) would lead to basin formation typical for greenstone belt formation. Apart from the sediment fill of the basin, the intercalated layered extrusives grading from komatiite at the base to mafic lavas in the higher parts, appear to confirm the foregoing scenario of the formation of greenstone belts. Thus, greenstone belts are characterized by outflow of komatiites as the first product of reheating and thus destabilizing the upper thick basalt-komatiite layer by heating from below. Cooling from the surface results in overall low temperature of this upper part of the upper mantle. This should explain the relatively low temperature of the high grade terrains, as a result of the low temperature regime in the upper mantle below.

ARCHAEAN CONTINENTAL LITHOSPHERE

The occurrence of low grade greenstone belts and high grade granulite facies terrains side by side within Archaean cratons poses a problem concerning the geothermal state of the earth. The high temperature of komatiite extrusives in the greenstone belts indicates high (> $1750°C$) mantle temperatures, whereas the high grade metamorphic terrains only give petrologically derived temperatures of $600-900°C$ at pressures of 7 - 13 kbar (Tarney and Windley, 1977). Thus, temperatures at these pressures, at immediate crustal levels of between 25 and 40 km depth, though higher than modern continental crustal values, can not be considered to be in accordance with radioactive heat production in the Archaean which we take to be three times larger than now (O'Nions et al., 1978). An upper mantle temperature of some $1750°C$ at 100 km depth appears to be conflicting with the relatively low temperatures of the high grade terrains. If heat production would be neglected, the temperature of the upper mantle would cause stationary thermal gradients of some $18\ °K\ km^{-1}$ for a constant value of heat conductivity. This gradient is not essentially lower than the one pertaining for high grade terrains. Considering that radioactive heat generation in the Archaean was threefold modern values, above stationary gradient does not comply with the Archaean geothermal state.

Archaean thermal regimes are usually considered to have reached equilibrium in analogy with studies of the modern continental geotherm. It should be noted, however, that the latter is to be considered to be quasi-stationary and undergoing secular change with a very large time constant, due to progressive cooling of the upper mantle and the decay of radioactive heat sources. This quasi-stationary state, of course, may be perturbed by tectonic events and erosion.

A one-dimensional stationary temperature field may be obtained by integrating the heat conduction equation, resulting in

$$q(z) = K\, dT/dz = -\int_0^{z'} Q(z') + q_o$$

$$T(z) = \int_0^z \frac{dz'}{k(z')} [q_o - \int_0^{z'} Q(z'')dz'']$$

where z is the depth coordinate, T(z) the temperature, K the coefficient of thermal conduction, q(z) the heatflow at depth z, and q_o the heat-flow through the earth's surface and where T(0) = 0.

Assuming for the moment that both the depth distribution of heat producing elements and the thermal conductivity are equal to modern values, the threefold heat generation in the Archaean and hence threefold heatflow, causes the temperature T(z) to be three times the present continental temperatures. Mercier and Carter (1975) give an experimentally derived temperature of 500°C at 40 km depth in a modern continental setting. In the present model this would result in 1500°C at the same depth in Archaean high-grade terrains, a value some 600°C in excess of those petrologically determined.

If we concentrate all heat producing elements at the shallow level of less than 40 km we may arrive at the required low temperatures of the granulite terrains. However, in this case, heat flow from below 40 km depth vanishes, which obviously is conflicting with the high temperature upper mantle requirement. If, in contrast, heat producing elements are concentrated below 40 km depth and assuming constant K for the upper 40 km, a temperature close to 2000°C at 40 km depth results. This also conflicts with metamorphic temperatures and pressures of the granulite facies rocks. Petrological data can only be met by assuming extremely high thermal conductivity in the upper 40 km, and an extremely low one below this level, preferably in combination with low heat production in the crust. England's (1979) results demonstrate that indeed in those extreme circumstances a stationary geotherm can be found to match these constraints.

However, as we find no reason to believe that thermal conductivity in the Archaean continent differed drastically from modern values, we prefer to rule out the possibility of an equilibrium thermal regime which meets petrological constraints. We, therefore, are bound to search for transient events during which the temperature remained low. A possible transient phenomenon affecting the thermal state of the crust is mantle diapirism. If, as a consequence, metamorphism of crustal material results, determined metamorphic temperatures must be higher than crustal temperatures before metamorphism. Thus, in this case, metamorphic temperatures constitute an upper bound for crustal temperatures before the onset of diapirism. Crustal layers in which mantle diapirs penetrated had to be cooler than is compatible with an equilibrium thermal state complying with experimental p/T conditions.

Therefore, the succession of events having led to Archaean high-grade terrains must have taken place in a cool, transient thermal regime, which paradoxically was accompanied by intrusion of mantle diapirs. A cool transient regime could only be achieved by a combination or succession of one of the following processes:
(1) Rapid sedimentation on a cool oceanic crust
(2) 'Cold' thickening of the continental protocrust
(3) Rapid erosion of the continental crust

The first possibility should be ruled out because, as is shown in the foregoing, cool oceanic crust could not exist.

The second possibility should be associated with processes such as large scale thrusting and nappe tectonics. However, as this usually is associated with thermal perturbation, the question arises how a thrust sheet stack could remain cool.

The third alternative might be possible only if the exhumed high-grade terrain was rather cool to start with, which can not be the case. However, if komatiite melting temperature is indicative of the shallow mantle thermal regime, above requirements can not be met. The cooling effect of the transient events of thrust sheet emplacement and erosion should have required a cool crust in advance or no subsequent reheating.

The only way to maintain this state is by assuming a relatively cool shallow upper mantle, which episodically may be thermally perturbed by deeper mantle diapirism. An upper mantle as suggested in the foregoing could meet above requirements.

As was shown before, a slightly stratified dominantly basaltic thick layer with ultramafics at its bottom, though on the average being stagnant at its base, could episodically give rise to rising blobs of hot basaltic and komatiitic magma. These blobs, when underplating the Archaean tonalitic protocontinent may result in crustal stretching and thinning, as is required for greenstone belts (McKenzie et al., 1980). Crustal thinning caused by the hot mantle blob is accompanied by extrusion of mafic and ultramafic lavas and simultaneous sedimentation from adjacent terrains. As has been argued before, the komatiites would be expected to extrude first.

Crustal stretching may lead to incipient rifting and spreading of the protocontinent, though the instable state of the oceanic upper mantle does not allow oceanisation to be intitiated.

The underplating of hot mantle blobs underneath the non-stretched and non-rifted part of the Archaean tonalitic protocontinent which must have been in a state of compression due to adjacent rifting in the future greenstone belts, should have several consequences. The underplating dominantly involved basalts with temperatures far above the liquidus. This should have given rise to widespread anatexis of the lower crust of the protocontinent and associated magmatism and intracrustal gravity tectonics. The cooling of the blob would lead to a succession of felsic and mafic intrusions and extrusions. Thus, the blob would be incorporated in the lower crust, leading to considerable crustal thickening from below and rising of the crustal column. The latter, in turn would give way to large scale erosion.

The scenario described above would easily cause granitization and granite diapirism by anatexis of the lower tonalitic crust. The style of Archaean geodynamics does not favour recycling of continental crustal material and subsequent enrichment of continental crust in L.I.L. and R.E.E. elements. For the latter we require plate tectonics and deposition of sediments over longer periods of time on an established and mechanically strong oceanic crust and lithosphere.

PROTEROZOIC PLATE TECTONICS

The onset of some form of plate tectonics had to wait until geothermal circumstances were favourable for a stably stratified oceanic crust and upper mantle to be generated. This, in turn, depends on the upper mantle temperature being sufficiently low to allow basalt to be dominantly in the solid state at low pressures and to be converted to eclogite at higher pressures. In order to segregate a basalt layer from lherzolite source material, a residue of depleted peridotite must be left behind. Basalt, when being recycled back into the upper mantle can react with depleted peridotite to form lherzolite, which in turn may serve again as new source material for basalt generation. The latter takes place above the peridotite solidus, the depth of which increases with increasing temperature. For a modern oceanic spreading ridge, according to Sleep and Windley (1982), large scale partial melting of undepleted peridotite starts at the eutectic, at some 30 km depth, resulting in a basaltic crust of 6 km thickness at partial melting between 5% and 30%. This leaves a depleted peridotite residue of some 25 km thickness. The extrusion temperature of basalt is near 1300°C. When assuming an extrusion temperature of 1400°C, that is between modern and Archaean values, partial melting up to 30% and segregation of the basalt fraction would take place between about 70 and 40 km depth within the rising undepleted peridotite column. This would give rise to a 12 km thick basaltic layer resting on a 28 km thick depleted peridotite layer if the basalt separated completely from the peridotite.
If this were not the case, a thinner basaltic layer would result. Therefore, the 12 km and 28 km are to be considered as an upper and lower bound of the thicknesses of the basaltic and depleted peridotite layers, respectively. At any rate, decisive for stable stratification is the total column of undepleted peridotite involved in fractional segregation, i.e. 30 km and 40 km for extrusion temperatures of 1300°C and 1400°C, respectively. The compositional defect thickness δ_{comp} would accordingly differ by a factor 4/3 and compositional buoyancy of the oceanic lithosphere, if existing, increases towards the past. The higher mantle temperature T_m on the other hand counteracts lithospheric buoyancy. The factor δ_{comp}/T_m, being a measure of lithospheric buoyancy, would be larger by $(4/3) \times (13/14) \approx 1.25$, if extrusion temperatures was 1400°C compared to 1300°C for the present.

Making use of (3) for the lithospheric age at transition from stable to unstable layering, we arrive at an age of (25/16) of modern values. Taking this transition to occur at an age of 30 Ma in modern oceanic lithosphere, this age is to be set at 47 Ma when the extrusion temperature at an oceanic spreading center was 1400°C.

We infer that the buoyancy of the oceanic lithosphere increases from now towards the past, due to higher mantle temperatures, shifting the basalt-peridotite eutectic to greater depth. The compositional effect far outweighs the cooling effect and, therefore, the first is the most decisive. We may extrapolate to the past until the plate tectonic mechanism is prohibited at higher mantle temperature. This is the case when this temperature is above the liquidus of basalt, and the latter can not be recycled into the upper mantle and, therefore, is bound to accumulate in the shallow upper mantle. In the foregoing we have argued that this situation pertains to the Archaean. Prior to and during the Archaean, cooling of the earth must have been very effective, upper mantle temperature in the Archaean being relatively low compared to radiogenic heat production. We ascribe this to the absence of a stable oceanic thermal boundary layer exerting a blanketing effect.

We now make the assumption that Archaean heat loss was large enough to result in an extrusion temperature of some 1400°C by the early Proterozoic, thus favouring some form of plate tectonics to exist at that time. In later Proterozoic and during the Phanerozoic, plate tectonics governed the geotectonic style. However, mantle temperature, being secularly decreasing, turns out to be the decisive parameter which critically determines the character of plate interactions through the influence of lithospheric buoyancy.

PROTEROZOIC PLATE INTERACTION

Bickle (1978) estimates that the present rate of plate production at oceanic ridges is 3 $km^2 a^{-1}$ and that the average age of oceanic lithosphere when being subducted at plate convergence zones is 60 Ma. Vlaar and Wortel (1976) demonstrated that the behaviour of subducted oceanic lithosphere is crucially dependent on lithospheric age. The 'classical', steeply dipping and deeply penetrating Benioff subduction only pertains to lithosphere with ages in excess of some 90 Ma. At this age, modern oceanic lithosphere has acquired sufficient negative buoyancy to evoke Benioff subduction. However, younger oceanic lithosphere when being subducted, only reaches shallower upper mantle levels, and very young lithosphere (< 30 Ma) can only be subducted subhorizontally underneath an older oceanic or continental lithosphere. At present, only a small fraction of subducted oceanic lithosphere is older than 90 Ma and thus 'classical' Benioff subduction at present, rather is the exception than the rule.

According to the foregoing, Proterozoic oceanic lithosphere starts to be unstably layered with respect to the underlying upper mantle at an age of 47 Ma. This value is far in excess of the 30 Ma for modern plate tectonics.

Assuming that oceanic plate creation was not less than now, this implies that a larger part of the oceanic lithosphere was buoyant in the Proterozoic than at present. Oceanic plate creation rate depends on spreading rate and hence on the plate driving forces. The main driving forces are the ridge push and the slab pull. The ridge push increases roughly linearly with mantle temperature T_m (Turcotte and Schubert, 1982), and hence, may be assumed to be slightly larger in the Proterozoic compared to now. With dominantly buoyant oceanic lithosphere in the Proterozoic, slab pull, associated with 'classical' Benioff subduction virtually must have been absent. This should have been more pronounced even by virtue of a higher upper mantle temperature and consequently shorter absorption time of a descending slab and also by a shift of possible phase transitions (olivine-spinel) to greater depth.

In total, driving forces may have been smaller in the Proterozoic than at present, and resistive forces at subduction zones, due to subduction of young and buoyant lithosphere may have been larger. However, it should be ruled out that plate tectonics was absent in the Proterozoic, as even a heat loss of $100°C/B.a$ of the earth would result in an extra heat flow of 1 H.F.U. (Sleep and Windley, 1982), leaving apart the higher radiogenic heat production in the past. This value of the heat flow contradicts the geothermal state of the Proterozoic crust. Moreover, in modern plate tectonics sufficient evidence is presented for subduction of young and very young lithosphere, only driven by ridge push, for example at the west coast of W.North America and of S.Chile. We, therefore, infer plate tectonics to be operative during the Proterozoic and to have involved the subduction of young and buoyant oceanic lithosphere caused by ridge push within the converging plates.

However, it is not to be discarded that periods of regional or global stagnation of plate tectonic processes could alternate with periods of activity.

LITHOSPHERIC DOUBLING AND CONTINENTAL ACCRETION

A main implication of the compressive regime of the Proterozoic oceanic lithosphere, is that continents could not easily be fragmented. We, therefore, deduce that continental drift could only start when buoyancy of the oceanic lithosphere had decreased sufficiently to allow compression at convergence zones to have decreased in order to allow for 'classical' Benioff subduction. This must have been not earlier than in the Mesozoic. However, the Proterozoic state of compression would not rule out motion along shear zones and transform faults. The average state of stress in the Proterozoic continent must have been more compressive than now.

The mechanical strength and thickness of lithospheric plates is strongly temperature dependent and is determined by the depth of a specific isotherm. The temperature being assumed to be slightly higher in the Proterozoic compared to now, $1400°C$ versus $1300°C$, at an oceanic ridge center, Proterozoic lithospheric plates must have been slightly thinner than modern ones. Therefore, lithospheric plates were sufficiently thick to act as guides for large stresses.

One of the consequences of subduction of young oceanic lithosphere and the resulting state of compression at convergence zones is the off-scraping of oceanic sediments to form accretionary wedges (Wortel and Cloetingh, 1983). Part of the growth of continents during the Proterozoic and later must be ascribed to accretion at a compressive convergence zone, not unlike modern 'melanges'. Subduction of young oceanic lithosphere and particularly of an active spreading ridge results, as has been stated before, in horizontal layering of the obducted upper cool lithosphere and the subducted young hot and buoyant lithospheric plate. This process has been termed lithospheric doubling (Vlaar, 1983). Lithospheric doubling can be seen as a mechanism causing a thermal perturbation beneath the upper cool lithosphere. Assuming a geothermal gradient not deviating much from the present we propose a temperature of $1000^{o}C$ at the base of a 80 km thick Proterozoic continental lithosphere. Lithospheric doubling results in a hot layer below the depth of 80 km. The extrusion temperature at the former spreading center (being $1400^{o}C$) will cause the temperature to rise from $1000^{o}C$ tot $1400^{o}C$ over a small depth interval below 80 km. The strongest rise occurs at the site of the subducted spreading ridge, whereas this rise is spread over a larger depth interval with increasing distance from this site.

Lithospheric doubling thus introduces a gravitationally unstable, hot layer below a cool continent. As such, the subducted ridge with its adjacent young flanks is a potential source for a suite of magmas and for ascending diapirs. As a consequence, underplating of the continental crust by hot diapirs may result in anatexis of the lower crust, crustal thinning and subsidence of earth's surface to form broad intracontinental ensialic basins. The compressive state of the Proterozoic continent may not facilitate the rising of magma to the surface. However, crustal thinning by the foregoing process may be accompanied by the intrusion of dike swarms into the crust. The former oceanic basaltic layer, probably partly hydrated, being now situated at a depth of 80-100 km, at a temperature of $1400^{o}C$, is above or close to the basalt liquidus. These circumstances appear to be ideally suited for the formation of anorthosite by the segregation of pyroxene megacrysts from the basaltic liquid (Emslie, 1975).
If the anorthositic magma rises diapirically to underplate the existing continental crust, considerable crustal thickening may result. As such, the Proterozoic is characterized by intrusive plutonism of mainly basaltic character and crustal growth by anorthosite diapirs.

Proterozoic plate tectonics and lithospheric doubling, in contrast to the thermal and geodynamic regime of the Archaean, should have led to a sudden increase of the enrichment of L.I.L., R.E.E. and other incompatible elements in the crust.

The Proterozoic plate tectonic regime is particularly suited for the transport (into the continental crust) of enriched oceanic crust basalt and sediments, by means of lithospheric doubling and subsequent magmatism.

PROTEROZOIC OROGENY

The standard concept of mountain building involves the so-called Wilson cycle. A phase of oceanic spreading separating continents is followed by closing of the ocean accompanied by subduction in Benioff style of oceanic lithosphere underneath one of the continents bordering the ocean. The last phase of this process involves 'continental collision' at the destructive margin, whereby accumulated sediments are squeezed between the colliding continental blocks. This then should result in the creation of fold belts as they are described in geology.

Above scheme has up till now resulted in partly conflicting mechanisms, which defer an understanding complying with geological and geophysical data.

As far as recent mountain building is concerned, the Wilson cycle approach involving the creation of large oceans may, at first sight, seem justified. However, when we turn to the Precambrian, this type of spreading, as is suggested by palaeomagnetic data, should be ruled out. This, in turn, makes the Wilson cycle approach invalid. In a recent paper (Vlaar and Cloetingh, 1984) an alternative theory for mountain building, with reference to the Alps, was presented. In particular, it was shown that mountain building had to be placed in the setting of the opening and closing of small and young oceanic basins. A full scenario for Alpine orogeny could be developed by means of new tectonophysical concepts. These concepts, which appear to be valid for young fold belts, are also applicable to earlier Phanerozoic mountain building periods and to Proterozoic orogenics.

In the following a condensed version of the above theory is given. A more detailed description is presented by Vlaar and Cloetingh (1984).
(1) The initial stage of the formation of small ocean basins usually takes place in a shear zone setting. Along with rifting and spreading, lateral movement along transform faults also takes place. A period of spreading is followed by closing of the young ocean basin. Vlaar and Wortel (1976) and Wortel and Vlaar (1978) have demonstrated that lithospheric age is the key parameter governing the subduction process. Upon closing a young and small ocean basin, young lithosphere and also ridge segments should be overridden by adjacent continental margins. The mechanism of lithospheric doubling should become effective and consequently thermally perturbed upper mantle should result. Cloetingh et al. (1983) demonstrated that the transformation from a passive to an active continental margin is most effective when young and mechanically weak oceanic lithosphere is involved, particularly when loaded by a thick sediment pile. During this compressive stage of basin closing, low-grade melanges may be formed by off-scraping of sediments from the underlying oceanic lithosphere.

The foregoing is particularly suited for the Proterozoic. The larger compressive state of the continent did not favour the Wilson cycle to take place. Transcurrent displacement and extension accompanied by pull apart basins could provide the creation of small basins. Closing of these young basins then constituted the initial stage of mountain building. It should be borne in mind that the creation of small basins implies oceanic spreading on oceanic ridges or ridge segments. This causes, by the very process of diapirism underneath the ridges, the upper mantle below the young lithosphere to be thermally perturbed. The events to follow are dominated by the fact that, in closing the basins, young and buoyant lithosphere is involved and that the underlying mantle acts as a 'hot spot', or rather as a 'hot line'. This is typical for the process of lithospheric doubling.

(2) The closing of the small basin may or may not be accompanied by apparent migration of the 'hot line' or 'hot area' with respect to the overlying continental platforms originally bordering the basin. The apparent migration of a thermally perturbed upper mantle region, originated by a former spreading event, has been termed 'lithospheric shifting' by Vlaar (1983).

As the initial tectonic setting is rift formation and possibly transcurrent displacement, the platform lithosphere is transsected by subparallel faults. These faults may act as effective feeders of magma well below and into the crust.

We may distinguish the cases where after basin closing no considerable migration takes place, and where the sutured platforms are displaced together.

In the first case the perturbed upper mantle will stagnate beneath the suture, which may act as a conduite for the transport of magma to crustal levels. This results in metamorphism and mobilisation of the lower crust and also in uplift of the area involved. This uplift could lead to bivergent gravitational decollement of superficial platform sediment layers and consequently to the creation of fold belts: a Pyrenean type orogeny.

If, on the contrary, apparent migration of the mantle perturbation takes place, old transcurrent faults may be reactivated and may act as magma feeders to shallow levels. This then may result in sequential stacking of nappes in a 'piggy back' fashion (Elliott, 1976) away from the apparently moving metamorphic source area.

(3) The general absence of ophiolite thrust sheets in Proterozoic fold belts can be attributed to the overall compressive regime of the continent which counteracted the creation of larger basins and hence, of young oceanic crust without appreciable sediment cover. As has been argued by Vlaar and Cloetingh (1984), Alpine ophiolite thrust sheets had to involve the decollement of very young, solidified, thin oceanic crust (< 10 Ma) from its underlying ductile mantle.

Due to limited spreading and more abundant sedimentation, circumstances can have prohibited the Alpine setting with respect to ophiolite emplacement to take place.

CONCLUSIONS

1. Temperature is the most important parameter governing the geodynamic regime during the earth's history. The temperature of the earth's interior and particularly of the upper mantle determines the degree of partial melting of undepleted peridotite to generate basalt and depleted peridotite. The solidus of peridotite shifts to greater depth with increasing temperature and, therefore, determines the thickness of the basaltic oceanic crust and the thickness of the depleted peridotite layer underneath. Increasing mantle temperature from now towards the past gives rise to increasing thickness of these layers. We have demonstrated that this mechanism has been working from early Proterozoic up till the present. However, the mechanism becomes selfprohibitive when the upper mantle temperature rises to above the basalt liquidus and/or the basalt-eclogite phase transition. The basalt then is bound to accumulate as a light fraction in the upper part of the upper mantle. Plate tectonics then is ruled out because this mechanism requires recycling of the oceanic basalt layer back into the mantle. We gave arguments that showed that the Archaean was characterized by a geodynamical regime that involved an (episodical) turn over of a relative thick shallow upper mantle basaltic layer. This process promoted a rapid heat loss of the earth and a relatively low upper mantle temperature. At the same time it favoured the existence side by side of greenstone belts and high-grade terrains during the Archaean.

2. Thickening of the oceanic basaltic crust and the depleted peridotite residual layer enhances the compositional buoyancy of the past oceanic lithosphere. It has been demonstrated that increased stable stratification of the Proterozoic oceanic lithosphere has governed the tectonic regime. This circumstance must have led to a compressive state of the continental lithosphere and to subduction of stably stratified oceanic lithosphere. The latter process, called lithospheric doubling, can govern Proterozoic tectonics and magmatism. Arguments are presented that these events have led to massive crustal thickening by anorthosites. Anorthosite is assumed to be the product of melting of basalt at near liquidus temperatures at upper mantle pressures. Mountain building during the Proterozoic is caused by lithospheric doubling due to the destruction of small and young oceanic sedimentary basins.

REFERENCES

Arndt, N.T., Geology, 11, 1983, 372-375.

Baer, A.J., Tectonophysics, 72, 1981, 203-227.

Bickle, M.J., E.P.S.L., 40, 1978, 301-315.

Cloetingh, S.A.P.L., M.J.R. Wortel, N.J. Vlaar, AAPG Memoir, 34, 1983, 717-723.

Condie, K.C., Tectonophysics, 105, 1984, 29-41.
Elliott, D., J. Geophys. Res., 81, 1976, 949-963.
Emslie, R.F., Can. Mineral., 13, 1975, 138-145.
England, P.C., Nature, 277, 1979, 556-558.
Goodwin, A.M., Science, 213, 1981, 55-61.
Green, D.H., I.A. Nicholls, M. Viljoen and R. Viljoen, Geology, 3, 1975, 11-14.
Green, D.H., Geology, Vol. 3, 1, 1975, 15-17.
Jarvis, G.T. and I.H. Campbell, Geophysical Res. Lett., 10, 1983, 1133-1136.
McKenzie, D., E. Nisbet and J.G. Sclater, E.P.S.L., 48, 1980, 35-41.
Mercier, J.C. and N.L. Carter, J. Geoph. Res., 80, 1975, 3349-
Nisbet, R.W., S.S. Sun and A.C. Purvis, Can. Mineral., 17, 1979, 165-186.
Nisbet, E.G. and C.M.R. Fowler, Geology, 11, 1983, 376-379.
Nisbet, E.G., in: Komatiites (Eds. N.T. Arndt and E.G. Nisbet), 1982, 501-520.
O'Nions, R.K., N.M. Evensen, P.J. Hamilton and S.R. Carter, Phil. Trans. R. Soc. Lond., A258, 1978, 547-559.
Oxburgh, E.R. and E.M. Parmentier, J. Geol. Soc. Lond., 133, 1977, 343-355.
Sleep, N.H. and B.F. Windley, Journal of Geology, 90, 1982, 363-379.
Tarney, J. and B.F. Windley, J. Geol. Soc., 134, 1977, 153-172.
Turcotte, D.L. and G. Schubert, Geodynamics, John Wiley and Sons Inc., 1982.
Vlaar, N.J., in: Progress in Geodynamics, North Holland Publishing Coy., 1975.
Vlaar, N.J. and M.J.R. Wortel, Tectonophysics, 32, 1976, 331-351.
Vlaar, N.J., E.P.S.L., 65, 1983, 322-330.
Vlaar, N.J. and S.A.P.L. Cloetingh, Geologie en Mijnbouw, 2, 1984, 159-164.
Weaver, B.L. and J. Tarney, Nature, 279, 1979, 689-692.
Wells, P.R.A., E.P.S.L., 46, 1980, 253-265.
Wortel, M.J.R. and N.J. Vlaar, P.E.P.I., 17, 1978, 201-208.
Wortel, M.J.R. and S.A.P.L. Cloetingh, Am. Assoc. Petr. Geol. Mem., 34, 1983, 793-801.

SEISMIC REFLECTION RESULTS FROM PRECAMBRIAN CRUST

Scott B. Smithson, William R. Pierson, Sharon L.
Wilson, and Roy A. Johnson
Department of Geology and Geophysics
Program for Crustal Studies
University of Wyoming
Laramie, Wyoming 82071 U.S.A.

ABSTRACT. Results of crustal reflection profiling in the
Precambrian can provide decisive information on such features as
faulting, extension vs. compression and crustal scale deformation
in general, vertical vs. horizontal movements, intrusions, mafic
differentiates, underplating, and plate tectonics even though
targets in the Precambrian are complex. Mylonite zones are probably the best crustal reflectors so that a moderately dipping
reflector separating the ancient Archean gneiss terrain from
younger Archean greenstone belts in Minnesota is probably a thrust
fault marking the suture zone between the two terrains. This
structure implies the operation of plate tectonics at least by the
late Archean. The 3.6 b.y. Minnesota gneiss terrain is underlain
by a thick (10 km) sequence of layered deformed rocks, whose presence restricts early Archean events. A proterozoic suture in
southeastern Wyoming is marked by a thick mylonite zone and a change
in crustal structure that has persisted through time. The
Adirondack anorthosite and granulite terrain is underlain by a thick
layered sequence in the lower crust. Formation of a late
Proterozoic basin filled with 10-12 km of interlayered sedimentary
rocks in Texas and Oklahoma may have been accompanied by crustal
underplating.

INTRODUCTION

The development of Precambrian crust remains one of the big
questions in earth science, and a major unknown concerning
Precambrian crust is the nature of the deep crust. In particular,
the character of crust beneath granulite terrains is especially
fascinating because granulites are generally interpreted to represent a deep crustal section. Crustal reflection profiling has
recently become a technique that is broadly applied to obtain new
information about the deep crust (Oliver et al., 1983; Phinney and
Odom, 1983). We discuss those data on Precambrian crustal structure that have primarily been provided by seismic reflection

profiling; we start with results from the oldest areas in the Archean and progress through to the young Proterozoic.

NATURE OF REFLECTIONS IN CRYSTALLINE ROCKS

The seismic reflection method is applied routinely to map structures in horizontal to sub-horizontal sedimentary rocks and still with success in strongly deformed sedimentary rocks such as fold-and-thrust belts. The situation is, however, more complex in crystalline rocks, mainly for the following three reasons: 1) reflection coefficients are much smaller in crystalline rocks (about 0.13 vs. 0.30 for maximum values in crystalline and sedimentary rocks, respectively), 2) interfaces in crystalline rocks are far more contorted and discontinuous from effects of deformation and intrusion because these rocks generally deformed much more ductilly than sedimentary rocks, 3) lack of drilling or exposure of objectives in crystalline rocks means lack of "ground truth" to confirm, modify, or reject interpretations. As a result of the last reason, interpretations are poorly constrained and use of the reflection method in crystalline terrains is in its infancy. At the same time seismic reflection results have the potential to provide the highest-resolution image of the deep crust that we can expect to obtain.

The geometry of Precambrian rocks is generally complex, and the seismic response of such rocks may be complex or weak or both. The strongest reflections are generated by constructive interference from layers involving velocity reversals (Fuchs, 1969; Fountain et al., 1984) so that layered metamorphic sequences and layered igneous intrusions conceivably could cause good reflections. If reflections are generated from layering then the reflection wavelet will be multicyclic instead of a single wavelet, and interference effects will produce a complex reflection pattern (Smithson et al., 1977a). Because reflections from layers may be two to four times stronger than reflections from a single interface (a simple single wavelet), these complex reflections will tend to be more common in crustal reflection sections.

Complex geometry such as folded, irregular, or corrugated contacts tends to interrupt the continuity of reflections and produce arcuate reflections and diffractions (Smithson, 1979; Smithson et al., 1980; Wong et al., 1982). Such data must be migrated or moved updip on the seismic section in order to achieve a "proper" image of the subsurface geometry, and a good image is not necessarily achieved for complex targets. Steep dips may be difficult to record and process, and the steepest likely to be determined depends on the velocity distribution in the crust and the recording parameters used. Although in industry applications of the reflection method, "sideswipe" or seismic events out of the plane of the recording profile are commonly considered noise, in the complex geometries found in the Precambrian, out-of-the-plane events may be the rule rather than the exception and should be considered signal unless these events follow a near-surface path.

Layered planar rock units should thus provide the best reflections, and these reflections are generated by differences in composition or anisotropy or both between the layers. At present, the importance of anisotropy is difficult to evaluate. Mylonites are layered planar zones commonly composed of a variety of rocks. Because they may be thick and relatively planar and dip moderately, mylonites are probably the best reflectors in the crust (Smithson et al., 1979; Fountain et al., 1984), and recent reflection profiling over mylonites demonstrates that they may be remarkably reflective (Hurich et al., submitted for publication). Thus mylonites may be mapped in seismic sections and used to resolve crustal scale deformation of the Precambrian crust.

ARCHEAN AREAS

Minnesota

Some of the oldest Archean crust in North America is found in the Precambrian of Minnesota, which has dates of 3.6-3.8 b.y. (Goldich and Hedge, 1974). Here the ancient Minnesota Valley gneiss terrain is bounded on the north by the Great Lakes tectonic zone (Sims et al., 1980) and a series of late Archean greenstone belts. The Great Lakes tectonic zone was presumed to be a steeply dipping Archean border but COCORP crustal reflection data (Gibbs et al., 1984; Pierson, 1984) shows a gently dipping reflection beneath the zone (Fig. 1). Because some of the best crustal reflectors may be

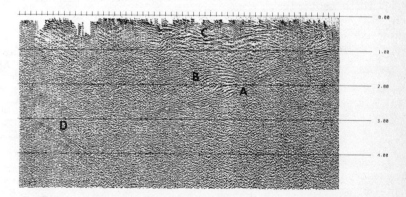

Figure 1. Seismic section from COCORP Minnesota line 3 reprocessed at the University of Wyoming. Time in seconds on vertical scale. Time can be converted to approximate depth by multiplying by 3 km/s. Reflection (A) is dipping multicyclic event believed to be generated by a mylonite zone along a thrust fault separating late Archean greenstone belts (hanging wall) from early Archean gneiss terrain (footwall). Reflection pattern (B) interpreted as a recumbent fold hinge on the basis of a migrated section (Pierson, 1984). Reflections (C) from Proterozoic sedimentary basin. Reflections (D) from folded layered rocks in greenstone terrain.

fault zones along which mylonites have formed within the ductile regime (Smithson et al., 1979; Fountain et al., 1984), the reflection could be caused by a mylonitic thrust fault. It could equally well be interpreted as a mylonitic listric normal fault, especially since younger rocks lie on older across the fault. The indication of a major recumbent fold above the fault reflection (Pierson, 1984, Figs. 1 and 2) suggests that this event is coming from a thrust fault; mylonitization along the fault zone is the postulated cause of the strong, multicyclic reflection (Fountain et al., 1984). Magnetic data from this area can be best modeled as a magnetized wedge of older Archean rocks (granulitic gneisses) underlying the younger Archean greenstone terrain (Fig. 2). The dip of the boundary is the same as the dip of the postulated thrust-fault reflection. Thus several lines of evidence indicate that the younger Archean greenstone belt terrain is thrust above the ancient Minnesota Valley gneiss terrain, presumably as the greenstone belt was accreted to the gneiss terrain, so that the dipping reflection in Figure 1 represents a suture zone. This dipping reflection, however, with a thickness of about 1-2 km, seems remarkably simple to represent an Archean suture. Its apparent simplicity may just be a function of resolution in the seismic survey and the postulated suture may be more geologically extensive and broader.

Other dipping events beneath the greenstone terrain may also be reflections from thrusts. Thus by at least late Archean time, a horizontal tectonic regime was dominant. This may reflect the operation of plate-tectonics mechanisms, and if greenstone belts are formed around island arcs, then their presence is more evidence for plate tectonics.

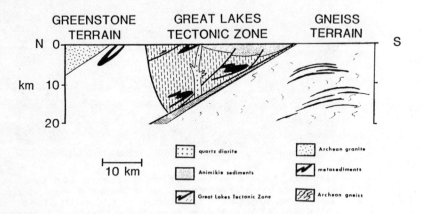

Figure 2. Interpretation of seismic sections showing structural relationships between different terrains in Minnesota Archean separated by a mylonitic thrust zone. Seismic section from Figure 1 in center and from Figure 3 at right.

Seismic data from underneath the granulite-facies Minnesota gneiss terrain shows abundant reflections between 3 and 6 s or about 9 to 20 km (Fig. 3). These are arcuate or dipping multicyclic events indicative of layering. The gneisses consist of granodioritic to tonalitic gneisses with mafic layers or schlieren and garnets (Grant, 1972). Foliation in the gneiss generally dips moderately. Although the layered sequences underlying the gneiss could be mylonites their arcuate geometry and general geologic setting suggest that these events are caused by layered gneisses that could represent a supracrustal sequence (although with transposed layering). This indicates that no mafic residuum from anatexis closely underlies the gneiss unless it is highly heterogeneous. The significance of layering may rather be to indicate a large-scale migmatite terrain. This reflection data furnishes important information concerning what underlies a granulite terrain. The rocks appear to be a heterogeneous, layered deformed sequence of rocks with moderately dipping foliation. The zone corresponds to a deep lower crustal level.

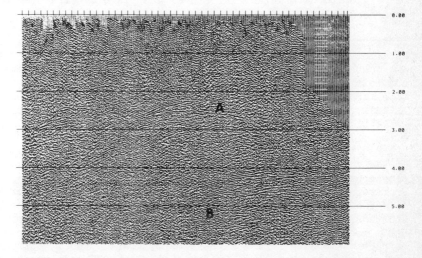

Figure 3. Seismic section from COCORP Minnesota line 3 showing arcuate multicyclic reflections (A and B) in the 3.6 b.y. Archean gneiss terrain. Time in seconds on vertical scale.

Wyoming

Seismic data is available from two areas within the Archean Wyoming province, the Wind River Mountains consisting of some of the oldest part of the Wyoming province, and the Laramie Range. The COCORP Wind River lines that imaged the Laramide Wind River thrust fault crossed the flank of a greenstone belt at South Pass (Condie, 1972). Except for one strong event at 2.5 s, the seismic section doesn't show many reflections from the upper crust. This is somewhat surprising because of the different velocity contrasts expected from within the greenstone belt rocks; however, the explanation for this might be the steep dips present near the surface.

The crust beneath the Wind River Mountains shows discontinuous but strong reflections all through the crust and complex arcuate events at 8 s on line 1A (Smithson et al., 1980). Interpretation of these events on the basis of fold structures exposed up-plunge in the hanging-wall block of the Wind River uplift suggests that the deep crust is not significantly different from the folded high grade metamorphic rocks intruded by granites exposed in the core of the range. Smithson et al. (1980), furthermore, suggest that a crustal thickness of 35-40 km was attained by 3 b.y. based on the interpretation of complex fold structure. These interpretations provide an example of how seismic sections may be used indirectly to infer aspects of deep crustal genesis in the sense that seismic events at 30 km depth can be related to deformational features and place constraints on lower crustal development.

A Proterozoic suture zone in southeastern Wyoming marks the border between the Archean Wyoming province, which includes the basement rocks of the Wind River Mountains, and 1.7 b.y. crust to the south (Karlstrom and Houston, 1984). The suture zone itself consists of a steeply dipping mylonite zone from 1 to 7 km wide (Karlstrom and Houston, 1984). This area has been the site of extensive geophysical studies including acquisition of seismic reflection and gravity data by the University of Wyoming (Smithson et al., 1977b; Johnson et al., 1984) and COCORP seismic lines (Allmendinger et al., 1982). These data have provided constraints on the interpretation of surficial geological features as well as deep crustal structure. Most of the seismic data come from the Proterozoic terrain south of the suture zone. Gravity interpretation indicated that charnockitic syenite associated with the Laramie anorthosite complex was underlain at several kilometers depth by mafic rock (Hodge et al., 1973), and later seismic work showed refractions coming from a high-velocity (mafic) zone. Both seismic reflection interpretation (Smithson et al., 1977b; Allmendinger et al., 1982) and gravity interpretation suggest that the Laramie anorthosite itself is about 6 km thick. Reflections from depths of 4 to 18 km are found beneath the vast 1.4 b.y. Sherman granite and indicate the presence of heterogeneities that could represent anything such as xenoliths or the base (or far below the base) of the batholith (Smithson et al., 1977b).

Many ambiguities still exist in the interpretation of deep crustal reflections. Complexities are illustrated by a zone of reflections resembling an unconformity on COCORP line 5 (Fig. 4). When this data is migrated to move dipping events into proper geometric position, the reflections then form the shape of a dipping saucer (Fig. 4). This could come from a layered mafic intrusion (Johnson et al., in press), and modeling, which is

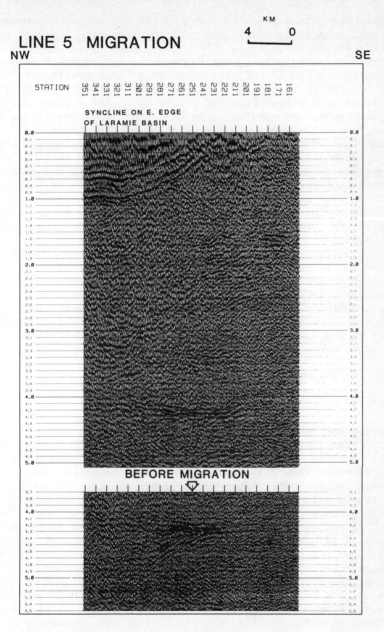

Figure 4. Seismic section from COCORP Laramie Mountains line 5 showing seismic section before migration (below) and after migration (above). Time in seconds plotted on vertical scale. Geometry of migrated reflection at 4.2 s suggests a layered mafic intrusion. Figure illustrates how distorted dipping reflections are before migration.

generally the only check on a deep crustal reflection interpretation, indicates that this is a plausible conclusion (Wong et al., 1982; Fig. 4, p. 96).

The shear zone marking the Proterozoic suture may have been detected by the COCORP reflection survey which suggests that it dips about $50°$ southeast (Allmendinger et al., 1982). Although Moho depths were picked from the COCORP data, the seismic data are highly ambiguous and unreliable on this question. Gravity data, however, indicates a change in crustal structure across the Archean-Proterozoic suture (Johnson et al., 1984). The gravity field decreases to the southeast at the suture zone (Fig. 5) and the interpretation of this observation is that crust thickens to the south and/or crustal density decreases to the south (Johnson et al., 1984). Because of Laramide effects, some long-wavelength thickening of the crust occurs to the south, but an abrupt change in crustal thickness or change in crustal density or both occurs at the boundary and has apparently persisted since middle Proterozoic time (Johnson et al., 1984). The southern Proterozoic province is believed to consist of deeply eroded roots of a migrating chain of island arcs and a continental margin that evolved from an Early Proterozoic Atlantic-type passive margin to a convergent margin in the Middle Proterozoic with the accretion of island arc terrains (Karlstrom and Houston, 1984. This is somewhat similar to the early Paleozoic history of the Appalachians.

THE PROTEROZOIC

The Adirondacks, New York

Reflection response is highly variable in the Proterozoic granulite-facies terrain of the Adirondacks. COCORP lines 8 and 10 show very few reflections from the deep crust (Brown et al., 1983), however, line 7 located over anorthosite shows some of the best deep crustal reflections yet recorded on land (Fig. 6). These events extend from 5 to 10 s as straight, gently dipping reflections suggesting a layered sequence of rocks at least 16 km thick. The reflections could be caused by a deformed supracrustal sequence, a layered mafic intrusion, a migmatite zone, a mylonite zone or numerous gabbroic sills. Because of the geometry, a tectonized supracrustal packet or migmatite zone seems likely, but either of these rock types could have undergone homogeneous ductile strain and now be mylonitized. Wide-angle reflections are found from 9 to 24 km depth near line 8 (Kubichek et al., 1984) where none were found at vertical incidence; it is important to understand whether this difference in seismic sections is due to actual structural variations within a short distance or to recording technique. This sub-granulite facies deep crust could range from homogeneous (or so strongly and steeply folded that it is non-reflective) to strongly layered within a few kilometers of the base of the crust.

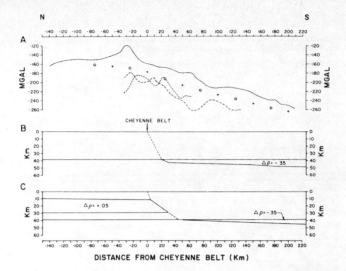

Figure 5. Gravity profiles (A) over Proterozoic suture (Cheyenne belt) in southeastern Wyoming. Profiles are from three different areas and show a general decrease in gravity across the Cheyenne belt. Gravity model B explains the decrease in gravity solely on the basis of a change in crustal thickness and gravity model C combines a change in crustal density with a small change in crustal thickness as an alternative interpretation (After Johnson et al., 1984).

Figure 6. Seismic section from COCORP Adirondack line 7 showing numerous gently dipping reflections beneath anorthosite. Time in seconds on vertical scale. Reflections are some of the best deep reflections found in COCORP data and indicate a thick layered succession of the lower crust.

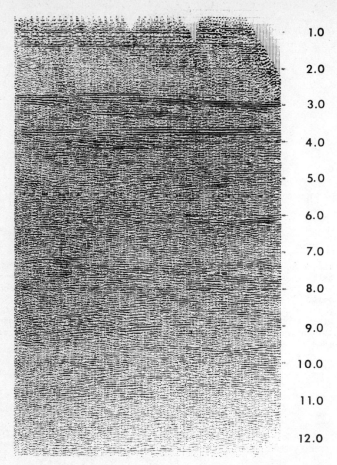

Figure 7. Seismic section from COCORP Hardeman County, Texas line 2. Time in seconds on vertical scale. Base of Phanerozoic sedimentary rocks at 1.4 s. Strong reflections from within Proterozoic basin at 2.8 s to 3.8 s. Dipping criss-crossing seismic events from 5 to 12 s.

Oklahoma

COCORP data from Hardeman County, Texas (Oliver et al., 1976) and adjacent areas in Oklahoma (Brewer et al., 1981) are unique in showing the strongest, most continuous crustal reflections yet found (Fig. 7). These strong reflections resemble those from the sedimentary rocks and mafic igneous rocks in the Witwatersrand Basin (Fatti, 1972). These come from a layered succession in the Hardeman basin believed to be a 10-15 km-thick sequence of interlayered volcanic and sedimentary rocks. Extraction of reflection coefficients from the amplitude of reflections indicates a sequence of quartzofeldspathic rocks interlayered with basalt or dolomite. Basalt seems to be the more likely choice. Here, the deep crust generates many arcuate events (Schilt et al., 1981) of the type that would come from any complex structure (Smithson, 1979; Smithson et al., 1980). These have been interpreted as originating from point diffractors between depths of 10 and 22 km (Schilt et al., 1981), but reflections from synformal features might be a more likely explanation because of the relatively high amplitude of these events. The deep crust slightly to the north, however, seems to be transparent. Interpretation of a crustal refraction line (Mitchell and Landisman, 1970) indicates unusually thick crust with a high velocity of 7.2 km/s in the lower crust. These data have been interpreted to represent a zone of crustal underplating (Fig. 8) which may have accompanied subsidence that allowed the accumulation of the thick supracrustal sequence (Wilson and Smithson; submitted for publication).

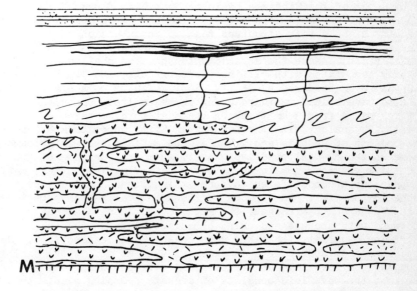

Figure 8. Interpretation of geological and geophysical data for crustal structure in southwestern Oklahoma. Phanerozoic sedimentary rocks (dotted) above basin in which basalts (black) and rhyolitic flow are interbedded with sedimentary rocks. Lower crust is underplated by coalescing gabbroic intrusions (V-pattern) that serve as source for basalts. As crust is underplated it subsides to allow deposition of 10-15 km of supracrustal rocks in the Proterozoic basin. M = Moho.

Michigan and Kansas

COCORP crustal reflection profiling has helped delineate the extent of Proterozoic rifts in Michigan and Kansas (Brown et al., 1982; Serpa et al., 1984). Both of these areas are correlated with exposed Keweenawan rifting in Northern Michigan. The seismic reflection profiles across the Paleozoic Michigan Basin show a series of subhorizontal reflections extending from 1.5 to 5 s beneath the Paleozoic sedimentary rocks in the Basin, and a large (about 50 mgal) gravity anomaly coincides with the zone of deep reflections (Brown et al., 1982). A deep well penetrated into the top of the pre-Paleozoic rocks and found redbeds and interlayered mafic rocks similar to the Keweenawan rocks. The seismic data indicates that the rift-infilling deposits are at least 9 km thick. These deposits are interpreted to consist of sandstones, shales, conglomerates and interlayered mafic flows and sills (Brown et al., 1982). The mafic rocks would represent a mass excess and contribute to the positive gravity anomaly. Whether the positive gravity anomaly is caused entirely by dense mafic rocks emplaced in the rift deposits or whether it is partially caused by a deep mafic intrusion in the root of the rift is not known. The deep crust shows no reflections outside the rift, and evidence concerning the nature of the deep Proterozoic rift structure or the reason for the Paleozoic subsidence of the Michigan Basin has not been found.

The COCORP seismic profiles in Kansas cross the Mid-Continent gravity high, a rift structure that is an extension of the Keweenawan, and these profiles contrast sharply with the Michigan seismic data. Here the seismic data show two shallow asymmetric basins associated with rifting (Serpa et al., 1984). Both basins appear to be bordered by gently dipping faults on their west flank and to have tilted westward along these faults. Deposits in these basins range from 3 km thick on the east to 8 km thick on the west. By analogy with the Keweenawan Rift, deposits are inferred to consist of redbeds and interlayered mafic flows, which cause some of the 80-mgal positive gravity anomaly. Because of the positive gravity anomaly and the high stacking velocities, mafic intrusions are postulated to form a significant part of the crust beneath the rift. As opposed to the seismically transparent deep crust underneath Michigan, the seismic sections in Kansas are covered with arcuate and criss-crossing events which are in striking contrast with other deep crustal seismic sections. Seismic sections from COCORP data in Kansas (Serpa et al., 1984) exhibit a complicated wavefield at travel-times corresponding to lower crustal depths; the wavefield consists of arcuate crossing events that cannot successfully be migrated to separate domains. This suggests that at least some of these events are energy arriving out of the plane of the section, but whether these events represent deep signal or shallow side-arriving energy (noise) is uncertain. This is a fundamental problem in the interpretation of many such seismic sections. If these events are coming from depth, then this seismic section represents a classic, complicated (not plane layered) deep crustal section. The arcuate and crossing events are, in fact, quite similar to complex patterns from seismic models generated for typical igneous and tectonic features in crystalline rocks (Smithson, 1979; Wong et al., 1982). Does this mean that, for the first time, we are seeing typical complex crustal structure in the seismic data from Kansas? Possibly, but not necessarily! Readers should be warned of the possibility of side-arriving seismic events which certainly cause some of these crossing or arcuate events. Serpa et al. (1984, p. 373) suggest that at least one of these events is arriving vertically from a reflector in the upper mantle because of its behavior on crossing seismic lines. Further information on the crust is provided from the xenolith population in diatremes. The xenoliths are predominantly gabbro and mafic granulites with relatively small amounts of more felsic rocks. With the predominantly mafic xenolith population, it is somewhat perplexing that the crust appears so reflective. Nevertheless, <u>if side-arriving events are minimal</u>, then the seismic sections from Kansas represent the seismic response of typical heterogeneous crystalline crust consisting of folded metamorphic rocks and igneous intrusives.

Canada

Canadian geophysicists have conducted a number of seismic reflection experiments aimed at a better understanding of the Precambrian (Overton, 1972; Clee et al., 1974; Hajnal and Stauffer, 1975; Mair and Lyons, 1976). A wide-angle reflection study (Clee et al., 1974) helped resolve the geometry of "greenstones" and adjacent granitic rocks but, most importantly, showed a highly variable and complex reflection response in a crustal section that had appeared simple from a crustal refraction profile. The reflection results indicate that the crust consisted of strongly layered rock units and that the layering varied greatly laterally, a view consistent with surface exposures.

INTRUSIONS

Igneous intrusions in the crust may be detected in seismic profiles. Seismically transparent zones have typically been interpreted as intrusions, but readers should be aware that seismic sections may appear transparent for many other reasons (Smithson, 1979). Because they are not generally layered, intrusions will not be as good reflectors as mylonites and layered metamorphic sequences, but results suggest that reflection data will give information about intrusions, particularly their lower contacts. A number of seismic data sets indicate that reflections come from the bottom of or beneath intrusions ranging from granites to anorthosites to layered intrusions. Reflections have been found from beneath granite batholiths (Smithson et al., 1977b; Lynn et al., 1981), from the base of Laramie anorthosite (Smithson et al., 1977b; Allmendinger, 1982), from beneath anorthosite (Brown et al.,

1983) and from within a layered intrusion in the Wichita Mountains (Widess and Taylor, 1959; Wilson and Smithson, submitted for publication). In general, the reflectivity of layered mafic intrusions is uncertain; however, the interlayered granites and gabbros of the Wichita Mountains in Oklahoma generate rather good reflections. Reflections from the tops of batholiths have generally not been verified although these features constitute important geologic targets. Roof zones of intrusions will probably require special recording and processing techniques before they can be imaged routinely.

THE MOHO

The Moho is at present generally unidentified as a discrete reflection in seismic reflection profiles at vertical incidence across Precambrian terrains. For example, in the Kansas COCORP data set, a change in seismic character may occur at approximate Moho depth (Serpa et al., 1984). A Moho reflection has been picked at 16 s (about 48 km) under the Proterozoic crust of the Laramie Range, Wyoming, but because of the ringy nature of the seismic section at this depth, there is little to recommend it as a true Moho reflection. The Moho appears to be resolved more frequently in wide-angle reflection profiles (Meissner, 1973). Thus we know little about the nature of the Moho beneath the Precambrian in the U.S. from the results of crustal reflection profiling. This may be because of the nature of the Precambrian Moho, and special recording experiments may have to be designed to resolve the question. Soviet DSS-type studies (wide angle reflections) have mapped the Moho beneath the Baltic shield, the Ukrainian shield, (Sollugub et al., 1973) and the Indian shield (Chowdhury and Hargraves, 1981). In Canada, wide-angle Moho reflections suggest a complex transitional Moho and no significant change in crustal thickness from the Yellowknife to the Churchhill Province (Mair and Lyons, 1976).

CONCLUSIONS

The relatively detailed picture of the third dimension provided by seismic reflection interpretation can greatly alter previous conclusions about the geometry and genesis of Precambrian rocks. Processes such as vertical vs. horizontal movements, extension vs. compression, plate tectonics, intrusion, mafic differentiates, underplating, crustal scale deformation and evolution, nature of the Moho, and basin formation may be distinguished in seismic reflection data. The technique even has the potential to determine crustal composition by means of detailed velocity measurements although much experimentation and development is necessary to achieve this goal. Because no geophysical technique can approach the resolution of the human eye, the best knowledge of deep Precambrian crust will be provided by recognizing exposed crustal cross-sections (Fountain and Salisbury, 1981; Percival and Card, 1983). Verification of such tilted sections through the Precambrian crust could be a major contribution obtained from the application of seismic reflection methods. The use of reflection techniques in crystalline crust is so new that its potential is far from realized. The full significance of many seismic events in the deep crust (Fig. 3) is simply not understood, but such reflection patterns must clearly be carrying information critical to the genesis of that part of the Precambrian crust.

ACKNOWLEDGMENTS

This research was supported by U.S. National Science Foundation Grants EAR-8300659 and EAR-8306542. We especially thank Barbara Cox for help with the research and manuscript. Processing was carried out on the DISCO VAX 11/780 computer system of the Program for Crustal Studies.

REFERENCES

Allmendinger, R.W., Brewer, J.A., Brown, L.D., Kaufman, S., Oliver, J.E., & Houston, R.S. 1982. 'COCORP profiling across the Rocky Mountain Front in southern Wyoming, part 2: Precambrian basement structure and its influence on Laramide deformation'. Geol. Soc. Am. Bull. 93, 1253-1263.

Brewer, J.A., Brown, L.D., Steiner, D., Oliver, J.E., Kaufman, S. & Denison, R.E. 1981. 'Proterozoic basin in the southern midcontinent of the United States revealed by COCORP deep seismic reflection profiling'. Geology, 9, 569-575.

Brown, L., Jensen, L., Kaufman, S., and Steiner, D. 1982. 'Rift structure beneath the Michigan Basin from COCORP profiling'. Geology, 10, 645-649.

Brown, L.D., Ando, C., Klemperer, S., Oliver, J., Kaufman, S., Czuchra, B., Walsh, T. & Issachsen, Y.W. 1983. 'Adirondack-Appalachian crustal structure: The COCORP Northeast traverse'. Geol. Soc. Am. Bull. 94, 1173-1184.

Chowdhury, K.R. & Hargraves, R.B. 1981. 'Deep seismic soundings in India and the origin of continental crust'. Nature, 291, 648-650.

Clee, T.E., Barr, K.G., & Berry, M.J. 1974. 'The fine structure of the crust near Yellowknife'. Can. J. Earth Sci. 11, 1534-1549.

Condie, K.C. 1972. 'A plate tectonics evolutionary model of the South Pass Archean greenstone belt, southwestern Wyoming'. 24th Int'l. Geol. Congr. Sect. 1. 104-112.

Fatti, J.L. 1972. 'The influence of dolerite sheets on reflection seismic profiling in the Karroo Basin'. Trans. Geol. Soc. South Africa, 75, Part 2, 71-76.

Fountain, D.M. & Salisbury, M.H. 1981. 'Exposed cross-sections through the continental crust: Implications for crustal structure, petrology, and evolution'. Earth Planet. Sci. Lett. 56, 263-277.

Fountain, D.M., Hurich, C.A & Smithson, S.B. 1984. 'Seismic reflectivity of mylonite zones in the crust'. Geology, 12, 195-198.

Fuchs, K. 1969. 'On the properties of deep crustal reflections'. J. Geophys. 35, 133-149.

Gibb, A.K., Payne, B., Setzer, T., Brown, L.D., Oliver, J.E. & Kaufman, S. 1984. 'Seismic-reflection study of the Precambrian crust of central Minnesota'. Geol. Soc. Am. Bull. 95, 280-294.

Goldich, S.S. & Hedge, C.E. 1974. '3,800 myr granitic gneiss in southwestern Minnesota'. Nature, 252, 467-468.

Grant, J.A. 1972. 'Minnesota River Valley, southwestern Minnesota'. in Sims, P.K. and Morey G.B., eds. Geology of Minnesota-A centenial volume. Minnesota Geological Survey, 177-198.

Hajnal, Z. & Stauffer, M.R. 1975. 'The application of seismic reflection techniques for subsurface mapping in the Precambrian shield near Flin Flon, Manitoba'. Can. J. Earth Sci. 12, 2036-2047.

Hodge, D.S., Owen, L.B. & Smithson, S.B. 1973. 'Gravity interpretation of Laramie anorthosite complex, Wyoming'. Geol. Soc. Amer. Bull. 84, 1451-1464.

Hurich, C.A., Smithson, S.B., Fountain, D.M., & Humphreys, M.C. 'Mylonite reflectivity: evidence from the Kettle dome, Washington'. Submitted for publication to Geology.

Johnson, R.A., Karlstrom, K.E., Smithson, S.B., & Houston, R.S. 1984. 'Gravity profiles across the Cheyenne Belt, a Precambrian crustal suture in southeastern Wyoming'. J. Geodynamics, 1, in press.

Johnson, R.A., Pierson, W. & Smithson, S.B. 'Reprocessing of crustal reflection data'. Int'l Symp. on Deep Structure of the Continental Crust, submitted to American Geophysical Union, Geodynamics Series.

Karlstrom, K.E. & Houston, R.S. 1984. 'The Cheyenne belt: Analysis of a Proterozoic suture in southern Wyoming'. Precambrian Research, 25, 415-446.

Kubichek, R.F., Humphreys, M.C. Johnson, R.A. & Smithson, S.B. 1984. 'Long-range recording of VIBROSEIS data: Simulation and experiment'. Geophys. Res. Lett. 11, 809-812.

Mair, J.A. & Lyons, J.A. 1976. 'Seismic reflection techniques for crustal structure studies'. Geophysics, 41, 1272-1290.

Oliver, J. Dobrin, M., Kaufman, S., Meyer, R. & Phinney, R. 1976. 'Continuous seismic reflection profiling of the deep basement, Hardeman County, Texas'. Geol. Soc. Am. Bull. 87, 1537-1546.

Overton, A. 1972. 'Seismic experiments on the Quill Lake syncline, Ontario'. Geol. Assoc. Can Proc. 24, 55-68.

Percival, J.A. & Card, K.D. 1983. 'Archean crust as revealed in the Kapuskasing uplift, Superior Province, Canada'. Geology, 11, 323-326.
Phinney, R.A. & Odom, R.I. 1983. 'Seismic studies of crustal structure'. Rev. Geophys. Space Phys. 21, 1318-1332.
Pierson, W.R. 1984. A geophysical study of the contact between the greenstone-granite terrain and the gneiss terrain in central Minnesota. Unpubl. M.S. Thesis, University of Wyoming, Laramie, 84 p.
Schilt, F.S., Kaufman, S. & Long, G.H. 1981. 'A three-dimensional study of seismic diffraction patterns from deep basement sources'. Geophysics, 46, 1673-1683.
Serpa, L., Setzer, T., Farmer, H., Brown, L., Oliver, J., Kaufman, S. & Sharp, J. 1984. 'Structure of the southern Keweenawan rift from COCORP surveys across the Midcontinent geophysical anomaly in northeastern Kansas'. Tectonics, 3, 367-384.
Sims, P.K., Card, K.D., Morry, G.B., & Peterman, Z.E. 1980. 'The Great Lakes tectonic zone - A major crustal feature in North America'. Geol. Soc. Amer. Bull. 91, 690-698.
Smithson, S.B. 1979. 'Aspects of continental structure and growth: targets for scientific deep drilling'. University of Wyoming Contributions to Geology. 17, 65-75.
Smithson, S.B., Shive, P.M., & Brown, S.K. 1977a. 'Seismic velocity, reflections, and structure of the crystalline crust'. The Earth's Crust, Amer. Geophys. Union Mono. 20, 254-270.
Smithson, S.B., Shive, P.N., & Brown, S.K. 1977b. 'Seismic reflections from Precambrian crust'. Earth Planet. Sci. Lett. 35, 134-144.
Smithson, S.B., Brewer, J.A., Kaufman, S., Oliver, J., & Hurich, C. 1979. 'Structure of the Laramide Wind River uplift, Wyoming, from COCORP deep reflection data and from gravity data'. J. Geophys. Res. 84, 5955-5972.
Smithson, S.B., Brewer, J.A., Kaufman, S., Oliver, J.E. & Zawislak, R.L. 1980. 'Complex Archean lower crustal structure revealed by COCORP crustal reflection profiling in the Wind River Range, Wyoming'. Earth Planet. Sci. Lett. 46, 295-305.
Sollugub, V. B. et al. 1973. 'New DSS data on the crustal structure of the Baltic and Ukrainian shields'. Tectonophysics, 20, 67-84.
Widess, M.B. & Taylor, G.L. 1959. 'Seismic reflections from layering within the basement complex'. Geophysics, 24, 417-425.
Wilson, S.L. & Smithson, S.B. 'Structure and nature of a Proterozoic basin in southwestern Oklahoma as revealed by COCORP surveys'. Submitted for publication in Geology.
Wong, Y.K., Smithson, S.B. & Zawislak, R.L. 1982. 'The role of seismic modeling in deep crustal reflection interpretation, Part I'. University of Wyoming Contributions to Geology, 20, 91-109.

PROTEROZOIC ANORTHOSITE MASSIFS

R.F. Emslie
Geological Survey of Canada
601 Booth Street
Ottawa, Ontario K1A OE8
Canada

ABSTRACT. Major anorthositic massifs intruded stable cratonic crust in the North Atlantic region about 1.4 to 1.7 Ga ago. Similar (meta-) anorthosite massifs (0.9-1.6 Ga) in Grenville and Sveconorwegian terranes exhibit varying degrees of deformational and metamorphic overprint and consequently have ambiguous tectonic settings. Close correspondence of rock and mineral associations in both types points to common origins. Mineral assemblages, chemical, and isotopic data for pristine and meta-anorthosite massifs strongly imply varying degrees of interaction between mantle-generated magmas and deep continental crust occurred prior to final emplacement. Olivine-bearing massifs consistently show less evidence of crustal assimilation than olivine-free massifs. Parent magmas of anorthosite massifs were LREE-enriched with moderately high Fe/Mg, consistent with fractionated basic magmas or their derivatives. Recently reported Nd and Sr isotopic relationships imply that the parent magmas were not simply partial melts of undepleted subcontinental mantle. Appropriate dry, plagioclase-saturated or -supersaturated magmas can be generated on a large scale only within a depth interval $\sim$35-50 km, embracing the lowermost continental crust and upper mantle. Fractionation of pyroxenes ±olivine was an important factor in determining the chemistry of parent magmas and in decreasing their densities prior to intrusion into the continental crust. The high level intrusions of Fennoscandia and the East European platform are not obviously related to crustal rifting. Widespread non-orogenic granitic suites, including rapakivi types, span a similar time interval and provide supporting evidence for local heat sources perhaps related to subcratonic mafic magma ponding. This unique period of geological history may mark a transition to Phanerozoic tectonic regimes in which large scale magmatism became increasingly confined to plate margins and intracontinental rift zones.

1. INTRODUCTION

The abundance of large anorthosite massifs in the Middle Proterozoic is one of the unique features of the Earth's geological record. Understanding the significance of these intrusions in the evolution of the Proterozoic crust hinges primarily on correctly deciphering the tectonic implications of their presence and the conditions of generation, intrusion, and crystallization of the magmas. Recent reviews (1,2,3,4,5) emphasize various aspects of the geology, petrology, and geochemistry of massif anorthosites from different viewpoints and summarize directions of current research. Outstanding problems remain in explaining apparent differences between some meta-anorthosite massifs in Grenville and Sveconorwegian terranes and pristine massifs elsewhere (5).

Efforts to explain massif anorthosites and their associated silicic rocks as stemming from a single parent magma, by invoking fractionation, hybridization, liquid immiscibility, or other processes, have not been notably successful. There has been a growing body of data to indicate that many of the attributes of magmas parental to anorthosite massifs were acquired prior to intrusion of the magmas into the continental crust and that a number of different magmas existed in larger complexes.

Only two source materials capable of producing dry, aluminous magmas parental to anorthosite massifs now receive serious consideration, basic granulites of the deep crust (6) and mantle peridotite. The tectonic setting in which the magmatism occurred is an important element in assessing preferences for one source or the other.

Mostly because of their distinctive compositions, massif anorthosites have long been recognized as representing an unusual kind of Proterozoic magmatism. As isotopic dating progressed, it became increasingly evident that anorogenic granites, many of which resemble those associated with anorthosite massifs, span a similar time interval (1,7). It is virtually certain that the ultimate control for both types of magmatism is the same and constitutes either a unique episode unrelated to previous or subsequent events, or forms part of the normal evolutionary development of the crust and mantle.

Rather than attempt an exhaustive review, my intent is to concentrate on those aspects of massif anorthosites that, in my opinion, bear most directly on their genesis and relationship to the Proterozoic continental crust.

2. DISTINCTIVE FEATURES OF MASSIF ANORTHOSITE

Aspects of massif anorthosites widely considered to be significant in assessing how they formed and the nature of the crust they intruded are summarized briefly below:

2.1 Ages

Major massifs lie in the age range 0.9 to 1.7 Ga. Some older examples may occur but are not well documented; Archean examples are unknown. Small Paleozoic massifs are very rare (8,9).

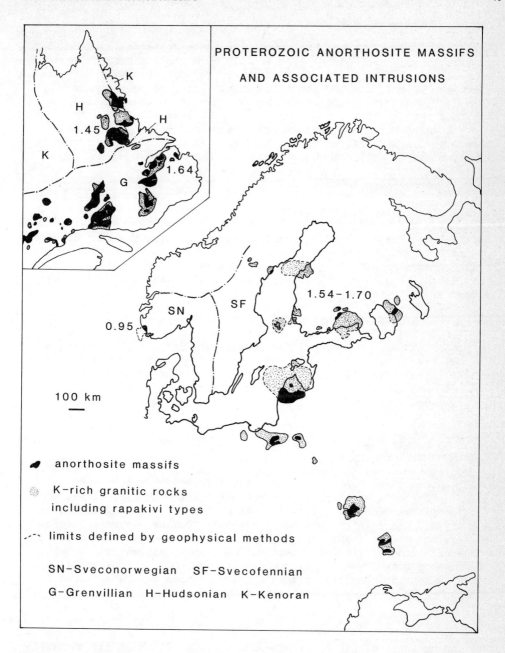

Fig. 1. Distribution of major Proterozoic anorthosite massifs and associated granitic rocks in the North Atlantic region. European data modified from (10). Inset and main map are at the same scale. Approximate ages (Ga) of igneous activity are indicated where U-Pb zircon data are available.

2.2 Geological Settings

Massifs typically, but not exclusively, intrude Proterozoic terranes that previously have undergone relatively high grade metamorphism and deformation. Pristine massifs of central Labrador, the East European platform, Finland, and the Ukrainian Shield (Fig. 1) demonstrate intrusion into nonorogenic environments. Massifs with variable overprints of metamorphism and deformation, as in Grenville and Sveconorwegian terranes, have ambiguous tectonic settings because of these superimposed effects whose absolute ages with respect to massif magmatism are largely unknown.

2.3 Associated Rocks

A distinctive suite of rock types, present in variable amounts, accompanies and typically shows intrusive relations toward massif anorthosite. The dominant compositions of this suite are ferrodiorite (jotunite) which is normally volumetrically small, and a K-rich group ranging from monzonite to granite (mangerite to charnockite). A most notable feature of the entire assemblage is the <u>absence</u> of significant volumes of mafic rocks.

2.4 Depths of Intrusion

Depths of intrusion not exceeding a few km are indicated for massifs in Finland, East European platform, and the Ukrainian Shield (10,11,12). Central labrador massifs were intruded to depths of about 10 to 12 km (13). In the Grenville Province and Rogaland, depths of 15 to 25 km are indicated for some massifs but examples of shallower depths of intrusion may also exist.

2.5 Parent Magma Compositions

Mineral compositions in massifs imply equilibrium with dry liquids having intermediate Mg/Mg+Fe and Ca/Ca+Na (ferrogabbro to ferrodiorite range). Plagioclase, olivine, and orthopyroxene compositions are much less 'primitive' (more evolved) in massifs than in early cumulates of many layered basic intrusions. The high plagioclase content of massifs requires operation of some process to concentrate plagioclase crystals or the existence of very high temperature liquids approaching anorthosite in composition (14). Occurrences of aluminous orthopyroxene megacrysts in massifs suggests polybaric fractionation may have played an important role in genesis of the parent magmas.

2.6 Trace Elements and Isotopes

Very low contents of all trace elements incompatible with plagioclase is consistent with the adcumulate nature of most massif anorthosite. Very high Sr contents are characteristic of plagioclase in massifs. LREE-enriched patterns are typical of massif plagioclase, implying crystallization from relatively evolved liquids. Initial $^{87}Sr/^{86}Sr$ is variable within individual, and between associated, massifs implying variable crustal contamination. Trace elements and isotopic data are consistent with crustal derivation of parent magmas of the K-rich monzonite-granite suite (15,16).

2.7 Physical Character of Anorthosite Parent Magmas

There is little agreement about the physical nature of the parent magmas of massifs. The choice between largely liquid magmas and liquids heavily charged with plagioclase crystals has not yet been successfully resolved. Crystal mush magmas have been favoured for several massifs in Grenville and Sveconorwegian terranes. Pristine massifs in central Labrador, Fennoscandia, and the Ukraine show much less obvious evidence of textures and structures that have been interpreted as protoclastic.

2.8 Contrasting Features of Archean Anorthosites

Unlike Proterozoic massifs, most Archean anorthosites form stratiform parts of basic layered intrusions. The plagioclase cumulates are highly calcic with magnesian augite or hornblende as typical mafic silicates. Archean anorthositic rocks commonly consist of coarse, equant, euhedral to subhedral plagioclase megacrysts in a finer grained mafic matrix and are associated with greenstone belts in which there is evidence to support a genetic relationship with the associated mafic volcanic rocks (17). Voluminous, K-rich, high Fe/Mg felsic intrusions are not associated with Archean anorthosites.

3. TECTONIC SETTINGS OF MASSIF COMPLEXES

Constraints on possible petrological processes that may lead to development of anorthosite massif complexes are imposed by the tectonic regime of the magmatism. Synorogenic and nonorogenic interpretations have been proposed but some of these have depended heavily on assumed models. Much evidence supports continental interiors as principal sites of intrusion; host rocks are most commonly gneiss terranes developed in an earlier (sometimes a few hundred Ma earlier) Proterozoic orogenic event.

Several meta-anorthosite massifs in Grenville and Sveconorwegian terranes have been interpreted as synorogenic intrusions (18,19). In much of the Grenville Province relative sequences of deformation, intrusion, and metamorphism can be established in some detail but absolute ages are typically unknown or uncertain and subject to several interpretations. Although degrees of metamorphism and deformation in these massifs vary widely, the fact that Grenvillian and Sveconorwegian terranes are themselves not well understood makes them fragile foundations on which to base tectonic interpretations of the magmatism. This does not imply that proposed tectonic models for these terranes are incorrect, but they are not easily defended.

Pristine massifs lacking overprints of deformation and metamorphism are characteristic of the Ukrainian Shield, East European platform, Fennoscandia, and central Labrador. These are widely regarded as examples of nonorogenic magmatism in stabilized cratonic crust (1,10,11,12,20); the interpretations are strongly supported by geological and isotopic evidence. Rock and mineral assemblages in these massifs are so alike in essential features to Grenville and Sveconorwegian examples that closely similar processes must have controlled their development. Rare Paleozoic anorthosite massifs in Niger (9) and Sept Iles, Quebec (8) are small, associated with alkalic basic magmatism, and clearly were intruded into cratonic settings.

For any magmatism in cratonic settings, mantle-derived heat or magma, or both, necessarily plays a central role. Continental rift settings for massif anorthosite magmatism have been suggested by Bridgwater and Windley (21) and by Berg (22). Currently favoured models for active continental rifting consider that the subcontinental lithosphere becomes thinned to zero at the base of the crust establishing a high geothermal gradient (23). Heat transfer to the crust may be largely conductive, causing significant, but slow, partial melting of the lower crust or, if mantle magmas develop, more efficient convective heat transfer from the magmas can cause crustal melting. An alternative scenario would maintain a relatively cool, thick lithosphere through which mantle magmas are delivered to the base of the crust via fracture systems or diapirs. In this case, a relatively low geothermal gradient is maintained and the magmas are not in equilibrium with the uppermost subcontinental mantle; they may cool, crystallize, and fractionate. Again, convectively transferred latent heat from the crystallizing ponded magma can provide abundant heat for crustal partial melting.

Mid-Proterozoic nonorogenic basic and silicic magmatism spans several hundred Ma or more in Fennoscandia and in central Labrador but clear evidence of fault-controlled continental rifting is absent even in the subvolcanic intrusions of Fennoscandia. Large anorthosite massifs have not been convincingly shown to be directly associated with known continental rifts or paleorifts.

4. MINERALOGY AND GEOCHEMISTRY

Plagioclase compositions in massifs lie almost exclusively within the labradorite to andesine range; $An_{50\pm10}$ encompasses most known examples. Some massifs contain rocks spanning a considerable part of the total range and it is clear that no rigid division occurs within the range. Nevertheless, a significant correlation exists (cf. Figs. 2 and 3) between olivine-bearing facies and more calcic plagioclase (generally labradorite), and between orthopyroxene-bearing facies and more sodic plagioclase (commonly andesine antiperthite).

Plagioclase much more calcic than about An_{60} is rare in massif anorthosite proper although more calcic compositions may occur in locally associated troctolite or leucotroctolite. Plagioclase more sodic than about An_{40} is also rare; notable exceptions include Labrieville (24) and Roseland (25) both of which are also unusually potassic.

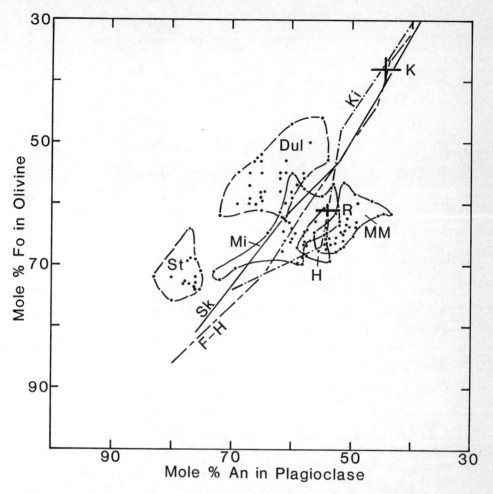

Fig. 2. Coexisting olivine and plagioclase compositions in anorthositic massifs compared to layered intrusions. Sk-Skaergaard, Ki-Kiglapait, F-H-Fongen-Hyllingen, St-Stillwater, Dul-Duluth, K-Korosten, R-Riga, Mi-Michikamau, H-Harp Lake, MM-Mealy Mountains. Data from (1,10,26,27,28,29,30) and unpublished.

An important feature of massif anorthosites emphasized by Morse (4) is their characteristically low content of augite. Where augite occurs, it is nearly always a minor, intercumulus mineral. Only in rocks with relatively ferrous-rich mafic silicates does augite become more abundant. This clearly implies that the parent magmas were undersaturated in augite component during most of their crystallization. Marked augite undersaturation relative to olivine and low-Ca pyroxene is evidently an intrinsic property of the parent magmas that requires explanation.

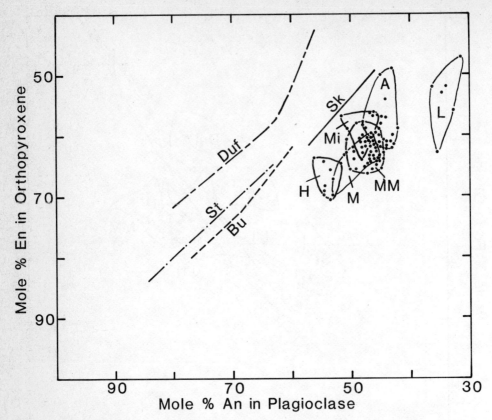

Fig. 3. Coexisting orthopyroxene and plagioclase compositions in anorthositic massifs compared to layered intrusions. Sk-Skaergaard, Bu-Bushveld, St-Stillwater, Duf-Dufek, A-Adirondacks, M-Morin, L-Labrieville, MM-Mealy Mountains, Mi-Michikamau, H-Harp Lake. Data from (1,24,31,32) and unpublished.

Olivine compositions in massif anorthositic rocks (excluding ferrodiorites, jotunites, and more K-rich rocks) lie chiefly in the range Fo_{70} to Fo_{58} but compositions of Fo_{40} to Fo_{36} are reported from Korosten anorthosites (10). Low-Ca pyroxene is typically hypersthene En_{70} to En_{55}. Inverted pigeonite occurs in some massifs and its composition generally is in the more Fe-rich part of the low-Ca pyroxene range.

Compositions of coexisting plagioclase with olivine, and plagioclase with orthopyroxene in massifs show consistent differences from comparable assemblages in layered basic intrusions (Figures 2 and 3). Data for olivine-bearing anorthositic rocks are not very abundant but there is a tendency for massifs to have more sodic plagioclase at equivalent olivine compositions relative to tholeiitic layered intrusions. The three most calcic pairs from Michikamau are from layered troctolite that forms a very small part of the whole complex. Kiglapait pairs, over part of their range, are similar to the massif assemblages and it should be kept in mind that Kiglapait is closely related to, and may be considered a late member of, the Nain complex. Relative to layered intrusions, the tendency for massif plagioclase to be more sodic at equivalent orthopyroxene compositions is much more pronounced in Figure 3. There is a heavy concentration approximately in the range An_{43-51}, En_{56-66}. The Nain complex also shows a maximum in this compositional region (4).

Compared to layered basic intrusions, massif mineral assemblages have a restricted range of mineral compositions. There is a marked truncation that excludes more Mg- and An-rich pairs; ferrodiorites or jotunites closely associated with many massifs have mineral compositions which would extend the ranges to considerably more Fe- and Ab-rich pairs.

Aluminous orthopyroxene megacrysts occur widely in massif anorthosites (33,34,35,36) and commonly have bulk Al_2O_3 contents in the range 4 to 9 weight percent. Exsolved calcic plagioclase (common), spinel (less common), and garnet (rare) have been reported from different localities. Megacrysts tend to have similar or higher Mg/Mg+Fe than matrix orthopyroxenes of their host rocks. Megacrysts may also have Ni and Cr contents several times higher than matrix orthopyroxenes (2,36). Megacrysts range from rounded, nodular shapes, sometimes with granular reaction rims containing low-Al orthopyroxene, to subophitic intergrowths with plagioclase.

The intermediate ranges of Mg/Mg+Fe in mafic silicates and of Ca/Ca+Na in plagioclase of massif anorthosites indicate parent magmas that were much less 'primitive' than typical MORB for example, but resemble some continental tholeiites. Minor and trace element chemistry is also consistent with relatively evolved parent magmas, possibly modified by crustal contamination, a characteristic also shared by many continental tholeiites.

REE patterns of massif anorthosite whole rocks, and of plagioclase separates from them, imply that the liquids with which they equilibrated were relatively enriched in light REE suggesting Ce ~40 to 60 times chondrites and Ce/Yb ~3 to 8 (6,36). Ferrodiorites and jotunites have REE patterns that most closely approach the requirements for equilibrium with massif plagioclase (3,15,37) and are also enriched in Fe^{2+}, Ti, P, and K, all consistent with advanced fractionation of basic magma.

Sr contents of massif plagioclase are invariably much higher than those of plagioclase crystallized from common tholeiitic magmas as shown in Figure 4. These data also show that An content of plagioclase is not a principal factor controlling $D_{Sr}^{plag/liq}$. The persistently high Sr concentrations in anorthosite plagioclase are evidence that significant extraction of plagioclase from the parent magmas did not occur prior to anorthosite crystallization. Intermediate massif plagioclase compositions cannot therefore, be attributed to prior plagioclase fractionation. Strong positive Eu anomalies in plagioclase provide

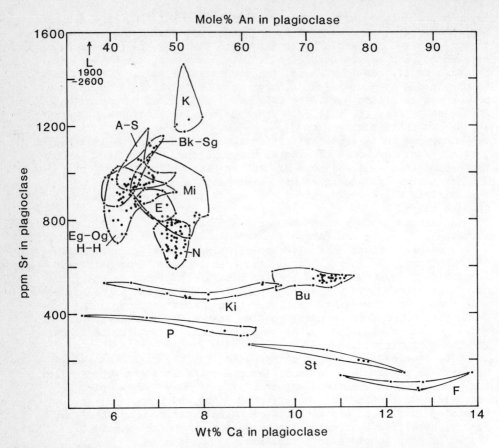

Fig. 4. Sr and Ca in plagioclase of anorthosite massifs compared to other basic intrusions. Eg-Og-Egersund-Ogna, H-H-Haaland-Helleren, Bk-Sg-Bjerkreim-Sogndal, A-S-Ana-Sira, K-Kenemich, E-Etagaulet (Mealy Mtns.), L-Labrieville, Mi-Michikamau, N-Nain, Ki-Kiglapait, Bu-Bushveld, P-Palisades, St-Stillwater, F-Fiskanaesset. Data from (38,39,40,41,42,43,44,45,46) and unpublished.

similar evidence but variable valence states of Eu are a function of fO_2 in equilibrium with the magmas and are known to affect partitioning.

The marked, regular inverse correlation between Sr and Ca in plagioclase crystallized from tholeiitic basic magmas of layered intrusions is due to the fact that plagioclase was a cumulus mineral and its abundant cotectic crystallization with mafic silicates prevented residual liquids from becoming notably enriched in Sr. High Sr in massif plagioclase on the other hand, rules out prior extraction of plagioclase from parent magmas and suggests previous abundant clinopyroxene fractionation which would be consistent with depletion of Ca

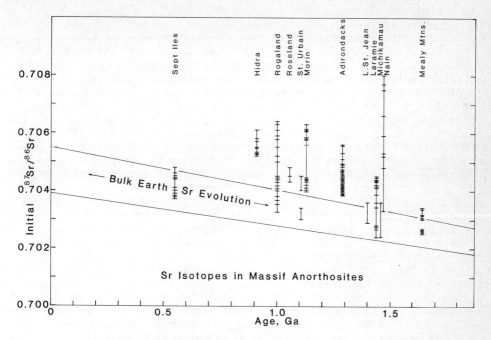

Fig. 5. Sr isotopes in massif anorthosites. Bulk earth Sr evolution envelope is shown for reference only. Data from (37,47,48,49,50,51,52, 53,54,55,56).

combined with Sr enrichment in the liquids. Olivine and orthopyroxene fractionation would also help boost Sr in liquids but pressure effects on $D_{Sr}^{plag/liq}$ may be the most important factor in explaining high Sr in massif plagioclase. The Sr^{2+} ion is substantially larger than Ca^{2+} (~12 percent larger radius) implying that $SrAl_2Si_2O_8$, the strontium analogue of anorthite, has a larger molar volume than anorthite. If plagioclase at higher pressures (sub-crustal) has lower density than liquid (i.e. it floats) then Sr should be concentrated in plagioclase (the lower density phase) relative to liquid. This effect would be enhanced if the liquid were denser than normal basalt (ferrogabbro to ferrodiorite for example).

All trace elements incompatible with plagioclase are typically low in massif anorthosites. This, combined with exceedingly high K/Rb in whole rocks (locally >10,000, e.g. (43)) and lack of significant zoning in plagioclase, is consistent with the extreme adcumulate nature of the rocks.

Sr isotopic data for massif whole rocks have been compiled in Figure 5. Isotopic data for ferrodiorites, jotunites, and K-rich and silicic rocks have been excluded from the compilation. Lower initial $^{87}Sr/^{86}Sr$ ratios (Sr_i) are associated with olivine- and labradorite-bearing massifs. This is particularly evident for the Mealy Mountains where the olivine-labradorite Kenemich massif has lower Sr_i than the adjacent orthopyroxene-andesine Etagaulet massif and at St. Urbain where labradorite anorthosite with lower Sr_i is intruded by andesine anorthosite with higher Sr_i. Nain, Michikamau, Laramie, and Rogaland massifs all contain olivine-bearing rocks in varying amounts. Sept Iles anorthositic rocks contain olivine to the exclusion of orthopyroxene. Within individual massifs or massif complexes, the range of apparent Sr_i is consistent with contamination by crustal Sr. The high Sr levels of massif anorthosites combined with Rb/Sr commonly in the range 0.01 to 0.001 requires that contamination took place at the magmatic stage, prior to, or during plagioclase crystallization and is not a product of secondary alteration. Sufficiently detailed data do not exist for any single massif to demonstrate that contamination was progressive but that remains a plausible interpretation.

Studies of Sr and Nd isotopes in anorthosite samples from a number of massifs by Ashwal and Wooden (54) have shown a consistent areal distribution pattern related to position within or outside the Grenville Province, suggesting either different mantle reservoirs, different crustal reservoirs, or both.

The limited, intermediate ranges of Mg/Mg+Fe in mafic silicates, and of Ca/Ca+Na in plagioclase in massifs is consistent with crystallization from a strongly fractionated derivative of basalt or more mafic magma. Assuming a fairly primitive basalt starting composition, the Mg/Mg+Fe of mafic silicates in many massifs would be consistent with 70-80 percent crystallization or more compared to tholeiitic layered intrusions crystallized in the crust. If the starting liquid were picritic or more magnesian, a correspondingly greater amount of crystallization would be necessary. The implied large degree of crystal fractionation took place prior to intrusion of the massifs into the crust as indicated by the general lack of earlier differentiates or geophysical evidence for them. Aluminous orthopyroxene megacrysts in massifs offer support for an earlier, higher pressure crystallization stage but other interpretations have been suggested for them (4,35). Subophitic intergrowths of some of these pyroxenes with plagioclase implies cotectic crystallization that could have taken place within a deep crustal chamber or prior to intrusion into the crust. Ca-Sr relationships in massif plagioclase suggest earlier, high pressure clinopyroxene fractionation also took place.

5. ANORTHOSITE PARENT MAGMAS AND THEIR GENERATION

Parent magmas of anorthosite massifs continue to be a central problem. If the parent magmas approached the bulk compositions of massifs, they were either very high temperature liquids (14), or lower temperature mixtures of crystals plus liquid. Dry liquids approximating anorthosite compositions require special source materials, special melting conditions, or both, and the intermediate Mg/Mg+Fe and Ca/Ca+Na of mafic silicates and plagioclase respectively do not favour unusually high crystallization temperatures. Such liquids, if fractionally crystallized, should produce residual liquids with large negative Eu anomalies which have not been observed. Mineral assemblages of massif anorthosites have commonly been regarded as early-formed crystallization products of a parent magma but the mineral compositions are more in keeping with advanced stages of fractional crystallization of basic magma. The general lack of evidence for a mafic complement to anorthosite within the crust implies that any such fractionation must have taken place largely below the continental crust.

Simmons and Hanson (6) have argued that the required intermediate Mg/Mg+Fe, high Al_2O_3, and dry character of anorthosite parent magmas do not permit derivation from peridotite source rocks and they favour lower crustal tholeiitic compositions as source materials. This assumes however, that the parent magmas must be direct melts from a peridotitic mantle.

Relatively low Mg/Mg+Fe, high Al_2O_3, dry magmas can be derived from peridotite mantle sources but only by cooling, crystallizing, and fractionating initially more 'primitive' liquids. As discussed in the previous section, mineral assemblages and geochemical features of massif anorthosite are consistent with crystallization from considerably evolved basic magmas.

In order to generate an aluminous magma (anorthositic, plagioclasic) the liquid component must have been saturated with plagioclase at its source or, more likely at some intermediate stage prior to intrusion into the crust. No other condition will permit plagioclase to be the sole liquidus phase for some crystallization interval at lower pressures and promote adcumulus growth in plagioclase cumulates. Of liquids that may equilibrate with the principal aluminous phases in the mantle, those in equilibrium with plagioclase reach maximum enrichment in Al_2O_3 (18 to 20 weight percent), but because of upper stability limits on plagioclase such liquids cannot exist at pressures much greater than about 17 kb (~50 km depth). In addition to plagioclase these liquids coexist at equilibrium with orthopyroxene, clinopyroxene, and spinel, but not with olivine (reaction points A_3, A_4, Fig. 6). If the subcontinental mantle is peridotite (i.e. olivine-bearing), aluminous liquids can never be in equilibrium with it. This limits generation of aluminous liquids to two processes; they may form by direct partial melting of basic granulites of the lower crust or they may develop by crystal-liquid fractionation from more primitive mantle magmas whose ultimate source was at greater depths. For reasons discussed in the next section, the first alternative may have contributed but is unlikely to have been a major factor in the genesis of massif anorthosite parent magmas.

Several aspects of polybaric basic magma generation and crystallization that bear directly on anorthosite production are summarized in Figure 6. Natural liquids do not behave exactly like the system $CaO-MgO-Al_2O_3-SiO_2$ but the same principles apply and the correspondence is close, especially for more 'primitive' compositions (58).

Projections from Di show the progressive shrinkage of the anorthosite volume with increasing pressure; this is accompanied by separation of reaction points (pseudoinvariant points in natural systems) A and U above ~9-10 kb and their migration in opposite directions, A toward the An apex, U toward the Fo apex. Above about 17 kb the garnet liquidus volume pierces the surface of projection and with a small further

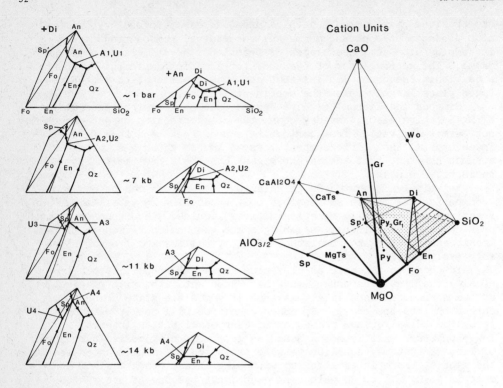

Fig. 6. Projected liquidus relations in the system $CaO-MgO-Al_2O_3-SiO_2$. Projection method and Di projections after Irvine and Sharpe (57). An projections constructed from the same data base and derived largely from (58). Barbs on liquidus boundary curves indicate down temperature direction. Reaction points designated A and U (A-aluminous assemblage, U-ultramafic assemblage) correspond to pseudoinvariant points in natural systems and constitute primary controls on melt compositions during partial melting and fractional crystallization of mafic and ultramafic solids and liquids.

increase of pressure the anorthite volume disappears entirely so aluminous A liquids can no longer exist.

Projections from An illustrate phase relations of liquids in equilibrium with plagioclase and show the remarkable expansion of the diopside liquidus volume with increasing pressure. Above ~9-10 kb, reaction points U_3, U_4 have migrated behind the An liquidus surface and are not visible in An projections. At pressures above about 17 kb increasing stability of garnet begins to encroach upon the Di volume.

Subcontinental mantle liquids such as U_3 or U_4, in equilibrium with peridotite, must cool, fractionate with loss of olivine by reaction with liquid and precipitate orthopyroxene, clinopyroxene, and spinel to reach reaction points A_3 or A_4; this can only happen if the uppermost mantle is well below solidus temperatures. Decreased solubility of clinopyroxene with increased pressure is due to the breakdown of $CaAl_2Si_2O_8$ (An) to form $CaAl_2SiO_6$ (CaTs) and occurs only in liquids in equilibrium with plagioclase. In natural magmas precipitation of an aluminous clinopyroxene component reduces the An/Ab ratio of the liquids.

Once liquids have reached reaction points like A_3 or A_4 they will have their minimum density just as plagioclase begins to precipitate. Prolonged crystallization may occur at such reaction points, in which case precipitation of mafic silicates of the assemblage would cause continuing decrease in Mg/Mg+Fe of liquids. If plagioclase remained in suspension or floated and mafic silicates sank, plagioclase could accumulate in the upper part of the chamber. Even with relatively Fe-enriched liquids the bulk density of a plagioclase-liquid suspension (magma) may be considerably less than that of the liquid alone. Decompression of such a magma by rising into the crust would cause immediate adcumulus growth of plagioclase from liquids undersaturated in clinopyroxene, orthopyroxene, and spinel. Depending upon the degree of contamination of the liquid by siliceous, crustal-derived melts during subcrustal crystallization, olivine or orthopyroxene would first crystallize cotectically with plagioclase after decompression.

The strongest evidence to support polybaric crystallization in generation of massif anorthosite magmas is the presence of aluminous orthopyroxene megacrysts. Aluminous clinopyroxene megacrysts are only rarely observed (33) and this may be due to their greater sensitivity to pressure change and silica contamination as the following reactions suggest:

$$CaAl_2SiO_6 + SiO_2 = CaAl_2Si_2O_8 \quad \Delta V_{298,\ 1\ bar} = +14.6\ cc/mole \quad (1)$$

CaTs cpx   liq   plag

$$MgAl_2SiO_6 = MgAl_2O_4 + SiO_2 \quad \Delta V_{298,\ 1\ bar} = +0.3\ cc/mole \quad (2)$$

MgTs opx   spinel   liq

The much larger ΔV of reaction (1) implies that decompression causes greater displacement from equilibrium and leads to clinopyroxene breakdown, other things being equal. Note also that increase in silica activity has opposite effects on the two reactions. In reaction (1) increase in silica activity promotes plagioclase formation at the expense of CaTs clinopyroxene whereas in reaction (2) increased silica activity stabilizes MgTs orthopyroxene relative to spinel. Silica contamination of mantle magmas by crustal melts may therefore contribute to preservation of high pressure aluminous orthopyroxene megacrysts and at the same time cause decomposition of high pressure aluminous clinopyroxene. Similar effects may help to account for the absence of clinopyroxene phenocrysts in some continental basalt suites that seem to require high pressure clinopyroxene fractionation at some pre-eruption stage of their evolution (59,60). Silica contamination of anorthositic magmas may also play a role in causing exsolution of calcic plagioclase from aluminous orthopyroxene megacrysts.

6. ROCKS ASSOCIATED WITH ANORTHOSITE MASSIFS

The suite of rocks associated with anorthosite massifs has important petrological and geological implications that require integration into any successful model. The marked bimodal character of the magmatism associated with massif anorthosites (dominated by anorthosite and related plagioclase-rich rocks as one end-member, and K-rich monzonite-granite or mangerite-charnockite as the other) implies the presence of a crustal component that yielded approximately granitic minimum melts. Trace element and isotopic evidence generally supports that interpretation (15,16,48). If both granitic and anorthositic parent magmas were products of crustal melting, it would be difficult to understand why bimodal distributions resulted instead of a continuum of compositions.

Ferrodiorites and jotunites, although subordinate members of the associations, appear to form an important link between massif anorthosites and the K-rich group. Ferrodiorites occur as both cumulate rocks (3,41,61,62) and as dykes and other fine grained rocks that approach probable liquid compositions (48,62,63). The latter group has bulk Mg/Mg+Fe and trace element contents generally in the appropriate range for liquids in equilibrium with mafic silicates and plagioclase of massifs as well as resembling residua from crystallization of basic magmas.

Mafic dykes and cumulate layers with negative Eu anomalies occur in the Adirondack Marcy massif and have been suggested to represent residual liquids or differentiates from anorthosite crystallization (3). Apatite is an abundant accessory mineral in all of the samples whose whole rock REE patterns are consistent with a strong apatite influence, including the marked negative Eu anomalies characteristic of apatite (64). For these reasons, the rocks cannot be used to argue that residual liquids must have had negative Eu anomalies. Although the rocks have similar REE patterns, they show wide variations in mineral modes which makes it difficult to accept them as plausible liquids and is supporting evidence that apatite exerts the dominant control on the patterns.

A notable feature of Proterozoic anorthosite massifs is that rocks resembling possible residual liquids and having strong negative Eu anomalies have never been identified. Marked negative Eu anomalies would be required in any residual liquids remaining after crystallizing large amounts of plagioclase alone. Fine grained ferrodiorites have only small positive or negative Eu anomalies at most (3,37,48). If ferrodiorites approach residual liquid compositions from anorthosite crystallization, the absence of significant Eu anomalies in them implies that plagioclase was never the sole liquidus phase of the parent magma over a large crystallization interval. If most of the plagioclase crystallized cotectically with mafic silicates as suggested in the last section, development of noticeable negative Eu anomalies in liquids would not be possible until more than about 80 percent of the liquid had fractionally crystallized (65).

There are a number of reasons for believing that sources of the magmas parental to the K-rich suite lie in the lower crust (1,2,7,66). If the heat source for partial melting is latent heat from subcrustal ponded mafic magma, granitic or charnockitic melts may be expected to form a sizeable body immediately overlying it. Densities of these K-rich melts would be substantially lower than any liquid or liquid-crystal mixture in the underlying mafic chamber. It seems probable that the granitic melt must reach some critical mass before gravitational instability sets in and buoyant diapirism upward into the crust can begin. This in turn implies that the main driving force for intrusion of anorthosite-granite (-charnockite) complexes lies in the thermally-generated buoyancy of the K-rich melt component; the less-buoyant anorthosite component may be unable to penetrate the crust on its own.

Geological evidence for intrusion driven principally by buoyant granitic magma can be found in complexes intruded to different crustal levels. Those complexes intruded to relatively high crustal levels — Fennoscandia, East European platform, Ukrainian Shield — contain large volumes of granite whereas lesser amounts of granite are associated with massifs in central Labrador, Rogaland, and the Grenville Province (Figure 1). Many anorogenic granites that lie within the same time interval as massif anorthosites (1,2,7) attest to the effectiveness of buoyant granitic magmatism in cratonic settings and may owe their origins to similar thermal driving mechanisms.

Nearly contemporaneous intrusion of anorthosite massifs and granitic plutons can only be accomplished if the two end members are unable to mix in any significant degree. Mixing would be hindered if the anorthosite magma were a partly crystalline mass and the dry granitic melt a viscous liquid. Continuing crystallization of the anorthosite component, as the composite diapir rose in the crust and after emplacement, could supply heat to maintain the granitic melt at high temperature. Common intrusive relations of granite toward anorthosite can be explained if granite magmas remained mobile after the anorthosite was essentially solid.

Local evidence of modal, chemical, and isotopic transition between ferrodiorite compositions and granite (charnockite) has been described (5,41,62,67). If hybridization were possible on a larger scale, the granitic melts might provide a 'sink' for residual liquids expelled from a slowly crystallizing anorthosite massif; monzonitic members of many complexes are relatively mafic (C.I. ≥ 20), consistent with mixing of ferrodiorite liquids with granitic melts (16,43). The systematics of Pb, Sr, and O isotopic data for Hidra massif (68) may also be explained by incomplete mixing of such melts. Mixing of similar end members has been suggested as an explanation for hybrid marscoites of Skye (69).

7. DISCUSSION AND SUMMARY

Pristine anorthosite massifs were intruded into stabilized cratonic continental crust. Evidence bearing on massif settings in Grenville and Sveconorwegian terranes is mostly equivocal, but not inconsistent with a similar interpretation.

Primary source materials for anorthosite parent magmas cannot be proven to be subcontinental mantle peridotite but bimodal magmatism of typical complexes and evidence for lesser crustal contamination of olivine- than of orthopyroxene-bearing assemblages provides considerable support for mantle origins. Subordinate contributions from basic granulites of the lower crust seem likely. The correlation of increased Sr_i with increased silica activity (as indicated by the olivine-orthopyroxene reaction) in massif mineral assemblages suggests operation of Bowen's assimilation-fractional crystallization process (70,71,72).

Plagioclase and mafic silicate compositions in massif anorthosites are evidence that the magmas from which they crystallized had undergone large degrees of fractional crystallization before intrusion into the crust. Intermediate plagioclase with high Sr contents and substantial LREE enrichment points to parent liquids from which major clinopyroxene extraction had occurred, presumably during subcrustal fractionation of a higher pressure crystal assemblage from mafic magma. The Egersund-Ogna massif contains two distinct plagioclase populations; core zone plagioclase An_{40-53} has 800-1000 ppm Sr and gneissic marginal zone plagioclase An_{48-68} has 400-600 ppm Sr (37,42,73). Marginal zone plagioclase was not included in Fig. 4 because of the possibility it might have been metamorphically re-equilibrated. New interpretations regard the core as having largely crystallized at higher pressures prior to emplacement and suggest that deformation and metamorphism in the massif is related to diapiric emplacement of a crystal mush and is not necessarily linked to a regional deformation (73,74).

Inverse changes in solubilities of plagioclase and aluminous clinopyroxene in equilibrium with basic liquids take place with pressure changes. These effects may largely account for development of anorthositic magmas and for adcumulus growth of plagioclase from clinopyroxene-undersaturated liquids during decompression of the magmas.

An appropriate physical model to initiate generation of anorthosite massif magmas may be provided by ponded basaltic, picritic, or more ultramafic magma beneath a lower density cratonic lid (1,60,75). Regional Moho temperatures must have been well below the subcontinental mantle solidus. In these conditions, ponded mafic magma is able to crystallize and fractionate, forming a thick layered sequence of olivine, olivine+orthopyroxene, olivine+orthopyroxene+clinopyroxene and spinel+orthopyroxene+clinopyroxene cumulates. When plagioclase reached saturation in the overlying liquid (plagioclase+orthopyroxene+clinopyroxene+spinel in equilibrium with liquid), it remained in suspension or floated, as mafic silicates sank or precipitated on the floor. In volume, these ultramafic cumulates would be at least equal to any derived anorthosite massifs and perhaps much larger. Contemporaneous progressive partial melting of the lower crustal roof by latent heat from crystallizing mafic magma formed a body of granitic (charnockitic) melt. When this melt reached some critical volume it was able to rise diapirically into the overlying crust, closely accompanied by the anorthositic magma substrate. The general features of the physical model are similar to those suggested by Morse (76) to explain derivation of an anorthosite massif from a body of basaltic magma.

Crystal-liquid proportions in the anorthosite magma cannot be easily estimated and would probably vary. Widespread layered structures in pristine massifs suggest that the liquid portion was substantial. The transition in rheological behaviour from a crystal-liquid suspension to the beginning of grain-to-grain interactions has been estimated from experimental data at about 30 to 35 percent liquid (77). On this evidence it seems possible that up to 65 percent crystals could be present during transport of anorthositic magmas without inducing significant protoclastic effects.

In nonorogenic conditions it may safely be assumed that rise of magmas in the crust is due mainly to buoyancy and that tectonic forces play no more than a minor role. Diapiric intrusion is regarded as the most efficient method for transferring magmas to higher crustal levels with minimum heat loss (78,79). This is an important consideration in explaining how some massif complexes are able to rise to shallow depths in the crust.

Diapiric intrusion of low density anorthositic crystal mush has been suggested to apply to the Morin and Lac St. Jean massifs (80,81). In both cases, there exists considerable evidence of solid state strain, explained as a consequence of continued buoyant rise after consolidation. By contrast, although the existence of dispersed, bent, warped, and kinked crystals has been widely observed in pristine massifs of central Labrador and elsewhere, evidence for large strain components in near solid anorthosite is not abundant. It is unlikely that intrusion to different crustal levels can account for the differences because, other things being equal, larger transport distances to higher crustal levels would be expected to lead to enhanced development of protoclastic textures. Longhi and Ashwal (82) have recently favoured diapirism as the principal mode of intrusion for terrestrial massifs and lunar anorthosites.

The wider implications of subcratonic magma ponding during the Mid-Proterozoic may be important. It could reflect a unique episode of enhanced heat dissipation from the mantle or one in which larger continental masses resulted in more abundant intracontinental magmatism. The fact that anorthosite-granite magmatism is largely confined to Proterozoic gneiss terranes may mean that only they were sufficiently rich in low-melting granitic components to permit generation of large, buoyant bodies of granitic melt. If subcontinental magma ponding was widespread as early as the Archean (75), it is possible that it diminished by the Late Proterozoic as plate tectonic regimes characteristic of the Phanerozoic became dominant and sites of major mantle magmatism were increasingly confined to plate margins and intracontinental rift zones.

ACKNOWLEDGEMENTS. Constructive suggestions toward improvements in the manuscript were made by L.D. Ashwal, J.-C. Duchesne, S.A. Morse, and R.A. Wiebe, to all of whom I am grateful. It should be unnecessary to add that this does not imply concurrence with all of the ideas presented.

REFERENCES

1. R.F. Emslie, Prec. Res., **7**, 61, 1978.
2. R.F. Emslie, Geol. Surv. Can. Bull., **293**, 136 p., 1980.
3. L.D. Ashwal and K.E. Seifert, Geol. Soc. Am. Bull., **91**, Pt. II, 659, 1980.
4. S.A. Morse, Am. Min., **67**, 1087, 1982.
5. J.-C. Duchesne, in: Feldspars and Feldspathoids, ed. by W.L. Brown, NATO ASI in Rennes, France, 1984, p. 411.
6. E.C. Simmons and G.H. Hanson, Contr. Mineral. Petrol., **66**, 119, 1978.
7. J.L. Anderson, in: Proterozoic Geology: Selected Papers from an International Symposium, ed. by L.G. Medaris, Jr., C.W. Byers, D.M. Mickelson, and W.C. Shanks, Geol. Soc. Am. Mem., **161**, 133, 1983.
8. M.D. Higgins and R. Doig, Can. Jour. Earth Sci., **18**, 561, 1981.
9. J.M. Husch and C. Moreau, Jour. Volc. Geotherm. Res., **14**, 47, 1981.
10. D.A. Velikoslavinskii and others, The Anorthosite-Rapakivi Granite Formation of the East European Platform, Acad. Sci. U.S.S.R., Nauka, Leningrad, 294 p., 1978.
11. A. Vorma, Geol. Surv. Finl. Bull., **285**, 98 p., 1976.
12. I. Haapala, Geol. Surv. Finl. Bull., **286**, 128 p., 1977.
13. J.H. Berg, Geol. Assoc. Can. Ann. Mtg., **4**, 39, 1979 (abstract).
14. R.A. Wiebe, Nature, **286**, 564, 1980.
15. D. Demaiffe, J.-C. Duchesne and J. Hertogen, in: Origin and Distribution of the Elements, ed. by L.H. Ahrens, Pergamon Press, Oxford, 417, 1977.
16. J.C. Fountain, D.S. Hodge and F.S. Hills, Lithos, **14**, 113, 1981.
17. L.D. Ashwal, D.F. Morrison, W.C. Phinney and J. Wood, Contr. Mineral. Petrol., **82**, 259, 1983.
18. A.R. Philpotts, Jour. Petrol., **7**, 1, 1966.
19. P. Michot, in: Anorthosites and Related Rocks, ed. by Y.W. Isachsen, New York Museum and Science Service Mem. **18**, 411, 1969.
20. M. Vaasjoki, Geol. Surv. Finl. Bull., **294**, 64 p., 1977.
21. D. Bridgwater and B.F. Windley, Geol. Soc. S. Africa. Sp. Publ. **3**, 307, 1973.
22. J.H. Berg, Jour. Petrol., **18**, 399, 1977.
23. P. Morgan and B.H. Baker, Tectonophys, **94**, 1, 1983.
24. A.T. Anderson, Jr., Am. Min., **51**, 1671, 1966.
25. N. Herz, Geol. Soc. Am. Sp. Paper **194**, 200, 1984.
26. L.R. Wager and G.M. Brown, Layered Igneous Rocks, Oliver and Boyd, Edinburgh, 588 p., 1968.
27. J.L. Snyder, Geochim. et Cosmochim. Acta, **16**, 243, 1959.
28. L.D. Raedeke and I.S. McCallum, Geochim. et Cosmochim. Acta, Suppl. **12**, 133, 1980.
29. S.A. Morse, Jour. Petrol., **20**, 555, 1979.
30. J.R. Wilson, K.H. Esbensen and P. Thy, Jour. Petrol., **22**, 584, 1981.

31. G.R. Himmelberg and A.B. Ford, Jour. Petrol., **17**, 219, 1976.
32. B.T.C. Davis, Geology of the St. Regis quadrangle, New York. Unpubl. Ph.D. thesis, Princeton Univ., 170 p., 1963.
33. R.F. Emslie, Can. Mineral., **13**, 138, 1975.
34. S.A. Morse, Earth Planet. Sci. Lett., **26**, 331, 1975.
35. R.F. Dymek and L.P. Gromet, Can. Mineral., **22**, 297, 1984.
36. R. Maquil, I. Roelandts, J. Hertogen and J.-C. Duchesne, Soc. Geol. France, 8e Reunion Ann. Sci. de la Terre, 238, 1980 (abstract).
37. J.-C. Duchesne and D. Demaiffe, Earth Planet. Sci. Lett., **38**, 249, 1978.
38. K.K. Turekian and J.L. Kulp, Geochim. et Cosmochim. Acta, **10**, 245, 1956.
39. K.R. Walker, Geol. Soc. Am. Spec. Paper **111**, 178 p., 1969.
40. P. Henderson, S.J. Fishlock, J.C. Laul, T.D. Cooper, R.L. Conrad, W.V. Boynton and R.A. Schmitt, Earth Planet. Sci. Lett., **30**, 37, 1976.
41. R.A. Wiebe, Can. Jour. Earth Sci., **15**, 1326, 1978.
42. J.-C. Duchesne, Ann. Soc. Geol. Belg., **90**, 643, 1967.
43. J.-C. Duchesne, Contr. Mineral. Petrol., **66**, 175, 1978.
44. R. Zeino-Mahmalat and H. Krause, Norsk Geol. Tids., **56**, 51, 1976.
45. S.A. Morse, Geochim. et Cosmochim. Acta, **46**, 223, 1982.
46. J. Ferguson and I.H. Wright, Geol. Soc. S. Africa Spec. Publ. **1**, 59, 1970.
47. M.D. Higgins, Age and origin of the Sept Iles anorthosite complex, Quebec. Unpubl. Ph.D. thesis, McGill Univ., 127 p., 1979.
48. D. Demaiffe and J. Hertogen, Geochim. et Cosmochim. Acta, **45**, 1545, 1981.
49. H.S. Pettingill, A.K. Sinha and M. Tatsumoto, Contr. Mineral. Petrol., **85**, 279, 1984.
50. P.L. Gromet and R.F. Dymek, Geol. Soc. Am. Ann. Mtg., **12**, 438, 1980 (abst.).
51. J.M. Barton, Jr., and R. Doig, Contr. Mineral. Petrol., **61**, 219, 1977.
52. S.A. Heath and H.W. Fairbairn, in: Orogin of Anorthosite and Related Rocks, ed. by Y.W. Isachsen, New York Museum and Science Service Mem. **18**, 99, 1969.
53. L.D. Ashwal and J.L. Wooden, Geochim. et Cosmochim. Acta, **47**, 1875, 1983.
54. L.D. Ashwal and J.L. Wooden, Nature, **306**, 679, 1983.
55. G.V. Subbarayudu, The Rb-Sr isotopic composition and the origin of the Laramie anorthosite-mangerite complex, Laramie Range, Wyoming. Unpubl. Ph.D. thesis, Univ. of New York at Buffalo, 109 p., 1975.
56. R.F. Emslie, W.D. Loveridge and R.W. Stevens, Geol. Assn. Can., Ann. Mtg., **8**, A20, 1983 (abst.).
57. T.N. Irvine and M.R. Sharpe, Carnegie Inst. Wash. Yb., **81**, 294, 1982.
58. D.C. Presnall, S.A. Dixon, J.R. Dixon, T.H. O'Donnell, N.L. Brenner, R.L. Schrock and D.W. Dycus, Contr. Mineral. Petrol., **66**, 203, 1978.

59. W.P. Leeman, C.J. Vitaliano and M. Prinz, Contr. Mineral. Petrol., **56**, 35, 1976.
60. K.G. Cox, Jour. Petrol., **21**, 629, 1980.
61. L.D. Ashwal, Am. Min., **67**, 14, 1982.
62. R.A. Wiebe and T. Wild, Contr. Mineral. Petrol., **84**, 327, 1983.
63. J.-C. Duchesne, I. Roelandts, D. Demaiffe, J. Hertogen, R. Gijbels and J. DeWinter, Earth Planet. Sci. Lett., **24**, 325, 1974.
64. I. Roelandts and J.-C. Duchesne, in: Origin and Distribution of the Elements, ed. by L.H. Ahrens, Pergamon Press, Oxford, 199, 1977.
65. L.A. Haskin, Petrogenetic modelling — use of rare earth elements, in: Rare Earth Element Geochemistry, ed. by P. Henderson, Elsevier, Amsterdam, 1984.
66. R.F. Wendlandt, Am. Min., **66**, 1164, 1981.
67. R.I. Kalamarides, E.C. Simmons, D.D. Lambert and R.A. Wiebe, Geol. Soc. Am. Ann. Mtg., **14**, 524, 1982.
68. D. Weis and D. Demaiffe, Geochim. et Cosmochim. Acta, **47**, 1405, 1983.
69. B.R. Bell, Nature, **306**, 323, 1983.
70. N.L. Bowen, The Evolution of the Igneous Rocks, Princeton Univ. Press, 251 p., 1928.
71. F. Barker, D.R. Wones, W.M. Sharpe and G.A. Desborough, Prec. Res., **2**, 97, 1975.
72. D.J. DePaolo, Earth Planet. Sci. Lett., **53**, 189, 1981.
73. J.-C. Duchesne and R. Maquil, in: Excursion Guide of the South Norway Geological Excursion ed. by C. Maijer, NATO ASI, 84, 1984.
74. R. Maquil and J.-C. Duchesne, Ann. Soc. Geol. Belg., **107**, 27, 1984.
75. C.T. Herzberg, W.S. Fyfe and M.J. Carr, Contr. Mineral. Petrol., **84**, 1, 1983.
76. S.A. Morse, in: Origin of Anorthosite and Related Rocks, ed. by Y.W. Isachsen, New York Museum and Science Service, Mem. **18**, 175, 1969.
77. I. van der Molen and M.S. Patterson, Contr. Mineral. Petrol., **70**, 299, 1979.
78. E.R. Oxburgh and T. McRae, Phil. Trans. Roy. Soc. Lond. Ser. A, **310**, 457, 1984.
79. B.D. Marsh, Am. Jour. Sci., **282**, 808, 1982.
80. J. Martignole and K. Schrijver, Geol. Soc. Finl. Bull., **42**, 165, 1970.
81. G. Woussen, E. Dimroth, L. Corriveau and P. Archer, Contr. Mineral. Petrol., **76**, 343, 1981.
82. J. Longhi and L.D. Ashwal, Proc. 15th Lunar Planet. Sci. Conf. (in press).

SM-ND ISOTOPIC STUDIES OF PROTEROZOIC ANORTHOSITES: SYSTEMATICS AND
IMPLICATIONS

Lewis D. Ashwal　　　　　　　Joseph L. Wooden
Lunar and Planetary　　　　　U.S. Geological Survey
　　Institute　　　　　　　　Isotope Geology Branch
3303 NASA Road 1　　　　　　 345 Middlefield Road
Houston, TX 77058　　　　　 Menlo Park, CA 94025

ABSTRACT. Sm-Nd isotopic studies of anorthosites can be used to provide information on their ages of crystallization and metamorphism, contamination history, and mantle sources. Proterozoic anorthosites in the Grenville and Nain Provinces of eastern North America crystallized between about 1100 and 1600 Ma, and some were metamorphosed at about 1000 Ma. Grenville Province anorthosite massifs were derived from depleted mantle. It is not clear whether massifs and related mafic intrusions throughout the Nain Province of Labrador were derived from enriched mantle, or were contaminated by early Archean (greater than 3500 Ma) silicic crustal materials, heretofore thought to be restricted to coastal Labrador.

1. INTRODUCTION

　　　It is generally assumed that massif-type anorthosites are a phenomenon unique to the Proterozoic (e.g. Anderson, 1969), and have, in fact, been discussed in terms of a distinct, catastrophic anorthosite "event" (Herz, 1969). There has always been some concern, however, that the apparent temporal restriction of anorthosite ages may be an artifact of the widespread high grade metamorphic event about one billion years ago that many massifs have been subjected to. The situation is further complicated by the difficulty in obtaining <u>direct</u> anorthosite ages because of their generally small variations in parent/daughter ratios and lack of zircons. As a result, essentially all anorthosite ages have come from spatially associated charnockites, mangerites, and other silicic rocks, assuming them to be coeval with the anorthosites. There has been much controversy, however, regarding the relationship between the anorthosites and the silicic rock suite: some favor the hypothesis that the silicic rocks and anorthosites are comagmatic and related by differentiation (e.g. Philpotts, 1969; Demaiffe and Hertogen, 1981) others are persuaded that the charnockites and mangerites are chemically and petrologically independent from the anorthosites (e.g. Buddington, 1969; Ashwal and Seifert, 1980).

A direct method of anorthosite age determination would be of value, therefore, in addressing a number of these important issues. The Sm-Nd isotopic system offers great promise in this regard for two reasons: (a) pyroxene-rich materials should, in principal, exhibit sufficiently high Sm/Nd to yield reasonably precise isochrons, and (b) Sm-Nd is more resistant to metamorphic resetting compared to other isotopic systems (e.g. Rb-Sr, K-Ar). In addition, the Sm-Nd system has been shown to be a reliable geochemical tracer to possible source materials, and may also provide information on processes involving contamination with "crustal" materials. Thus far, Sm-Nd isotopic data are available only for anorthosite massifs and possibly related mafic intrusives of similar age from eastern North America. In this paper we summarize these data and discuss their possible implications.

2. GEOLOGICAL SETTING

The Grenville Province of the Canadian Shield (Fig. 1) is predominantly a deeply eroded mid- to late-Proterozoic orogenic belt, characterized by upper amphibolite to granulite facies metamorphic

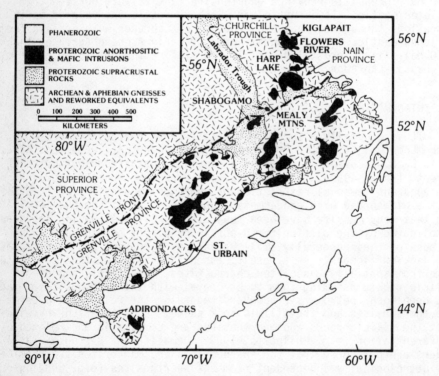

Fig. 1. Geologic map showing locations of major Proterozoic anorthosite massifs and mafic intrusions.

grade, and a predominance of K-Ar ages between 900 and 1100 Ma (Harper, 1967). In the most simplified sense, the Province consists of about 60% quartzofeldspathic gneiss, which in some places can be demonstrated to be reworked late Archean material, 20% supracrustal rocks, and 20% anorthosite and related plutonic rocks (Wynne-Edwards, 1972). The anorthosite massifs pre-date the metamorphism, and many, but not all, have been metamorphosed and deformed. The Grenville Province is separated from the older Provinces to the northwest (Superior, Churchill, and Nain) by the Grenville Front (Fig. 1), a prominent linear feature with variable properties along its 1600 km length (Wynne-Edwards, 1972), including marked geophysical anomalies (Gibb et al., 1983). In contrast to the Grenville Province, the Nain Province of Labrador has not been affected by major metamorphism since the Archean. Anorthosites transect the Grenville Province in Labrador, and extend into the Nain Province. Here, and also in large areas of the northeastern Grenville Province, the anorthosites are completely unmetamorphosed and undeformed, and retain primary igneous textures, structures, and mineralogy.

3. ISOTOPIC SYSEMATICS OF ANORTHOSITIC AND RELATED ROCKS

Direct whole-rock isochrons for anorthosite massifs should be possible if pyroxene-rich rocks which crystallized from the same parental magma as the anorthosites can be identified. Many anorthosite massifs contain minor volumes of mafic to ultramafic rocks, which occur as conformable layers, interpreted as cumulates, and as dikes interpreted as late-stage residual magmas (Ashwal and Seifert, 1980; Ashwal, 1982). These mafic rocks are associated with Fe-Ti oxide ore deposits, whose association with Proterozoic anorthosite massifs is well known (Rose, 1969). Ashwal and Wooden (1983b) have shown that a pyroxene-rich ultramafic cumulate layer from the Adirondack Marcy massif (Fig. 1) has sufficiently high Sm/Nd, and associated leuconorite sufficiently low Sm/Nd, such that a reasonably precise Sm-Nd whole-rock isochron can be determined. In the case of some massifs, such as Marcy, which has been metamorphosed to high pressure granulite facies (max. P = 8 kbar; T = 750°C), the situation is complicated by the presence of abundant metamorphic mineralogy in rocks of appropriate bulk composition. Thus in the Marcy massif layered sequence, which grades over several meters from basal oxide-rich pyroxenite to leuconorite, it is only the intermediate gabbroic and noritic rocks which contain abundant metamorphic garnet and/or hornblende (Ashwal, 1982; Ashwal and Wooden, 1983b). Sm-Nd isotopic data from these intermediate, recrystallized rocks and from a garnet mineral separate plot along an isochron of distinctly younger age (973 $\pm$ 19 Ma) compared to data for the rocks of extreme composition from the base and top of this layered sequence (Ashwal and Wooden, 1983b). These rocks, which have largely retained their primary texture as well as mineralogy plot along an older isochron (1288 + 36 Ma), which has been interpreted as the age of primary crystallization of the Marcy massif (Ashwal and Wooden, 1983b). It is perhaps better considered as a <u>minimum</u> age of crystallization. Rb-Sr isotopic data for all samples from this layered sequence, including whole rocks, and igneous and metamorphic mineral separates give an isochron age of 919 + 12 Ma, illustrating that this isotopic system has been completely disturbed.

The isotopic data from the Marcy anorthosite massif described above demonstrate that even the whole-rock Sm-Nd system is resettable, at least partially, at granulite metamorphic conditions. Samples with the best chance of retaining their primary crystallization age are those which still possess their original igneous texture, with minimal development of metamorphic mineralogy. For unmetamorphosed anorthosite massifs, it should be possible, considering the precision of modern Sm-Nd isotopic measurements, to obtain internal isochrons by separating and analysing plagioclase and pyroxene mineral pairs. Isotopic data from a labradorite anorthosite inclusion in andesine anorthosite of the unmetamorphosed St. Urbain massif, Quebec (Fig. 1) (Gromet and Dymek, 1982) confirm this. A three-point Sm-Nd isochron, including whole-rock, plagioclase, and pyroxene (orthopyroxene + clinopyroxene) corresponds to an age of 1079 ± 22 Ma. This age is more than 100 Ma older than the Rb-Sr isochron age (Gromet and Dymek, pers. comm.) of the same samples (926 ± 24 Ma), again illustrating the unsuitability of the Rb-Sr isotopic system as a direct geochronometer even for reportedly unmetamorphosed anorthosite massifs. Apparently better results have been obtained from mafic layered intrusions. DePaolo (1981a) reported mineral isochrons (plagioclase + augite + whole rock) for a ferrodiorite from the Kiglapait intrusion in the Nain Province of Labrador (Fig. 1). In this case, the Sm-Nd internal age (1416 ±50 Ma) and Rb-Sr internal age (1413 ±54 Ma) are concordant.

Giant pyroxene megacrysts (up to 50 cm) with <u>exsolved</u> plagioclase lamellae and relatively primitive Mg/(Mg + Fe) up to 0.8 have been found in many anorthosite massifs. These aluminous megacrysts are thought to be products of either (a) crystallization at high pressure (Emslie, 1975), or (b) rapid crystallization from hyperaluminous magmas (Morse, 1975). Available isotopic measurements of a few of these megacrysts show them to have relatively primitive isotopic compositions, indicating that they may have been somehow protected from high level crustal contamination effects (discussed below). For example, an aluminous orthopyroxene megacryst from the Adirondack Marcy massif has the highest initial Nd ratio (epsilon Nd = +5) of any material measured thus far from that massif (Ashwal and Wooden, 1983b). For the large unmetamorphosed massifs of Labrador, it may be possible to utilize isotopic data from pyroxene megacrysts to obtain age information. For the Mealy Mountains massif (Fig. 1), a two-point isochron (aluminous pyroxene megacryst and two isotopically similar anorthosites) gives an age of 1576 Ma, which confirms the relatively old age of this massif as indicated by the U-Pb zircon age of 1640 +3 Ma for a crosscutting monzonite-granite unit (Emslie et al., 1983). Samples from the Harp Lake massif (Fig. 1) yield a three-point isochron (pyroxene megacryst, gabbro, and two anorthosites) corresponding to an age of 1663 ± 45 Ma, which is significantly older than the 1460 Ma U-Pb zircon age from crosscutting granite (Krogh and Davis, 1973).

4. CRUSTAL CONTAMINATION EFFECTS

Many anorthosite massifs show evidence of interaction with crustal materials, particularly near their margins. Hybrid rocks are common, consisting of silicic suite rocks with xenoliths of anorthosite or xenocrysts of anorthositic plagioclase (Buddington, 1939). Where anorthosite intruded carbonate-rich units, complex skarns are developed (Buddington, 1950). Inclusions of country rocks can be found in the anorthosites, and in the Marcy massif of the Adirondacks, large xenoliths of metasedimentary material, including marbles are present (Valley and Essene, 1980). Evidence is accumulating for extensive fluid circulation, particularly at anorthosite-country rock contacts (Valley and O'Neil, 1982). The isotopic effects of such interactions should be pronounced, considering the relatively low concentrations of trace elements (except Sr) in typical anorthositic rocks. This indeed appears to be the case in the Marcy massif, where anorthositic rocks from near the margins of the massif have high Rb contents (up to 38 ppm) and initial $^{87}Sr/^{86}Sr$ (Sr_o up to 0.70483) compared to interior samples, which typically have 5 ppm Rb and Sr_o = 0.7039 - 0.7040 (Ashwal and Wooden, 1983b). Variability in initial Nd isotopic ratios can also be reliably attributed to contamination effects; a spread of about 5 epsilon Nd units (as defined in DePaolo, 1981b) appears to be typical of anorthosite massifs measured thus far (Ashwal and Wooden, 1983a). The isotopic signature of at least two contaminants has been recognized in the Adirondack anorthosite (Ashwal and Wooden, 1983b), which is not surprising, considering the diverse lithologies available in that terrane.

5. REGIONAL ISOTOPIC SIGNATURE OF ANORTHOSITES IN THE EASTERN CANADIAN SHIELD

Proterozoic anorthosite massifs and mafic layered intrusions in eastern North America (Figs. 1, 2; Table 1) can be divided into two types based on location and Nd isotopic signature (Ashwal and Wooden, 1983a). This is possible despite age uncertainties of as much as 200 Ma, caused by metamorphism and incomplete isotopic studies, and the variations in initial isotopic ratios caused by contamination effects. All of the Labrador intrusions northwest of, or within close proximity to the Grenville Front have negative epsilon Nd values (Table 1). These include data for the Harp Lake massif (Ashwal and Wooden, 1983b and this paper), the Kiglapait layered intrusion (DePaolo, 1981), the Flowers River gabbro (Brooks, 1982), and the Shabogamo gabbro (Zindler et al., 1981). Massifs well within the Grenville Province (i.e. southeast of the Grenville Front) have positive epsilon Nd values (Table 1), suggesting derivation from LREE depleted sources. These include the Adirondack Marcy (Ashwal and Wooden, 1983b), Mealy Mountains, Labrador, and St. Urbain, Quebec massifs (Ashwal and Wooden, 1983a and this paper), and the Roseland anorthosite complex of Virginia (Pettingill et al., 1984). With the exception of two massifs (Adirondacks and Roseland), initial Sr ratios (Sr_o) can also be correlated with geographic position (Fig. 3). The Labrador massifs northeast of the Grenville Front all have Sr_o of 0.703 or greater, whereas the Grenville Province massifs have Sr_o of 0.703 or less (Ashwal and Wooden, 1983a). It may be noted that the two exceptions have both been metamorphosed, and both also have anomalously high $^{18}O/^{16}O$ compared to most other anorthosites (Taylor, 1969). For the Adirondack anorthosite, the oxygen isotopic anomaly has been discussed in terms of exchange with pervasive metamorphic fluids (Taylor, 1969) or assimilation of metasedimentary units by the anorthosite (Valley and O'Neil, 1982). Possibly the enriched Sr_o and $^{18}O/^{16}O$ of the

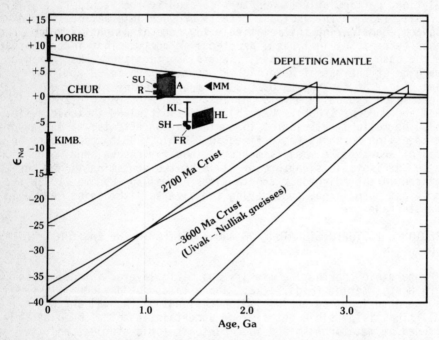

Fig. 2. Plot of age vs. epsilon Nd for Proterozoic anorthositic and mafic intrusions from the eastern Canadian shield. Refer to Fig. 1 for locations. Massifs derived from sources with a depleted signature plot above the chondritic evolution line (CHUR), and include St. Urbain (SU), Roseland (R), Adirondacks (A), and Mealy Mountains (MM). It is not certain whether intrusions plotting below the CHUR line, including Kiglapait (KI), Shabogamo (SH), Flowers River (FR) and Harp Lake (HL) were derived from "enriched" mantle or were affected by crustal contamination. Data for some massifs are shown as polygons because of age uncertainties and contamination effects. Curve representing depleted mantle is from (DePaolo, 1981b). Field showing evolution of 2700 Ma crust is from unpublished data of J.L. Wooden; that for evolution of 3600 Ma crust (Uivak-Nulliak gneisses) is from (Collerson et al., 1983). Ranges for modern-day MORB (BVSP, 1981) and Australian kimberlites (McCulloch et al., 1983) are also shown.

Adirondack and Roseland massifs are correlated, and attributable to the same general process (Ashwal and Wooden, 1983a).

The positive epsilon Nd values of the Grenville Province anorthosite massifs require that they were derived from sources with long-term depletion in light rare earth elements (LREE) compared to chondrites or "bulk earth." This implies a source with similar isotopic characteristics to the mantle source of mid-ocean ridge basalts. It cannot be used, necessarily, to conclude that oceanic-type mantle existed under eastern North America during the Proterozoic, but this is a possibility. Evidence is accumulating for depleted mantle of regional extent under the North American continent, at least as far back as the late Archean (Tilton, 1983; Wooden et al., 1983). In this regard, it is the Labrador massifs with negative epsilon Nd values that are anomalous. Two possible explanations can account for their isotopic signature: (a) crustal contamination, or (b) enriched mantle. Zindler et al. (1981) suggested that the epsilon Nd of the Shabogamo gabbro, Labrador may have been lowered from 0 to -5.3 by 20% assimilation of a granitic component derived from late Archean ($\sim$2700 Ma) silicic crust. Significantly larger amounts of contamination would be required if the original Shabogamo magma had an Nd isotopic signature of depleted mantle (positive epsilon Nd), as for the Grenville massifs. There is little independent evidence for such extensive contamination for the Shabogamo gabbro or for the other Labrador massifs. A more plausible contaminant would be early Archean (greater than 3500 Ma) silicic crustal materials, which would have evolved to extremely low epsilon Nd values by the late

TABLE 1. Isotopic compositions of Proterozoic anorthosites and mafic intrusions

	Age (Ma)	Sr_o	ϵ_{Nd}	Reference
Depleted source:				
Adirondacks, New York	1288 ± 36*	0.7039–0.7048	+0.4 to +4.1	Ashwal and Wooden (1983a)
	1113 ± 10†	0.7040–0.7055	−0.3 to +4.4	Silver (1969)
St. Urbain, Quebec	1079 ± 22*	0.7029	+2.2	Gromet and Dymek (1982)
				Ashwal and Wooden (1983b)
Mealy Mtns., Labrador	1576*		+2.0	Ashwal and Wooden (1983b)
	1640 ± 3†	0.7026–0.7032	+1.3 to +2.8	Emslie et al. (1983)
Roseland, Virginia	1045 ± 44*	0.704–0.7055	+1.0 ± 0.3	Pettingill et al. (1984)
Enriched source:				
Harp Lake, Labrador	1663 ± 45*	0.7040–0.7066	−5.3 to −2.0	Ashwal and Wooden (1983b)
	1460†	0.7040–0.7068	−6.1 to −3.4	Krogh and Davis (1973)
Kiglapait, Labrador	1416 ± 50*	0.7039–0.7066	−4.9 to −1.1	DePaolo (1981); Simmons et al. (1981); Morse (1983)
Shagbogamo, Labrador	1379 ± 65*	0.7041	−5.3	Zindler et al. (1981); Brooks et al. (1982)
Flowers River, Labrador	1411 ± 48*		−5.6	Brooks (1982)
	1200 ± 210‡	0.7065		

*From Sm-Nd whole-rock and mineral isochron
†U-Pb zircon age of crosscutting granitic intrusion
‡Rb-Sr errorchron

Proterozoic. The Uivak and related gneisses (~3600 Ma), which would have had epsilon Nd values between -20 and -40 at 1400 Ma ago (Fig. 2) (Collerson and McCulloch, 1982), would be more suitable contaminants in that less assimilation would be required. However, known surface exposures of these old Archean gnesises are limited to a narrow strip along the eastern coast of Labrador approximately 600 km long, and reaching a maximum width of about 100 km near the Grenville Front (Collerson et al., 1976). If crustal contamination with these ancient gneisses is required to explain the isotopic composition of the Labrador anorthosite massifs and mafic intrusions, then the extent of old Archean basement must be much more widespread than currently recognized, extending at least as far west as the Labrador Trough.

If contamination effects are not responsible for altering the isotopic compositions of the Labrador massifs, then the possibility must be considered that they were derived from sources with long-term

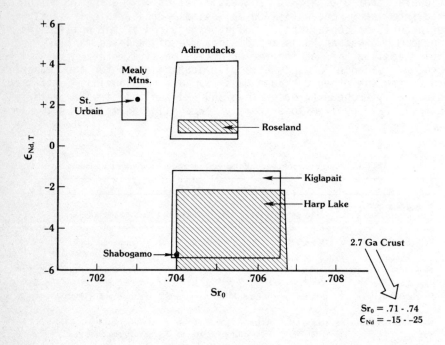

Fig. 3. Plot of epsilon Nd at time of crystallization vs. initial Sr ratio for Proterozoic anorthosites and mafic intrusives. The large fields are presumably caused by contamination with crustal materials, which would plot below, and to the right of the diagram. Labrador massifs close to or northwest of the Grenville Front have negative epsilon Nd and Sr_o of 0.704 or greater. Grenville Province massifs have positive epsilon Nd, but variable Sr_o. Possibly the Sr_o of the Adirondack and Roseland anorthosites was elevated by metamorphic processes.

enrichment in LREE compared to chondrites. Evidence is accumulating for the existence of enriched mantle sources; some appear to have given rise to kimberlites and associated alkalic volcanics in subcontinental settings (Menzies and Murthy, 1981; Collerson and McCulloch, 1983; Erlank et al., 1982). Thus, the anorthosites in eastern North America may have been derived from isotopically and geographically distinct mantle sources. If the two-source model is correct, these reservoirs must have been juxtaposed by at least 1.6 Ga to give rise to the Harp Lake and Mealy Mountains complexes, but the fractionation event(s) responsible for their isotopic compositions must be much older, particularly if the sources originally developed from a single reservoir with chondritic Sm/Nd (Ashwal and Wooden, 1983a). Although the isotopic discontinuity discussed here has been identified only in Labrador, it is possible that the Grenville Front represents the surface expression of a deep planar boundary which separates lithospheres of different affinity. Evaluation of this possibility awaits further regional isotopic studies.

6. SUMMARY

Isotopic work on Proterozoic anorthosites from eastern North America shows:
1. The Sm-Nd isotopic system can give direct anorthosite ages if pyroxene-rich mafic or ultramafic facies are present. Pyroxenes have sufficiently high Sm/Nd to yield reasonably precise internal isochrons for unmetamorphosed massifs. Rb-Sr isotopic systematics are largely disturbed, even in reportedly unmetamorphosed complexes.
2. Ages of anorthosite crystallization can be distinguished from later metamorphic events. In the Adirondacks these may be as much as 300 Ma apart, indicating that metamporphism was not a causative factor in anorthosite genesis.
3. Effects of crustal contamination are evident, particularly in the border regions of the massifs. The most isotopically primitive materials are coarse, igneous-textured anorthosites and giant aluminous pyroxene megacrysts.
4. Anorthosite massifs in the Grenville Province, including Marcy (Adirondacks, N.Y.), St. Urbain (Quebec), Mealy Mountains (Labrador), and Roseland (Virginia) have positive epsilon Nd, suggesting derivation from sources with long-term depletion in light REE, probably depleted mantle.
5. Anorthosites and mafic intrusives in Labrador, close to or northwest of the Grenville Front (including: Kiglapait, Flowers River, Harp Lake, and Shabogamo intrusions) have negative epsilon Nd. This can be attributed either to (a) contamination of the parental magmas of the Labrador massifs with ancient (greater than 3500 Ma) silicic crust, or (b) derivation of the parental magmas from "enriched" mantle. If (a) is the explanation, then suitable contaminants must be sought. The Uivak gneisses of coastal Labrador ($\sim$3600 Ma) are good candidates, and if they are responsible for lowering the epsilon Nd of the Labrador massifs, then old Archean basement must be more extensive than currently recognized. If (b) is the case, this would have important implications as to the nature and origin of subcontinental mantle heterogeneity. Pb

isotopes may be able to distinguish (a) from (b). In either case, presently available isotopic data indicate that the Grenville Front is a deep feature, which separates lithospheres of different affinity. This must be accounted for in regional tectonic models.

ACKNOWLEDGMENTS

This work is supported by a NASA grant to L.D. Ashwal. We thank L.E. Nyquist for access to the JSC mass spectrometry laboratory, H. Wiesmann, C.-Y. Shih, and B. Bansal for help with the isotopic analyses, and R.F. Emslie, L.P. Gromet, R.F. Dymek, and H.W. Jaffe for providing samples. R.K. O'Nions, J.R. Touret, and G. Ryder reviewed the manuscript. This is LPI contribution no. 545.

REFERENCES

Anderson, A.T., Jr. (1969) Massif type anorthosite: a widespread Precambrian igneous rock. In, Origin of Anorthosite and Related Rocks, N.Y. State Museum and Sci. Serv. Memoir 18, pp. 47-55.

Ashwal, L.D. (1982) Mineralogy of mafic and Fe-Ti oxide-rich differentiates of the Marcy anorthosite massif, Adirondacks, New York. Amer. Mineralogist 67, pp. 14-27.

Ashwal, L.D. and Seifert, K.E. (1980) Rare earth element geochemistry of anorthosite and related rocks from the Adirondacks, New York, and other massif-type complexes. Bull. Geol. Soc. Amer. 91, pp. 105-107; 659-684.

Ashwal, L.D. and Wooden, J.L. (1983a) Isotopic evidence from the eastern Canadian shield for geochemical discontinuity in the Proterozoic mantle. Nature 306, pp. 679-680.

Ashwal, L.D. and Wooden, J.L. (1983b) Sr and Nd isotope geochronology, geologic history, and origin of the Adirondack anorthosite. Geochim. Cosmochim. Acta 47, pp. 1875-1885.

Basaltic Volcanism Study Project (BVSP) (1981) Basaltic Volcanism on the Terrestrial Planets, Pergamon, N.Y., pp. 974-1031.

Brooks, C., quoted in Hill, J.D. (1982) Geology of the Flowers River - Notakwanon area, Labrador. Newfoundland Dep. Mines Energy Rept. 82-6, p. 11.

Brooks, C., Wardle, R.J., and Rivers, T. (1981) Geology and geochronology of Helikian magmatism, western Labrador. Can. Jour. Earth Sci. 18, pp. 1211-1227.

Buddington, A.F. (1939) Adirondack igneous rocks and their metamorphism. Geol. Soc. Amer. Memoir 7, 354 pp.

Buddington, A.F. (1950) Composition and genesis of pyroxene and garnet related to Adirondack anorthosite and anorthosite-marble contact zones. Amer. Mineralogist 35, pp. 659-670.

Buddington, A.F. (1969) Adirondack anorthosite series. In, Origin of Anorthosite and Related Rocks (Y.W. Isachsen, Ed.) N.Y. State Museum and Sci. Serv. Memoir 18, pp. 215-231.

Collerson, K.D. and McCulloch, M.T. (1982) The origin and evolution of Archean crust as inferred from Nd, Sr, and Pb isotopic studies in Labrador. Fifth Int. Symp. Geochron., Cosmochron. and Isotope Geol., pp. 61-62.

Collerson, K.D. and McCulloch, M.T. (1983) Nd and Sr isotope geochemistry of leucite-bearing lavas from Gaussberg, East Antarctica. Proc. Fourth Int. Symp. Antarctic Earth Sci. 18, pp. 676-680.

Collerson, K.D., Jesseau, C.W. and Bridgwater, D. (1976) Crustal development of the Archean Gneiss Complex: Eastern Labrador. In, The Early History of the Earth, (B.F. Windley, Ed.) Wiley, N.Y., pp. 237-252.

Demaiffe, D. and Hertogen, J. (1981) Rare earth element geochemistry and strontium isotopic composition of a massif-type anorthosite-charnockite body: the Hidra Massif (Rogaland, SW Norway) Geochim. Cosmochim. Acta 45, pp. 1545-1561.

DePaolo, D.J. (1981) Age, source, and crystallization-assimilation history of the Kiglapait intrusion as indicated by Nd and Sr isotopes (abstr.) Geol. Soc. Amer. Abstr. with Prog. 13, p. 437.

DePaolo, D.J. (1981) Neodymium isotopes in the Colorado Front Range and crust-mantle evolution in the Proterozoic. Nature 291, pp. 193-196.

Emslie, R.F. (1975) Pyroxene megacrysts from anorthositic rocks: new clues to the sources and evolution of the parent magmas. Canadian. Mineral. 13, pp. 138-145.

Emslie, R.F., Loveridge, W.D., Stevens, R.D., and Sullivan, R.W. (1983) Igneous and tectonothermal evolution, Mealy Mountains, Labrador (abstr.). Geol. Assoc. Canada - Mineral. Assoc. Canada Prog. Abstr 8, p. A-20.

Erlank, A.J., Allsopp. H.L., Hawkesworth, C.J. and Menzies, M.A. (1982) Chemical and isotopic characterization of upper mantle metasomatism in peridotite nodules from the Bulfontein kimberlite. Terra Cognita 2, pp. 261-263.

Gibb, R.A., Thomas, M.D., LaPointe, P.L., and Mukhopadhyay, M. (1983) Geophysics of proposed Proterozic sutures in Canada. Precambrian Res. 19, pp. 349-384.

Gromet, L.P. and Dymek, R.F. (1982) Petrological and geological characterization of the St. Urbain anorthosite massif, Quebec: summary of initial results. In, Workshop on Magmaric Processes of Early Planetary Crusts: Magma Oceans and Stratiform Layered Intrusions (D. Walker and I.S. McCallum, Eds.), Lunar and Planetary Inst. Tech. Rept. 82-01, pp. 72-74.

Harper, C.T. (1967) On the interpretation of Potassium-Argon ages from Precambrian shields and Phanerozoic orogens. Earth Planet. Sci. Lett. 3, pp. 128-132.

Herz, N. (1969) Anorthosite belts, continental drift, and the anorthosite event. Science 164, pp. 944-947.

Krogh, T.E. and Davis, G.L. (1973) The significance of inherited zircons on the age and origin of igneous rocks- an investigation of the ages of the Labrador adamellites. Yearbook, Carnegie Inst. Washington 72, pp. 610-613.

Menzies, M. and Murthy, V.R. (1981) Enriched mantle: Nd and Sr isotopes in diopsides from kimberlie nodules. Nature 300, pp. 634-636.

McCulloch, M.T., Jaques, A.L., Nelson, D.R., and Lewis, J.D. (1983) Nd and Sr isotopes in kimberlites and lamproites from Western Australia: an enriched mantle origin. Nature 302, pp. 400-403.

Morse, S.A. (1975) Plagioclase lamellae in hypersthene, Tikkoatokhakh Bay, Labrador. Earth and Planet. Sci. Lett. 26, pp. 331-336.

Pettingill, H.S., Sinha, A.K. and Tatsumoto, M. (1984) Age and origin of anorthosites, charnockites, and granulites in the Central Virginia Blue Ridge: Nd and Sr isotopic evidence. Contrib. Mineral. Petrol. 85, 279-291.

Philpotts, A.R. (1969) Parental magma of the anorthosite-mangerite suite. In, Origin of Anorthosite and Related Rocks (Y.W. Isachsen, Ed.), N.Y. State Museum and Sci. Serv. Memoir 18, pp. 207-212.

Rose, E.R. (1969) Geology of titanium and titaniferous deposits of Canada. Geol. Surv. Canada Econ. Geol. Rept. No. 25, 177pp.

Silver, L.T. (1969) A geochronologic investigation of the Adirondack complex, Adirondack Mountains, New York. In, Origin of Anorthosite and Related Rocks (Y.W. Isachsen, Ed.), N.Y. State Museum and Science Service Mem. 18, pp. 233-251.

Taylor, H.P., Jr. (1969) Oxygen isotope studies of anorthosites, with particular reference to the origin of bodies in the Adirondack Mountains, New York. In, Origin of Anorthosite and Related Rocks (Y.W. Isachsen, Ed.), N.Y. State Museum and Sci. Serv. Mem. 18, pp. 111-134.

Tilton, G.R. (1983) Archean crust-mantle differentiation. In Workshop on A Cross Section of Archean Crust (L.D. Ashwal and K.D. Card, Eds.), Lunar and Planetary Inst. Tech Rept. 83-03, pp. 92-94.

Valley, J.W. and Essene, E.J. (1980) Akermanite in the Cascade Slide xenolith and its significance for metamorphism in the Adirondacks. Contrib. Mineral. Petrol. 74, pp. 143-152.

Valley, J.W. and O'Neil, J.R. (1982) Oxygen isotope evidence for shallow emplacement of Adirondack anorthosite. Nature 300, pp. 497-500.

Wooden, J.L., Bansal, B. and Wiesmann, H. (1983) Archean crust-mantle reservoirs: Sr and Nd evidence from the Superior Province (abstr.) EOS 64, p. 340.

Wynne-Edwards, H.R. (1972) The Grenville Province. In, Variations in Tectonic Styles in Canada (R.A. Price and R.J.W. Douglas, Eds.), Geol. Assoc. Canada Spec. Paper no. 11, pp. 264-334.

Zindler, A., Hart, S.R., and Brooks, C. (1981) The Shabogamo Intrusive Suite, Labrador: Sr and Nd isotopic evidence for contaminated mafic magmas in the Proterozoic. Earth and Planet. Sci. Letters 54, pp. 217-235.

TEMPERATURE, PRESSURE AND METAMORPHIC FLUID REGIMES IN THE AMPHIBOLITE FACIES TO GRANULITE FACIES TRANSITION ZONES

Robert C. Newton

Department of the Geophysical Sciences
University of Chicago
Chicago, Illinois 60637, USA

ABSTRACT

Regional metamorphic progressions from amphibolite facies to granulite facies may be fossil profiles of the lower crust or may outline ancient "hot-spots" or collision zones. Paleotemperatures in different transition zones range from 650°C to 800°C, with pressures of below 5 to nearly 8 kbar, based on the most reliable geothermometry-geobarometry. Water activities were 0.35 and lower, based on calculations of orthopyroxene stability, fluid inclusions, and volatiles in cordierite. Oxygen fugacities were remarkably variable. High activities of F, CO_2 and SO_2 are recorded in fluorbiotite and scapolite.

Paleotemperatures increase across some transition zones, as in the Adirondack Highlands, the Willyama Complex (S. Australia), and the West Uusimaa Complex of South Finland. Some areas show increase of pressure towards higher grade, compatible with the depth-zone concept, as in southern India. Others record nearly isobaric increase of grade, as in Namaqualand, South Africa, and West Uusimaa. Granulites of the Nain Area, Labrador, are spatially distributed about the anorthosite bodies and show no correlation of pressures and temperatures, as might be expected of contact metamorphism at variable depth. Many transition zones show progressively lower H_2O activity with increasing grade: decreasing water pressure seems to have been the most important factor in producing granulite terrains.

Opening and closing of an infracontinental ocean basin could provide a plausible, but possibly not unique, explanation of some features of the granulite transition zones, including overthrusts, involvement of anorogenic igneous rocks and evaporites, and pressures of 5 to 8 kbar.

GRANULITE FACIES TERRAINS

Importance to Study of the Deep Crust

Granulite facies terrains are commonly regarded as representative of the deep portion of the continental crust (Fountain and Salisbury, 1981). The densities of granulite facies rocks are generally higher than for most chemically equivalent rocks of lower metamorphic grade, because of the presence of pyroxenes and garnets, even in acid compositions, and the seismic velocities of intermediate granulites are considered by some as appropriate for the deep crust (Smithson and Brown, 1977). The characteristic depletion of large ion lithophile (LIL) elements such as Th and U relative to average upper-crustal rocks make granulites suitable for heat flow modelling of the crust (Heier, 1973); low heat production and dryness may be key factors in long-term survival of large continental blocks (Hamilton et al., 1979).

Granulite facies terrains invariably contain abundant to dominant lithologies of supracrustal origin in addition to the intermediate to acid calc-alkaline pyroxene-bearing gneisses which are characteristic of many terrains. The presence of these lithologies requires consideration of processes of profound burial and subsequent uplift in crustal evolution. The antiquity of most large granulite terrains poses a problem: either the same crustal processes were not active on the same scale in later geologic times or uplift and erosion have not had sufficient time to expose large tracts of younger granulites.

Nature of Granulite Metamorphism

Granulite facies metamorphism is often ascribed to the action of unusually high temperatures, high pressures, low-H_2O mineralizing volatiles, partial anatectic melts, or some combination of these. Many granulite terrains are bounded by transitional zones of progressive metamorphism. The deep crustal granulite model implies that the high grade terrains formed at greater pressures than adjoining lower grade terrains, i.e. that granulite terrains and their transition zones represent a lower-crustal profile.
Evidence for increase of pressure, hence depth, with increasing grade of metamorphism was found by Phillips and Wall (1981) for the Broken Hill area (Willyama Complex) of S.E. Australia and Wells (1979) for the Buksefjorden region of S.W. Greenland. Both of these terrains also record substantial increase of metamorphic temperatures, from about 650°C to 800°C, through the transition zones. In contrast, Touret (1971) inferred that the controlling parameter in prograde metamorphism in the Bamble region of southern Norway was decreasing water pressure, with little evidence for large changes of temperature and pressure going from amphibolite facies to granulite facies. Mineralogic evidence for steady decrease of $P(H_2O)$ with increasing grade was found also by Phillips

(1980) for Broken Hill. Waters (1984) concluded that the
amphibolite facies to granulite facies transition in Namaqualand,
South Africa, is essentially isobaric, with increase in temperature
and decrease in water activity of the metamorphism, and Schreurs
(1984) reached a similar conclusion for the West Uusimaa area of
southern Finland.

Mechanisms that have been invoked to decrease H_2O pressure
below rock pressure are strength of rocks (Thompson, 1955),
absorption of H_2O into anatectic melts (Nesbitt, 1980), vapor-
deficient metamorphism (Valley and Essene, 1977), and dilution of
pore fluids with anhydrous volatiles, most notably CO_2 (Touret,
1971). Since Touret's (1970) discovery of CO_2-rich fluid
inclusions in granulites from several large terrains, numerous
similar findings have been reported (e.g. Coolen, 1981; Hollister,
1982). Some authors have attached great importance to streaming of
CO_2 through the lower crust in creating granulites. For example,
Harris et al. (1982) envisioned that vast quantities of CO_2 of
mantle origin passed through the South Indian crust in the late
Archaean, drying it out and transferring heat upwards to give an
"advective geotherm". In contrast, Valley et al. (1983) and
Valley and O'Neil (1984) argued from mineral stabilities and oxygen
isotope patterns that CO_2 and oxygen exchange were locally
controlled in the Adirondacks, with little evidence of pervasive
flux of low-$P(H_2O)$ volatiles. Here, initial dryness or H_2O
absorption by anatectic melts may have been more important than
removal of H_2O in a vapor phase.

TRANSITIONAL GRANULITE ZONES

General Features

Few transitional granulite terrains have been studied in any
detail; nevertheless, a few generalizations may be made. The
transitional terrains of southern Karnataka (South India), the
Adirondack Lowlands and the Buksefjorden region are rich in
supracrustals (paragneisses, marbles, quartzites, etc.).
Transitional regions of charnockitic terrains are often migmatitic
(see Quensel, 1951, for a review of migmatitic charnockites). A
zone of mixed amphibolite facies and granulite facies rocks in
intimate association, including "patchy" charnockite, is known from
southern Karnataka (Janardhan et al., 1982), southwest Australia
(Wilson, 1978), S.W. Sweden (Hubbard, 1978), and S.W. Greenland
(Wells, 1979). Although this relation can also be one of partial
retrogression from granulite facies, the above examples were
interpreted as arrested prograde granulite formation. The
incipient charnockites of the transition zone are not generally
depleted in LIL elements (Field et al., 1980; Janardhan et al.,
1982; Condie et al., 1982), in contrast to their higher-grade
neighbors.

Relations to High Grade Terrains ("Massifs")

Many transitional granulite terrains are separated from adjacent highest-grade granulites by some sort of dislocation, either a profound shear zone or a thrust. Examples include the Inari complex of northern Finland (Kröner, 1980), the Adirondack Highlands relation (Romey et al., 1980) and the northern and southern margins of the Limpopo Belt of Zimbabwe (Du Toit et al., 1983). Some transitional granulite terrains apparently advance to high grade without major interruptions, including the Broken Hill area, S.W. Greenland, and southern Karnataka (Hansen et al., 1984). In regions of continuous transition to granulite facies, regional isograds have been mapped, such as muscovite out at Broken Hill and orthopyroxene in at Bamble, the Adirondacks and southern India.

Geochemical Profiling

Geochemical traversing of the transitional zones promises to yield important data concerning the origin of granulites, especially in those zones which are apparently without major structural interruption. Of especial interest are the study of isograd reactions, $P(H_2O)$ and the nature of the metamorphic fluids, and the mechanisms of LIL depletion. Discontinuous metamorphic progression may possibly be detected by steps in the paleopressure profile. A few geochemical surveys of various types in transitional granulite terrains have been carried out, by Hörmann et al. (1980) in the north Finland complex, Schreurs (1984) in the West Uusimaa area, Condie et al. (1982) in India southeast of Bangalore, Field and Clough (1976) and Field et al. (1980) in the Bamble area, and Hansen et al. (1984) in southern Karnataka. In the Schreurs (1984) and Hansen et al. (1984) studies, geobarometric profiling showed regional continuity. The Bamble and southern Karnataka areas were the only traverses which penetrated a very high grade, LIL-depleted terrain.

GEOTHERMOMETRIC METHODS

Choice of Geothermometers

Important advances in geothermometry of high grade rocks have been made in recent years; however, lack of precision remains, as indicated by intercomparison of different methods in the same rocks and by relatively large scatter of results using the same thermometers on closely-related rocks. These problems result from inadequacy of experimental and theoretical calibrations and of varying degrees of retrogressive reaction in many granulites. The continuous, or distribution, thermometers most used are the Fe-Mg partitioning between garnet and pyroxene and garnet and biotite, the two-pyroxene method based on clinopyroxene-orthopyroxene solid solubility, the two-feldspar solvus thermometer, and coexisting magnetite and ilmenite. The old-fashoined petrogenetic grid method of discontinuous reactions is still useful in providing broad constraints.

Petrogenetic Grid Constraints

Experimental rock-melting P-T curves and the Al_2SiO_5 diagram provide the most useful temperature information, shown in Fig. 1.

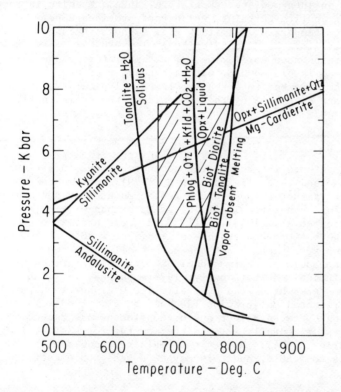

Fig. 1. Experimental univariant equilibria bearing on the amphibolite facies to granulite facies transition zones. Al_2SiO_5 relations from Holdaway (1971). Mg-cordierite breakdown from Newton et al. (1974). Vapor-absent biotite tonalite melting from Wyllie (1977). Vapor absent biotite diorite melting from Brown and Fyfe (1970). Phlogopite-quartz melting under CO_2-H_2O vapors from Wendlandt (1981). H_2O-saturated tonalite solidus from Wyllie (1977). Diagonal-ruled area is probable P-T region of transitional granulite facies.

Many transitional granulite terrains are migmatitic. Therefore, H_2O-saturated solidus curves for representative compositions (for example, tonalitic gneisses) provide lower temperature limits, since $P(H_2O)$ lower than $P(total)$ raises the solidus. Fig. 1 shows the H_2O-saturated solidus of a typical biotite tonalite (Wyllie, 1977). The vapor-absent solidus in the same composition can be expected to provide upper temperature limits, since biotite is almost always present in transitional charnockitic gneisses. A vapor-absent biotite-out curve for diorite composition (Brown and Fyfe, 1970) is shown for comparison. The reaction of biotite with K-feldspar, quartz, and CO_2-H_2O vapor to form orthopyroxene and liquid is a useful reference curve, which should apply to migmatitic charnockites, in an intermediate temperature range. Kyanite is rare in transitional granulite terrains, which fact provides upper pressure and lower temperature limits. Fig. 1 shows the upper pressure stability limit of Mg-cordierite at low $P(H_2O)$; many transitional terrains have cordierite. The temperature range to be deduced from Fig. 1 is 675°-800°C, with attendant pressure conditions of from 7.5 kbar to 4 kbar or lower. Typical transition-zone conditions are suggested by the centroid of the frame in Fig. 1 near 6 kbar and 740°C.

Garnet-Pyroxene K_D^{Fe-Mg}

A number of authors intercomparing geothermometers in various terrains have endorsed the Ellis and Green (1979) experimental calibration of the garnet-clinopyroxene geothermometer as the most reliable method for granulites involving Fe-Mg exchange (Harris et al., 1982; Johnson et al., 1983). Ellis and Green estimated the uncertainty of their temperature scale at ± 5%, or ± 40°C at 800°C.

Recent attempts to recalibrate the garnet-clinopyroxene thermometer on the basis of thermodynamic theory using various assumed solid solution models (Ganguly, 1979; Saxena, 1979) have generated thermometers which read generally 100°C higher than the Ellis-Green experimental calibration (Johnson et al., 1983).

Fe-Mg exchange between garnet and orthopyroxene has recently been calibrated experimentally as a geothermometer for basic compositions (Harley, 1984). Harley's formula reproduces his experimental data to ± 60°C, and this may be a reasonable estimate of its accuracy.

Garnet-Biotite K_D^{Fe-Mg}

The Fe-Mg distribution between garnet and biotite was determined experimentally as a function of temperature by Ferry and Spear (1978) and more recently by Perchuk and Lavrent'eva (1983). Their experiments did not include systems with Ti or Ca, which are important components of granulite-grade biotite and garnet, respectively, and hence, the simple-system calibrations may not be suitable for higher-grade reactions (Ghent et al., 1982).

Two-Pyroxene Method

The experimentally-determined intersolubility of pyroxenes in the simple system $CaMgSi_2O_6-Mg_2Si_2O_6$ was treated by Wood and Banno (1973) as an ideal solution, and extension to more complex natural systems was made assuming ideal ion-site substitutions. A more extensive but essentially identical calibration was made by Wells (1977), using more recent experimental work. Several authors, intercomparing different thermometers in several granulite terrains, have stated that the Wood-Banno and Wells thermometers commonly yield temperatures about 100°C too high in the temperature range around 800°C (Bohlen and Essene, 1979; Harris et al., 1982).

Lindsley (1983) constructed a new two-pyroxene thermometer for igneous and metamorphic rocks based on all available reliable experimental work in the $MgSiO_3-FeSiO_3-CaSiO_3$ pyroxene system. The thermometer is used graphically, and requires projection of non-quadrilateral components into the experimental system. Lindsley suggested a projection procedure, but emphasized that the effects of Na, Al, Fe^{3+}, Ti, etc. in pyroxenes are not known well enough yet to allow absolute confidence in the projection scheme. The estimated uncertainty, for non-quadrilateral components less than two mol percent, is ± 50°C.

Two-Feldspar and Oxide-Mineral Methods

Barth (1951) empirically calibrated a thermometer scale based essentially on the albite-potassium feldspar solvus. This scale has been up-dated by thermodynamic analysis of experimental work (Stormer, 1975). Recent thermodynamic and theoretical work suggests revision of Stormer's scale upwards in temperature by 40°-100°C for the crustal pressure range (Brown and Parsons, 1981; Haselton et al., 1983). Coexisting magnetite and hemo-ilmenite provide, in theory, a combination thermometer and oxygen pressure barometer (Buddington and Lindsley, 1964).

The above two methods have been criticized in application to granulite metamorphism because of relative ease of back-reaction, and hence, tendency to record only closure temperatures during the cooling period (Stoddard, 1980; Lambert, 1983, p. 160). However, Bohlen and Essene (1977) got consistent results from these methods in regional geobarometry of the Adirondacks, by reintegrating the compositions of exsolved perthites and antiperthites and of exsolved magnetite and ilmenite, assuming that no mass loss occurred in local systems in cooling. Rollinson (1980) extracted detailed information from the oxides about temperature and oxygen fugacity histories in granite sheets from the Scourie terrain, including data for igneous, granulite metamorphic, and retrogressive events. Apparently, it is possible to use the feldspar and oxide thermometers in favorable situations of very dry metamorphism and subsequent cooling.

GEOBAROMETRIC METHODS

Comparison of Geobarometers

The principles of geobarometry of high-grade rocks are summarized in Newton (1983). Among the most useful barometric reactions are those in which a mineral of low density, such as feldspar or cordierite with four-coordinated Al, breaks down under pressure to a dense assemblage with minerals of six-fold Al coordination, as in garnet. An example is the reaction of anorthite to grossular, kyanite and quartz, which was first shown to be the basis of a feasible geobarometer for metapelites by Ghent (1976). Newton (1983) compared geobarometers for several different terrains for which abundant mineral analyses are available, and found much consistency among several barometers in common use.

The Ghent Barometer

The anorthite breakdown reaction has been calibrated theoretically in a number of studies (Newton and Haselton, 1981; Ganguly and Saxena, 1984). The several thermodynamic calibrations based on experimental determinations of the end-member reaction and thermodynamic properties of the garnet and plagioclase solid solutions determined from calorimetry or theoretically are in moderately good agreement. A semi-empirical field calibration by Ghent et al. (1979) also gives consistent results, at least for the middle-grade metamorphic range.

Olivine-Orthopyroxene-Quartz Barometer

The reaction of ferrosilite to fayalite + quartz:

$$Fe_2Si_2O_6 = Fe_2SiO_4 + SiO_2 \qquad (A)$$

was calibrated experimentally by Bohlen and Boettcher (1981) with high precision as a geobarometer. The barometer is applicable to assemblages with high Fe/Mg ratio, where olivine and quartz are stable. This is a somewhat rare situation, but mafic assemblages of very high Fe/Mg occur in Precambrian metaironstones (Berg, 1977) and in some Proterozoic anorthosite associations (Jaffe et al., 1978).

"Charnockite Barometers"

A useful geobarometric reaction for high-grade pyroxene gneisses, first suggested by Wood (1975), is:

$$CaAl_2Si_2O_8 + Mg_2Si_2O_6 = 2/3 Mg_3Al_2Si_3O_{12} + 1/3 Ca_3Al_2Si_3O_{12}$$
anorthite   enstatite   pyrope   grossular
$$+ SiO_2 \qquad (B)$$
quartz

The substances in the reaction are principle components of the phases plagioclase, orthopyroxene and garnet. A similar reaction involving $CaMgSi_2O_6$ (diopside) in clinopyroxene is well-suited for two-pyroxene granulites, commonly metabasic. The charnockite barometers were calibrated by Newton and Perkins (1982) almost entirely on the basis of thermodynamic data. The major advantages of these barometers are that the assemblages are very common and that the calibrations are almost temperature-independent.

Another useful geobarometric reaction for charnockites and metabasites is the reaction of anorthite + fayalite to garnet:

$$CaAl_2Si_2O_8 + Fe_2SiO_4 = Fe_2CaAl_2Si_3O_{12} \quad\quad (C)$$

This reaction is pseudounivariant in the system $CaO-FeO-Al_2O_3-SiO_2$ and has been worked out experimentally with high precision by Bohlen et al. (1983a). When combined with reaction A), the resulting reaction:

$$CaAl_2Si_2O_8 + Fe_2Si_2O_6 = 2/3 Fe_3Al_2Si_3O_{12} + 1/3 Ca_3Al_2Si_3O_{12}$$
anorthite   ferrosilite   almandine   grossular
$$+ SiO_2 \quad\quad (D)$$
quartz

can be calibrated as a geobarometer. Bohlen et al. (1983b) found fairly close agreement using this barometer with the Newton-Perkins barometer of reaction B) in the same rocks, although somewhat different solid solution properties were used. Newton (1983) and Johnson and Essene (1982) calibrated barometers from reaction C) for olivine-garnet gabbros, of which numerous examples occur in the Adirondacks. The agreement between the two barometers is fairly good although the calibration methods are somewhat different.

Garnet-Cordierite-Sillimanite-Quartz Barometer

This is based on the reaction of Mg-cordierite to pyrope, sillimanite and quartz:

$$3Mg_2Al_4Si_5O_{18} = 2Mg_3Al_2Si_3O_{12} + 4Al_2SiO_5 + 5SiO_2 \quad\quad (E)$$

and its Fe^{2+} analog. This reaction, together with the Fe-Mg exchange equilibrium between garnet and cordierite, constitutes a simultaneous geothermometer-geobarometer, most useful for metapelites. Experimental calibrations by Currie (1971), Hensen and Green (1973), Holdaway and Lee (1977) and Aranovich and Podlesskii (1983) exist. The latter study is the most comprehensive, though it refutes some of the earlier work. A closely related geobarometric assemblage is garnet-cordierite-hypersthene-quartz, calibrated experimentally as a geobarometer by Hensen and Green (1973).

The pressure indication of these barometers is sensitive to the hydration state of cordierite, and thus, to $P(H_2O)$ of the metamorphism. Numerous theoretical calibrations of reaction E) have been attempted, some of which explicitly consider $P(H_2O)$ as a variable in addition to P(total). For a review of these calibrations, see Newton (1983). Unless some independent knowledge of $P(H_2O)$ is obtainable this barometer serves only to bracket the pressures of metamorphism between extremes of $P(H_2O) = 0$ and $P(H_2O) = P(total)$.

CO_2 Equation of State Barometer

Touret and Bottinga (1979) constructed analytic expressions and graphs of the P-T isochores (lines of constant volume) of CO_2 based on theoretical extrapolation of experimental measurements. Their isochores in the pressure range 5-10 kbar and temperature range 600°-900°C have relatively flat dP/dT slopes, so that, if the temperature of metamorphism can be approximately characterized, the pressure of CO_2 entrapment in fluid inclusions can be estimated. The assumptions must be made that the entrapment occurred during peak metamorphic conditions and that no inelastic deformation of the host crystal or loss of material occurred during uplift.

REGIONAL GEOTHERMOMETRY AND GEOBAROMETRY

Transitional Granulite and High-Grade Terrains

Fig. 2 shows temperatures and pressures deduced for a number of granulite terrains, mostly Precambrian. The temperatures are generally those advanced by the authors of mineral analyses, and the pressures are based on the charnockite geobarometers. Sources of data are given in Newton and Perkins (1982) and Newton (1983). The most noteworthy features are:
1) The pressures and temperatures satisfy the experimental Al_2SiO_5 diagram (Holdaway, 1971). The few terrains having kyanite as the regional Al_2SiO_5 polymorph plot in the kyanite field.
2) The characteristic pressure range of the high grade granulites is 8.5 ± 2 kbar. This contrasts with earlier estimates of 15 kbar (O'Hara and Yarwood, 1978) to 2 kbar (Saxena, 1977).
3) The transitional granulite terrains, so classified because of position in well-defined transition zones, or because they are of the undepleted type, show lower pressures, as well as lower temperatures, than the high-grade granulites. The spread of temperature and pressure conditions shown by the transitional granulites is almost exactly the same as deduced in Fig. 1 from the petrogenetic grid.

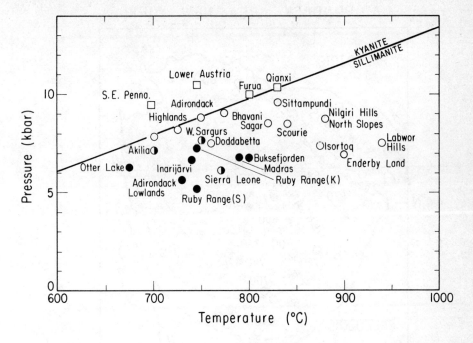

Fig. 2. Average P-T conditions of granulite terrains. Temperatures from a variety of geothermometers. Pressures from the garnet-plagioclase-pyroxene-quartz geobarometers (Perkins and Newton, 1981). Squares: kyanite-bearing terrains. Circles: sillimanite-bearing terrains. Shaded symbols: transitional granulite terrains. Open symbols: high-grade granulite terrains. Half-shaded symbols: indeterminate status. S.E. Pennsylvania Taconic granulites: M.E. Wagner (1982) and personal communication; Lower Austria Variscan granulites: H.G. Scharbert, unpublished data.

Adirondack Highlands

Fig. 3 shows the temperatures deduced for Adirondack Highlands granulites by Bohlen and Essene (1977) in a large-scale regional survey based on the two-feldspar and magnetite-ilmenite methods. As is evident, there is a high degree of consistency between the two methods. Results using other thermometers were much more scattered. The "bull's-eye" pattern centered around the Adirondack anorthosite body is enigmatic. Sm-Nd and Rb-Sr systematics suggest that there was a 300 million year interval between the intrusion of the anorthosite and subsequent metamorphism

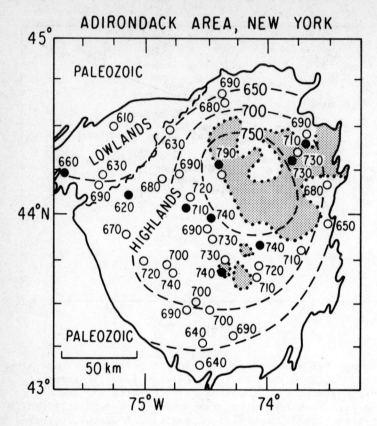

Fig. 3. Paleotemperatures in the Adirondack area, from Bohlen and Essene (1977). <u>Open circles</u>: two-feldspar method (Stormer, 1975). <u>Filled circles</u>: magnetite-ilmenite method. Area outlined in dots is outcrop of Adirondack anorthosite.

1000 million years before present (Ashwal and Wooden, 1983). The reason for the coincidence of highest temperatures with the anorthosite outcrop is therefore not evident, as it is not likely that residual heat from the igneous event could have significantly influenced the metamorphism 300 million years later. A possible explanation derives from the suggestion of Martignole and Schrijver (1970) that the Grenville anorthosites rose in the solid state by bouyancy through surrounding rocks. Under this hypothesis, the deeper isotherms could have been dragged upwards after they had been frozen in (post-metamorphic uplift) or hotter anorthosite from a deeper level could have thermally perturbed higher levels during metamorphism (syn-metamorphic uplift).

Fig. 4 shows the distribution of pressures calculated by Newton (1983) from the reaction B) and C) barometers. Analytical data were

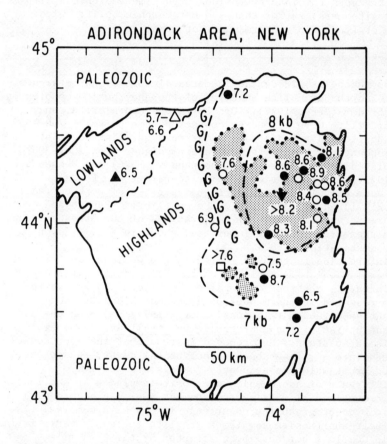

Fig. 4. Paleopressures in the Adirondacks area, from Newton (1983). Open circles: garnet-pyroxene-plagioclase-quartz metabasites. Filled circles: olivine-garnet-pyroxene-plagioclase metabasites. Open triangle: garnet-plagioclase-orthopyroxene-quartz (Newton and Perkins, 1981). Filled triangle: sphalerite-pyrite; Filled inverted triangle: presence of ferrosilite-quartz; Square: presence of kyanite; all after Johnson and Essene (1982). Row of G's: garnet isograd in charnockites after Buddington (1963). Outcrop of Adirondack anorthosite outlined in dots.

taken from Bohlen and Essene (1979, 1980) and Johnson and Essene (1982), and the Bohlen and Essene (1977) isotherms were assumed. A few other pressure data are plotted, as noted in the caption to Fig. 4, as well as approximate locations of a regional garnet-in isograd in acid rocks, according to Buddington (1963). There is a high pressure region, coinciding approximately with the anorthosite outcrop and with the Bohlen and Essene (1977) temperature maximum. This result favors the interpretation of post-metamorphic uplift of the anorthosite mass, rather than syn-metamorphic uplift, with excavation of frozen-in isobars and isotherms.

Nain Complex, Labrador

Orthopyroxene is widespread in metasedimentary and metaigneous assemblages in the granulite aureoles of the anorthosite complex in central Labrador (Berg, 1977). Orthopyroxene, cordierite, and, locally, osumilite were formed by contact metamorphism at 700°-950°C when anorthosite and associated igneous rock were intruded into a rift environment at initial temperature no higher than 200°-300°C. Berg calculated temperatures from Thompson's (1976) and Hensen and Green's (1973) calibrations of cordierite-garnet K_D (Fe-Mg). Pressures were calculated from earlier experimental work on reaction A) (Smith, 1971) and garnet-cordierite equilibria (Hensen and Green, 1973). More recent experimental calibrations of the pressure scales would give pressures 1-2 kbar lower than shown in Fig. 5 (J. Berg, personal communication, 1982).

The interesting feature of Fig. 5 is the well-defined elongate pressure maximum which is centered on the Nain anorthosite body. Also noteworthy is the fact that paleotemperatures do not correlate with paleopressures in the manner of a metamorphic depth-zone profile, but show random relations. Berg (1977) rejected the idea that bouyant uprise of the anorthosite _by itself_ was responsible for the paleoisobar pattern. He proposed instead that the anorthosite was intruded into a rift-graben in continental crust, thus creating a thickened crust which subsequently rose isostatically.

The inference of anorogenic anorthosite emplaced at shallow levels in a continental rifting environment seems in accord with the aureole type of granulite occurrence in the Nain Complex. The Grenville Belt anorthosite complexes are lithologically very similar to the Nain Complex, but have been affected by the Grenville orogeny, 1.2-1.0 billion years ago. According to Morse (1982), anorogenic and orogenic events recorded in the Proterozoic anorthosite-bearing terrains need have no connection and should not be confused. However, the very similar anorthosite kindred and associated granulites of S.W. Norway were both considered by Falkum and Petersen (1980) to be part of an orogenic cycle 1.2 to 0.85 billion years old and to be the direct correlatives of the Grenville Belt across the Atlantic. The presence of recumbent folds in the Rogaland area along with the regional distribution of osumilite spatially related to the Egersund anorthosite

(Maijer et al., 1981), indicates that this has characteristics of both the Nain and Grenville anorthosite associations.

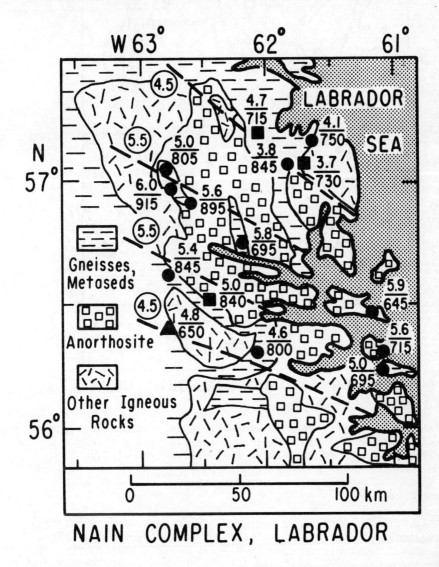

Fig. 5. Nain Complex, Labrador, simplified from Berg (1977). Paleopressures in kbar and paleotemperatures in °C. <u>Filled circles</u>: garnet-cordierite-hypersthene-quartz (Hensen and Green, 1973). <u>Filled squares</u>: olivine-orthopyroxene-quartz (Smith, 1971). <u>Filled triangle</u>: garnet-sillimanite-plagioclase-quartz. Four isobars of regional pressures shown. All pressures would be reduced by two kbar by more recent calibrations.

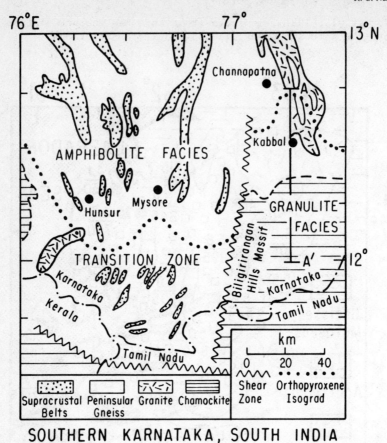

Fig. 6. Amphibolite facies to granulite facies transition zone in southern Karnataka, India. Geology after Pichamuthu (1965) and Janardhan et al. (1982). Section A-A' is traverse of Fig. 7.

Southern India

Fig. 6 shows an area in southern Karnataka, India, straddling the regional late-Archaean amphibolite facies to granulite facies transition. Temperatures along a N-S traverse, normal to the orthopyroxene isograd, were calculated by Hansen et al. (1974) for two-pyroxene and garnet-pyroxene granulites by the methods of Lindsley (1983), Ellis and Green (1979) and Harley (1984). The

thermometers are broadly consistent and indicate nearly the same
temperature range as was deduced from the petrogenetic grid of
Fig. 1, namely 670°-800°C. The scatter of ± 60°C is probably
representative of the present accuracy obtainable, and is too great
to discriminate any certain increase in temperature over the
transition zone. The representative temperature of 750° ± 50°C is
a sufficient base for geobarometry, since the methods used are
relatively temperature insensitive.

Fig. 7 shows a geobarometer profile along the N-S axis,
assuming a uniform temperature of 750°C. There is a clear

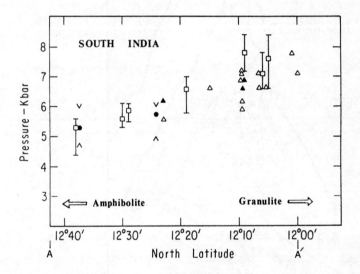

Fig. 7. Paleopressures in the granulite facies transition zone
along A-A' of Fig. 6, from Hansen et al. (1984). Opposing
arrowheads: garnet-cordierite-sillimanite-quartz for $P(H_2O)$ =
$P(total)$ and $P(H_2O)$ = 0. Filled circles: garnet-sillimanite-
plagioclase-quartz. Open triangles: garnet-orthopyroxene-
plagioclase-quartz. Filled triangles: garnet-clinopyroxene-
plagioclase-quartz. Squares: pressures from densities of CO_2
inclusions in quartz (Touret and Bottinga, 1979).

indication of a pressure increase going southwards towards higher-
grade granulites, and no indication is evident of any discontinuity.
The entrapment pressures of CO_2 fluid inclusions in quartz are
shown for seven localities in the transition zone. The agreement
of the CO_2 pressures with the mineralogic pressures indicates that
the CO_2 was trapped at peak metamorphic conditions, and that nearly

pure CO_2 is representative of the fluids of granulite metamorphism in the region.

Willyama Complex, South Australia

The early Proterozoic Willyama Complex contains a variety of lithologies with a metamorphic gradation from high amphibolite grade to granulite grade (Phillips and Wall, 1981). Fig. 8 shows the

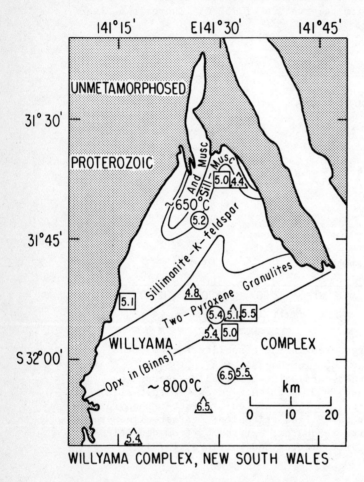

Fig. 8. Metamorphic zones in Willyama Complex, New South Wales, after Phillps and Wall (1981). Paleopressures in kbar. <u>Squares</u>: garnet-plagioclase-sillimanite-quartz. <u>Circles</u>: garnet-plagioclase-orthopyroxene-quartz. <u>Triangles</u>: garnet-cordierite-sillimanite-quartz. Orthopyroxene isograd in basic granulites according to Binns (1964).

metamorphic zones and the pressures calculated by Phillips and Wall using their own formulations of barometers based on reaction D) and the Fe analog of reaction E). Here, as in southern India, pressure increases towards higher metamorphic grade, indicating a depth-zone arrangement of the metamorphic facies. Phillips and Wall inferred a concomitant temperature increase of about 650°C to 800°C over the transition zone, which range is similar to other transition zone estimates made in this paper.

H_2O FUGACITIES AND METAMORPHIC FLUIDS

Biotite and Amphibole Stability

Fig. 9 shows the experimental determination of Luth (1963)

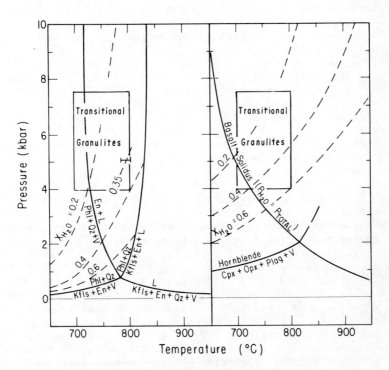

Fig. 9. Mol fractions of H_2O in CO_2-rich fluids which could coexist with enstatite, K-feldspar, phlogopite and quartz, calculated by Newton and Hansen (1983) from experimental data of Luth (1963), with experimental bracket at $X(H_2O) = 0.35$ of Bohlen et al. (1983c); mol fractions of H_2O for which orthopyroxene, clinopyroxene and plagioclase could be stable relative to hornblende and quartz, calculated by Wells (1979). The figure shows that $X(H_2O)$ for transition-zone charnockites was probably less than 0.35.

of the equilibrium phlogopite + quartz = K-feldspar + orthopyroxene + vapor in the system $K_2O-MgO-Al_2O_3-SiO_2-H_2O$. The reaction line is terminated by melting at about 700 bars $P(H_2O)$; at higher pressures the four-mineral assemblage can coexist with a vapor phase only if $P(H_2O)$ is less than P(total). Fig. 9 gives the calculated position of this equilibrium when H_2O is diluted with CO_2, assuming ideal mixing. A modified Redlich-Kwong equation of state (Flowers, 1979; Holloway, 1977) calculation gives very similar results. The lines of constant vapor composition are somewhat sensitive to the choice of experimental data for the basis reaction at $P(H_2O)$ = P(total), and there is some conflict among several determinations. However, the Luth (1963) curve yields good agreement with the reversed bracket of Bohlen et al. (1983c) at 5 kbar for a H_2O-CO_2 mixture of $X(H_2O)$ = 0.35.

The region of 4-7.5 kbar and 700°-800°C believed to be characteristic of most transitional granulite zones corresponds to vapor isopleths of $X(H_2O)$ = 0.35, to below 0.20. That this range should be considered as an upper limit is shown by the probable effects of natural-system components outside the 5-component model system. Fe^{2+} is the major extra component, but will be almost equipartitioned between biotite and orthopyroxene, and will therefore not affect the equilibrium substantially (Bohlen et al., 1983c). The other components which can have a significant effect are Ti, F and Fe^{3+}. All of these will partition strongly into biotite, effecting some stabilization of biotite to lower $P(H_2O)$ relative to pyroxene. In particular, F and Ti are potent stabilizing factors in biotite (Westrich, 1981; Forbes and Flower, 1974). Ti content of biotite increases regularly and dramatically across the transition zones in southern India (E.C. Hansen and R.C. Newton, unpublished data) and in the West Uusimaa area (Schreurs, 1984). High F biotites are characteristic of southern Karnataka (Janardhan et al., 1982).

A similar exercise was performed by Wells (1979) on amphibole stability at reduced $P(H_2O)$ (Fig. 9). He found $X(H_2O)$ of 0.1 to 0.3 at probable granulite P and T conditions for the critical stability of two-pyroxene-plagioclase assemblages in S.W. Greenland.

Phillips (1980) showed, by calculations on the stability of biotite relative to cordierite in pelitic compositions, that $X(H_2O)$ decreased steadily from $\sim$ 1.0 to 0.35 from the amphibolite facies to the granulite facies in the Willyama Complex. The pelite method gave $X(H_2O)$ $\sim$ 0.5 at the regional grade where orthopyroxene appears in metabasic assemblages. Calculations by Phillips (1980) on the stability of hornblende in metabasic rocks gave lower water activities for the first appearance of orthopyroxene.

Fluid Inclusions

Many recent studies have found that CO_2 is the dominant species in fluid inclusions in granulites, with H_2O secondary (e.g., Wells, 1979). Several studies have found quantitative agreement between

CO_2 entrapment pressures and mineralogic geobarometry (Coolen, 1981; Hollister, 1982; Hansen et al., 1984; Schreurs, 1984). The critical amount of H_2O for the first appearance of orthopyroxene in gneisses in southern Karnataka was estimated from fluid inclusions by Hansen et al., (1984). A few large irregular CO_2-rich inclusions in quartz from the incipient charnockite localities in southern Karnataka show immiscible H_2O, estimated to be about 20% of an inclusion, adsorbed on cavity walls and filling capillaries. Roedder (1972) estimated that as much as 20 mol percent of H_2O might be present in a CO_2-rich inclusion, adsorbed on the walls, and escape optical detection. It seems probable that many of the fluid inclusions from the South India incipient charnockites have about 20% H_2O, immiscible at laboratory conditions but miscible with CO_2 at granulite facies conditions.

In contrast to the above observations, Henry (1978) found that typical fluid inclusions in Adirondacks charnockites were mixed H_2O-CO_2, with H_2O often greater than 0.5. This much H_2O is not compatible with the low $X(H_2O)$ deduced here for orthopyroxene-K-feldspar stability. The reasons for this discrepancy are not apparent. The low fluid densities found by Henry, with estimated entrapment pressures of 2-4 kbar, are much lower than those deduced from mineralogic geobarometry, suggesting that the high-H_2O inclusions are not representative of peak metamorphic conditions and perhaps belong to a retrogressive event.

Volatiles in Cordierite

Hörmann et al. (1980) found abundant CO_2 in cordierites from granulites in the N. Finland transition zone, with H_2O/H_2O+CO_2 varying from 0.77 to 0.65 with increase in metamorphic grade. According to the experiments of Johannes and Schreyer (1981), this range corresponds to a change in the vapor phase from $X(H_2O) = 0.35$ to $X(H_2O) = 0.25$ at 5 kbar total pressure, which agrees with the range for transitional granulites deduced from biotite stability.

Oxidation Conditions

The oxygen fugacity states recorded by oxides from the Scourie granulites are low, near the quartz-fayalite-magnetite curve (Rollinson, 1980). Glassley (1982) inferred common low $f(O_2)$ conditions, near graphite stability, determined by Fe^{2+} minerals in the deep crust in the presence of CO_2. In contrast, Dymek (1983) found highly oxidized assemblages from the granulite facies of S.W. Greenland. The best generalization at present seems to be that $f(O_2)$ is highly variable in granulite facies metamorphism and probably is often established by local mineralogy and inherited oxygen contents of precursor rocks (Valley et al., 1983).

LIL Element Depletion

   A major problem in granulite metamorphism is to determine at what stage in the amphibolite facies to granulite facies transition the major LIL depletion occurs, and what the mechanisms of depletion are. Geochemical profiling data presently available are insufficiently definitive. As several authors have noted, the incipient granulites are not depleted in Rb, but have K/Rb ratios similar to amphibolite facies gneisses. In the southern Karnataka traverse, large Rb depletion is found in a few rocks at about the position where the granulite facies becomes dominant (i.e., entering the massif area). However, several rocks sampled from the massif area have K/Rb ratios not much higher than for upper crustal rocks. A similar sudden onset of Rb depletion was found by Field et al. (1980) in the highest-grade zone of the Bamble terrain. It is not yet apparent whether volatile processes can produce extreme depletion, as advocated by Rollinson and Windley (1980) and Okeke et al. (1983) or whether partial melting and removal of a granitic fraction is necessary to produce large depletion of Rb.

ORIGIN OF GRANULITE TERRAINS

Summary of T, P, and Fluid Regimes

   The preceeding review has shown that the temperature range of granulite metamorphism is from near 700°C to above 900°C. The lower temperature cutoff may reflect a characteristic Precambrian geotherm, with higher temperatures the result of magmatism or other thermally perturbing factors. Pressures varied from near 5 kbar to above 10 kbar, with the characteristic range of high grade granulites being 8 ± 2 kbar. These pressures indicate burial under nearly a full continental thickness. Transitional granulites, so classified because of location in well-defined metamorphic gradational tracts or because they are undepleted in LIL elements, characteristically show lower pressures, in the range 5-8 kbar. A few well-defined transitional zones show pressure and temperature gradients across them, suggestive of a depth-zone arrangement. Others show a nearly isobaric increase of grade. The water activity when orthopyroxene formed in acid gneisses was low, no higher than 0.35, and possibly lower than 0.10. Biotite and amphibole survived to very high grades in some lithologies because of high F and Ti contents. The presence of F, as well as SO_4 in scapolite, has suggested common evaporite involvement in some Proterozoic granulite metamorphism (Ortega-Gutierrez, 1984). Some very dry granulites may have been metamorphosed in the absence of a vapor phase.

The Dryness of Granulites

   Low H_2O activity in high grade metamorphism may be explained by

several mechanisms, as follows:
1) CO_2 streaming. CO_2 of deep crustal or subcrustal origin has been suggested as the major agent of dehydration and, sometimes, of LIL depletion in a number of terrains, including Scourie (Okeke et al., 1983; Rollinson and Windley, 1980), the Bamble region (Touret, 1971) and southern India (Janardhan et al., 1982; Condie et al., 1982). The source of abundant CO_2 could be a degassing mantle (Harris et al., 1982), deeply-buried carbonate rocks (Glassley, 1983), or exsolution from basaltic intrusions (Touret, 1971) or tonalitic intrusions (Wells, 1979).
2) Drying-out of terrains in a shallow zone prior to metamorphism. Prolific shallow intrusions could have baked out the Adirondacks area in a low-pressure environment which allowed H_2O to escape without forming hydrates (i.e., the pyroxene hornfels facies). This concept is in accord with the model of shallow igneous emplacement of the Proterozoic massif anorthosites (Morse, 1982) and is also indicated by oxygen isotope patterns in the contact area of the Adirondack anorthosite (Valley and O'Neil, 1984). Subsequent high pressure metamorphism thus operated on very dry rocks. The preservation of pre-metamorphic igneous textures in the Adirondack Highlands, including ophitic and rapakivi textures (Brock, 1980), igneous olivine rimmed by coronal garnet (Johnson and Essene, 1983) and relict chilled margins of gabbros (Gasparik, 1980) all point to a dry style of metamorphism. Pervasive CO_2 action in the Adirondack Highlands is refuted by local mineralogic control of oxygen isotope distributions and CO_2 (Valley et al., 1983; Valley and O'Neil, 1984).
3) Removal of H_2O-undersaturated anatectic melts. This is conceptually a feasible mechanism which has been invoked many times (McCarthy, 1976; Nesbitt, 1980). Many authors have called attention to the existence of granulite-facies granitic migmatites, however, which, if they were the products of anatexis in high grade metamorphism, were not removed, but were themselves dehydrated by some other mechanism (Janardhan et al., 1982; Rollinson and Windley, 1980; Weaver, 1980). Thus, H_2O absorption into granitic melts may have been an adjunct drying agency, and, perhaps, the explanation of severe LIL depletion, but cannot be a sufficient mechanism to account for all granulites.
It is certainly possible that a combination of the above three agencies acted together in a given terrain.

Possible Tectonic Mechanisms of Granulite Metamorphism

A regional high-level granulite terrain in an anorogenic setting must be the result of some large-scale thermal anomaly in the crust which could only have been an expression of subcrustal activity. Subsequent uplift without tilting could give isobaric metamorphic gradients. Such large intracontinental thermal effects are widely believed to result from incipient or early continental rifting (Windley, 1983). The anorthosite kindred magma suite, including anorthosites, peralkaline and rapakivi granites, and magmatic charnockites (Hubbard and Whitley, 1979), is believed by some to be a feature of continental rifting. Some granulites associated with the anorthosite suite, as in the Nain, Labrador aureoles, probably represent shallow baking out of the crust in the pyroxene hornfels facies during early rifting. Drying out at a deeper level might occur in such a setting if CO_2 is abundant, perhaps released by basic magmas, or if escape of H_2O-undersaturated granitic melts leaves refractory residues. Such mechanisms could account for the Ivrea Zone granulites (Schmid, 1979) and the West Uusimaa granulites (Schreurs, 1984).

Most well-described granulite terrains display orogenic features. Early-orogenic large-scale thrusting and nappe stacking occurred in the Scourie (Khoury, 1968), Adirondack (McLelland and Isachsen, 1980), South India (Drury et al., 1983) and many other terrains. However, metasediments are most often of the shelf or passive margin type, including marbles, quartzites and K-pelites (Shackleton, 1976). Evaporites have been reported in several terrains, including the Adirondack Lowlands (Brown and Engel, 1956) and the Oaxaca, southern Mexico, region (Ortega-Gutierrez, 1984). Marine evaporites are believed to be characteristic of rifted passive margins (Emery, 1976). Overthrusts and nappes in shelf sediments suggest closure of an ocean basin. This type of tectonic event has been suggested as the cause of metamorphism in the Norwegian Basal Gneisses (Cuthbert et al., 1983), the Adirondacks (Seyfert, 1980), and the Wopmay Orogen of N.W. Canada (Hoffman, 1980). Rocks of the collapsed basin, including anorogenic plutons and volcanics, may be overriden in the collision. It has been postulated that H_2O-laden shelf sediments and evaporites can provide a lubricated horizon which localizes continental-scale overthrusting (Behr and Horn, 1982; Hodges et al., 1982). Abundant CO_2 may be supplied to the metamorphic site by decarbonation of shelf sediments or by degassing of a mantle diapir previously carbonated by subducted limestone. Closure of a basin probably supposes some type of subduction which consumes a small oceanic plate. Therefore, calc-alkaline plutonic rocks may be involved in the collisional orogenesis, and these may penetrate into, and be hard to distinguish from, the anorogenic suite. Careful selection of samples for dating has, in a few areas, shown a crystallization age of anorthosites 50-300 million years older than an overprinted metamorphic age (Ashwal and Wooden, 1983; Cliff et al., 1983; Foland and Muessig, 1978).

An amphibolite facies to granulite facies transition might, in an orogenic model, be expected to correspond to a critical depth in the overthrust sequence, and thus show an increase of pressure with increasing grade. A monotonous increase of temperature with increasing pressure may not always be expected, however. If the collision process was fast, there could have been an initial thermal overturn immediately after thrusting (England and Richardson, 1977), which might not have been erased during subsequent uplift and erosion (Wagner, 1982). Uplift might, more often than not, have taken place along zones of petrologic contrast, so that the transitional regions may be separated from the highest-grade areas by faults. Such vertical faults could plausibly coincide with the surface traces of earlier overthrusts, since density contrasts, and hence, isostatic adjustments, would tend to be greatest along these lines.

The foregoing interpretation is not presented as a universal solution to the granulite problem, but merely to show that a number of features of different granulite terrains can be explained by tectonic models under current discussion. Another mechanism to account for some granulites may be emplacement of hot mantle slices into continental crust along a major transcurrent fault as postulated by Vielzeuf and Kornprobst (1984) for the northern Pyrenees Mesozoic granulites. Metasomatic charnockite emplacement along vertical shears which allowed access of CO_2 to the upper crust was suggested by Drury and Holt (1980) for southern India. It may be that change of tectonic style from dominantly overthrusting to transcurrent shearing was a common consequence of embrittlement as the lower crust was dried out and converted to granulite (Chadwick and Nutman, 1979).

IMPLICATIONS FOR THE DEEP CONTINENTAL CRUST

The above speculations interpret granulite terrains as the result of continental rifting and collision processes. Many of the lithologies caught up in such processes were supracrustal, and others were high-level plutons. The greatest burial of these rocks was to middle depths in a thickened continental crust. Rocks underlying the supracrustal-plutonic assemblage may rarely or never be seen in Precambrian granulite terrains.

Granulite-grade regional metamorphism seems very much analogous to young metamorphic belts, in showing zones of progressive metamorphism with more or less well-defined isograds, nappes, overthrusts, and synorogenic plutons. The major differences between the older type of regional metamorphism and younger metamorphism is that H_2O activity was lower and temperatures were somewhat higher in the earlier type. If this analogy of ancient granulite metamorphism with younger metamorphism in orogenic belts is correct, there seems no reason to suppose that the metamorphism need be directly related to crustal accretion or that the present lowermost continental crust is commonly a metasedimentary-plutonic assemblage bearing the scars of ancient continental collision.

ACKNOWLEDGEMENTS

The author benefitted from discussions and correspondence with many people, including, especially, Lew Ashwal, Jon Berg, Ed Hansen, Neil Phillips, Heinz Scharbert, Jacques Touret, John Valley, Mary Emma Wagner, and Dave Waters. The research of Robert Newton is supported by National Science Foundation grants #EAR 84-11192 and EAR 82-19248.

REFERENCES

Aranovich, L.Ya. and Podlesskii, K.K. 1983, in Kinetics and Equilibrium in Mineral Reactions (S.K. Saxena, ed.), Springer-Verlag, New York, pp. 173-198.
Ashwal, L.D. and Wooden, J.L. 1983, Geochim. Cosmochim. Acta 47, pp. 1875-1885.
Barth, T.F.W. 1951, Neues Jahrb. Mineral. 82, pp. 143-154.
Behr, H.F. and Horn, E.E. 1982, Chem. Geol. 37, pp. 173-190.
Berg, J.H. 1977, J. Petrol. 18, pp. 399-430.
Binns, R.A. 1964, J. Geol. Soc. Australia 11, pp.283-330.
Bohlen, S.R. and Boettcher, A.L. 1981, Amer. Mineral. pp. 951-964.
Bohlen, S.R. and Essene, E.J. 1977, Contr. Mineral. Petrol. 62, pp. 153-169.
Bohlen, S.R. and Essene, E.J. 1979, Lithos 12, pp. 335-345.
Bohlen, S.R. and Essene, E.J. 1980, Geol. Soc. Amer. Bull. 91, pp. 685-719.
Bohlen, S.R., Wall, V.J. and Boettcher, A.L. 1983(a), Contr. Mineral. Petrol. 83, pp. 52-61.
Bohlen, S.R., Wall, V.J. and Boettcher, A.L. 1983(b), in Kinetics and Equilibrium in Mineral Reactions (S.K. Saxena, ed.), Springer, Verlag, New York, pp. 141-171.
Bohlen, S.R., Boettcher, A.L., Wall, V.J. and Clemens, J.D. 1983(c), Contr. Mineral. Petrol. 83, pp. 270-277.
Brock, B.S. 1980, Geol. Soc. Amer. Bull. 91, pp. 93-97.
Brown, G.C. and Fyfe, W.S. 1970, Contr. Mineral. Petrol. 28, pp.310-318.
Brown, J.S. and Engel, A.E.J. 1956, Geol. Soc. Amer. Bull. 67, pp. 1599-1622.
Brown, W.L. and Parsons, I. 1981, Contr. Mineral. Petrol. 76, pp. 369-377.
Buddington, A.F. 1963, Geol. Soc. Amer. Bull. 74, pp. 1155-1182.
Buddington, A.F. and Lindsley, D.H. 1964, J. Petrol. 5, pp. 310-357.
Chadwick, B. and Nutman, A.P. 1979, Precamb. Res. 9, pp. 199-226.
Cliff, R.A., Gray, C.M. and Huhma, H. 1983, Contr. Mineral. Petrol 82, pp. 91-116.
Coolen, J.J.M.M.M. 1981, Chem. Geol. 37, pp. 59-77.
Condie, K.C., Allen, P. and Narayana, B.L. 1982, Contr. Mineral. Petrol. 81, pp. 157-167.
Currie, K.L. 1971, Contr. Mineral. Petrol. 33, pp. 215-226.

Cuthbert, S.J., Harvey, M.A. and Carswell, D.A. 1983, J. Meta. Geol. 1, pp. 63-90.
Drury, S.A., Harris, N.B.W., Holt, R.W., Reeves-Smith, G.J. and Wightman, R.T. 1983, J. Geol. 92, pp. 3-20.
Drury, S.A. and Holt, R.W. 1980, Tectonophys. 65, pp. T1-T15.
Du Toit, M.C., Van Reenen, D.D. and Roering, C. 1983, Spec. Publ. Geol. Soc. South Africa 8, pp. 121-142.
Dymek, R.F. 1983, Gronlands Geol. Unders. 112, pp. 83-94.
Ellis, D.J. and Green, D.H. 1979, Contr. Mineral. Petrol. 71, pp. 13-22.
Emery, K.O. 1977, Amer. Assoc. Petrol. Geol. Short Course Notes No. 7.
England, P.C. and Richardson, S.W. 1977, J. Geol. Soc. Lond. 134, pp. 201-213.
Falkum, T. and Petersen, J.S. 1980, Geol. Rundschau 69, pp. 622-647.
Ferry, J.M. and Spear, F.S. 1978, Contr. Mineral. Petrol. 66, pp. 113-117.
Field, D. and Clough, P.W.L. 1976, J. Geol. Soc. Lond. 132, pp. 277-288.
Field, D., Drury, S.A. and Cooper, D.C. 1980, Lithos 13, pp. 281-289.
Flowers, G.C. 1979, Contr. Mineral. Petrol. 69, pp. 315-318.
Foland, K.A. and Muessig, K.W. 1978, Geology 6, pp. 143-146.
Forbes, W.C. and Flower, M.F.J. 1974, Earth. Plan. Sci. Lett. 22, pp. 60-66.
Fountain, D.M. and Salisbury, M.H. 1981, Earth, Plan. Sci. Lett. 56, pp. 263-277.
Ganguly, J. 1979, Geochim. Cosmochim. Acta. 43, pp. 1021-1029.
Ganguly, J. and Saxena, S.K. 1984, Amer. Mineral. 69, pp. 88-97.
Gasparik, T. 1980, Geol. Soc. Amer. Bull. 91, pp. 78-88.
Ghent, E.D. 1976, Amer. Mineral. 61, pp. 710-714.
Ghent, E.D., Robbins, D.B. and Stout, M.Z. 1979, Amer. Mineral. 64, pp. 874-885.
Ghent, E.D., Knitter, C.C., Raeside, R.P. and Stout, M.Z. 1982, Canad. Mineral. 20, pp. 295-305.
Glassley, W.E. 1982, Nature 295, pp. 229-231.
Glassley, W.E. 1983, Contr. Mineral. Petrol. 84, pp. 15-35.
Hamilton, P.J., Evensen, N.M., O'Nions, R.K. and Tarney, J. 1979, Nature 277, pp. 25-28.
Hansen, E.C., Newton, R.C. and Janardhan, A.S. 1984, in Archaean Geochemistry (A. Kröner, A.M. Goodwin and G.N. Hanson, eds.), Springer-Verlag, Heidelberg (in press).
Harley, S.L. 1984, Contr. Mineral. Petrol. (in press).
Harris, N.B.W., Holt, R.W. and Drury, S.A. 1982, J. Geol. 90, pp. 509-527.
Haselton, H.T., Hovis, G.L., Hemingway, B.S. and Robie, R.A. 1983, Amer. Mineral. 68, pp. 398-413.
Heier, K.S. 1973, Phil. Trans. R. Soc. Lond. A-273, pp. 429-442.
Henry, D.L. 1978, Unpub. Senior Thesis, Dept. of Civil Engineering, Princeton Univ.

Hensen, B.J. and Green, D.H. 1973, Contrib. Mineral. Petrol. 38, pp. 151-166.
Hodges, K.V., Bartley, J.M. and Burchfiel, B.C. 1982, Tectonics 1, pp. 441-462.
Hoffman, P.F. 1980, Geol. Assoc. Canada Spec. Paper 20, pp. 523-549.
Holdaway, M.J. 1971, Amer. J. Sci. 271, pp. 97-131.
Holdaway, M.J. and Lee, S.M. 1977, Contr. Mineral. Petrol. 63, pp. 175-198.
Hollister, L.S. 1982, Canad Mineral. 20, pp. 319-332.
Holloway, J.R. 1977, in Thermodynamics in Geology (D. Fraser, ed.), D. Reidel, Dordrecht, pp. 161-181.
Hörmann, P.K., Raith, M., Raase, P., Ackermand, D. and Seifert, F. 1980, Finland Geol. Surv. Bull. 308, pp. 1-95.
Hubbard, F.H. 1978, Geol. För. Stock. Förh 100, pp. 31-38.
Hubbard, F.H. and Whitley, J.E. 1979, Nature 217, pp. 439-440.
Jaffe, H.W., Robinson, P. and Tracy, R.J. 1978, Amer. Mineral. 63, pp. 1116-1136.
Janardhan, A.S., Newton, R.C. and Hansen, E.C. 1982, Contr. Mineral. Petrol. 79, pp. 130-149.
Johannes, W. and Schreyer, W. 1981, Amer. J. Sci. 281, pp. 299-317.
Johnson, C.A. and Essene, E.J. 1982, Contr. Mineral. Petrol. 81, pp. 240-251.
Johnson, C.A., Bohlen, S.R. and Essene, E.J. 1983, Contr. Mineral. Petrol. 84, pp. 191-198.
Khoury, S.G. 1968, Scot. J. Geol. 4, pp. 109-120.
Kröner, A. 1980, in Mobile Earth (H. Closs, K.V. Gehlen, H. Illies, E. Kuntz, S. Neumann and E. Seibold, eds.), Harald Boldt Verlag, pp. 225-234.
Lambert, R.St.J. 1983, Geol. Soc. Amer. Memoir 161, pp. 155-165.
Lindsley, D.H. 1983, Amer. Mineral. 68, pp. 477-493.
Luth, W.C. 1963, Ph.D. Thesis, Pennsylvania State Univ.
Maijer, C., Andriessen, P.A.M., Hebeda, E.H., Jansen, J.B.H. and Verschure, R.H. 1981, Geol. en Mijnb. 60, pp. 267-272.
Martignole, J. and Schrijver, K. 1970, Tectonophys. 10, pp. 403-409.
McCarthy, T.S. 1976, Geochim. Cosmochim. Acta 40, pp. 1057-1068.
McLelland, J. and Isachsen, I. 1980, Geol. Soc. Amer. Bull. 91, pp. 68-72.
Morse, S.A. 1982, Amer. Mineral. 67, pp. 1087-1100.
Nesbitt, H.W. 1980, Contr. Mineral. Petrol. 72, pp. 303-310.
Newton, R.C. 1983, Amer. J. Sci. 283-A, pp. 1-28.
Newton, R.C. and Hansen, E.C. 1983, Geol. Soc. Amer. Memoir 161, pp. 167-178.
Newton, R.C. and Haselton, H.T. 1981, in Thermodynamics of Minerals and Melts (R.C. Newton, A. Navrotsky and B.J. Wood, eds.), Springer-Verlag, New York, pp. 129-145.
Newton, R.C. and Perkins, D. 1982, Amer. Mineral. 67, pp. 203-222.
Newton, R.C., Charlu, T.V. and Kleppa, O.J. 1974, Contr. Mineral. Petrol. 44, 295-311.
O'Hara, M.J. and Yarwood, G. 1978, Phil. Trans. R. Soc. Lond. A-228, pp. 441-456.

Okeke, P.O., Borley, G.D. and Watson, J. 1983, Mineral. Mag. 47, pp. 1-10.
Ortega-Gutiérrez, F. 1984, Precamb. Res. 23, pp. 377-393.
Perchuk, L.L. and Lavrent'eva, I.V. 1983, in Kinetics and Equilibrium in Mineral Reactions (S.K. Saxena, ed.), Springer-Verlag, New York, pp. 199-239.
Perkins, D. and Newton, R.C. 1981, Nature 292, pp. 144-146.
Phillips, G.N. 1980, Contr. Mineral. Petrol. 75, pp. 377-386.
Phillips, G.N. and Wall, V.J. 1981, Bull. Minéral. 104, pp. 801-810.
Pichamuthu, C.S. 1965, Indian Mineral. 6, pp. 119-126.
Quensel, P. 1951, Arkiv. for Min. Geol. 1, pp. 229-332.
Roedder, E. 1972, U.S. Geol. Surb. Prof. Pap. 440JJ, pp. 1-164.
Rollinson, H.R. 1980, Mineral. Mag. 43, pp. 623-631.
Rollinson, H.R. and Windley, B.F. 1980, Contr. Mineral. Petrol. 72, pp. 257,263.
Romey, W.E., Elberty, W.T., Jacoby, R.S., Christoffersen, R., Shrier, T. and Tietbohl, D. 1980, Geol. Soc. Amer. Bull. 91, pp. 97-100.
Saxena, S.K. 1977, Science 198, pp. 614-617.
Saxena, S.K. 1979, Contr. Mineral. Petrol. 70, pp. 229-235.
Schmid, R. 1978, Mem. di Schien. Geol. 33, pp. 67-69.
Schreurs, J. 1984, J. Meta. Petrol. (in press).
Seyfert, C.K. 1980, Geol. Soc. Amer. Bull. 91, pp. 118-120.
Shackleton, R.M. 1976 in Early History of the Earth (B.F. Windley, ed.) John Wiley and Sons, New York, pp. 317-322.
Smith, D. 1971, Amer. J. Sci. 271, pp. 370-382.
Smithson, S.B. and Brown, S.K. 1977, Earth, Plan. Sci. Lett. 35, pp. 134-144.
Stoddard, E.F. 1980, Geol. Soc. Amer. Bull. 91, pp. 100-102.
Stormer, J.C. 1975, Amer. Mineral. 60, pp. 667-674.
Thompson, A.B. 1976, Amer. J. Sci. 276, pp. 425-454.
Thompson, J.B. 1955, Amer. J. Sci. 253, pp. 65-103.
Touret, J. 1970, Comptes Rend. Acad. Sci. Paris 271, Ser. D, pp. 2228-2231.
Touret, J. 1971, Lithos 4, pp. 239, 249; pp. 423-436.
Touret, J. and Bottinga, Y. 1979, Bull. Minéral. 102, pp. 577-583.
Valley, J.W. and Essene, E.J. 1977, Geol. Soc. Amer. Abstr. w/ Prog. 9, pp. 260-261.
Valley, J.W. and O'Neil, J.R. 1984, Contr. Mineral. Petrol. 85, pp. 158-173.
Valley, J.W., McLelland, J. and Essene, E.J. 1983, Nature 301, pp. 226-228.
Vielzeuf, D. and Kornprobst, J. 1984, Earth, Plan. Sci. Lett. 67, pp. 87-96.
Wagner, M.E. 1982, Geol. Soc. Amer. Abstr. w/ Prog. 14, p. 640.
Waters, D.J. 1984, Proc. of the Conf. on Middle to Late Proterozoic Lithosphere Evolution, Cape Town, July 1984.
Wells, P.R.A. 1977, Contr. Mineral. Petrol. 62, pp. 129-139.
Wells, P.R.A. 1979, J. Petrol. 20, pp. 187-226.

Wendlandt, R.F. 1981, Amer. Mineral. 66, pp. 1164-1174.
Westrich, H.R. 1981, Contr. Mineral. Petrol. 78, pp. 318-323.
Wilson, A.F. 1978, in Archaean Geochemistry (B.F. Windley and S.M. Naqvi, eds.), Elsevier, Amsterdam, pp. 241-268.
Windley, B.F. 1983, Geol. Soc. Amer. Memoir 161, pp. 1-10.
Wood, B.J. 1975, Earth, Plan. Sci. Lett. 26, pp. 299-311.
Wood, B.J. and Banno, S. 1973, Contr. Mineral. Petrol. 42, pp. 109-124.
Wyllie, P.J. 1977, Tectonophys. 43, pp. 41-71.

FLUID ENHANCED MASS TRANSPORT IN DEEP CRUST AND ITS INFLUENCE ON ELEMENT ABUNDANCES AND ISOTOPE SYSTEMS

W.E. Glassley[1] and D. Bridgwater[2]

[1]Dept. of Geology, Middlebury College, Middlebury, Vt., 05753 USA, [2]Geologisk Museum, Øster Voldgade 5-7, DK-1350, København K, Danmark

ABSTRACT. We report here the results of two studies designed to establish the chemical and isotopic compositional makeup of fluids in deep crustal settings. One study concentrates on a small shear zone where fluids associated with retrogression of granulites were probably aqueous, were Cl- and N-rich and had significant sulfate and potassium contents. REE abundances in the shear zones are fractionated (La/Yb ~160), relative to the country rock (La/Yb ~22), with a 50% increase in the Nd/Sm ratio. The second study, dealing with granulite facies rocks in a large (> 40 km wide) shear zone, demonstrates that Rb-Sr systems are reset to varying degrees, on the scale of tens to thousands of sq meters. In some cases, complete resetting occurs of Archaean gneisses to Proterozoic ages with low Sr87/86 initial ratios. Other areas of the same region experience only weak disturbance of the original Archaean Rb-Sr systematics. The degrees of resetting are correlated with the degree of mineral homogenization, and suggest that fluid volume varies widely from region to region, although no clear lithologic characteristics delineate the different regions. Theoretical consideration of the chemical characteristics of the fluids which modified these two rock systems demonstrates 1) nitrogen may significantly modify CO_2/H_2O ratios, 2) geothermal gradients strongly influence equilibrium fluid composition and solvent properties of the fluid, and 3) the relative proportions of the ionic species in the fluid are strongly dependent on fluid pH.

1. INTRODUCTION

Fluids in deep crustal environments are important equilibrium parameters in two respects. First, the chemical potentials of fluid components contribute to the overall free energy of any assemblage, and therefore affect which mineral assemblages will occur at any given set of P-T conditions. Because of the dramatically different free energies of different fluid species, small changes in fluid composition can lead to significant changes in mineral assemblages and/or mineral compositions. This will alter Kd rock/fluid parameters as minerals change composition or become unstable. Second, fluids provide effective communication pathways which encourage element mobility and mass transport. Thus fluids may act as vehicles whereby the composition of a particular rock system is modified as a result of its local chemical environment.

However, we remain ignorant of the thermodynamic behavior of the highly complex fluids likely to occur at the high temperatures and pressures of deep crustal rocks. It is therefore difficult to predict the features to be expected in regions where fluid activity has been high. We also do not know what fluid volumes to expect in such settings, nor do we know the scale over which chemical communication can be expected.

In this paper we present results of two studies undertaken in high grade metamorphic rocks typical of Greenland granulites. The purpose of these studies was to determine in high grade rocks what characteristics may be observed which record fluid-rock interaction. We looked at major, minor, trace element and isotope features, and used phase relationships to compute fluid compositions. We carried out these studies on the scale of 100s of sq km, and on the scale of cm, to see if limits could be placed on the scale of fluid effects. We report first the observations from the two sites, and then discuss the theoretical aspects of the work.

2. STORØ EAST GREENLAND

The Storø site ($66.1°N$, $35.9°W$) is a region of granulite facies gneisses which are cut by transcurrent shear zones ~.5 m wide (1). The shear zones formed at about the time of the granulite facies metamorphism (2), and locate regions of fluid migration; weakening of the rock due to fluid activity probably led to shear zone growth.

The composition of fluids in the shear zone and in the country rock were determined using several methods. Fluid inclusions were used as an approximate indicator of CO_2/H_2O ratios. The inclusions in the country rock are primarily CO_2-rich and are interpreted to represent the pre-shear zone fluid composition (3). Inclusions in the shear zone are primarily late stage, saline and aqueous. Fluids were also studied using phase equilibria, as described below. These results
are consistent with the fluid inclusion work, in that shear zone fluids were probably aqueous, while the country rocks equilibrated with more CO_2-rich fluids. In addition, the shear zone rocks experienced H_2 fugacities twice that of the enclosing rock. We have also carried out bulk rock analyses of volatile species, including F, Cl, S (total sulfur, sulfate and sulfide) and N_2. These results document strong enrichment of Cl and N_2 in the shear zone rocks; F exhibited no change between shear zone and country rock. Sulfur occurs primarily as sulfate in the shear zone, but as sulfide in the country rock.

These contrasts in fluid composition resulted in significant mineralogical changes across the shear zone boundaries. Quartzo-feldspathic country rock gneisses consistently contain biotite-orthopyroxene-hornblende-pyrrhotite, minor perthitic microcline and rare garnet. The shear zone rocks lack hornblende and pyrrhotite, but contain, instead, scapolite, magnetite and abundant microcline, with minor garnet. Using the observed phase compositions, the following reaction relates the assemblages:

$$Na_{.45}K_{.55}Ca_{1.8}Mg_{2.5}Fe_{2.5}Al_{2.5}Si_{6.0}O_{22}(OH)_2 + 1.86SiO_2 +$$
$$+ .038S_2 + .113Cl ===$$
$$.188Na_{2.4}Ca_{1.6}Al_{4.2}Si_{7.8}O_{24}Cl_{.6}SO4_{.4} + .581Mg_{.88}Fe_{2.12}Al_2Si_3O_{12} +$$
$$+.298KAlSi_3O_8 + .252KMg_{2.2}Fe_{.8}AlSi_3O_{10}(OH)_2 + 1.5CaMg_{.95}Fe_{.05}Si_2O_6 +$$
$$+.33Fe_3O_4 + .75H_2 + .1O_2$$

In this reaction, hornblende and scapolite are written using observed mineral compositions; the other phases are written as solid solutions using the expected Fe-Mg partitioning for these minerals at granulite facies conditions (7,8).

This reaction is confirmed by the occurrence of biotite-microcline-garnet-magnetite clusters replacing hornblende; scapolite is usually in the immediate vicinity of these clusters. The importance of the activity of fluid constituents other than those limited to the C-O-H system is emphasized by this reaction.

It is also apparent that zircons, which are trace constituents in all of the rocks, are chemically rounded in the shear zone samples, but remain subidiomorphic in the country rocks. This modification may account for the REE pattern changes discussed below.

The mineralogical changes observed between the shear zone and country rock reflect, in addition to changes resulting from the above reaction relationship, a change in bulk composition (1). The modal abundances of scapolite, microcline and quartz must reflect a change in the bulk composition of the shear zone rocks, relative to the country rock. This is confirmed by bulk rock analyses, which clearly show increases in SiO2, K2O, Pb and K2O/Na2O, and decreases in CaO, MgO, Al2O3, Cu and Zn (Fig. 1 and ref 1).

Analyses of rare earth elements from regional granulites and the shear zone indicate strong depletion of the HREEs, reflecting interaction of the shear zone fluids with the rock systems, and breakdown of hornblende and zircon (Fig. 2). Within the shear zone the light REEs appear to increase slightly, while the heavy REEs are depleted by 10-20 times, relative to the country rock. This modification of the REE abundances results in a change in the Ce/Yb ratio from 6.3-50 in the regional granulites, to 70-250 in rocks within 5 meters of the shear zone. Nd/Sm ratio increases with increasing alkali content (from ~1.8 to ~ 2.6). No change in Rb-Sr systematics could be identified (Fig. 3).

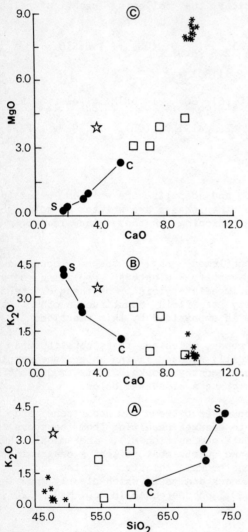

Figure 1. SiO_2 vs K_2O (A), CaO vs K_2O (B) and CaO vs MgO (C) for shear zone (S) and country (C) rocks at the Storø site. Adjacent samples are connected by lines. Shown for comparison are the ranges for acid and basic andesites (squares), Agto gabbros (7; filled stars) and average metapelites from Marranguit (9; star).

3. NORDRE STROMFJORD SHEAR ZONE

On the West coast of Greenland a major shear zone (~40 km wide) developed under granulite facies conditions (4-9) about 1.8 b.y.a. (10-12). Within the shear zone, fluid heterogeneity was common, with CO_2-rich domains occurring in metapelitic units where graphite is abundant, and H_2O-rich fluids occurring in quartzo-feldspathic tonalitic gneisses (8). Unlike the E. Greenland site, where sampling was restricted to

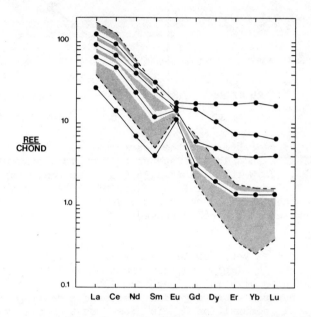

Figure 2. REE variation diagram comparing the regional granulites (solid lines and symbols; 4 samples indicate the range of patterns and abundances observed in 11 spectra) to rocks within 5 meters of the shear zone (shaded field; defined by 9 spectra). Eu is constant for all samples at 10-17 times chondrite, suggesting Eu is little affected by either partial melting or fluid processes (cf. 31).

one lithologic unit, the West Greenland location covers large tracts of land containing a variety of lithologies. It is thus not possible to establish that progressive changes in bulk chemistry occur as a consequence of fluid interaction, although a correlation between bulk composition and fluid composition can be demonstrated (8). In addition, Rb-Sr isotopic data and mineralogical characteristics suggest that fluid-dominated processes strongly influenced the chemical features of the rocks (13). Two features are particularly striking in these shear zone rocks. First, the degree of resetting of the Rb-Sr system varies substantially, with some areas (on the scale of sq. kms.) preserving an Archaean age with no evidence of a 1.8 by. event, while other areas of similar size exhibit complete resetting of the Rb-Sr system to a 1.8 by. age (Fig. 3). One of the reset areas

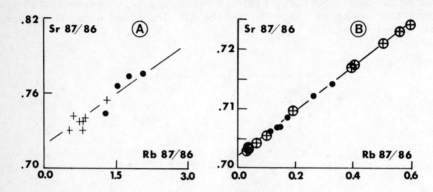

Figure 3. 87Rb/86Sr vs 87Sr/86Sr for the Marranguit (13;A) and Storø (2;B) sites. The line in A is a reference 2050 my isochron, with an initial ratio of .721. In B the line is a 2630 my isochron, with a two sigma uncertainty of 65 my and an initial ratio of .7016. In A, solid dots are banded gneisses, crosses are granitic gneisses. In B, solid dots are country rock samples, circled crosses are shear zone samples.

(Marranguit) also preserves very low initial ratios of .7037, despite the fact that this crust is simply reworked crust of >2.7 by. age, and is derived from crust with high initial ratios (13). Second, a strong correlation exists between the degree of isotopic resetting and the homogeneity of mineral compositions. In the reset rocks mineral compositions tend to cluster around a common value, despite wide variations in rock type, while in the less reset rocks mineral compositions scatter widely, with a clear distinction between rock types (13).

Because of the scale involved, diffusion cannot be considered an important influence in the homogenization which is characteristic of the reset rocks. Instead, the implied element mobility virtually requires the presence of a fluid phase; differences in degree of homogenization/resetting can be attributed to differences in the total fluid volume within a particular area, and perhaps to differences in fluid composition (13).

4. C-O-H Fluid Rock Interaction

By restricting our considerations to metapelitic systems which contain graphitic quartzo-feldspathic gneisses bearing magnetite-biotite-potassium feldspar-sillimanite, we can model fluid rock interaction which may be expected under granulite facies conditions. This assemblage buffers hydrogen partial pressures (14,15) and, in the presence of graphite, becomes an invariant assemblage if all solid phase compositions are known. Other rock systems can be qualitatively examined by comparison with the results from this system, since it is usually possible to establish, simply through examination, the relative changes in one or more fluid species fugacities and to deduce how these will modify the overall fluid composition. The equilibria and computational procedure used to evaluate fluid-rock interaction in this system have been outlined in (8).

The results demonstrate that the solubility of alkali metal ions (Na^+, K^+, and Ca^{++}) is dependent on acidity and CO_2/H_2O ratio. The equilibrium activity of the ions in solution increases logarithmically as pH decreases, but increases exponentially at high XCO_2 as CO_2/H_2O increases. However, the relative solubilities of the ions ($Na^+>K^+>>Ca^{++}$) will lead to an apparent enrichment in Ca^{++} relative to the other ions in the rock residue remaining after dissolution. The theoretical implication derived from this is that CO_2-rich fluids which migrate through deep crust will deplete the rock in sodium and potassium, and will enrich it in Ca^{++}. It can be expected that elements which behave in an analogous manner to these elements will mimic their analogues. Therefore, such a fluid will be expected to leave behind Ca-, Sr-, Mg-enriched, Na-, K-, Rb-, Th-, Pb-, U-depleted crust. These "depleted" elements will be redeposited at some other location along the migration path of the fluid. Experimental studies suggest that Si will also be an important ionic species in this migrating fluid, and respond to the equilibrium parameters in a fashion similar to that of Na^+ and K^+ (16-19). It is important to remember that the difference between H_2O-dominated and CO_2-dominated fluids is primarily in the equilibrium abundances of the ionic species; although the equilibrium activity ratios of the ions will change as acidity and CO_2/H_2O change, Na^+ will remain an ion with the highest solubility, and Ca^{++} an ion with the lowest solubility, in any CO_2-H_2O fluid.

The results from Storø are perfectly consistent with these results; an H_2O-rich fluid, with high potassium and silica abundances, migrated along the shear zone pathways, resulting in recrystallization of the rock along the shear zone as the chemical environment changed. The source for this fluid may well have been underlying continental crust, of low Na content, which was experiencing high grade metamorphism (3). However, the chemistry of the migrating fluid was more complex than that in the model system, and requires consideration of sulfur and nitrogen as equilibrium parameters (see below).

The Nordre Stromfjord study area also suggests migration of fluids from one crustal level to another crustal level, but the fluid compositions at the different levels do not appear to have been strikingly different. Instead, this area suggests the presence of different, time-integrated, fluid volumes in different parts of the shear zone. The fluid compositions were probably variable from one region to another, but in all cases were not dramatically different from the in situ fluids.

5. THE C-O-H-N-S SYSTEM

Although the Nordre Stromfjord region does not require the presence of highly complex fluids, the Storø site demonstrates the need to understand systems containing sulfur, nitrogen and chlorine, in addition to C-O-H. Sufficient data are available to incorporate nitrogen and sulfur (20,21) into model systems. In the metapelitic system described above, addition of nitrogen and sulfur can be accommodated by considering, in addition, pyrrhotite-bearing assemblages and the following equilibria (ref 3):

$$.5N_2 + 3H_2O = NO_3^- + H^+ + 5/2H_2$$

$$.5N_2 + 1.5H_2O + H^+ = NH_4^+ + .75O_2$$

$$4H_2O + HS^- = SO_4^= + H^+ + 4H_2$$

$$.5S_2 + OH^- = HS^- + .5O_2$$

$$H_2S = H_2 + .5S_2$$

$$Fe_3O_4 + 3H_2S = 3FeS + 3H_2O + .5O_2$$

The computations were carried out along several geothermal gradients; only the results of the gradients 150 C/kb and 100 C/kb are discussed here. These gradients were selected because they represent, respectively, gradients which may be likely in areas experiencing elevated temperatures during thermal perturbations, and areas experiencing "normal" thermal activity.

The results for the neutral fluid species demonstrate that neither H2S nor S2 occur in abundances of greater than 1 mole %. In all cases, the ratio PH2S/PS2 >10. This reflects the effect of graphite on the oxygen fugacity, which leads to the high abundance of reduced sulfur.

Addition of N2 dramatically effects the CO2/H2O ratio. Addition of N2 to a fluid will lead to a drop in CO2/H2O, reflecting the fact that the fluid must become more reducing as N2 abundance increases. This is accomplished by the magnetite-biotite-potassium feldspar assemblage which buffers fH2 to constant values; addition of N2 to such a fluid reduces the total oxygen content of the fluid and will thus favor more reduced fluid species over more oxidized fluid species. This is consistent with the Storø results, in which high water activities correlate with high nitrogen abundances.

In addition, as temperature increases, the CO2/H2O value decreases at any given PN2. As a result of this, the PN2 increment necessary to accomplish a given change in CO2/H2O decreases, thus making the system more sensitive to N2 at low temperature, than it is at high temperature (Fig. 4).

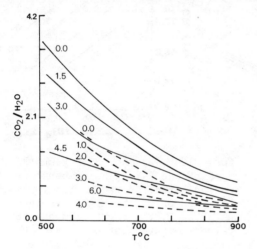

Figure 4. Temperature vs ratio of CO2/H2O partial pressures, plotted as a function of nitrogen partial pressure in kilobars. Each line defines the CO2/H2O ratio at the indicated N2 partial pressure (kbars). Solid lines are for a geothermal gradient of 100C/kb, dashed lines for a geothermal gradient of 150C/kb.

Examination of Figure 4 demonstrates that a change in the geothermal gradient causes dramatic changes in the fluid CO2/H2O ratio. At any given Ptotal and/or PN2, an increase in geothermal gradient will result in a significant decrease in the CO2/H2O ratio. Thus, upon heating of any crustal segment containing the assemblage biotite-magnetite-potassium feldspar- graphite, fluid evolution must occur which will result in an increase in PH2O and a drop in PCO2. Conversely, cooling of crust at constant P must lead to a progressively more CO2-enriched fluid. The petrological implications of this are discussed below.

The ionic equilibria are less well constrained because they are direct functions of pH, which is a variable that cannot yet be quantitatively evaluated in most rock systems. The computational scheme, therefore, determines ionic activities as a function of pH. Nevertheless, it is apparent that HS- and NH4+ will be the dominant ionic species, and that NO3- and SO4= will be relatively low in abundance, with SO4= much more abundant than NO3- (Fig. 5).

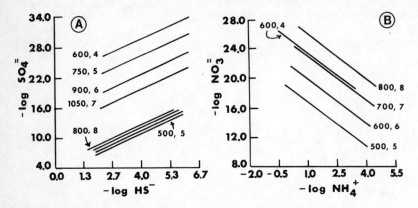

Figure 5. Variation diagrams for sulfur (A) and nitrogen (B) single ion activities, as a function of pH and P-T. The temperature (C) and pressure (kb), respectively, are indicated for each line. The left terminus of each line in A is at a pH of 6, the right terminus is at a pH of 2; in B the left terminus pH is 2, the right terminus pH is 6. The lines define the range of values computed between these pHs, at the indicated P and T.

The most striking aspect of the ionic species is their relative sensitivities to geothermal gradients. At low geothermal gradients the sulfur species exhibit only a limited range of activity values at a given pH. At high gradients, however, SO4= becomes very sensitive to temperature and/or pressure changes, while HS- is virtually insensitive to T or P change. Thus, at any given pH, HS- activity values are restricted to a very narrow range, regardless of the P and T, while SO4= activity can change by orders of magnitude.

NH4+ ions exhibit a behavior very similar to that of HS-, with only a slight sensitivity to P and T values. NO3- activity, on the other hand, spans several orders of magnitude as T or P change along a given geothermal gradient; as T changes at a fixed P, the changes in nitrate ion activity are even more pronounced.

6. DISCUSSION

These results emphasize the importance of complex fluids as equilibrium parameters. Although the CO_2/H_2O ratio has long been recognized as an important variable, the influence of other fluid constituents, although previously noted (22,23,26-28) has not been systematically evaluated in N_2-bearing systems. The results presented here demonstrate that a variety of features of deep crustal rocks may be significantly affected by these other fluid species. We note, in particular, the following:

1) The role of nitrogen in metamorphic fluids may, in some systems, be much more than that of a dilutant. Because of competing ionic equilibria in a C-O-H-N-S fluid, changes in nitrogen partial pressures must result in changes in CO_2/H_2O ratios. As demonstrated previously, CO_2/H_2O ratio is an important variable when considering fluid-rock interaction. N_2 has the potential to play a significant role in establishing the CO_2/H_2O character of deep crustal fluids.

2) The addition of such fluid constituents as sulfur and chlorine may encourage dehydration reactions similar to reaction (1), even in the presence of a fluid that is H_2O-rich. The occurrence of scapolite-bearing granulites (24,29) may be a reflection of this process. This implies that granulites may reflect something more than CO_2 enrichment or H_2O depletion of segments of deep crust; the possibility must be considered that releases of chlorine or sulfur occur in some regions of the crust, resulting in the development of scapolite-bearing granulites.

3) The influence of complex fluids on Nd-Sm, Pb-Pb and Rb-Sr systematics needs careful evaluation. The data presented here demonstrate that fluid activity may significantly modify those isotope systems with the consequence that, at least in some cases, the apparent crustal residence times may be dramatically decreased. Clearly, the implication is that very old crustal segments may not preserve evidence of their antiquity if large volumes of complex fluids pass through them.

4) Thermal evolution of deep crustal material affects more than solid phase relationships. In the system described here many of the fluid species activities are dramatically modified by changes in P/T. Two contrasting behaviors are evident; one is a cooling process which leads to an increase in the CO_2/H_2O ratio, the other is a heating process which decreases the CO_2/H_2O ratio. It has been argued (25) that stabilization of the continents reflects the presence of significant H_2O in the lithosphere. Although this may be true, most reported deep crustal segments either equilibrated with no fluid phase, or equilibrated with a fluid dominated by CO_2. The possibility must be entertained that stabilization of continental crust at 3.8 Ga reflected an increase in CO_2/H_2O which may reflect nothing more than a relaxation of geothermal gradients. Furthermore, in already stabilized crust, thermal events may result in complex reequilibration processes which

involve both recrystallization of preexisting assemblages through continuous and discontinuous reactions, and fluid evolution toward more H2O-rich conditions.

These processes, when considered together, characterize the complexity of metamorphic events at deep crustal levels. This suggests that careful re-examination of high grade terrains be encouraged, with emphasis being placed on development of criteria for establishing a) the time-integrated fluid volume present during any particular metamorphic event, b) the activities of fluid species in addition to those in the C-O-H system, and c) the significance of isotopic signatures in systems where fluid activity has been high.

Acknowledgements: The REE analyses were carried out by INAA at Risø Atomic Research Reactor Center and was controlled by isotope dilution by Bohr Ming Jahn at the University of Rennes. Permission to publish was provided by the Greenland Geological Survey.

REFERENCES

(1) Bridgwater D (1979) U.S. Geol. Surv. Open File Rept. 79-1239: 505-512.
(2) Bridgwater D and Pedersen S (1979) U.S. Geol. Surv. Open File Rept. 79-1239: 512-526
(3) Glassley WE, Bridgwater D and Konnerup-Madsen J (1984) Earth Planet. Sci. Letters, in press
(4) Sørensen K (1970) Rapp Grønlands geol Unders 27: 1-32
(5) Bak J, Sørensen K, Grocott J, Korstgaard JA, Nash D and Watterson J (1975a) Nature 254: 566-569
(6) Bak J, Korstgard JA and Sørensen K (1975b) Tectonophysics 27: 191-209
(7) Glassley WE and Sørensen K (1980) J Petrol 21: 69-105
(8) Glassley WE (1983a) Geochim Cosmochim Acta 47:597-616
(9) Glassley WE (1983b) Contrib. Mineral. Petrol. 84:15-24.
(10) Hickman MH (1979) Rapp Grønlands geol Unders 89: 125-128
(11) Kalsbeek F (1979) Rapp Grønlands geol Unders 89: 129-132
(12) Pedersen S and Bridgwater D (1979) Rapp Grønlands geol Unders 89: pp 133-146
(13) Hickman MH and Glassley WE (1984) Contrib. Mineral. Petrol., in press
(14) Wones DR (1972) Amer. Min. 57: 316-317.
(15) Glassley WE (1982) Nature 295: 229-231
(16) Anderson GM and Burnham CW (1965) Am. J. Sci. 263: 494-511
(17) Walther JV and Helgeson HC (1977) Am. J. Sci. 277: 1315-1351
(18) Eggler DH and Rosenhauer M (1978) Amer. J. Sci. 278: 64-94.
(19) Eggler DH and Kadik AA (1979) Amer. Min. 64: 1036-1048
(20) Rhyzhenko BN and Volkov VP (1971) Geochem. Intl. 8: 468-481.

(21) Robie RA Hemingway BS and Fisher JR (1978) U. S. Geol. Surv. Bull. 1452
(22) Blattner P (1980a) Proceeds. 26th Intl. Geol. Congr. Paris
(23) Blattner P (1980b) Neues Jb. Mineral Monatsh: 283-288
(24) Goldsmith JR (1976) Geol. Soc. Amer. Bull. 87: 161-168.
(25) Campbell IH and Taylor SR (1983) Geop. Res. Letters 10: 1061-1064.
(26) Kreulen R and Schuiling RD (1982) Geochim. Cosmochim. Acta 46: 193-203.
(27) Valley JW, Petersen EU, Essene EJ and Bowman JR (1982) Amer. Mineral. 67: 545-557.
(28) Ohmoto H and Kerrick DM (1977) Amer. J. Sci. 277: 1013-1044.
(29) Hoefs J, Coolen JJM and Touret J (1981) Contrib. Mineral. Petrol. 78: 332-336.
(30) Rosing, M (1983) Thesis, Univ. of Copenhagen, unpubl.
(31) Moller P and Muecke GK (1984) Contrib. Mineral. Petrol. 87: 242-250.

C-O-H FLUID CALCULATIONS AND GRANULITE GENESIS

William M. Lamb and John W. Valley

Department of Geology and Geophysics
The University of Wisconsin, Madison, WI 53706, USA

ABSTRACT

Pervasive flooding of CO_2 into the deep crust has been proposed as the cause of granulite facies metamorphism. One effect of CO_2 infiltration would be the dilution of H_2O and the stabilization of orthopyroxene-bearing assemblages. Calculations in the C-O-H system indicate that infiltration of CO_2 into rocks whose oxygen fugacities are buffered to within the stability of graphite will cause graphite to precipitate. Oxygen fugacities determined from coexisting magnetite and ilmenite from three granulite facies terranes are sufficiently low so that it can be shown that most of these samples have not been flooded by sufficient CO_2 to form granulites as graphite is not reported. Instead, most of these fO_2 values are consistent with either H_2O rich or vapor absent conditions assuming the fraction of non C-O-H components is relatively small. If these values of fO_2 are common in rocks which were metamorphosed under conditions of low fH_2O then vapor absent metamorphism may be common in granulites.

INTRODUCTION

It has been proposed that pervasive flooding of CO_2 is an important, perhaps even essential, mechanism for forming many features which characterize the granulite facies. These features include reduced water activities, orthopyroxene-bearing assemblages, and depletion of large-ion lithophile elements (Touret, 1971; Newton et al., 1980; Condie et al., 1982; Glassley, 1982; Janardhan et al., 1982; Glassley, 1983). This model requires that CO_2 infiltrate a rock and mix with H_2O reducing its activity and stabilizing orthopyroxene bearing assemblages. The calculation of fluid fugacities in the C-O-H system can place important constraints on models which call upon the infiltration of CO_2 to form granulites. These calculations show that some granulite facies rocks did not form via the infiltration of CO_2 (Lamb and Valley, 1984) and that some other process such as partial melting or recrystallization of an already dry rock was probably involved (Fyfe, 1973; Valley and O'Neil, 1984).

FLUID CALCULATIONS

At granulite facies pressures and temperatures calculations in the C-O-H system show that there are six important fluid species (H_2O, CO_2, CH_4, CO, H_2, O_2) which can be related by four independent equations (French, 1966; Ohmoto and Kerrick, 1977; Valley et al., 1982).

$$C + O_2 = CO_2 \tag{1}$$

$$CO + 0.5O_2 = CO_2 \tag{2}$$

$$H_2 + 0.5O_2 = H_2O \tag{3}$$

$$CH_4 + 2O_2 = CO_2 + 2H_2O \tag{4}$$

At any given lithostatic pressure (P) and temperature (T) there are 8 variables in this system; the fugacity of the six fluid species, the activity of carbon (αC), and the fluid pressure (P_F) which may be less than the lithostatic pressure. The fluid pressure may be related to the partial pressures of the fluid species via equation (5).

$$P_F = PH_2O + PCO_2 + PCH_4 + PCO + PH_2 + PO_2 \tag{5}$$

At fixed P and T there are 5 equations and 8 unknowns so it is necessary to fix three variables to specify the system. If a free fluid phase is present then P_F = P and it is then necessary to fix 2 variables to specify the system.

All fluid calculations were performed using the free energy of the fluid phase (Robie et al., 1979). Fugacity coefficients were calculated for H_2O, CO_2, and CH_4 from a hard sphere modified Redlich Kwong equation of state which considers mixing of these three fluid species (Kerrick and Jacobs, 1981; Jacobs and Kerrick, 1981). Fugacity coefficients for H_2 and CO were calculated from equations based on the law of corresponding states (Ryzhenko and Volkov, 1971).

C-O-H calculations are commonly made for the graphite-bearing system in which the activity of carbon is defined as one. If the fugacity of one of the fluid species is known and there is a free fluid phase then it is possible to solve for the fugacity of the five remaining fluid species (French, 1966; Ohmoto and Kerrick, 1977). Figure 1a shows the results of such calculations for a graphite bearing system at 800°C and 8 kbar in which the mole fractions of the major gas species have been plotted versus oxygen fugacity. CO and H_2 have been omitted as they never amount to more than 5% of the total fluid at any oxygen fugacity. This diagram shows that under relatively oxidizing conditions, near the maximum stability of graphite, CO_2 is the predominant species. At intermediate conditions H_2O becomes the dominant species and under relatively reducing conditions CH_4 dominates.

The effect of varying the activity of carbon is shown by comparison of figures 1a and 1b. Figure 1b is constructed in the same manner as figure 1a, but for graphite absent conditions where the activity of carbon is less than one ($\alpha C = 0.1$). The relative positions of the major gas species have not changed but the range of oxygen fugacities at which H_2O is the predominant species has expanded at the expense of CO_2, CH_4, and CO. Thus reducing the activity of carbon reduces the stability of all the carbon bearing fluid species.

Figures 1a and 1b also serve to illustrate the point that a granulite facies fluid phase cannot be composed of CO_2 and CH_4 as the two most abundant species. In fact the relationships illustrated in figures 1a and 1b are true for a wide range of temperatures and pressures. A number of workers, however, have reported the presence of fluid inclusions composed primarily of CO_2 and CH_4 in high grade rocks (Hollister and Burruss, 1976; Wilkins, 1977; Crawford et al., 1979; Kreulen and Schuiling, 1982; Rudnick et al., 1983; Touret and Dietvorst, 1983; Touret, 1984). Some of these inclusions appear to be primary, but this interpretation is at variance with the relationships illustrated in figures 1a and 1b. In light of this conclusion it is difficult to interpret the presence of these fluid inclusions in high grade rocks. Perhaps such inclusions indicate subsolvus, post metamorphic entrapment of fluids.

THE CO_2 INFILTRATION MODEL

Calculation of fluid fugacities in the C-O-H system places constraints on models which call upon infiltration of CO_2 as an important agent in the formation of granulite facies rocks (Lamb and Valley, 1984). Results of these calculations are shown as a function of log fCO_2 vs. log fH_2O in figures 2a-d (calculated at 800°C, 8 kbar, 2a,d; 800°, 6 kbar, 2b; and 600°C, 8 kbar, 2c). The solid curve in each of these figures delineates the stability field of graphite. Along this invariant curve $\alpha C = 1$ and $P_F = P$. A free fluid phase in the C-O-H system is not stable inside of (below and to the left) the solid curve as $P_F < P$. The dotted lines in this region are isopleths of fluid pressure which are calculated by varying P_F independently of P in equation (5) while P is used to calculate the fugacity coefficients of the fluid species. These isopleths (calculated at $\alpha C = 1$) represent the maximum possible fluid pressure for a given point on the diagram as reducing the activity of carbon will reduce the calculated fluid pressure. The dashed curves in figures 2a-d are isopleths of constant oxygen fugacity. Outside the graphite stability curve these contours are calculated by varying αC at $P_F = P$ while within the stability field of graphite they are calculated by varying P_F at $\alpha C = 1$.

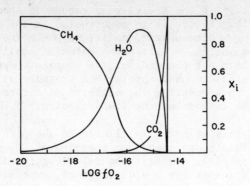

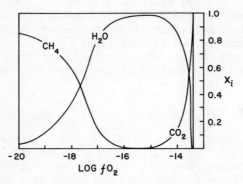

Figures 1a and 1b. Mole fractions of CO_2, H_2O, and CH_4 plotted against oxygen fugacity at 800°C and 8 kbar in both the graphite bearing system ($\alpha C = 1$, figure 1a, top) and a system in which αC is reduced ($\alpha C = 0.1$, figure 1b, bottom). The vertical line in figure 1a represents the maximum stability of graphite ($\alpha C < 1$ to the right of this line) whereas the analogous line in figure 1b is placed at the fO_2 above which αC must be < 0.1. These figures show that high grade metamorphic fluids cannot consist only of CO_2 and CH_4. Instead, there is a range of oxygen fugacities in which H_2O is the predominant species and this range is intermediate between fO_2's which are compatible with either CO_2 or CH_4 rich conditions. Comparison of these figures shows that reducing the activity of carbon expands the stability of H_2O at the expense of the carbon-bearing fluid species. In either case infiltrating CO_2 is more oxidized and therefore out of equilibrium with rocks buffered to intermediate H_2O-CO_2 compositions.

It should be noted that Lamb and Valley (1984) have calculated a phase diagram which is similar to figure 2, but a notable difference is that when performing calculations where $P_F < P$ Lamb and Valley (1984) utilized the same pressure (P_F) to satisfy both equation (5) and to calculate fugacity coefficients. In the C-O-H system both approaches show that no free fluid phase is stable within the stability field of graphite as shown on figures 2a-d. The calculations presented here, however, more accurately reflect the partial pressures of C-O-H fluids if there is some non C-O-H component in the fluid (e.g. N, S, F, Cl). This possibility has been proposed for some terranes, for example at the Dome de l'Agout, France some fluid inclusions contain up to 72 mol% N_2 (Kreulen and Schuiling, 1982). Only in the presence of a fluid with non C-O-H components could a free fluid phase exist within the graphite stability field as shown in figure 2.

The petrologic significance of these phase diagrams can be demonstrated for a rock in which log fO_2 is buffered to -15 at 800°C and 8 kbar. Figure 2d (an enlargement of the center portion of figure 2a) shows that at this fO_2 and at low fCO_2 (point A) graphite is unstable but as fCO_2 increases the system will move along the dashed isopleth from point A to point B intersecting the graphite stability curve, at this point graphite becomes stable and $XCO_2 \approx 0.15$. The system is then invariant in the presence of graphite. Infiltration of additional CO_2 can not then change the fluid composition, instead the CO_2 will break down precipitating graphite and liberating oxygen. The oxygen will be absorbed by the rock buffer and as long as the buffering capacity is not exhausted CO_2 will be converted quantitatively to graphite. The CO_2 to rock ratio (CO_2/R) for infiltration can then be calculated from the amount of graphite that is precipitated.

The amount of CO_2 flooding which is necessary to form a granulite has been estimated by different methods. Newton et al. (1980) estimate that roughly 0.3 rock volumes are needed to produce the conversion from amphibolite to granulite facies. While this estimate is a generalization it is possible to calculate CO_2/R by considering how much CO_2 would be required to dilute the H_2O evolved by the breakdown of a given amount of a hydrous mineral such as a mica or amphibole. Lamb and Valley (1984) calculated that a CO_2/R of 0.1 on a molar oxygen basis is required to form 10% orthopyroxene in a particular rock unit from the Adirondack Mountains of New York in which XH_2O was buffered to a value of 0.12 (at the metamorphic P-T for these rocks a CO_2/R of 0.1 on a molar oxygen basis is approximately equivalent to 0.12 rock volumes).

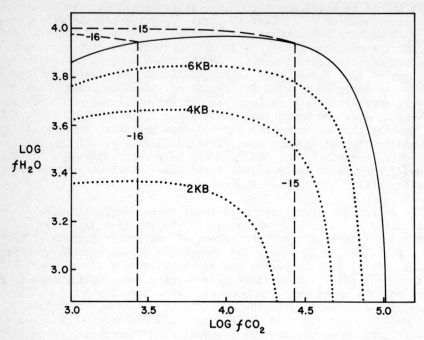

Figure 2a

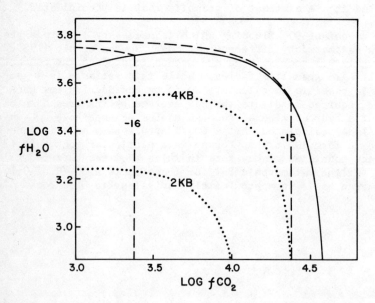

Figure 2b

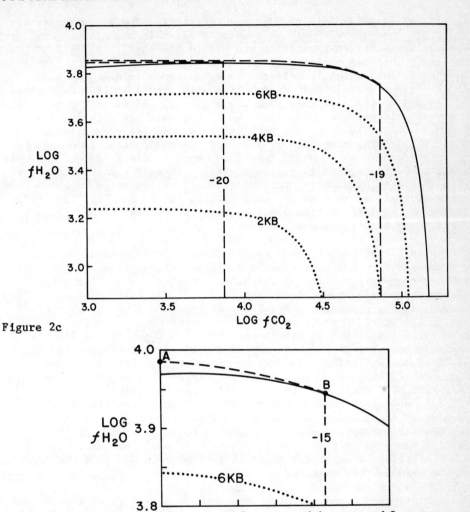

Figure 2c

Figure 2d

Figures 2a, 2b, 2c, and 2d. C-O-H fluid equilibria plotted as log fH_2O vs. log fCO_2 at 800°C, 8 kbar (2a and 2d); 800°C, 6 kbar (2b); and 600°C, 8 kbar (2c). Along the solid curve graphite is stable and $P_F = P$. Outside of this curve (above to the right) $\alpha C < 1$ at $P_F = P$. Inside of this curve $P_F < P$ $\alpha_c = 1$ as shown by the dotted isopleths. Isopleths of constant oxygen fugacity (dashed) are calculated assuming $\alpha C = 1$ within the graphite stability curve and $P_F = P$ outside this curve. Within the graphite stability field values of P_F are maxima at $\alpha_c = 1$ and are reduced if $\alpha_c < 1$. Thus no free fluid phase is stable in the C-O-H system inside the solid curve suggesting vapor-absent metamorphism or non C-O-H fluid species.

If such large amounts of CO_2 have infiltrated the granulite facies as a whole then there should be readily observable amounts of graphite in rocks whose oxygen fugacities are within the stability field of graphite. This may be illustrated for a given volume of rock (in this example 100 cm^3) which is infiltrated by 0.1 rock volumes CO_2 under the conditions described in conjunction with figure 2d (800°C, 8 kbar and log fO_2 = -15). If this rock has a porosity of 1% (a figure which is probably large) then of the 10 cm^3 of CO_2 which infiltrates the rock only 0.15 cm^3 could enter the rock prior to graphite precipitation (as $XCO_2 \approx 0.15$). The remaining 9.85 cm^3 of CO_2 would break down and precipitate graphite which would then form about 1 mode % of the rock. As little as 0.1% graphite is easily identifiable either macroscopically or in thin section allowing reliable identification of rocks which have undergone small amounts of CO_2 infiltration ($CO_2/R > 0.01$). Thus the effects of CO_2 infiltration would be readily observable even in rocks experiencing relatively small amounts of infiltration.

The applicability of these calculations depends on what values of fO_2 are buffered during granulite facies metamorphism. If fO_2 is within the stability field of graphite then large amounts of graphite must be present in order to satisfy models of CO_2 infiltration. However, if fO_2 exceeds the stability of graphite then the absence of graphite provides no information concerning CO_2 infiltration. The presence of graphite does not demonstrate that CO_2 infiltration has occurred, however, as graphite may have a biogenic precursor and be unrelated to CO_2 infiltration. A CO_2/R value calculated from the abundance of graphite would, therefore, represent a maximum value and the presence of small amounts of graphite may be restrictive. As an example the maximum possible CO_2/R in a rock which contains 0.01% graphite is about 0.001 rock volumes. This assumes that all the graphite was precipitated by infiltrating CO_2, but if any of the graphite predated infiltration then the CO_2/R would be less. If all of the graphite is unrelated to precipitation from CO_2 then the CO_2/R must be zero in this example.

Coexisting Fe-Ti-oxides have been used in a variety of igneous and metamorphic rocks to determine both the temperature and oxygen fugacity (Buddington and Lindsley, 1964; Anderson and Lindsley, pers. comm., 1984). Values of temperature and oxygen fugacity from three granulite facies terranes are plotted in figure 3: Adirondacks (Bohlen et al., 1980a), Quebec (Perkins et al., 1982), and Scourie (Rollinson, 1980). Nearly all of these samples fall within the stability field of graphite thus if they were infiltrated by CO_2 they should contain graphite. Furthermore, if a fluid phase was present during metamorphism the oxygen fugacity of many of the samples is only consistent with an H_2O-rich fluid (in the C-O-H system). These samples are not reported to contain graphite and further examination of the Adirondack samples shows no graphite at the 0.01% level.

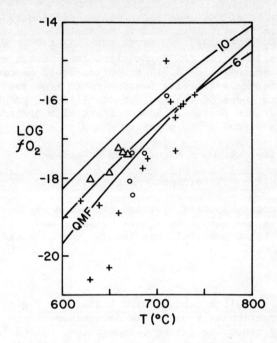

Figure 3. Values of fO_2 and T for granulite facies samples estimated from coexisting magnetite and ilmenite calculated from Anderson and Lindsley (pers. comm., 1984). The maximum stability of graphite is shown at 6 and 10 kbar and QMF, the quartz + magnetite + fayalite buffer is plotted at 1 bar. Open circles represent samples from Scourie (Rollinson, 1980), triangles from Quebec (Perkins et al., 1982), and crosses from the Adirondacks (Bohlen et al., 1980). The samples from Scourie are averaged analyses of Rollinson's (1980) grain boundary pairs. Graphite was not reported in any of these samples indicating that there was little or no CO_2 infiltration in most of these rocks (see text).

While infiltration of CO_2 into rocks with sufficiently low fO_2 will precipitate graphite it will also tend to exhaust the oxygen buffering capacity of the rock. In fact, the buffering capacity of a rock may be a very sensitive indicator of CO_2 infiltration. Small amounts of CO_2 infiltration, precipitating 0.01% graphite, would be capable of oxidizing most magnetite + ilmenite bearing rocks from the values shown in figure 3 to those of the maximum stability of graphite unless other oxygen buffers are present. Such buffers with higher capacity may be found in granulites. Two possible examples are:

$$3 \text{ferrosilite} + 0.5 O_2 = \text{magnetite} + 3 \text{quartz} \tag{6}$$

$$\text{annite} + 0.5 O_2 = \text{K-feldspar} + \text{magnetite} + H_2O \tag{7}$$

While the buffering capacity of these reactions may be larger than magnetite + ilmenite the oxidizing effect of large amounts of CO_2 infiltration could be so powerful that low fO_2 values would not be preserved.

VAPOR ABSENT METAMORPHISM

Rocks from the granulite facies which buffer the fugacity of two fluid species are particularly useful in determining fluid composition and movement during granulite facies metamorphism. Given the fugacity of two fluid species (for example H_2O and O_2) it is possible to calculate the activity of carbon for a given fluid pressure or the fluid pressure for a given activity of carbon. The applicability of these relationships can be seen for an Adirondack charnockite in which the assemblage biotite + K-feldspar + magnetite + ilmenite buffered both fO_2 and fH_2O. The estimated values are log fO_2 = -17.2 ± 0.4 and log fH_2O = 2.65 ± 0.65 at 700°C and 7 kbar (Bohlen et al., 1980b). Assuming $\alpha C = 1$ the calculated fluid pressure is about 1 kbar and reducing αC further reduces the fluid pressure (this calculated fluid pressure differs from that of Lamb and Valley (1984) due to the difference in the calculation procedure discussed earlier). These calculations show that no free fluid phase could have coexisted with this rock during the peak of metamorphism unless the metamorphic fluid deviated substantially from the C-O-H system.

It should be emphasized that for pressures and temperatures found in many granulite terranes oxygen fugacities which are sufficiently below the maximum fO_2 stability of graphite (> 0.25 log unit) preclude the possibility of a CO_2-rich fluid. Such oxygen fugacities are only consistent with four possibilities: 1) an H_2O rich fluid, 2) a CH_4 rich fluid, 3) a fluid phase which deviates substantially from the C-O-H system, or 4) vapor absent metamorphism. While it is possible that there may be some granulite facies rocks which were metamorphosed under conditions of high water activities, this is not the case for most rocks. Thus the first possibility could be ruled out if low oxygen fugacities are found in rocks with demonstrably low fugacities of H_2O. Most granulite facies rocks are probably not sufficiently reducing to be compatible with a methane-rich fluid as indicated by the fO_2 values plotted in figure 3. The data which are currently available, therefore, indicate that the second possibility is unlikely. The third possibility may be an important factor on a local scale as nitrogen has been found in fluid inclusions in granulite facies rocks (Touret, 1984) but there is no evidence suggesting that pervasive flooding of non C-O-H fluids in the amount required to form granulites has occurred. More data are needed to further restrict these four possibilities, but still it is indicated that many granulite facies rocks with low oxygen fugacities were metamorphosed under vapor-absent conditions. This conclusion applies to most samples in figure 3 if XH_2O was low (< 0.5) during granulite facies metamorphism. In fact, some of the samples from the Adirondack Mountains, New York which are plotted on figure 3 contain orthopyroxene + K-feldspar, an assemblage which, according to Janardhan et al, (1982), places an upper limit of P_{H_2O} = 700 bars. Other samples, also from the Adirondack Mountains, combine relatively reducing fO_2's ($<$ QMF) with low water activities (Powers and Bohlen, 1984). These samples are a further indication of vapor-absent granulite facies metamorphism.

DISCUSSION

The results of this study show that some common granulite facies rocks have not been infiltrated by sufficient CO_2 to explain their genesis. Studies of certain quarries in Southern India, however, suggest that CO_2 infiltration was important in forming those granulite facies rocks (Newton et al., 1980, Janardhan et al., 1982, Condie et al., 1982). The value of oxygen fugacity buffered in the rocks from southern India is so high that infiltration of CO_2 would not precipitate graphite (R. Newton, pers. comm. 1984) and so the results of these studies are not necessarily contradicted by this study.

Extreme gradients in fCO_2 and lack of pervasive fluid movement is indicated for some localities in the Adirondacks by calculations of mineral equilibria and by sharp gradients in oxygen isotope ratios (Valley and O'Neil, 1984). This suggests that low fH_2O may form by a number of processes including extraction of magmas, channelized CO_2 infiltration, or metamorphism of already dry rocks (i.e. orthogneisses) and these processes may operate in close proximity. All these processes are yet to be demonstrated in the same terrane, however, and it may be possible that real differences in fluid conditions correlate with age or tectonic style.

ACKNOWLEDGMENTS

This study was supported by grants from the National Science Foundation (EAR81-21214 and 83-11772), the Gas Research Institute (5083-260-0852), the Geological Society of America Penrose Fund (3256-83 and 3017-82) and the Lewis G. Weeks Fund. We thank P. E. Brown and J. Touret for helpful conversations and reviews.

REFERENCES

Bohlen SR, Essene EJ, Hoffman KS (1980a) Update on feldspar and oxide thermometry in the Adirondack Mountains, New York. Geol Soc Am Bull 91:110-113.

Bohlen SR, Peacor DR, Essene EJ (1980b) Crystal chemistry of a metamorphic biotite and its significance in water barometry. Am Min 65:55-62.

Buddington AF, Lindsley DH (1964) Iron-titanium oxide minerals and synthetic equivalents. Jour Pet 5:310-357.

Condie KC, Allen P, Narayana BL (1982) Geochemistry of the Archean low-to high-grade transition zone, Southern India. Contrib Mineral Petrol 81:157-167.

Crawford ML, Kraus DW, Hollister LS (1979) Petrologic and fluid inclusion study of calc-silicate rocks, Prince Rupert, British Columbia. Am Jour Sci 9:1135-1159.

French BM (1966) Some geological implications of equilibrium between graphite and a C-O-H gas phase at high temperatures and pressures. Rev Geophys 4:223-253.

Fyfe WS (1973) The granulite facies, partial melting and the Archean crust. Phil Trans R Soc Lond A 273:457-461.

Glassley W (1982) Fluid evolution and graphite genesis in the deep continental crust. Nature 295:229-231.

Glassley W (1983) Deep crustal carbonates as CO_2 fluid sources: Evidence from metasomatic reaction zones. Contrib Mineral Petrol 84:15-24.

Hollister LS, Burruss RC (1976) Phase equilibria in fluid inclusions from the Khtada Lake metamorphic complex. Geochim Cosmochim Acta 40:163-175.

Jacobs GK, Kerrick DM (1981) Methane: an equation of state with application to the ternary system $H_2O-CO_2-CH_4$. Geochim Cosmochim Acta 45:607-614.

Janardhan AS, Newton RC, Hansen EC (1982) The transformation of amphibolite facies gneiss to charnockite in southern Karnataka and northern Tamil Nadu, India. Contrib Mineral Petrol 79:130-149.

Kerrick DM, Jacobs GK (1981) A modified Redlich-Kwong equation for H_2O, CO_2, and H_2O-CO_2 mixtures at elevated pressures and temperatures. Am Jour Sci 281:735-767.

Kreulen R, Schuiling RD (1982) N_2-CH_4-CO_2 fluids during formation of the Dome de l'Agout, France. Geochim Cosmochim Acta 46:193-203.

Lamb WM, Valley JW (1984) Metamorphism of reduced granulites in low-CO_2 vapour-free environments. Nature 312:56-58.

Newton RC, Smith JV, Windley BF (1980) Carbonic metamorphism, granulites and crustal growth. Nature 288:45-50.

Ohmoto H, Kerrick D (1977) Devolatilization equilibria in graphitic systems. Am Jour Sci 277:1013-1044.

Perkins D, Essene EJ, Marcotty LA (1982) Thermometry and barometry of some amphibolite-granulite facies rocks from the Otter Lake area, southern Quebec. Can Jour Earth Sci 19:1759-1774.

Powers RE, Bohlen SR (1984) A contact aureole in the granulite facies, Lake Bonaparte area, N.W. Adirondacks. GSA abstracts with programs 16-6:627.

Robie RA, Hemingway BS, Fisher JR (1979) Thermodynamic properties of minerals and related substances at 298.15 K and 1 bar (10^5 pascals) pressure and at higher temperatures. US Geol Sur Bull 1452.

Rollinson HR (1980) Iron-titanium oxides as an indicator of the role of the fluid phase during cooling of granites metamorphosed to granulite grade. Min Mag 43:623-631.

Rudnick RL, Ashwal LD, Henry DJ (1983) Metamorphic fluids and uplift-erosion history of a portion of the Kapuskasing structural zone, Ontario, as deduced from fluid inclusions. In Workshop on a cross section of Archean crust, LD Ashwal and KD Card eds. Lunar and Planetary Institute Technical Report 83-03:76-78.

Ryzhenko BN, Volkov VP (1971) Fugacity coefficients of some gases in a broad range of temperatures and pressures. Geochim Int 8:468-481.

Touret J (1971) Le facies granulite en norvege meridionale II. Les inclusions fluides. Lithos 4:423-436.

Touret J (1985) Fluid regime in southern Norway. This volume.

Touret J, Dietvorst P (1983) Fluid inclusions in high-grade anatectic metamorphites. Jour Geol Soc. London 140:635-649.

Valley JW, O'Neil JR (1984) Fluid heterogeneity during granulite facies metamorphism in the Adirondacks: stable isotope evidence. Contrib Mineral Petrol 85:158-173.

Valley JW, Petersen EU, Essene EJ, Bowman JR (1982) Fluorphlogopite and fluortremolite in Adirondack marbles and calculated C-O-H-F fluid compositions. Am Min 67:545-557.

Wilkins RWT (1977) Fluid inclusion assemblage of the stratiform Broken Hill ore deposit, New South Wales, Australia. Science 198:185-187.

TECTONIC FRAMEWORK OF THE GRENVILLE PROVINCE IN ONTARIO AND WESTERN
QUEBEC, CANADA

A. Davidson
Geological Survey of Canada
588 Booth Street
Ottawa K1A 0E4
Canada

ABSTRACT. The southwestern Grenville Province embraces three subprovinces of different character. Northwesternmost, the Grenville Front Tectonic Zone (GFTZ) exhibits ductile deformation in gneisses with margin-parallel structures inclined toward the orogen. At its northwest limit (Grenville Front), steeply southeast-dipping mylonitized rocks abut Archean supracrustal and plutonic rocks overlain by Lower Proterozoic (Huronian) sediments. Southeastward, gneisses in the Central Gneiss Belt (CGB) include younger deformed and metamorphosed plutonic rocks (1.5 – 1.35 Ga). Curved shear belts ($\sim$1.1 Ga) outline distinct lithotectonic domains. Farther southeast the Central Metasedimentary Belt (CMB) is underlain by Grenville Supergroup marble, metavolcanic and other metasedimentary rocks, intruded by plutons either pre- ($\sim$1.2 Ga) or syntectonic (1.15 – 1.0 Ga) with respect to Grenvillian Orogeny. A major, shallowly southeast-dipping tectonic zone separates the CMB and CGB. Shortening across the Province is implied by abundant evidence for northwestward overriding found in high strain zones. Granulite facies distribution points to a formerly much thicker crust. Grenvillian supracrustal and plutonic rocks are lacking in the CGB and GFTZ, in which Grenvillian Orogeny is registered solely as a major tectonothermal event. Whether the northwest CMB margin is a décollement or a telescoped unconformity remains undetermined.

1. INTRODUCTION

Terranes of gneissic and migmatitic rocks can be difficult to map and to interpret, and unrewarding for mineral exploration. For these reasons, as well as inaccessibility and lack of outcrop, many parts of the Grenville Province of the North American craton remain little known and poorly understood. The meagre results of sparse reconnaissance surveys coupled with local detailed knowledge and regional geophysical studies provide the data base from which evolutionary models have been proposed (e.g. Wynne-Edwards, 1972, 1976; Baer, 1976, 1981; for others, see Baer, 1974). In the southwest part of the exposed Province, two relatively well known regions, the 'marble belt' of southeastern

Ontario and the Grenville Front in the neighbourhood of Sudbury, are
separated by a 200 km wide tract of gneissic rocks, a large part of
which had received little more than cursory attention until as recently
as 1980, despite relative ease of access and plentiful outcrop. Recent
reconnaissance mapping in this gneiss terrane has resulted in recognition of features, of a structural nature in particular, that allow a
coherent evolutionary scheme to be presented for a broad cross-section
of the Province, although much detailed documentation is still to come.
Results of this work (Davidson and Morgan, 1981; Davidson et al., 1982;
Culshaw et al., 1983; Hanmer and Ciesielski, 1984) and its interpretation (Lindia et al., 1983; Hanmer, 1984; Davidson, 1984 and in press
(a); van Breemen et al., in press) have been or are to be published
elsewhere. This paper presents a summary overview of these recent findings. The region under consideration is shown in Figure 1.

2. REVIEW OF TERMINOLOGY

The name 'Grenville' has been used in many ways, at times loosely or
inconsistently, as pointed out by Engel (1956), leading to potential
confusion on the part of the reader. It has been applied to four basic
aspects of geology, namely rock, area, process and time, and it is
appropriate at this time to review the development of current usage and
how this relates to changing concepts of Grenvillian geology.

Logan (1847, 1863) first applied the name to a suite of highly
deformed and metamorphosed sedimentary gneisses and marble that he discovered north of the Ottawa River near the town of Grenville, situated
between Ottawa and Montreal. The term 'Grenville series' came to be
used for all similar rocks that were found to occur within a much
larger area of western Quebec and southeastern Ontario. Other 'series'
(Hastings, Bristol) were recognized within parts of the same area, but
dispute arose over whether or not the different appearances of these
and the 'Grenville series' were due merely to differences in degree of
metamorphic and structural modifications of the same rocks. The term
'series', with its time connotation, was subsequently dropped after
admonishment by Engel (1956) in favour of 'group' and then 'supergroup'
(Wilson, 1965; Wynne-Edwards, 1969, 1972). Based on unconformable
relationships, a distinctly younger succession, the Flinton Group, was
separated from the Grenville Supergroup in southeastern Ontario (Moore
and Thompson, 1972, 1980), and includes some of the rocks formerly
classified as 'Hastings series'. Although the Grenville Supergroup has
been divided locally into groups and formations (Hewitt, 1959; Lumbers,
1967), unresolved structural complexities limit correlations to within
the type area (Hastings Lowlands); a unified stratigraphic breakdown
cannot be made throughout the area in which these supracrustal rocks
occur. As such, therefore, the Grenville Supergroup is a catch-all
term used to describe a strongly deformed and generally highly metamorphosed succession of volcanic and sedimentary rocks whose base is not
defined, whose original thickness is not known, and within which stratigraphic principles can be applied only locally.

TECTONIC FRAMEWORK OF THE GRENVILLE PROVINCE

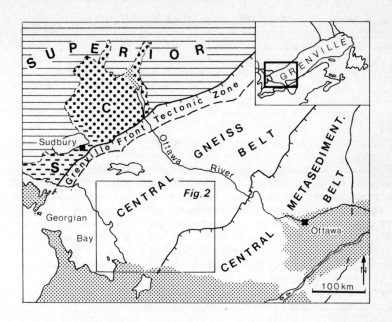

Figure 1. Subprovinces of the southwestern Grenville Province, after Wynne-Edwards (1972). S - Southern Province, C - Huronian cover on the Superior Province (Cobalt Plate). Paleozoic cover is stippled.

The current use of the term 'Grenville Province', or more correctly 'Grenville Structural Province' (Stockwell, 1982), developed from the original concept of defining an area in which certain rocks occur. Thus the area underlain in whole or in part by the 'Grenville series', characterized by marble, was referred to as the 'Grenville subprovince' by M.E. Wilson (1918, 1925), distinguishing it from the 'Timiskaming subprovince' to the northwest, characterized by different types of metasediments, greenstones and granites, all of which are overlain by the 'Huronian series'. These two subprovinces were included in the St. Lawrence province, primarily a geographic division of the Canadian Precambrian Shield, and were stated to be separated by "... an extensive belt composed chiefly of banded gneisses." (Wilson, 1927, p. 2). This gneiss belt was later included within the 'Grenville subprovince' by Wilson (1939, Fig. 2), even though he maintained its separate status (ibid., p. 238). Division of the Shield on structural grounds into provinces roughly as they are known today followed recognition of distinct tectonic breaks, on one side of which younger structures truncate those of the other side (Gill, 1948; J.T. Wilson, 1949). The Grenville Province thus became limited to that area southeast of the northeast-trending tectonic break, first referred to as the Grenville Front by Derry (1950), that truncates the east-trending earlier folds of the Southern and Superior provinces ('Timiskaming subprovince' of M.E. Wilson).

Recognition of superposition of regional structural elements gave rise to the concept of a succession of mountain-building events, or orogenies, in the Canadian Shield. Provinces thus became the regions affected by particular orogenies, that is, they were recognized as bearing the mark of deposition, deformation, metamorphism and plutonism related to a major episode of mountain-building. Hence the term Grenvillian (or Grenville) Orogeny was coined, having the Grenville Province as its type area; this region has also been referred to as the Grenville orogen or orogenic belt (McLaughlin, 1954).

Early workers did not have the benefit of radio-isotope geochronology to supply evidence for the absolute ages of the rocks they studied, and could only assign relative ages on the basis of rock relationships such as unconformities and crosscutting igneous intrusions. Almost invariably, little-deformed rocks were regarded as Late Precambrian in age, and highly deformed and metamorphosed ones as old (Archean). Correlation of rocks of various parts of the Shield were made on this basis; thus the well preserved 'Huronian series' was considered to be young, and although it was suggested that the 'Grenville series' might be its equivalent (e.g. Quirke, 1926), the latter was generally considered to be Archean (Wilson, 1939). The advent of increasingly reliable radiometric dating from the mid-1950s on has led to a complete reassessment of age relationships between various rock units, and in particular has allowed time constraints to be placed on orogenies, a major outcome of which has been the use of major, widespread orogenic events, such as the Grenvillian, as a basis for dividing Precambrian time (Stockwell, 1964, 1982). Ideally, isotopic geochronology has the potential for dating both the ages of formation of certain rocks and the ages of events that have subsequently affected them. Thus within the Grenville Province, Rb-Sr and U-Pb methods applied to volcanic rocks in the Grenville Supergroup yield ages between 1.3 and 1.25 Ga (Silver and Lumbers, 1966; Bell and Blenkinsop, 1980; J.R. Bartlett and D.W. Davis, personal communication, 1984), roughly 1 Ga younger than ages that constrain the Huronian Supergroup (Krogh et al., 1971; Van Schmus, 1965). Pegmatites, metamorphic minerals and late tectonic plutonic rocks within the Grenville Province generally give 'Grenvillian' ages (1.15 to 1.0 Ga). K-Ar ages are also Grenvillian, range to as young as 850 Ma, and are now considered to be cooling ages related to epi-orogenic uplift (Stockwell, 1982). Pre-Grenvillian ages have been determined for gneissic rocks in most parts of the Province, and reflect earlier plutonic and metamorphic events not related to Grenvillian Orogeny. In the Hastings Lowlands of southeastern Ontario, recognition that the Grenville Supergroup was deformed, metamorphosed and intruded by granitoid plutons (~1.2 Ga) before deposition of the Flinton Group, and that all these rocks were subsequently deformed and metamorphosed (1.1 to 1.0 Ga) caused Moore and Thompson (1980) to suggest the term 'Grenvillian orogenic cycle' to encompass all these events; they suggested the terms 'Elzevirian' and 'Ottawan' for the orogenies that respectively pre- and postdate deposition of the Flinton Group. Stockwell (1982), however, uses the term Grenvillian Orogeny for the later one. It seems likely that the Elzevirian and Ottawan orogenies were related in a continuum of crust-shaping events in their type area (Davidson, in press (a)), but at present it is not clear what, if any, were the effects of the Elzevirian orogeny in the vast remainder of the Province.

Various assumptions have been made concerning relationships between rocks and events within the Grenville Province. For example, there has been a tendency for any occurrence of crystalline limestone in the Province to be referred to as 'Grenville marble', causing the reader to make an unwarranted mental correlation with the Grenville Supergroup in which marble is a major component and has also been referred to loosely as 'Grenville marble'. Plutonic rocks in many parts of the Province were tacitly assumed to be the products of Grenvillian Orogeny until isotopic age determinations proved that most of them are older. Granitic rocks along the Grenville Front south of Sudbury are a case in point; formerly considered Grenvillian by association, they are now known to be $\sim$1.7 Ga old and likely related to events beyond the Grenville Province (Davidson, in press (b)). Plutonic rocks of Grenvillian age, it seems, are rather rare, except within the limited area underlain by the Grenville Supergroup. Misconceptions about the nature of the Grenville Province may have been generated by statements such as "The Grenville subprovince ... is named after the Grenville group of sedimentary and volcanic rocks which occur so extensively within its area." (Wilson, 1965, p. 347); the Grenville Supergroup, in fact, is limited in occurrence to an area that is less than 10 per cent of the exposed Province, and marble of any kind is a rare and minor component everywhere else. Again, in the Grenville Province, a "... feature worthy of note is the widespread occurrence of anorthosite and anorthositic gabbro." (Gill, 1948, p. 102), whereas the important point is that anorthositic massifs within the Grenville Province are more or less deformed, at least at their margins, but those north of the Grenville Front in Labrador are not; anorthosites in both regions are pre-Grenvillian (Duchesne, 1984). Gill (1948, p. 103) also states: "A most significant fact is that within this belt (Grenville Province) the dominant trend is north-eastward, not just near the line separating it from the Superior Province, but right down to the Shield border." The truth is that continuous northeast-trending structure is restricted to a relatively narrow zone adjacent to the Grenville Front; in the body of the Province there are large regions with north or northwest trends, as pointed out by J.T. Wilson (1949). Most of these discrepancies were dealt with in the last major published synthesis of the geology of the Grenville Province (Wynne-Edwards, 1972).

In summary, the names Grenville and Grenvillian are properly used as follows: Grenville Supergroup - a contiguous package of highly deformed metasupracrustal rocks in southwest Quebec and southeast Ontario, locally divided into groups and formations, whose base and original thickness are not defined, but which predate plutonic rocks of Elzevirian age (1.25 - 1.15 Ga); Grenvillian Orogeny - a polyphase tectonic event, manifest in highly deformed and metamorphosed rocks of several ages, that affected the southeastern Canadian Precambrian Shield, culminating between 1.15 and 1.0 Ga ago; Grenville Front - the northwest limit of severe transposition of older rocks during the Grenvillian Orogeny, usually marked by a zone along which mylonitized rocks locate thrusts or high angle reverse faults, and that truncates older regional structures of adjacent Shield provinces; Grenville Province - that part of the Canadian Shield southeast of the Grenville Front that has been affected by the Grenvillian Orogeny.

Wynne-Edwards (1972) divided the Grenville Province into a number of subprovinces on the basis of differences in structural and metamorphic style. This paper is concerned with the tectonic effects of Grenvillian Orogeny in the southwest part of the Province, namely within the Central Gneiss Belt, Central Metasedimentary Belt, and Grenville Front Tectonic Zone (Fig. 1). In brief, the Central Metasedimentary Belt corresponds roughly to the original restricted 'Grenville subprovince' of Wilson (1925), and the Central Gneiss Belt and Grenville Front Tectonic Zone to the extensive belt of banded gneisses that lies between it and his 'Timiskaming subprovince' (Superior and Southern provinces). Amphibolite facies gneisses in the Grenville Front Tectonic Zone exhibit a high degree of ductile deformation, with foliation, layering and map units aligned essentially parallel to the Grenville Front, dipping southeast toward the Grenville orogen, and usually carrying a well developed down-dip lineation. At the Front itself these gneisses are juxtaposed against Archean greenstone, metagreywacke and plutonic rocks, or against Lower Proterozoic (Huronian) sedimentary formations that form a relatively undisturbed cover on the Archean rocks (Superior Province) or are folded and metamorphosed in the Penokean fold belt (Southern Province). Some gneisses in the Tectonic Zone are reworked equivalents of the Archean rocks to the northwest. Unequivocal Huronian equivalents are restricted to a very narrow part of the Tectonic Zone adjacent to the Grenville Front south of Sudbury.

The Central Gneiss Belt is composed of a variety of gneissic rocks, many of which are migmatitic, that includes a suite of Middle Proterozoic metaplutonic units, dated in the range 1.5 to 1.35 Ga (van Breemen et al., in press). Country rocks include supracrustal gneisses and older metaplutonic complexes that cannot be correlated with rocks northwest of the Grenville Front. Orientation of major structures is variable, but in many places is at a high angle to the Tectonic Zone. It is in the Central Gneiss Belt that recent identification of large-scale ductile shear zones (Davidson, 1984) has led to reinterpretation of the tectonic history of this part of the Grenville Province.

3. CENTRAL GNEISS BELT

In the well exposed region east of Georgian Bay, aerial photographs, satellite images and shaded topographic maps show aligned low ridges and narrow valleys that reflect bedrock structures in gneissic rocks. Prominent in several places are continuous belts, up to a few kilometres in outcrop width, in which particularly well developed layering defines what look like large-scale folds. It can be seen, however, that these belts circumscribe areas within which structures of smaller amplitude display a more 'confused' pattern, and that there appear to be truncations of structure at the edges of some belts.

Reconnaissance fieldwork began in 1980 in this previously poorly known area (Davidson and Morgan, 1981; Davidson et al., 1982). It confirmed the continuity of the belts of gneissic rocks, and showed that in most places the layering is moderately to shallowly inclined. It also showed not only that different types of gneiss in these belts maintain a 'stratigraphy' for tens of kilometres along strike, but that in many cases the gneisses themselves, although usually well layered, are not obviously of supracrustal origin, and indeed that some of them are clearly highly attenuated metaplutonic rocks. In addition, some gneisses in these belts are mylonitic in character, and are associated with gneisses carrying a variety of other features suggestive of extreme ductile deformation. These features include severe disruption of competent rocks, detached isoclinal folds, sheath folds, networks of mesoscopic shears, well developed mineral stretching lineation, rotated, flattened, boudined and disaggregated pegmatites, C-and-S fabrics and shear band foliation (Davidson, 1984). Metamorphic mineral assemblages in all belts indicate middle to upper amphibolite facies conditions during deformation, in some belts accompanied by development of quartzofeldspathic leucosomes that now commonly have the form of lenses parallel to foliation. In contrast, areas between the gneiss belts, referred to as domains or subdomains, although containing highly deformed rocks, in general maintain a clear distinction between gneisses of supracrustal and plutonic origin. The domains and subdomains are distinct one from another, having different lithologic assemblages, structural orientations, metamorphic facies and geophysical signatures that set them apart. Their distinctive features are truncated on a regional scale by the intervening gneiss belts. A certain polarity is evident: gneiss belts that strike northeast truncate structures on their northwest sides; those that strike northwest usually, though not invariably, truncate structures in the underlying domains. Metaplutonic rocks within domains become progressively more deformed at the margins of the gneiss belts, and do not intrude them. Moreover, diagnostic map units cannot be traced across gneiss belts from one domain to another. The distribution of domains and gneiss belts is outlined in Figure 2.

These findings prompted the suggestion that the belts of layered gneiss cannot be interpreted wholly and simply as belts of supracrustal rocks now thrown into large folds, but are best ascribed to a tectonic origin involving displacement of deep crustal blocks and slices along broad zones of ductile deformation. Where present, kinematic indicators noted during reconnaissance within these gneissic tectonites imply a consistent component of displacement to the northwest, and thus have thrust sense where the belts are inclined to the southeast. Detailed studies have confirmed this scenario (Hanmer, 1984; Nadeau, 1984), which also explains anomalies in the distribution of metamorphic facies, for example at the west side of Parry Sound domain (Fig. 2), where structurally overlying granulite facies rocks are separated from middle amphibolite facies by a gneiss belt less than 2000 m thick.

Within the zones of gneissic tectonites some rocks can reasonably be termed mylonite or protomylonite, particularly where textural changes and marked grain refinement can be traced across strain gradients. This can be best documented in formerly massive plutonic rocks, especially where they carry strain markers such as flattened or elongated megacrysts or xenoliths. Where strain is inhomogeneous, strain markers may be observed to have become progressively distorted to the point that they are no longer discernible. In some places, well foliated and fine grained but otherwise nondescript gneissic rocks contain low strain lenses from which an igneous plutonic protolith for the whole assemblage can be reasonably deduced. In many places, however, this kind of evidence is lacking, and it becomes a guessing game as to what many of the rocks once were, and also how strained they are. The problem is compounded where recrystallization has resulted in a general increase in grain size, as well as growth of porphyroblasts and the formation of leucosomes, and has masked or obliterated evidence of previous accumulated strain. In such all too common instances some qualitative impression of the total strain may be given by mesoscopic and even map-scale features, such as the separation between boudins or blocks of competent rocks (e.g. mafic dykes), or the extreme attenuation of diagnostic rock types (e.g. anorthosite) to layers only a few metres thick that can be traced within the layering for tens of kilometres; these may themselves have been subsequently separated into slivers and small blocks strung out in the plane of foliation.

In summary, the gneiss belts between lithotectonic domains in the Central Gneiss Belt, taking together their form, attitude, regional distribution and the tectonic features of the rocks themselves, are interpreted as zones of high strain concentration formed during ductile deformation deep in the crust along which large crustal blocks and slices have been displaced with respect to one another; as such they are properly considered as ductile shear zones. Kinematic indicators at various scales imply dominance of northwest-directed displacement of structurally overlying blocks. That they have formed late in the regional tectonic history is intimated by the fact that all types of plutonic rocks are deformed within them, with the exception only of the youngest dykes of pegmatite. A recent attempt to date the deformation at Parry Sound has succeeded in showing that syntectonic pegmatites within the shear zones have Grenvillian ages (U-Pb on zircons 1160 and 1120 Ma), and that plutonic rocks within and marginal to the bordering domains are 200 to 300 Ma older (van Breemen et al., in press).

4. CENTRAL METASEDIMENTARY BELT

Stratigraphic analysis of the Grenville Supergroup in the Central Metasedimentary Belt has not proceeded far, principally because the internal structure of this subprovince is not understood. In most places the rocks are highly deformed, and metamorphism is predominantly amphibolite or granulite facies; both factors have obscured the original nature and disposition of most of the rocks. Only in the low-grade region in southeast Ontario, referred to as the Hastings 'metamorphic low' by Carmichael et al. (1978), are primary sedimentary and volcanic structures preserved, allowing determination of stratigraphic facing. It is thus only in this region that meaningful attempts have been made to divide the strata into groups and formations and, as already stated, these divisions cannot yet be traced with confidence into the higher grade neighbouring areas.

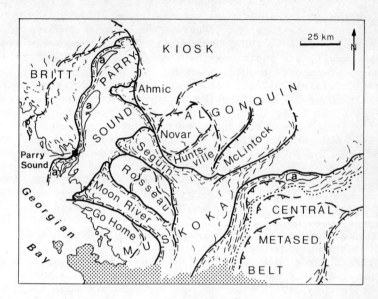

Figure 2. Distribution of lithotectonic domains and major ductile shear zones in part of the southwestern Grenville Province. Ticks on domain boundary shear zones (heavy lines) indicate dip direction. Also shown are high strain zones within domains (broken lines) and thrusts in the Metasedimentary Belt, in which the dotted line indicates the northwest limit of thick marble units; a - anorthosite. Paleozoic is stippled.

Study of the distribution of diagnostic sedimentary and volcanic facies, based on regional mapping, shows that certain rock types are more abundant in some parts than others, and may in fact be restricted to specific areas (Davidson, in press (a)). Whether this uneven distribution reflects different levels of the same succession or whether it is the result of regional facies changes has yet to be determined. Perusal of structural patterns on map compilations of the Central Metasedimentary Belt reveals that 'straight belts' dissect this region and in places lie along the boundaries of internal 'terranes' defined on the basis of differences in lithological assemblages (Brock and Moore, 1983; Bartlett et al., 1984). Disrupted and mylonitic rocks have been described from these belts (e.g. Hewitt, 1961). It may be, therefore, that major displacements of Grenville Supergroup stratigraphy are present within the Metasedimentary Belt.

5. BOUNDARY BETWEEN THE CENTRAL METASEDIMENTARY AND GNEISS BELTS

Along the northwestern part of the Central Metasedimentary Belt, interleaved marble and other gneissic rocks dip gently southeast, away from the Central Gneiss Belt. It has been suggested that the boundary between these two subprovinces is the site of an unconformity, albeit deformed, representing an erosion surface on an older plutonic terrane that forms the basement of the Grenville Supergroup (Appleyard, 1974; Schwerdtner and Lumbers, 1980; Lumbers, 1982). The rocks in the boundary zone, however, are replete with evidence that the two terranes are separated by an important ductile shear zone (Culshaw et al., 1983; Hanmer and Ciesielski, 1984). This zone of highly deformed rocks extends for at least 350 km along the northwest margin of the Metasedimentary Belt. Gneisses low in the 'tectonic succession' include flaggy 'straight gneiss', porphyroclastic and disrupted gneiss with associated anorthositic and gabbroic blocks and slivers. Above these, coarse marble units are flow-foliated and contain an assortment of disarranged silicate gneiss blocks. Hanmer and Ciesielski (1982, p. 121) interpreted this marble as "... a tectonic mélange which has played an important mechanical role in the tectonics of the boundary zone;". The silicate tectonites are directly analogous to the gneissic tectonites of the shear zones between domains in the Central Gneiss Belt, and contain the same kinds of kinematic indicators implying the same northwestward overthrust sense of tectonic transport. Nappe-like structures of orthogneiss thrust over marble mélange have been identified along the southeast side of part of the boundary zone (Easton and Bartlett, 1984). It is thus likely that the Grenville Supergroup, and also the Elzevirian plutonic rocks that intrude it, are at least parautochthonous with respect to the Central Gneiss Belt, and possible that they are wholly allochthonous.

6. GRENVILLE FRONT TECTONIC ZONE

As the Grenville Front Tectonic Zone is approached from the Central Gneiss Belt, structures within the latter curve into parallelism with the northeast-striking, southeast-dipping foliation that characterizes the Tectonic Zone. The boundary between these two subprovinces, however, is not sharply defined. Within the Tectonic Zone the rocks assume many of the same diagnostic features that characterize the gneissic tectonites of the ductile shear zones farther southeast, already described. Close to the Grenville Front, many rocks are conspicuously mylonitized; this is especially evident in granitoid rocks which show transitions through protomylonite to mylonite. At the Front itself, particularly where it can be defined as a single, linear tectonic break between gneissic rocks and recognizable Archean or Huronian formations to the northwest, mylonite and ultramylonite occur in narrow zones, and may be accompanied locally by pseudotachylite veins. Mylonitic foliation along the Front is generally steep, becomes shallower to the southeast, and is everywhere accompanied by a strong down-dip stretching lineation. Recent fieldwork in mylonitized granitoid rocks in the Tectonic Zone south of Sudbury (Pryer, 1985) has revealed rotated augen, C-and-S fabrics and shear band foliation indicating a consistent sense of steep northwestward thrusting. This is entirely in keeping with earlier interpretations (Lumbers, 1971; Wynne-Edwards, 1972) that Grenville Province rocks are in thrust contact with those of the older structural provinces to the northwest.

There has been some debate concerning how long the Grenville Front has been tectonically active. Stockwell (1982) argued that the Front has been in existence since Late Archean time. Lumbers (1978; Sims et al., 1981) required that it was the site of a deep trough at the south margin of an Archean craton during Early Proterozoic (Huronian) sedimentation. Krogh and Davis (1971) interpreted ages determined on deformed granitoid rocks near the Front to imply deformation within (and specific to?) the Grenville Province as long ago as 1700 Ma. The occurence of Grenvillian-age deformation along the Front is proven by isotopic ages, and inferred from the fact that late Middle Proterozoic rocks, for example the diabase dykes of the northwest-trending Sudbury swarm ($\sim$1250 Ma; Palmer et al., 1977), are deformed at the Front.

Stockwell appears to have based his argument largely on a consideration of up-dated isotopic ages related to increase in metamorphism toward the Front from the northwest; in fact, he located parts of the Front at a biotite isograd rather than defining it strictly on structural grounds. Wynne-Edwards (1972) pointed out that isograds parallel to the Front may well be up-turned 'fossil' isograds, related to earlier metamorphism that had nothing to do with orogeny within the Grenville Province. Lumbers interpreted biotite gneisses in the Tectonic Zone as coarse clastic deep water facies equivalents of the lower Huronian formations, mainly the Mississagi Formation of crossbedded feldspathic and quartz arenites. However, Davidson (1979 and in press (a)) considered it more likely that this gneiss unit (Red Cedar Lake formation) was derived from metawacke of the Archean Pontiac Group, a prominent unit in the Superior Province north of the Front in Quebec. In addition, paleocurrent analysis of the Mississagi Formation near the Front infers sediment transport away from the supposed marginal trough (Sims et al., 1981). Recent work along the Front southwest of Sudbury (Davidson, in press (b)) has led to identification of the Killarney granite as a high-level plutonic-volcanic association that is likely related to similar rocks that underlie much of mid-continental North America (Van Schmus and Bickford, 1981), far removed from the covered southwestern extension of the Grenville Province. Deformation and metamorphism of these mid-continental rocks, imprinted long before Grenvillian Orogeny, may be preserved in slices between Grenvillian mylonite zones in the Tectonic Zone, giving rise to older ages. These things considered, there is little evidence, none of it compelling, that tectonic activity in the Grenville Front Tectonic Zone, specific to the Grenville Province, was active for as long a period as has been suggested.

An important feature of the northwest edge of the Tectonic Zone is the progression down structural section from ductile to brittle deformation. Only very close to the Front are flinty ultramylonite and pseudotachylite found, and only here can some of the rocks be termed cataclasites in the sense of mechanical deformation. There is no doubt that metamorphic grade rises sharply across the Grenville Front. It is suggested that this was accomplished by the thrust emplacement of hot rocks over cold, with original thermal inversion. The present inclination of isograds along the Front is not known, but may now be 'normal', though probably still steep, allowing that re-equilibration followed overthrusting.

7. CONCLUSIONS

It has been shown that sinuous gneiss belts between distinctive lithotectonic domains in the Central Gneiss Belt of the southwestern Grenville Province are best interpreted as broad ductile shear zones of deep origin along which blocks and slices of crust have been displaced northwestward. Similarity in tectonic style at the northwest boundary of the Central Metasedimentary Belt led Culshaw et al. (1983, p. 251) to propose a working model that "... involves stacking of large crustal slices with major shear zones along the slice interfaces." In brief, Britt and Algonquin domains and Rosseau and Go Home subdomains (Fig. 2) are seen as the 'structural basement' over which Parry Sound domain and then Seguin and Moon River subdomains were emplaced. The northwest margin of the Metasedimentary Belt is the site of thrusting of Grenville Supergroup and other rocks over Muskoka domain of the Central Gneiss Belt, but the timing of this thrusting with respect to formation of the ductile shear zones in the Gneiss Belt cannot be deduced from field studies and must await isotopic geochronology. The same tectonic polarity is evident in the Grenville Front Tectonic Zone; its northwest margin, the Grenville Front, is the leading edge of Grenvillian orogenic effects. Again, precise timing of deformation in this zone relative to shear zone tectonics farther southeast is not known.

The surface distance perpendicular to the Grenville Front across the area under consideration is about 250 km, roughly seven times its present crustal thickness (Mereu and Forsyth, in press). Granulites now exposed at the surface were formed at depths greater than 20 and perhaps as much as 35 km (Davidson and Morgan, 1981). Abrupt metamorphic gradients associated with major ductile shear zones (including the Grenville Front Tectonic Zone), possibly with inverted isograds, suggest that large rock masses originating at different crustal levels have been brought to a common level during Grenvillian Orogeny. The proposed tectonic scheme of stacking large blocks and slices toward the northwest can explain the distribution of granulite facies in the Central Gneiss Belt, and allows that the crust was thickened during the process. Evidence for crustal thickening by ductile thrust stacking is not restricted to the southwestern Grenville Province. The Grenville Front is known to be the site of thrusting at several places along its length, and large blocks of high-grade granulite emplaced over amphibolite facies gneisses are now recognized in Labrador and eastern Quebec (Wardle et al., in press; Rivers and Nunn, this volume).

Lack of granitic magmatism of Grenvillian age in the Central Gneiss Belt and Grenville Front Tectonic Zone is somewhat enigmatic, as melting at depth might be expected in a thickened crust. It seems unlikely that the exposed rocks in this region have already divested themselves of a melt fraction that rose to levels now eroded; many of the gneisses have compositions suitable for producing minimum melt, and thus they are not restites. Accumulation and rise of magma formed from already metamorphosed rocks, thus at least partly dehydrated, may require some time for thermal readjustment in a thickened crust to bring about melting conditions beneath hot overthrust blocks. Deformed leucosomes and pegmatite in many of the ductilely deformed rocks suggest a possible relationship between the beginning of melting and the inception of ductile flow. Perhaps a balance was attained and maintained between these two processes during Grenvillian Orogeny in this region, inhibiting the accumulation of large bodies of magma. Lateral upward movement of blocks of crust along broad ductile shear zones might have effectively removed the blocks from the level of potential melting. It is suggested that crustal thickening, and the erosion that removed the ensuing insulating cover, must have been accomplished rapidly in order to prevent extensive magma generation, and to enable preservation of steep metamorphic gradients in high-grade rocks.

The formation of some tracts of low-dipping, well lineated gneisses has been attributed to extension tectonics; in the case of the Grenville Province, however, this explanation is refuted by the observed kinematic indicators and by evidence that granulites lie structurally above amphibolite facies rocks. In terms of invoking a plate tectonic mechanism to explain crustal thickening, it has been suggested that the Grenville Front might mark an ancient plate boundary (Krogh and Davis, 1971), that a former suture may lie within the body of the Province (Irving et al., 1974; Windley, in press), or that it lies southeast of the exposed Province (Dewey and Burke, 1973; Seyfert, 1980). At present, however, there is no firm field evidence to support the first two contentions, and if the third is true, the former suture now lies buried, and presumably reworked, within the Paleozoic Appalachian orogen. Nevertheless, subduction-related piling up of crustal blocks and slices is the most appealing mechanism to explain the deduced tectonic history outlined above. It is possible that one or more deformed cryptic sutures do exist within the exposed Grenville Province; the problem is one of recognition. It is conceivable that the ductile shear zones so far identified, while not themselves former sutures, are splays from such a structure.

8. REFERENCES

Appleyard, E.C., 1974, 'Basement/cover relationships within the Grenville Province of eastern Ontario'; Can. J. Earth Sci., 11, p. 369-379.

Baer, A.J., 1974, 'Grenville geology and plate tectonics'; Geoscience Canada, 1, p. 54-64.

————, 1976, 'The Grenville Province in Helikian times: a possible model of evolution'; Roy. Soc. London, Phil. Trans., Ser. A, 280, p. 499-515.

————, 1981, 'A Grenvillian model of Proterozoic plate tectonics'; in Precambrian Plate Tectonics, A. Kröner, ed., Elsevier, Amsterdam, p. 353-385.

Bartlett, J.R., Brock, B.S., Moore, J.M., Jr., and Thivierge, R.H., 1984, Grenville Traverse A, Cross-sections of Parts of the Central Metasedimentary Belt; Geol. Ass. Can., Field Trip Guidebook 9A/10A, 63 p.

Bell, K., and Blenkinsop, J., 1980, 'Whole rock Rb-Sr studies in the Grenville Province of southeastern Ontario and western Quebec - a summary report'; Can., Geol. Surv., Paper 80-1C, p. 152-154.

Brock, B.S., and Moore, J.M., Jr., 1983, 'Chronology, chemistry, and tectonics of igneous rocks in terranes of the Grenville Province, Canada'; Geol. Soc. Amer., Abstracts with Programs, 15, p. 533.

Carmichael, D.M., Moore, J.M., Jr., and Skippen, G.B., 1978, 'Isograds around the Hastings metamorphic "low"'; in Toronto '78, Field Trips Guidebook, A.L. Currie and W.O. Mackasey, eds., Geol. Ass. Can., p. 325-346.

Culshaw, N.G., Davidson, A., and Nadeau, L., 1983, 'Structural subdivisions of the Grenville Province in the Parry Sound - Algonquin region, Ontario'; Can., Geol. Surv., Paper 83-1B, p. 243-252.

Davidson, A., 1979, 'Regional synthesis of the Grenville Province of Ontario and western Quebec. Part I: Some observations on the Grenville Front, and on metamorphism and plutonic rocks in the southwestern Grenville Province'; Can., Geol. Surv., Paper 79-1B, p. 155-163.

————, 1984, 'Identification of ductile shear zones in the southwestern Grenville Province of the Canadian Shield'; in Precambrian Tectonics Illustrated', A. Kröner and R. Greiling, eds., Schweizerbart., Stuttgart, p. 263-279.

————, in press (a), 'The southwestern Grenville Province'; in New Perspectives on the Grenville Problem, J.M. Moore., Jr., A.J. Baer and A. Davidson, eds., Geol. Ass. Can., Spec. Pap.

————, in press (b), 'The Killarney granite and its relationship to the Grenville Front'; in J.M. Moore, Jr., A.J. Baer and A. Davidson, eds., Geol. Ass. Can., Spec. Pap.

Davidson, A., and Morgan, W.C., 1981, 'Preliminary notes on the geology east of Georgian Bay, Grenville Structural Province, Ontario'; Can., Geol. Surv., Paper 81-1A, p. 291-298.

Davidson, A., Culshaw, N.G., and Nadeau, L., 1982, 'A tectono-metamorphic framework for part of the Grenville Province, Ontario'; Can., Geol. Surv., Paper 82-1A, p. 175-190.

Dewey, J.F., and Burke, K.C.A., 1973, 'Tibetan, Variscan, and Precambrian basement reactivation: products of continental collision'; J. Geol., 81, p. 683-692.

Derry, D.R., 1950, 'A Tectonic Map of Canada'; Geol. Ass. Can., Proc., 3, p. 39-53.

Duchesne, J-C., 1984, 'Massif anorthosites: another partisan review'; in Feldspars and Feldspathoids, W.L. Brown, ed., Reidel, Dordrecht, p. 411-433.

Easton, R.M., and Bartlett, J.R., 1984, Precambrian geology of the Howland area, Haliburton, Peterborough and Victoria counties; Ont. Geol. Surv., Prelim. Map P.2699, geol. ser.

Engel, A.E., 1956, 'Apropos the Grenville'; in The Grenville Problem, J.E. Thomson, ed., Roy. Soc, Can., Spec. Publ. No. 1, p. 74-96.

Gill, J.E., 1948, 'Mountain building in the Canadian Precambrian Shield'; 18th Internat. Geol. Congr., pt. XIII, p. 97-104.

Hanmer, S.K., 1984, 'Structure of the junction of three tectonic slices; Ontario Gneiss Segment, Grenville Province'; Can., Geol. Surv., Paper 84-1B, p. 109-120.

Hanmer, S.K., and Ciesielski, A., 1984, 'Structural reconnaissance of the northwest boundary of the Central Metasedimentary Belt, Grenville Province'; Can., Geol. Surv., Paper 84-1B, p. 121-131.

Hewitt, D.F., 1959, Geology of Cardiff and Faraday Townships; Ont. Dept. Mines, Ann. Rept., 64, pt. 3, 82 p.

———, 1961, Nepheline Syenite Deposits of Southern Ontario; Ont. Dept. Mines, 59, pt. 8, 194 p.

Irving, E., Emslie, R.F., and Ueno, H., 1974, 'Upper Proterozoic paleomagnetic poles from Laurentia and the history of the Grenville Structural Province'; J. Geophys. Res., 79, p. 5491-5502.

Krogh, T.E., and Davis, G.L., 1971, 'The Grenville Front interpreted as an ancient plate boundary'; Carnegie Inst. Washington, Yearbook 70, p. 239-240.

Krogh, T.E., Davis, G.L., and Frarey, M.J., 1971, 'Isotopic ages along the Grenville Front in the Bell Lake area, southwest of Sudbury, Ontario'; Carnegie Inst. Washington, Yearbook 69, p. 337-339.

Lindia, F.M., Thomas, M.D., and Davidson, A., 1983, 'Geological significance of Bouguer gravity anomalies in the region of Parry Sound domain, Grenville Province, Ontario'; Can., Geol. Surv., Paper 83-1B, p. 261-266.

Logan, W.E., 1847, Report of Progress for the Year 1845-6; Can., Geol. Surv., p. 1-98.

———, 1863, Report on the Geology of Canada; Can., Geol. Surv., Report of Progress from its Commencement to 1863, 983 p.

Lumbers, S.B., 1967, 'Geology and mineral deposits of the Bancroft-Madoc area'; in Guidebook - Geology of Parts of Eastern Ontario and Western Quebec, S.E. Jenness, ed., Geol. Ass. Can., p. 13-29.

———, 1971, 'Some aspects of the northwestern margin of the Grenville Province between Sudbury and Lake Timiskaming, Ontario'; Geol. Ass. Can., Abstracts of Papers, p. 36-37.

———, 1978, 'Geology of the Grenville Front Tectonic Zone in Ontario'; in Toronto '78, Field Trips Guidebook, A.L. Currie and W.O. Mackasey, eds., Geol. Ass. Can., p. 347-371.

———, 1982, Summary of Metallogeny, Renfrew County Area; Ont. Geol. Surv., Rept. 212, 59 p.

McLaughlin, D.B., 1954, 'Suggested extension of the Grenville orogenic belt and the Grenville Front'; Science, 120, p. 287-289.

Mereu, R.F., and Forsyth, D.A., in press, 'A summary of the results of the 1982 COCRUST long range seismic experiment across the Ottawa-Bonnechere graben and Grenville Front'; in New Perspectives on the Grenville Problem, J.M. Moore, Jr., A.J. Baer and A. Davidson, eds., Geol. Ass. Can., Spec. Pap.

Moore, J.M., Jr., and Thompson, P.H., 1972, 'The Flinton Group, Grenville Province, Ontario'; 24th. Internat. Geol. Congr., Proc., Sect. 1, p. 221-229.

———, ———, 1980, 'The Flinton Group: a late Precambrian metasedimentary succession in the Grenville Province of eastern Ontario'; Can. J. Earth Sci., 17, p. 1685-1707.

Nadeau, L. 1984, Deformation of Leucogabbro at Parry Sound, Ontario;
   Unpubl. M.Sc. Thesis, Carleton University, Ottawa, Canada.
Palmer, H.C., Merz, B.A., and Hayatsu, A., 1977, 'The Sudbury dikes of
   the Grenville Front region: paleomagnetism, petrochemistry, and
   K-Ar age studies'; Can. J. Earth Sci., 14, p. 1867-1887.
Pryer, L.L., 1985 (in press), 'Preliminary report on the Grenville Front
   Tectonic Zone, Carlyle Township, Ontario'; Can., Geol. Surv.,
   Paper 85-1A.
Quirke, T.T., 1926, 'Huronian-Grenville relations'; Amer. J. Sci., 5th.
   Ser., XI, p. 165-173.
Seyfert, C.K., 1980, 'Paleomagnetic evidence in support of a middle
   Proterozoic (Helikian) collision between North America and Gond-
   wanaland as a cause of the metamorphism and deformation in the
   Adirondacks: summary'; Geol. Soc. Amer., Bull., Pt. 1, 91,
   p. 118-120.
Schwerdtner, W.M., and Lumbers, S.B., 1980, 'Major diapiric structures
   in the Superior and Grenville provinces of the Canadian Shield';
   in The Continental Crust and its Mineral Deposits, D.W. Strangway,
   ed., Geol. Ass. Can., Spec. Pap. 20, p. 149-180.
Silver, L.T., and Lumbers, S.B., 1966, 'Geochronological studies in the
   Bancroft-Madoc area of the Grenville Province, Ontario'; Geol. Soc.
   Amer., Spec. Publ. 87, p. 156.
Sims, P.K., Card, K.D., and Lumbers, S.B., 1981, 'Evolution of early
   Proterozoic basins of the Great Lakes region'; in Proterozoic
   Basins of Canada, F.H.A. Campbell, ed., Can., Geol. Surv., Paper
   81-10, p. 379-397.
Stockwell, C.H., 1964, Fourth Report on Structural Provinces, Orogenies,
   and Time-classification of Rocks of the Canadian Precambrian Shield;
   Can., Geol. Soc., Paper 64-17 (Part II), p. 1-21.
————, 1982, Proposals for Time Classification and Correlation of Precambrian Rocks and Events in Canada and Adjacent Areas of the
   Canadian Shield; Part I: A Time Classification of Precambrian
   Rocks and Events; Can., Geol. Surv., Paper 80-19, 135 p.
van Breemen, O., Davidson, A., Loveridge, W.D., and Sullivan, R.W.,
   in press, 'U-Pb zircon geochronology of Grenville tectonites,
   granulites and igneous precursors, Parry Sound, Ontario'; in New
   Perspectives on the Grenville Problem, J.M. Moore, Jr., A.J. Baer
   and A. Davidson, eds., Geol. Ass. Can., Spec. Pap.
Van Schmus, W.R., 1965, 'The geochronology of the Blind River - Bruce
   Mines area, Ontario, Canada'; J. Geol., 73, p. 755-780.
Van Schmus, W.R., and Bickford, M.E., 1981, 'Proterozoic chronology and
   evolution of the midcontinent region, North America'; in Precamb-
   rian Plate Tectonics, A. Kröner, ed., Elsevier, Amsterdam,
   p. 261-296.
Wardle, R.J., Rivers, T., Gower, C.F., Nunn, G.A.G., and Thomas, A., in
   press, 'The northeastern Grenville Province: new insights'; in
   New Perspectives on the Grenville Problem, J.M. Moore, Jr., A.J.
   Baer and A. Davidson, eds., Geol. Ass. Can., Spec. Pap.
Wilson, J.T., 1949, 'Some major structures of the Canadian Shield';
   Can. Inst. Mining Met., Trans., 52, p. 231-242.

Wilson, M.E., 1918, 'The subprovincial limitations of pre-Cambrian nomenclature in the St. Lawrence basin'; J. Geol., 26, p. 325-333.
———, 1925,'The Grenville pre-Cambrian subprovince'; J. Geol., 33, p. 389-407.
———, 1927, 'Some problems of classification in the Canadian Pre-Cambrian Shield'; Geol. Mag., 64, p. 1-7.
———, 1939, 'The Canadian Shield'; in Geologie der Erde: Geology of North America, Volume 1, R. Ruedemann and R. Balk, eds., Verlag von Gebrüder Borntraeger, Berlin, p. 232-311.
———, 1965, 'The Precambrian Shield'; in The Geologic Systems, K. Rankama, ed., Interscience, New York, p. 263-415.
Windley, B.F., in press, 'Comparative tectonics of the western Grenville and western Himalayas'; in New Perspectives on the Grenville Problem, J.M. Moore, Jr., A.J. Baer and A. Davidson, eds., Geol. Ass. Can., Spec. Pap.
Wynne-Edwards, H.R., 1969, 'Tectonic overprinting in the Grenville Province, southwestern Quebec'; in Age Relations in High-grade Metamorphic Terrains, H.R. Wynne-Edwards, ed., Geol. Ass. Can., Spec. Pap. 5, p. 163-182.
———, 1972, 'The Grenville Province'; in Variations in Tectonic Styles in Canada, R.A. Price and R.J.W. Douglas, eds., Geol. Ass. Can., Spec. Pap. 11, p. 263-334.
———, 1976, 'Proterozoic ensialic orogenesis: the millipede model of ductile plate tectonics'; Amer. J. Sci., 276, p. 927-953.

A 1650 MA OROGENIC BELT WITHIN THE GRENVILLE PROVINCE OF
NORTHEASTERN CANADA

A. Thomas, G.A.G. Nunn and R.J. Wardle
Department of Mines and Energy
Government of Newfoundland and Labrador
P.O. Box 4750, St. John's
Newfoundland, CANADA, AlC 5T7

ABSTRACT. Recent geological mapping and allied Rb-Sr and U-Pb isotopic work have revealed the existence of a circa 1650 Ma orogenic event within the Labrador component of the Grenville Structural Province which is known as the Labradorian Orogeny. The area affected by this event, the Labrador Orogen, is a belt composed largely of high-grade, metasedimentary, sillimanite and/or kyanite gneiss ± hypersthene, garnet, cordierite and sapphirine, which is intruded by a linear, posttectonic granitoid batholith. The Labrador Orogen is juxtaposed against rocks of the slightly older Lower Proterozoic orogen of the Churchill Province (circa 1800 Ma), but has been almost entirely remetamorphosed, together with the southern margin of the Churchill Province, during the Grenvillian Orogeny (circa 1000 Ma). Metamorphism during the Grenvillian event varied from greenschist to granulite facies, but had remarkably little effect on Labradorian Rb-Sr and U-Pb systems throughout much of the area.
  Regional correlations are made with other circa 1650 Ma areas of deformation, both within and outside of the Grenville Province.

1. INTRODUCTION

Vestiges of a suspected pre-Grenvillian gneiss terrane within the eastern Grenville Structural Province of Canada (Figure 1) have been found (1), (2) and (3). Preliminary findings of a geological surveying programme, initiated in 1978 by the Newfoundland Department of Mines and Energy, substantiated and documented these earlier suspicions. The study, expanded during the period 1979 to 1983, consisted of regional 1:100,000 scale geological mapping allied with comprehensive geochronological and bedrock geochemical programmes. To date 23,482 square kilometres of predominantly crystalline gneissic and granitoid rocks including areas mapped by Geological Survey of Canada personnel (3) and (4), have been covered within the northern part of the Grenville Province in central Labrador (Figures 1 and 2). Three isotopic methods were utilized; these involved mineral dating (U-Pb), whole rock determinations (Rb-Sr), or both mineral and whole rock dating

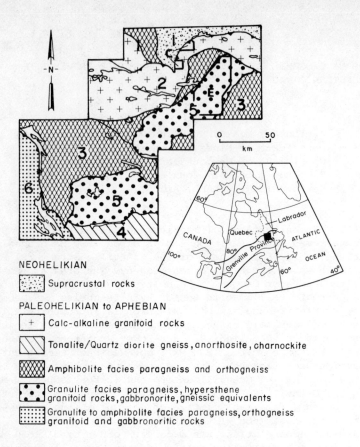

Figure 1. Location and general geology of the north-central Grenville Province, Labrador showing lithotectonic subdivisions (numbers 1 to 6). Areas marked by letters were mapped by Geological Survey of Canada personnel (3) C and (4) E.

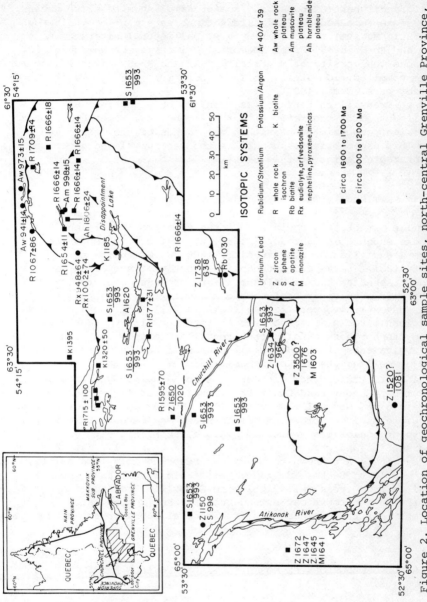

Figure 2. Location of geochronological sample sites, north-central Grenville Province, Labrador. Ages are in millions of years, single U-Pb numbers represent concordant ages, those given for discordant U-Pb ages represent upper and lower intercepts respectively. Structural Province divisions after (5), (6) and modified from (7).

(40Ar-39Ar) and, wherever possible, common sampling sites were chosen
for the different methods. U-Pb determinations were done by T.E. Krogh
(Royal Ontario Museum), 40Ar-39Ar by R.D. Dallmeyer (University of
Georgia), and Rb-Sr by B. Fryer (Memorial University of Newfoundland)
and C. Brooks (University of Montreal). K-Ar and Rb-Sr mineral isochron
dates were compiled from other published work (8), (9), (10), (11) and
K.L. Currie (pers. comm.,1983).

The pre-Grenvillian terrane was found to be widespread and to
contain elements of a separate, hitherto unrecorded, Lower Proterozoic
(Paleohelikian) orogenic event within the bounds of the Grenville
Province. This paper briefly summarizes the geological and geochronological characteristics of this event.

2. REGIONAL SETTING

The Grenville Structural Province encompasses approximately one half
of the rocks exposed in Labrador (Figure 2), which were tectonized
during the Grenvillian Orogeny at circa 1000 Ma. Granulite to
amphibolite facies para- and orthogneiss, mostly of suspected Lower
Proterozoic (Aphebian and/or Paleohelikian) or older age, predominate.
Migmatite, Middle to Upper Proterozoic mafic to felsic intrusions,
intrusive and extrusive peralkaline rocks as well as metasedimentary
and metavolcanic rocks are also present together with younger dykes.

Along its northern margin, the Grenville Province contains
reworked equivalents of the Superior, Churchill and Nain Structural
Provinces as well as the Makkovik Sub-province. In westernmost
Labrador and eastern Quebec, Archean Superior Province gneisses and
their previously undeformed Lower Proterozoic cover (12) extend as far
as 200 km into the Grenville Province (13). Throughout most of Labrador,
however, Proterozoic or older metasedimentary and metavolcanic rocks,
granitic to granodioritic gneiss, migmatite, amphibolite to granulite
facies paragneiss and granite belonging to the Churchill, Nain and
Makkovik Provinces are known to have been reworked only in a narrow,
approximately 10 to 50 km wide zone. Relict older ages have recently
been obtained from areas well within the Grenville Province indicating
that reworked older rocks may extend much farther south. Dates
representative of these older ages from central Labrador are documented
here.

3. GENERALIZED GEOLOGY OF THE NORTH-CENTRAL GRENVILLE PROVINCE,
   LABRADOR

North-central Grenville Province rocks in Labrador can be divided into
six lithotectonic blocks (Figure 1) on the basis of contrasts in lithology,
metamorphic grade and structure. Metamorphism in all blocks is regional
and associated with moderate to intense deformation of several ages. A
more thorough discussion of the structural and metamorphic features of
Grenvillian age, including lithotectonic block boundaries, is given
elsewhere (14).

Block 1 comprises predominantly Middle Proterozoic (Neohelikian) shallow marine to subaerial, arenaceous to argillaceous sedimentary rocks with intercalated basic volcanic rocks and an underlying sequence of comenditic volcanic and allied peralkaline intrusive rocks. Included within block 1 are minor amounts of Churchill Province gneisses that have been intruded by granitoid rocks, most commonly found in block 2, and which together form basement to the supracrustal sequence. The metamorphic grade of both the granitoid rocks and the cover sequences is greenschist facies which is superimposed on an amphibolite facies metamorphism in the Churchill Province gneisses. The boundary between blocks 1 and 2 is a zone of south-dipping thrust and high-angle reverse faults.

Block 2 is dominated by a suite of calc-alkaline granitoid rocks, ranging from diorite to granite in composition, that belong to the Trans-Labrador batholith (15). Biotite granite to granodiorite predominates and commonly contains xenoliths of older gneissic, tonalitic, dioritic and mafic material. Minor amounts of gabbronorite both pre- and postdate the granitoid suite. Two large rafts of older paragneiss, typical of that found in block 3, as well as patches of peralkaline volcanic and intrusive rocks are also present within the block. Metamorphic grade throughout most of the block 2 granitoid rocks is greenschist facies, increasing locally along the southwestern block boundary to middle to upper amphibolite facies. The boundary between blocks 2 and 3 is a modified intrusive contact, characterized along its eastern half by a moderate to high-angle, south-dipping reverse fault and along its western half by a ductile high-strain zone with an L-S fabric.

Block 3 comprises polydeformed quartzofeldspathic paragneiss with a variable mineral assemblage that may also include any or all of the following; sillimanite, kyanite, muscovite, biotite, garnet and opaques. Biotite ± garnet, foliated and/or gneissic equivalents of block 2 granitoid rocks are also present within this block. Relict fine-grained hypersthene is present, locally, in both the orthogneiss and paragneiss. Gabbro, norite, amphibolite and patches of hypersthene diorite also occur scattered throughout this block. The metamorphic grade of block 3, as demonstrated by the granitoid rocks and orthogneiss, ranges from greenschist facies in the northeast to upper amphibolite facies in the southwest and has been superimposed on a pre-existing middle to upper amphibolite (locally granulite) facies metamorphism of the paragneiss. The boundary between blocks 3 and 4, although not observed, lies beneath block 5 and is thought to be a fault.

Ortho- and clinopyroxene-bearing garnet tonalite to quartz diorite gneiss, charnockitic granitoid rocks and anorthosite predominate in block 4. Minor rock types include norite, amphibolite and sillimanite-garnet paragneiss. Granulite facies metamorphic conditions prevailed throughout this block.

Quartzofeldspathic paragneiss, containing hypersthene, sillimanite, cordierite, quartz, feldspar, magnetite, ilmenite and rare but widespread sapphirine, is the main constituent of block 5. The paragneiss is extremely uniform in character, and its average chemical composition is similar to that of the paragneiss in block 3, of which it is thought to be a higher grade equivalent. Small ultramafic lenses, as well as plutonic

bodies of gabbronorite, gabbro and hypersthene monzodiorite to quartz
monzonite also occur within this block. Block 5 rocks have been regionally metamorphosed under granulite facies conditions, the assemblage
hypersthene + sillimanite + sapphirine + quartz indicating the development, locally, of extremely high temperatures and pressures. Block 5 is
allochthonous; constituting a deep-seated crustal fragment that has
been detached and thrust over blocks 3 and 4 upon which it now rests as
a klippe (16), (17).

The sixth lithotectonic block comprises quartzofeldspathic sillimanite + biotite ± hypersthene and garnet paragneiss with minor metabasite, foliated adamellite and gabbronorite. In the northern part of
the block, granulite facies rocks structurally overlie amphibolite facies
equivalents on a syn- to late kinematic, internal, ductile shear zone
system. The southern portion of block 6 has undergone amphibolite facies
retrogression. This block is also allochthonous (12) and forms a thrust
nappe lying upon gneisses of block 3. Poor exposure along contacts with
blocks 4 and 5 masks their relationships with block 6.

4. AGES OF NORTH-CENTRAL GRENVILLE PROVINCE ROCKS IN LABRADOR

Geochronological results are illustrated in Figure 2. U-Pb data give
concordia and discordia dates ranging from 3500 Ma to 993 Ma. The data
from age determinations on all sphene samples were combined to produce
a single discordia plot which yielded an upper intercept of 1653 Ma and
a lower intercept of 993 Ma. The precision of 206Pb/238U vs 207Pb/235U
ratios is 2σ for all U-Pb dates except the 1738/638 Ma discordia. This
last date was compiled and its precision is not known. Rb-Sr results
are mostly isochrons and give dates ranging from 1666 Ma to 948 Ma. The
1709 ± 94 Ma (18) and 1715 ± 100 Ma (19) dates are based on errorchrons.
Precision of 87Sr/86Sr versus 87Rb/86Sr ratios is 1σ (18), (19) or 2σ
(8). 40Ar-39Ar dates range from 1806 Ma to 941 Ma with a reported
precision of 2σ.

Most of the dates from the area fall into two groups; an older group
ranging from 1700 Ma to 1600 Ma and a younger group ranging from 1200 Ma
to 900 Ma. Dates from lithotectonic block 1 generally range from 1100 Ma
to 900 Ma and record the timing of postorogenic cooling of metamorphic
rocks. The circa 1700 Ma to 1600 Ma ages from block 2 record primary
cooling and crystallization of various phases of granitoid rocks
belonging to the Trans-Labrador batholith. At present, the average age
of intrusion of these granitoid rocks is given by a composite Rb-Sr
isochron (18) as 1654 ± 11 Ma. The circa 1200 Ma to 900 Ma dates from
block 2 are the result of a generally low-grade metamorphic overprint
equivalent to that in block 1. Two K-Ar dates (10), (11) from near the
boundary of blocks 1 and 2 are thought to represent a mixing age of
these two events. Circa 1700 Ma to 1600 Ma dates from lithotectonic
blocks 3,5 and 6 predominantly represent recrystallization and cooling
in high-grade, regionally metamorphosed gneisses. Note that the 1666 ±
14 Ma dates in these blocks represent a composite isochron, and include
both amphibolite-grade paragneiss of block 3 and granulite-grade paragneiss of block 5. In addition, the 1666 ± 18 Ma date, from an amphibolite

facies paragneiss raft in block 2 granitoid rocks was also combined to form this isochron (18). 1150 Ma to 900 Ma dates from these blocks, together with block 4, result from a later low- to high-grade metamorphic overprint that has associated with it new zircon growth or overgrowth and/or partial to complete resetting of other isotopic systems. A single U-Pb upper intercept discordia age of 3500 Ma from block 5 may record the time of crystallization of zircon which was inherited from an ancient source terrane during formation of the paragneiss protolith. Due to the clustering of most of the points toward the lower intercept on the discordia plot however, the significance of the 3500 Ma date remains questionable at this time. Similar uncertainty applies to the 1520 Ma date from block 4. Zircon cores in adamellite from block 6 gave a U-Pb date of 1672 Ma, which may represent original crystallization and cooling, rather than metamorphism of the rock. The 40Ar-39Ar hornblende plateau age of 1806 Ma is enigmatic in that it records an older event and appears to have fortuitously escaped the high-grade circa 1700 Ma to 1600 Ma reworking.

5. DISCUSSION

The structural sequence in each lithotectonic block can be simplified into a three-stage cycle, namely i) deformation of older rocks, ii) emplacement of a plutonic suite and iii) further deformation. The earlier deformation is represented by a migmatitic gneissosity and the later deformation usually takes the form of folding and axial planar mineral recrystallization of the primary gneissosity, and development of foliated or gneissic fabrics in the plutonic suite. Despite the similarity of structural sequences recorded at outcrop scale throughout the area, the cycles are not necessarily of the same age (either between or within blocks). Therefore it has not been possible to use structural criteria as the principal orogenic discriminant. Based on geochronological and metamorphic characteristics however, there is evidence to conclude that three major events comprise the polyorogenic history of central Labrador.

The primary gneissosity in the Churchill Province gneisses contained within lithotectonic block 1 is known to be Hudsonian (circa 1800 Ma). The plutonic suite forms part of the Trans-Labrador batholith (circa 1650 Ma) and most of the postplutonic deformation consists of low-grade Grenvillian fabrics (circa 1000 Ma).

Paragneiss within blocks 2 and 3 contains a primary gneissosity which formed during stage i) of the cycle and has associated with it medium to high-grade preplutonic fabrics. Circa 1666 Ma Rb-Sr dates may record the time of formation of the gneissosity, or the medium to high-grade fabrics or both. On the other hand, the 40Ar-39Ar date of 1806 Ma may be local confirmation that the primary gneissosity actually formed during the Hudsonian Orogeny and the medium to high-grade fabrics followed at circa 1666 Ma. The plutonic suite of stage ii) is represented in blocks 2 and 3 by calc-alkaline granitoid rocks of the Trans-Labrador batholith. Stage iii) postplutonic deformation in block 2 and the northern and eastern parts of block 3 consists of greenschist facies foliations that

are known to be mostly Grenvillian (circa 1000 Ma) in age,with the
exception of a pre-Neohelikian, north-northeasterly-trending greenschist
facies fabric (20), (21) developed locally in blocks 1 and 2. Farther
south in block 3, the postplutonic deformation resulted in gneissic to
foliated, amphibolite facies fabrics in the Trans-Labrador batholith and
new migmatization in the paragneiss. Throughout blocks 2 and 3 the suite
of sphene samples from the batholith indicate an isotopically homogeneous
source at circa 1653 Ma and a single episode of lead loss at circa 993 Ma.
These are taken to represent repectively the emplacement of the batholith
and a postplutonic Grenvillian dynamothermal event.

Three possibilities exist for the relative timing of gneiss- and
fabric-formation in block 5. The primary gneissosity may be Hudsonian,
in which case circa 1676-1666 Ma dates on paragneiss from the block
represent the time of formation of later, high-grade regional metamorphic
fabrics,   during stage i) of the deformation - plutonism - deformation
cycle. The fabrics may have a temporal association with the gneissosity,
the circa 1676-1666 Ma dates representing the age of stage i) formation
of both (a view currently favoured by the authors). The emplacement of
the stage ii) plutonic suite in block 5 was followed by selective, high-
grade reworking, and a third possibility is that this postplutonic deformation may be what the ca 1676-1666 Ma dates have actually recorded.
The stage ii) plutonic suite is represented in block 5 by a charnockite-
gabbronorite assemblage (dated at ca 1675-1650 Ma, pers. comm. R.F.
Emslie, 1980), the hypersthene-bearing granitoid rocks of which may be
possible equivalents of parts of the Trans-Labrador batholith. Grenvillian dates representing stage iii) within block 5 have generally been
obtained only from systems with low blocking temperatures (e.g. unpublished 40Ar-39Ar preliminary results, pers. comm. R.D. Dallmeyer, 1984).
It is uncertain at present whether the deformation-plutonism-deformation
cycle in block 5 is an earlier diachronous equivalent to, or coeval with,
the cycle in blocks 2 and 3. We regard the cycle in block 5 as a temporally and probably genetically closely related trio of events and if the
ca 1676-1666 Ma dates do record a postplutonic, early stage iii) event
in block 5 (third possibility above) then the cycle is almost certainly
diachronous to its counterpart in blocks 2 and 3. If these dates record
preplutonic events (first or second possibilities above) then the block
5 cycle may be coeval with its block 2 and 3 counterparts.

Block 6 appears to contain a major postplutonic high-grade event at
circa 1646 Ma that may be equivalent to the later deformation in block
5. There is a paucity of data from this block and no preplutonic or
Grenvillian ages have yet been determined.

Block 4 has been thoroughly reworked in the granulite facies during
the Grenvillian Orogeny and at present insufficient data is available to
make any statement as to its pre-Grenvillian history.

The belt of rocks in lithotectonic blocks 2, 3, 5 and 6 that range
in age between 1676 Ma and 1620 Ma (14), (this study) and comprise mixed
para- and orthogneiss with a predominantly posttectonic granitoid
batholith, is known to extend southward from near the northern margin of
the Grenville Province for a width of at least 100 to 150 km. Its true
southerly limit is unknown, being constrained by the present limit of the
mapping and geochronological programmes. At present it is also not known

whether the rocks in this belt are reworked equivalents of any of the older structural provinces in Labrador or of a separate geographical affinity. In central Labrador, this belt of Lower Proterozoic rocks is now referred to as the Labrador Orogen (14) and the post-Hudsonian, pre-Grenvillian, high-grade deformation and metamorphism associated with the gneiss terrane has been assigned to the Labradorian Orogeny (17). Although the breakup of the area into lithotectonic blocks 1 to 6 is essentially a Grenvillian effect, Labradorian tectono-metamorphic, plutonic and geochronological effects dominate within blocks 2, 3, 5 and 6.

6. REGIONAL CORRELATIONS

The Labrador Orogen can be extended in an east-west direction across Labrador, where 1:100,000 scale mapping programmes have also been ongoing since 1977. Block 6 is the eastern margin of a large thrust nappe, the Lac Joseph allochthon (12) which appears to be of relatively homogeneous character throughout. Although the central part of it has not been mapped during these programmes, the allochthon extends to the western Labrador border (12). It is not yet possible to match the pre-Grenvillian tectonic histories of the lithotectonic blocks given here with those of the lithotectonic terranes established in eastern Labrador (22). However, the Trans-Labrador batholith is a dominant feature of both areas and structural relationships and radiometric data (23) suggest that the orogen can be extended to the coast. In eastern Labrador, the Labradorian equivalent tectonism may not be at such a high grade of metamorphism as that in central Labrador and is interpreted (23) to be the latest event of an orogenic sequence that may go back to circa 2100 Ma (23). A strong regional correlation of age, rock types, orogenic events and spatial relationships has been demonstrated (23), (24) between eastern Labrador and the Sveconorwegian orogenic belt of southwestern Sweden. In Sweden orogenic events of Labradorian age (1700-1600 Ma) have been noted (25), (26), (27) where they are known as the Gothian Orogeny (25), (27) and the 1000 km long Småland-Värmland granitoid belt is correlated directly with the Trans-Labrador batholith (23), (24).

In the United States, the Penokean Orogeny (circa 1850 Ma) of the north-midwest is succeeded to the southwest by younger orogenic activity (1780-1700 Ma) (28) and then both are overprinted along a southeastern margin by a continuous belt of tectonism and plutonism extending from Arizona to the Great Lakes (28), (29). In the southwest this belt was involved in the Mazatzal Orogeny (30) and throughout yields ages of 1680-1610 Ma, including foliated granitoid plutons at circa 1625 Ma (28). This belt clearly has similarities, at least in age with the Labrador Orogen, although poor exposure and younger cover sequences obscure whether or not the associated plutons are of regional batholithic proportions. The easternmost extension of the belt, some 1300 km southwest of Labrador, may be at Killarney on the north shore of Lake Huron where a Rb-Sr date of 1623 ± 74 Ma possibly represents an orogenic resetting of post-Penokean granitoid rocks (pers. comm. A. Davidson, 1984).

It is not known whether the Labrador Orogen is part of a once continuous orogenic belt stretching from Arizona to southern Sweden, but we can perhaps note that the gaps are partly filled by younger amphibolite to granulite facies orogenic segments. Representatives of these segments are in Norway, the 1600-1500 Ma Kongsbergian Orogeny (31) and in Canada, the 1500-1400 Ma Central Gneiss Belt (32), which may have obscured Labradorian-age events. Reports by workers in Scandinavia (25), (31), (33), (34) and North America (28), (35), (23), indicate that this margin of Laurentia has been the locus of repeated, local to extensive, often overlapping orogeny (a "protracted pre-Sveconorwegian history"(36)). There is a general younging of these events to the southeast in America and to the southwest in Scandinavia. In this scheme, temporal equivalents may not have been parts of continuous orogenic belts, and continuous belts may have been diachronous and/or selectively reworked or destroyed during subsequent events.

7. ACKNOWLEDGEMENTS

We are indebted to T. Krogh whose U-Pb data are used in this paper and with whom the authors have further work in progress on the geochronology and isotope systematics of these rocks. The manuscript was reviewed by P.C. Smalley, Institutt for energiteknikk, Norge and T. Rivers, Memorial University of Newfoundland, Canada. We express our appreciation to D. Gamberg and J. Morgan for typing the manuscript. This work is a direct consequence of both the Canada-Newfoundland Mineral Development Subsidiary Agreement (1977-1981) and the Canada-Newfoundland co-operative mineral program (1982-1984) carried out in conjunction with the Geological Survey of Canada. The paper is published with the permission of the Director, Mineral Development Division, Newfoundland Department of Mines and Energy.

8. REFERENCES

(1) Gandhi, S.S. (1970). Compendium of monthly repts. on Seal Lake copper area, Labrador. British Nfld. Exploration Ltd.
(2) Gandhi, S.S. (1971). 'Regional Geology of the Seal Lake area, Labrador', (unpublished 1:250,000 scale map, prepared for Brinex Corp.)
(3) Curtis, L.W. and Currie, K.L. (1981). Geol. Surv. Can. Bull. 294, 61p.
(4) Emslie, R.F., Hulbert, L.J., Brett, C.P. and Garson, D.F. (1978). Geol. Surv. Can., Paper 78-1A, Pt. A, pp. 129-134.
(5) Taylor, F.C. (1971). Can. Jour. Earth Sci., **8**, pp. 579-584.
(6) Wardle, R.J. (1983). Dept. Mines and Energy, Nfld. and Lab., Rept. 83-1, pp. 68-90.
(7) Smyth, W.R. and Greene, B.A. (1976). Dept. Mines and Energy, Nfld. and Lab., Map 764.
(8) Blaxland, A.B. and Curtis, L.W. (1977). Can. Jour. Earth Sci., **14**, pp. 1940-1946.
(9) Wanless, R.K., Stevens, R.D., Lachance, G.R. and Rimsaite, J.Y.H. (1966). Geol. Surv. Can., Paper 65-17.
(10) Wanless, R.K., Stevens, R.D., Lachance, G.R. and Delabio, R.N. (1972). Geol. Surv. Can., Paper 71-2.

(11) Lowden, J.A., Stockwell, C.H., Tipper, H.W. and Wanless, R.K. (1963). Geol. Surv. Can., Paper 62-17.
(12) Rivers, T. and Nunn, G.A.G. This volume.
(13) Rivers, T. and Chown, E.H. (in press). G.A.C. Spec. Paper, New Perspectives on the Grenville Problem, Proc. of Symp. at Joint G.A.C.-M.A.C. meeting, London, Ont., 1984.
(14) Thomas, A., Nunn, G.A.G. and Krogh, T.E. (in press). G.A.C. Spec. Paper, New Perspectives on the Grenville Problem, Proc. of Symp. at Joint G.A.C.-M.A.C. meeting, London, Ont., 1984.
(15) Wardle, R.J. and staff. (1982). Abst., Grenville Workshop, Prog. with Absts., Rideau Ferry Inn, Feb., 1982, pp. 11.
(16) Thomas, A. (in prep). 'Geol. of Red Wine Mtns. and surrounding area, north-central Grenville Province, Labrador'. Memoir, Nfld. and Lab. Dept. Mines and Energy.
(17) Nunn, G.A.G., Noel, N. and Culshaw, N.G. (1984). Dept. Mines and Energy, Nfld. and Lab. Rept. 84-1, pp. 30-41.
(18) Fryer, B.J. (1983). Unpub. summary of Geochron. results on samples collected from central Labrador between 1978 and 1980, 35p.
(19) Brooks, C. (1981). Unpub. rept. on geochron. to Dept. Mines and Energy, Nfld. and Lab., Open File LAB 519, 23p.
(20) Nunn, G.A.G. (1981). Dept. Mines and Energy, Nfld. and Lab. Rept. 81-1, pp. 138-148.
(21) Nunn, G.A.G. and Noel, N. (1982). Dept. Mines and Energy, Nfld. and Lab. Rept. 82-1, pp. 149-167.
(22) Gower, C.F. (1984). Dept. Mines and Energy, Nfld. and Lab. Rept. 84-1, pp. 68-79.
(23) Gower, C.F. and Owen, V. (1984). Can. Jour. Earth Sci., **21**, pp. 678-693.
(24) Gower, C.F. This volume.
(25) Berthelson, A. (1980). Colloque C6, Géologie de l'Europe du Précambrien aux bassins sedimentaires post-hercyniens. Bureau de Recherches Géologiques et Minières, Société Géologique du Nord, pp. 5-21.
(26) Samuelsson, L. and Åhäll, K.I. (1985). This volume.
(27) Verschure, R.H. (1985). This volume.
(28) Van Schmus, W.R. and Bickford, M.E. (1981). Precambrian plate tectonics, A. Kroner, (ed), Elsevier, Amsterdam, pp. 261-296.
(29) King, P.B. (1976). U.S.G.S. Prof. Paper 902, 85p.
(30) Silver, L.T. (1965). Geol. Soc. Amer. Abstr. for 1964, Spec. Paper 82, pp. 185-186.
(31) Starmer, I.C. (1985). This volume.
(32) Davidson, A. (1985). This volume.
(33) Field, D. and Råheim, A. (1981). Precam. Res., 14, pp. 261-275.
(34) Field, D., Smalley, P.C., Lamb, R.C. and Råheim, A. (1985). This volume.
(35) Baragar, W.R.A. (1981). Geol. Surv. Can. Bull. 314, pp. 47.
(36) Torske, T. (1985). This volume.

A REASSESSMENT OF THE GRENVILLIAN OROGENY IN WESTERN LABRADOR

T. Rivers[1] and G.A.G. Nunn[2]
1. Department of Earth Sciences, Memorial University,
   St. John's, Newfoundland, A1B 3X5, Canada.
2. Mineral Development Division, Newfoundland
   Department of Mines and Energy, St. John's,
   Newfoundland, A1C 5T7, Canada.

ABSTRACT. Recent mapping and radiometric dating in the Grenville Province of western Labrador indicate two major metamorphic events, one Early Proterozoic (circa 1650 Ma) and one Grenvillian (circa 1000 Ma). A threefold tectonic subdivision of the Grenville Orogen is proposed, consisting of an autochthon, a parautochthon and a number of allochthons. The parautochthon in the west consists of a previously unmetamorphosed sedimentary succession, whereas in the east it consists of paragneiss, metamorphosed in the Early Proterozoic, and a granitoid batholith. From the Grenville Front southwards, metasediments of both the autochthon and parautochthon increase in Grenvillian metamorphic grade from greenschist to upper amphibolite facies. These rocks are imbricated by northward-directed thrusts. In the interior Grenville Province, the parautochthonous sequences are tectonically overlain by Lower Proterozoic gneisses of upper amphibolite to granulite facies. These were thrust into place during the Grenvillian orogeny, when they were structurally reworked but only mildly metamorphosed. Upper amphibolite and granulite facies rocks of an Early Proterozoic metamorphism are thus adjacent to those at equivalent and lower grades of a Grenvillian metamorphism.

1. INTRODUCTION

It has long been recognized that the Grenville Province is in part composed of older tectonic provinces that were reworked during the Grenvillian orogeny (e.g. 1). For instance in western Labrador the Grenville Front truncates the southern margins of the Superior and Churchill provinces, and reworked extensions of both these provinces may be traced across the boundary into the northern Grenville Orogen. In the interior Grenville Orogen, however, the rocks are distinctly different from those in the adjacent provinces, and they were considered by Wynne-Edwards (1) to have been through a gneiss-forming event (or at least to have been strongly modified) during the Grenvillian orogeny.
    During the last seven years much of the area covered in Figure 1 has been remapped by geologists at Newfoundland Department of Mines and

Energy, and many units have been dated. In this paper we summarize some of the more important conclusions to date.

A feature dominating most of the new geologic maps of this region is the presence of numerous thrust faults, which indicates considerable tectonic stacking during the Grenvillian orogeny. The thrust faults can be divided into two groups: (a) those across which lithologic units may be traced, and on which tectonic transport is limited; and (b) those on which tectonic transport may be considerable and across which in many cases it is not possible to correlate lithologic units. In general the former are situated in the area immediately south of the Grenville Front, whilst the latter occur within the interior Grenville Orogen. Recognition of the importance of thrusting has led to the proposal of a new subdivision of the Grenville Orogen by Rivers and Chown (2), which is entirely tectonic in concept. These authors subdivide the orogen into three zones, an autochthon, a parautochthon and a number of allochthons (Figure 1). The autochthon consists principally of rocks of the adjacent, older structural provinces and their previously undeformed cover. These rocks were variably deformed _in situ_ during the Grenvillian orogeny. The intensity of deformation increases towards the south in the autochthon, and with the crossing of the first major thrust or reversed fault the autochthon gives way to a parautochthonous zone, much of which is interpreted to be underlain by thrust faults at depth and to be telescoped in an imbricate structure which resulted in considerable crustal thickening. A number of lithologic units may be traced from the autochthon into the parautochthon demonstrating an overall structural continuity between the two zones.

South of the parautochthon and structurally overlying it are the allochthons which are interpreted to have been emplaced as large thrust nappes. The Lac Joseph allochthon (Figure 1) is composed of high-grade gneisses and acid to basic intrusive rocks that from field relations both pre- and postdate the metamorphism in the surrounding gneisses. U/Pb radiometric data (Figure 1) show that these gneisses were formed about 1650 Ma in what is now known as the Labradorian orogeny. Since the data points plot on concordia (3,4), indicating no subsequent lead loss, the gneisses appear to have undergone minimal resetting and by extension minimal heating during the Grenvillian orogeny.

The picture that has emerged, therefore, is of a parautochthon characterized by Grenvillian metamorphism, for the most part in the amphibolite facies, tectonically overlain by allochthonous nappes with a negligible Grenvillian overprint.

The terminology of Wynne-Edwards (1) (e.g. Grenville Front Tectonic Zone, Baie Comeau Segment and Eastern Grenville Province) is not used in this paper as the boundaries of his subdivisions do not concur with the tectonic zonation used here. For instance, the Grenville Front Tectonic Zone contains parts or all of the elements of the parautochthon and allochthons as defined here. The tectonic subdivisions used in this paper are, however, compatible with a lithotectonic terrane scheme established in eastern Labrador by Gower and Owen (5) and Gower (6) and with the lithotectonic blocks of Thomas, Nunn and Wardle (7) in central Labrador. For example, block 1 (7) corresponds to the autochthon; the Groswater Bay Terrane (6) and blocks 2 and 3 (7) are

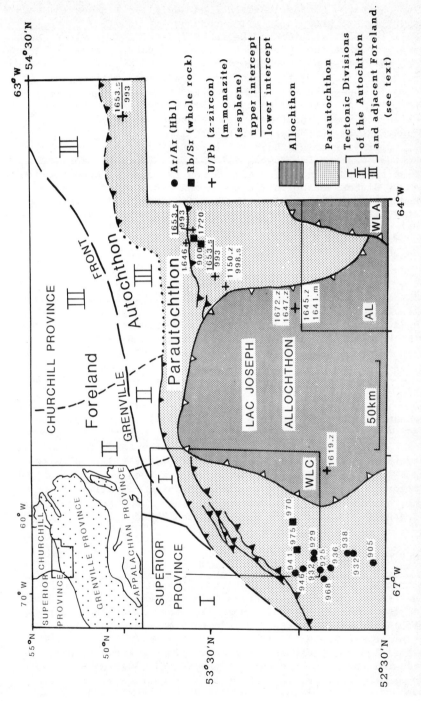

Figure 1. Tectonic subdivision of the northern Grenville Province of western Labrador, showing the locations of recent age data (3,4,8,9). WLA - Wilson Lake allochthon. Inset map shows location of the map area in the eastern Grenville Province. Boxes show locations of Figures 2 (WLC) and 3 (AL). Wilson Lake allochthon and Lac Joseph allochthon = respectively, blocks 5 and 6 of Thomas, Nunn and Wardle (7).

equivalent to the parautochthon; and the Lake Melville Terrane (6) and blocks 5 and 6 (7) are grouped as allochthonous terranes.

In Figure 1 a modification to the tectonic zonation of Rivers and Chown is shown, in that the autochthon has been separated from a foreland to the north, the latter being undeformed during the Grenvillian orogeny. The line separating the two, the Grenville Front, is a diffuse zone in the field, and in supracrustal rocks is marked by the northern limit of ductile deformation of Grenvillian age. The autochthon is considerably appressed or absent where the foreland is composed of anhydrous, high-grade gneisses as in the west of Figure 1.

In this paper we examine the characteristics of the autochthon, the parautochthon and the Lac Joseph allochthon in western Labrador. The central part of the Lac Joseph allochthon has not yet been remapped, and we base our discussion on work done on the western margin near Wabush-Labrador City (10,11) and on the eastern margin in the Atikonak River area (12, 13; Figure 1).

2. AUTOCHTHON

The autochthon and the foreland to the north, with which it is continuous, are divided into three tectonic elements (numbered I-III on Figure 1), each of which is composed of deformed rocks and a younger cover sequence. The western division (I) is an extension of the Superior Province, and consists of a basement of Archaean gneisses overlain by a Lower Proterozoic platformal sequence known as the Knob Lake Group (10, 14). The central and eastern divisions are extensions of the Churchill Province. In the central area (II), the Knob Lake Group was deformed and metamorphosed to greenschist facies during the Hudsonian orogeny (circa 1800 Ma), and forms the basement to minor Middle Proterozoic cover sequences; whereas the eastern division (III) consists of high-grade gneisses and syntectonic granitoid rocks from the core of the Hudson Orogen. These latter rocks are intruded by voluminous posttectonic granitoid rocks belonging to the Ungava and Trans-Labrador batholiths (15), and also contain minor areas of Middle Proterozoic cover sequences. None of the Middle Proterozoic sequences were involved in pre-Grenvillian deformations, and only the Lower Proterozoic Knob Lake Group is generally preserved south of the autochthon/parautochthon boundary.

3. PARAUTOCHTHON

In the map area of Figure 1 the parautochthon is composed of two elements with distinct pre-Grenvillian histories. In the west is the southerly extension of the Lower Proterozoic Knob Lake Group of the Labrador Trough, which was unmetamorphosed in this part of the Grenville Province prior to the Grenvillian orogeny (10); and in the east are high-grade gneisses and plutonic rocks which are now known to be pre-Grenvillian in age, and are part of the recently-defined Labradorian orogen (13, 16, 7). Both these sequences underwent metamorphism up to amphibolite facies during the Grenvillian orogeny and it is this imprint and the associated deformation, together with the position of the parautochthon with respect to the other tectonic zones, which unite these otherwise diverse rocks into a coherent tectonic unit. The contact between the Knob Lake Group and the Labradorian gneisses is obscured by the Trans-Labrador batholith and the Lac Joseph allochthon; its tectonic significance is unknown.

3.1 Wabush-Labrador City Area

The parautochthon in the Wabush-Labrador City area (Figure 2) has recently been described in some detail by Rivers (10, 11) and will be reviewed only briefly here. Since the rocks in this area were, for the most part, not metamorphosed prior to the Grenvillian orogeny, they provide sensitive indicators of Grenvillian thermotectonism. The rocks are polydeformed, and a sequence of three deformation episodes has been identified (10), which are all considered part of the Grenvillian orogeny. The zone south of the Grenville Front is dominated by northeast-trending thrust faults and overturned folds, both of which display northwest vergence and, together with the presence of a ubiquitous, southeast-trending (downdip) stretching lineation, indicate tectonic transport towards the northwest (i.e. towards the craton). Structural trends in the interior of the parautochthon are perpendicular to those in the Front zone (i.e. northwest-trending), and are dominated by later cross-folds which are superimposed on the earlier structures. A zone some 15-70 km wide south of the Grenville Front is characterized by the development of map-scale structural interference patterns which are well-defined by the thin platformal sequence of marble-quartzite-iron formation in the Knob Lake Group, and also by intrusive gabbro sills of the Middle Proterozoic Shabogamo Intrusive Suite (10). Rivers and Chown (2) relate the polyphase deformation in the parautochthon to the emplacement of the overlying allochthons.

Grade of metamorphism in the parautochthon varies from greenschist facies in the north, to middle to upper amphibolite facies in the south. Rivers (11) reported the occurrence of a progressive sequence of Grenvillian metamorphism which spanned zones characterized by the following assemblages in pelitic and quartzofeldspathic rocks; (a) muscovite-chlorite; (b) muscovite-chlorite-biotite; (c) muscovite-chlorite-biotite-garnet; (d) muscovite-staurolite-kyanite; (e) muscovite-kyanite-garnet-biotite; (f) muscovite-kyanite-biotite-garnet-granitic veins (interpreted to be the result of *in situ* anatexis; (g) kyanite-K.feldspar-granitic veins; and (h) sillimanite-K.feldspar-granitic veins. With the recognition of the new tectonic zonation, some of these assemblages are assigned to different tectonic zones, and the progressive nature of the metamorphism is reassessed.

The autochthon is underlain by the lowest grade rocks in the sequence, and corresponds approximately to the muscovite-chlorite zone in Figure 2. Metagreywackes in this area occur as phyllitic schists with a strong cleavage, and massive to weakly-cleaved psammites. With the passage into the parautochthon, the rocks become pervasively deformed and recrystallized, and a progressive rise in metamorphic grade is recognized, which culminated in widespread anatexis (granite veins

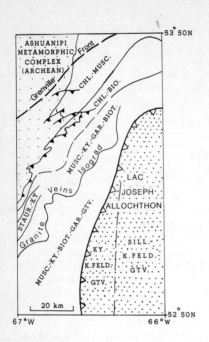

Figure 2. Wabush-Labrador City area (modified after 11). Metamorphic zones in the parautochthon and adjacent part of the Lac Joseph allochthon. CHL - chlorite, MUSC - muscovite, STAUR - staurolite, KY - kyanite, GAR - garnet, K.FELD - K.feldspar, BIOT - biotite, SILL - sillimanite, GTV - granite veins.

isograd, Figure 2). The occurrence of kyanite-bearing assemblages throughout much of the parautochthon, and especially in the zone spanning the granite veins isograd, implies pressures of about 6 Kb and temperatures of $650^{o}C$ (11). The two highest-grade metamorphic zones recognized by Rivers (11) are now known to be part of the Lac Joseph allochthon, and not the parautochthon as considered previously, and will be discussed in a later section.

3.2 Atikonak Lake Area

The Atikonak Lake area (Figure 3) spans the boundary between the parautochthon and the Lac Joseph and Wilson Lake allochthons. The parautochthon consists of pelitic paragneiss and foliated to gneissic granitoid rocks, the latter being regarded as deformed equivalents of the Trans-Labrador batholith (17, 12). The paragneiss is thought to be Labradorian in age on the basis of field observations in areas of low-grade Grenvillian metamorphism farther to the north (Figure 1) and northeast, where intrusive relationships of granitoid rocks into the paragneiss are seen (18). In the Atikonak Lake area the mineralogy and microstructure are considered to be a result of superposition of upper amphibolite facies metamorphism of Grenvillian age on Labradorian gneisses of similar metamorphic grade. The paragneiss is characterized by the assemblage quartz-two feldspars-sillimanite-magnetite-biotite and is extensively migmatitic. The leucosome is deformed into isoclinal folds with an associated penetrative axial planar fabric. Two subsequent deformations of this fabric resulted in moderate to tight folds and later

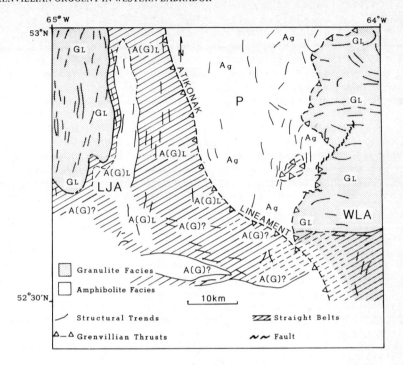

Figure 3. Atikonak Lake area (modified after 13). Map showing a boundary area between the parautochthon (P) and the Lac Joseph (LJA) and Wilson Lake (WLA) allochthons, the proposed facies and ages of metamorphism, and the ductile strain zones in the allochthons. A - amphibolite facies, G - granulite facies, (G) - relict granulite facies, 1 - Labradorian, g - Grenvillian, ? - unknown age of retrogression.

moderate to open folds, without associated fabrics, which developed under waning metamorphic conditions and at which time sillimanite was locally replaced by muscovite.

Deformation and metamorphism in the granitoid rocks of the Trans-Labrador batholith are considered to be Grenvillian in age, on the basis of age dates just north of the Atikonak Lake area (Figure 1), and a progressive metamorphic sequence has been recognized, which ranges from greenschist facies at the Grenville Front to granulite facies just southeast of this area (19, 7).

4. LAC JOSEPH ALLOCHTHON

The predominant rock-type in the Lac Joseph allochthon is pelitic to semipelitic paragneiss, similar to that in the parautochthon of the Atikonak Lake area, and intruded by large amounts of gabbronorite and minor granite. The paragneiss is in upper amphibolite to granulite facies and contains granitic veins composed of K.feldspar-quartz-plagioclase+magnetite+ biotite, and an aluminous restite assemblage of sillimanite-magnetite+biotite+garnet. In the west Rivers (11) has documented an amphibolite facies prograde sequence from kyanite- to sillimanite-bearing assemblages (Figure 2). In the east granulite facies paragneiss contains hypersthene, and intercalated metabasic layers are composed of two-pyroxene-bearing assemblages that are commonly net-veined or agmatitic, with plagioclase-quartz-biotite+ garnet+ orthopyroxene+K.feldspar leucosomes (12, 13). Hornblende-bearing rims on orthopyroxene characterize areas of amphibolite facies retrogression of unknown age in the south part of the Atikonak Lake area (Figure 3).

The gabbronorites are leucocratic, commonly layered and generally fine grained. They have a variably developed plagioclase and pyroxene mineral fabric, but may contain relict, pegmatitic, igneous textures. The granites are K.feldspar-megacrystic and possess a mineral foliation defined locally by orthopyroxene.

Along the eastern margin of the Lac Joseph allochthon, the structural history is very similar to that of the adjacent parautochthon (7). An early high-grade metamorphism and gneissosity was followed by the emplacement of plutonic rocks which themselves underwent further high-grade metamorphism at amphibolite to granulite facies. However, towards the end of this event, granulite facies rocks in the allochthon were thrust over their amphibolite facies counterparts along a series of ductile high-strain zones; shear deformation was accompanied by retrogression of the granulite facies rocks and the production of a very strongly lineated biotite-rich fabric (13). All this deformation is considered to be Labradorian since radiometric determinations on the granulite facies foliation yield dates of circa 1650 Ma (4, Figure 1), and structural elements are concordant between the granulite facies areas and the high-strain zones (12, 13). Hence it is interpreted that the metamorphism of these rocks is unrelated to the Grenvillian metamorphism in the parautochthon. However, preservation of Labradorian isotopic systematics during anhydrous Grenvillian recrystallization cannot be discounted at this stage.

The effects of Grenvillian deformation in the allochthon are poorly understood at present, but the regional set of late crossfolds with northwest to north-trending upright axial planes is considered to be a Grenvillian feature since it is superimposed on both the allochthon and the parautochthon.

4.1 Topographic and Geophysical Signatures

The boundary between the parautochthon and the allochthon is not exposed in the field, but is marked topographically by a break of slope with the result that the allochthonous rocks form a plateau some 150 m above the level of the parautochthon at its northwesterly leading edge. The difference in elevation is much less conspicuous 50 km farther south, perhaps indicating the location of the thinned or necked trailing edge of the nappe.

On aeromagnetic maps the parautochthon is clearly distinguished from

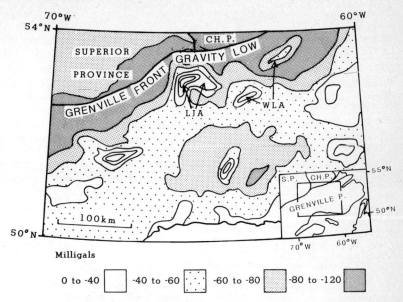

Figure 4. Bouguer gravity anomaly map of part of the eastern Grenville Province and adjacent foreland, showing the position of the Grenville Front Zone Low (adapted from 20). S.P. - Superior Province, CH.P. - Churchill Province, LJA - Lac Joseph allochthon, WLA - Wilson Lake allochthon.

the allochthon; the former has a subdued magnetic topography, contrasting strongly with the complex, swirly, high-relief topography of the allochthon. The boundary between the two is an aeromagnetic straight zone, which probably corresponds to a high-strain zone in the field: along the eastern margin of the Lac Joseph allochthon this feature has been called the Atikonak lineament by Nunn et al. (13). The magnetic relief of the allochthon is a result of the presence of abundant magnetite intergrown with sillimanite, which occurs as distinctive restite seams in the gneisses.

The lobate shape of the Lac Joseph allochthon (Figure 1) also conforms closely with the Bouguer gravity anomaly map (20, 21), and the allochthon is a dominant feature immediately south of the linear zone of gravity anomalies known as the Grenville Front Low (Figure 4). The allochthon itself is represented as a relatively positive element, consistent with the presence of higher density crust (composed of gabbronorite and quartzofeldspathic rocks in upper amphibolite and granulite facies). The profile of the allochthon in cross-section is a matter of conjecture at present, but the marked relief of the gravity anomaly suggests that it is broadly synformal in shape, and may have sunk into the underlying, less dense amphibolite-facies gneisses (cf. 22).

5. DISCUSSION

From the foregoing it is apparent that Grenvillian structural zones are transgressive across the pre-existing boundary between the Labrador and Hudson Orogens. Furthermore, in Figure 5 it can be seen that grade of metamorphism rises in the parautochthon from sub-biotite grade in the north to anatectic conditions some 20 - 40 km farther south. Since we have defined these isograds in sequences and units that were not previously metamorphosed, the metamorphism is ascribed to the Grenvillian orogeny. From the distribution of the isograds, it is clear that the grade rises southwards thereby indicating the direction of the axial zone of the orogen. No granulite facies rocks have been found in the parautochthon in the area of Figure 1, although Thomas et al. (16) interpret Grenvillian-age granulite facies assemblages south of the Wilson Lake allochthon to represent its southerly continuation.

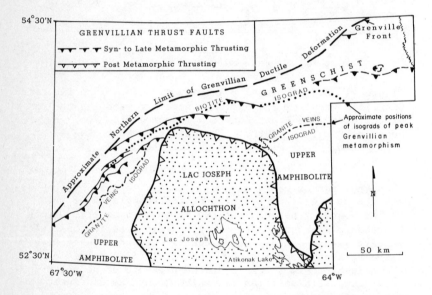

Figure 5. Map showing position and relative timing of thrust faults, and the location of the biotite and granite veins isograds of Grenvillian age. Compiled from mapping done for the Newfoundland Department of Mines and Energy by T. Rivers, G.A.G. Nunn, R.J. Wardle and N. Noel.

Figure 5 shows that the Grenvillian metamorphic zones are truncated by the post-metamorphic emplacement of the Lac Joseph allochthon, and it is clear that in 3-dimensions they are overlain by it. On the basis of concordant U/Pb data (3, 4) from the allochthonous rocks, it appears that the allochthon experienced only minor reworking during the Grenvillian orogeny. Thus it is suggested that the allochthon formed a tectonically-emplaced, suprastructural carapace over the parautochthon as the latter was undergoing penetrative tectonothermal reworking during the Grenvillian orogeny. The lack of observed Grenvillian metamorphic features in the allochthon may be ascribed to the anhydrous character of the previously deformed gneissic cover, which rendered it resistant to metamorphic reworking. However, retrograde Grenvillian shear zones might be expected (A(G)? in Figure 3 ?) to allow for upward fluid migration from the parautochthon beneath.

Emplacement of allochthons is considered to be partly responsible for the deep burial of the parautochthon during the Grenvillian orogeny, which resulted in its metamorphism and deformation. Hence it is likely that the allochthons themselves were reworked at least along their basal zones during their emplacement, and also later as the heat front migrated upwards across the parautochthon/allochthon boundary, although this has yet to be demonstrated. In the light of this hypothesis, it remains possible that the progressive metamorphic zonation (from kyanite-K.feldspar to sillimanite-K.feldspar) recorded along the western edge of the Lac Joseph allochthon (11) is indeed a Grenvillian feature. Further dating will be necessary to elucidate this problem. At present, lack of recent mapping in the central part of the allochthon holds back correlation of metamorphic history between the eastern and the western margins.

6. ACKNOWLEDGEMENTS

The authors wish to thank R.J. Wardle, E.H. Chown and A. Thomas for discussion and T.E. Krogh and C. Brooks for permission to use their preliminary radiometric data. The manuscript was edited by M.J. Murray and A. Davidson and typed by Mary Driscoll. L. Nolan, K. Staples and W. Marsh of the Memorial University of Newfoundland supplied drafting and photographic services. This work took place under the Canada-Newfoundland Mineral Development Subsidiary Agreement (1977-1981) and the Canada-Newfoundland Co-operative Mineral Program (1982-1984) in conjunction with the Geological Survey of Canada. The paper is published with the permission of the Director, Mineral Development Division, Newfoundland Department of Mines and Energy.

7. REFERENCES

(1) Wynne-Edwards, H.R. 1972, Geol. Assoc. Can., Special paper #11, pp. 264-334.
(2) Rivers, T. and Chown, E.H. in press, Geol. Assoc. Can., Special paper. New perspectives on the Grenville problem.
(3) Brooks, C. 1983, Rept. #4 (supp), Nfld. Dept. Mines and Energy, 16 p.
(4) Krogh, T.E. 1983, written communications.
(5) Gower, C.F. and Owen, V. 1984, Can. J. Earth Sci. **21**, pp. 678-693.
(6) Gower, C.F. this volume.
(7) Thomas, A., Nunn, G.A.G. and Wardle, R.J. this volume.

(8) Brooks, C., Wardle, R.J. and Rivers, T. 1981, Can. J. Earth Sci. **18**, pp. 1211-1277.
(9) Dallmeyer, R.D. and Rivers, T. 1983, Geochim, Cosmochim. Acta **47**, pp. 413-428.
(10) Rivers, T. 1983a, Precam. Res. **22**, pp. 41-73.
(11) Rivers, T. 1983b, Can. J. Earth Sci. **20**, pp. 1791-1804.
(12) Nunn, G.A.G. and Christopher, A. 1983, Geol. Surv. Can., Paper 83-1A, pp. 363-370
(13) Nunn, G.A.G., Noel, N. and Culshaw, N.G. 1984, Nfld. Dept. Mines and Energy, Min. Dev. Div., Rept. 84-1, pp. 30-41.
(14) Rivers, T. 1981, Can. J. Earth Sci. **17**, pp. 668-670.
(15) Wardle, R.J. and staff, 1982 Grenville Workshop, Ottawa-Carleton Centre for Geoscience Studies, Progr. + Abstr., p. 11.
(16) Thomas, A., Nunn, G.A.G. and Krogh, T., in press, Geol. Assoc. Can., Special paper. New perspectives on the Grenville problem.
(17) Wardle, R.J. and Britton, J.M. 1981, Nfld. Dept. Mines and Energy, Min. Dev. Div., Rept. 81-1, pp. 130-137.
(18) Thomas, A., Jackson, V. and Finn, G. 1981, Nfld. Dept. Mines and Energy, Min. Dev. Div., Rept. 81-1, pp. 111-120.
(19) Thomas, A., Culshaw, N.G., Mannard, G. and Whelan, G. 1984, Geol. Surv. Can., Paper 84-1A, pp. 485-493.
(20) Douglas, R.J.W. (ed.) 1970, Geol. Surv. Can., Econ. Geol. Rept. #1.
(21) Thomas, M.D. 1974, Energy, Mines and Resources, Earth Phys. Branch, Gravity Map Ser. 67.
(22) Lindia, F.M., Thomas, M.D., and Davidson, A. 1983, Geol. Surv. Can., Paper 83-1B, pp. 261-266.

GEOLOGICAL EVOLUTION OF THE ADIRONDACK MOUNTAINS: A REVIEW

J. M. McLelland
Department of Geology
Colgate University
Hamilton, New York 13346

Y. W. Isachsen
New York Geological Survey
State Museum and Science Service
Albany, New York 12230

ABSTRACT. The anomalous Tertiary(?) Adirondack dome exposes an oblique section across 10-12 km of middle to lower Proterozoic crust (from 17 km depth at 550° in the Northwest Lowlands to 25-30 km at 750-800°C in the Central Highlands). Platform sediments, migmatite, granitic to quartz monzonitic gneisses, and basal leucogranitic gneisses dominate the Lowlands, whereas granitic, charnockitic and anorthosite gneisses exceed metasediments by more than 2:1 in the Highlands. Anorthosite emplacement is pre- or early syntectonic. The Carthage-Colton mylonite zone separates these terranes, but the metamorphic gradient may be smooth across it.

A coherent stratigraphic section of folded rocks may cross the entire Adirondacks. Five phases of folding, from isoclinal to open, followed. Strong elongation lineations parallel early isoclinal fold axes and indicate regional ductile rotational strain with a tentatively assigned E over W sense of shear. Peak metamorphism (1.1 to 1.02 Ma) outlasted third phase folding in the Highlands and second phase folding in the Lowlands.

The 6-8 kb mineral equilibria in the Adirondacks and adjoining Central Granulite Terrane of Ontario, coupled with a normal crustal thickness today, suggest a double-thickened crust during the Grenville Orogeny, barring significant underplating. Three models for the Grenville Province have been proposed: 1) Continental collision and segmental underthrusting of a cold continental slab to produce double-thickened crust beneath a Tibetan-type plateau, convective heating by anorthosite, granite (sensu lato) and gabbro, gravity tectonics and formation of nappes, diapirs, domes and mylonites, erosion and isostatic recovery followed by inheritance of original plateau fracture pattern as rising mass enters brittle regime. 2) Continental arc magmatism, lower crustal level. 3) Crustal extension, anorogenic intrusion of anorthosite and mangerite-charnockite suite, northwestward continental overthrusting and mylonitization, vertical rise of rigid anorthosite and deflection of earlier structures. Crustal compression is supported by rock fabric and structures discussed in this paper. Preorogenic history involved deposition of a carbonate-clastic sequence on a wide platform, with local hypersaline environments indicated by anhydrite and stromatolites.

1. INTRODUCTION

The northeast trending, elliptical Adirondack dome forms a southeastern extension of the Grenville Province with which it connects via the Frontenac Arch (Fig. 1). The region has been physiographically subdivided into the Northwest Lowlands, underlain principally by metasedimentary rocks, and the anticlinorial Highlands, underlain principally by metaplutonic rocks with intervening synclines of metasediment. Separating the Lowlands from the Highlands is the 110 km-long Carthage-Colton mylonite zone (Geraghty and others, 1981) characterized by intense ductile strain and igneous intrusion (Figs. 1,2). Within the eastern Highlands several large bodies of anorthosite are exposed; the most notable being the Marcy massif that underlies the high peak region of the Adirondacks (Figs. 1,2).

This review describes the more salient aspects of Adirondack lithic types and discusses the possible correlation of layered sequences, as well as the evolution of metamorphosed igneous rocks and mineral deposits. Geochronologic results and interpretations are summarized and regional metamorphism is discussed. Descriptions of both ductile and brittle deformation are presented and several reconstructions of tectonic history are reviewed.

2. ADIRONDACK LITHIC SEQUENCE

Because of pervasive, intense deformation, the extent of tectonic stacking and concordant igneous intrusion remains unknown. However, lithotectonic sequences have been locally defined and regionally correlated (Fig. 3). Whether or not these represent true stratigraphies is debatable, but they may, nonetheless, be utilized for tracing out the structural geology of the region. In several instances along strike lithologic variations interpreted as primary facies changes may prove to be due to subtle tectonic truncations. However, this alternative remains to be demonstrated, and a facies change model will continue to be employed until tectonic truncations are reasonably well documented.

The areal extent of the principal Adirondack rock types is given in Table I. Complete descriptions of Adirondack lithologies and stratigraphic nomenclature are given in McLelland and Isachsen (1980) and Wiener and others (1984). Here we briefly summarize the salient aspects of the lithic sequence. The discussion will proceed according to the two major lithological subdivisions of the Adirondacks (Figs. 2,3,4): the basal Piseco Group, and the overlying layered sequence of

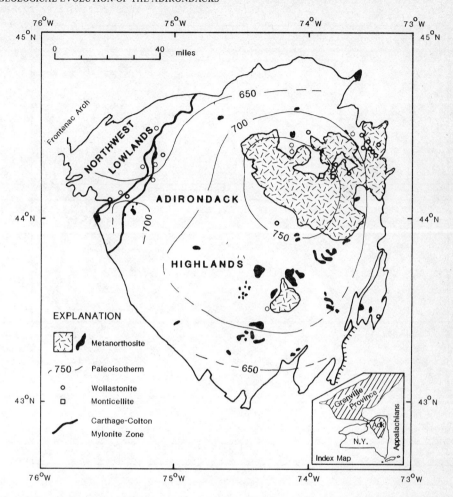

Figure 1. Location map of the Adirondacks showing the principal geologic subdivisions. Paleoisotherms are taken from Bohlen and others (1980). Wollastonite and monticellite localities are from Valley and Essene (1980) and Valley (this volume). Hachured line represents the McGregor border fault of the eastern Adirondacks.

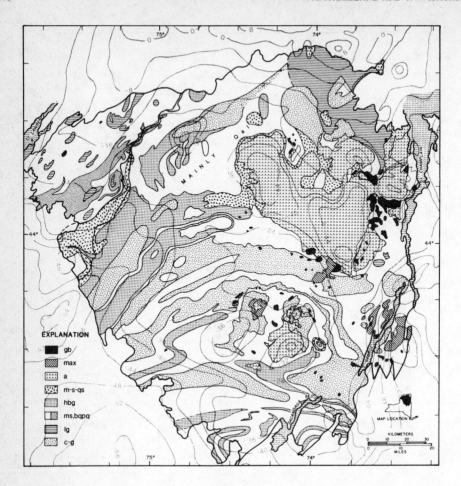

Figure 2. Generalized geologic map of the Adirondacks. Bouguer anomaly contours are from Simmons (1964). Strong negative anomalies associated with the Marcy massif center on values of −44 to −48 milligals. Legend: c-g: charnockitic and granitic gneisses assigned to basal Piseco Group including Brant Lake Gneiss in the Highlands. lg: in the Northwest Lowlands refers to the Alexandria Bay Leucogranitic Gneiss of the Piseco Group; in the northeastern Highlands refers to lithologically similar, but stratigraphically unassigned leucogranitic gneisses of the Lyon Mt. Fm. ms: undivided metasediments including marbles, quartzites, and metapelites; bqpq: within the Lowlands only biotite-quartz-plagioclase metapelites and migmatites of the Poplar Hill Fm. (Major Paragneiss). hbg: hornblende granitic gneiss; m-s-qs: mangeritic, syenitic, and quartz-syenitic gneiss; a: metanorthosite; max: charnockitic and mangeritic gneisses containing xenocrysts of blue-grey andesine plagioclase. In the Lake George region these occur in thrust sheets discussed in text; gb: olivine metagabbro.

TABLE 1. AREAL EXTENTS OF PRECAMBRIAN LITHIC UNITS IN THE ADIRONDACKS AS SHOWN ON GEOLOGIC MAP OF NEW YORK (FISHER AND OTHERS, 1971)

Lithologic type	Total Adirondacks			Adirondack Highlands			Northwest Lowlands		
	Area (%)	Area (km²)	Area (mi²)	Area (%)	Area (km²)	Area (mi²)	Area (%)	Area (km²)	Area (mi²)
Metagabbro	0.1	42	18	0.1	35	15	0.3	7	3
Leucogranitic gneiss	3.1	825	320	1.9	465	180	16.2	360	140
Hornblende-(biotite) granitic gneiss	23.8	6,410	2,470	24.1	5,970	2,300	19.8	440	170
Syenitic gneiss (mangerite in Highlands)	2.9	772	297	3.0	740	285	1.4	32	12
Charnockitic gneiss	23.8	6,413	2,475	25.9	6,400	2,470	0.5	13	5
Meta-anorthosite	14.0	3,750	1,450	15.2	3,750	1,450	--	--	--
Metasedimentary rocks, amphibolite, mixed gneisses	32.4	8,740	3,370	29.8	7,360	2,840	61.8	1,380	530
TOTALS	100.0	26,952	10,400	100.0	24,720	9,540	100.0	2,232	860

NOTE: Accuracy of planimetric measurements is slightly better than 1% (by summing areas).

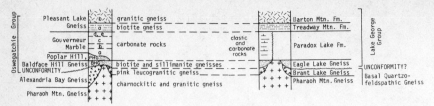

Figure 3. Correlation chart for the Adirondacks (after Wiener and others, 1984). Lowlands sequence shown on left, Highlands on right.

the Oswegatchie and Lake George Groups in the western and eastern Adirondacks respectively (Wiener and others, 1984).

To date, the only reliable facing-direction criterion in the Adirondacks are domal stromatolites described by Isachsen and Landing (1983) in the Northwest Lowlands. deWaard (1964) utilized the laccolithic shape of a metagabbro to assign tops. This limited evidence forms the basis for arranging the units to be described below in an order from oldest to youngest.

2.1 Piseco Group

Rocks of the Piseco Group have been identified throughout the Adirondacks (Fig. 2). In the Highlands they are represented by the lower Pharoah Mt. Gneiss consisting of charnockitic and granitic gneiss together with lesser amounts of mangerite. Locally these are overlain by leucogranitic gneisses with amphibolite interlayers known as the Brant Lake Gneiss. Occasional trondjemitic varieties occur and magnetite is an important accessory. The Pharoah Mt. Gneiss does not crop out within the Northwest Lowlands, but lithologies similar to the Brant Lake Gneiss, exposed within 14 domical culminations, are assigned to the basal Alexandria Bay Leucogranitic Gneiss (Fig. 2). Carl and Vandiver (1975) have interpreted these alaskitic bodies as metamorphosed rhyolite-dacite ash flow tuffs based on their compositions, but no relict volcanic textures have yet been recognized.

The Lyon Mt. Gneiss (Postel, 1952) of the northeastern Adirondacks is lithologically similar to the Alexandria Bay Leucogranitic gneiss and the Brant Lake gneiss, but its stratigraphic position has not yet been determined (Fig. 2).

2.2 Layered Sequence

2.2.1 Oswegatchie Group. This group constitutes the layered sequence overlying the Alexandria Bay Leucogranitic Gneiss within the Northwest Lowlands (Figs. 2,3,4). The Group is tentatively correlated with the Lake George Group of the Highlands, but since unambiguous documentation for this correlation does not yet exist, the two groups are treated separately herein.

The basal unit of the Oswegatchie Group is the Baldface Hill

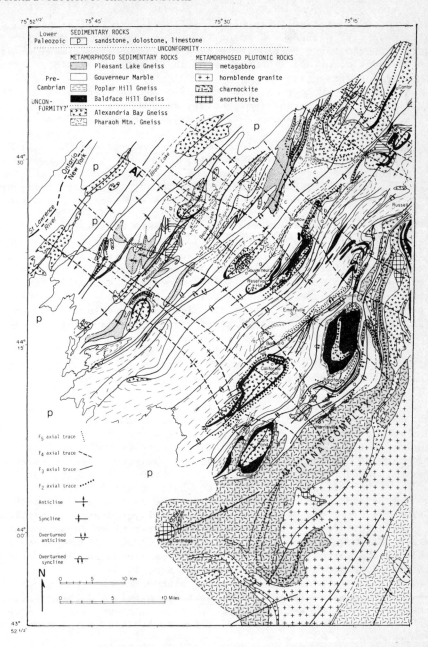

Figure 4. Geologic map of the Northwest Lowlands and parts of the adjoining Northwest Highlands. Diana Complex includes mangeritic as well as charnockitic gneiss (from Wiener and others, 1984).

Gneiss which consists principally of sillimanite-garnet-biotite-quartz-feldspar gneisses. Cordierite is locally present and magnetite occurs throughout the section. Thin, discontinuous sillimanite-garnet rich gneisses occur near the bottom of this unit and are interpreted as a metamorphosed regolith.

According to Wiener and others (1984) the Baldface Hill Gneiss is overlain by a biotite-quartz-oligoclase gneiss, designated Poplar Hill Gneiss. This unit corresponds to the Major Paragneiss of Engel and Engel (1953) which they, and others, placed higher in the sequence with a marble unit (Lower Marble) intervening between it and the sillimanitic rich gneisses of the Baldface Hill.

Wiener and others (1984) place the thick Gouverneur Marble above the Poplar Hill Gneiss. Earlier workers (Engel and Engel, 1953; Lewis, 1969) had divided the Gouverneur Marble into two separate formations, i.e. the Upper and Lower Marble, one above and one below the Poplar Hill Gneiss (Major Paragneiss). The one marble hypothesis requires either a tectonic slide (Fleuty, 1964) as proposed by Foose and Brown (1976) or a major isoclinal fold axis as hypothesized by Wiener and others (1984). Lithologically, the Gouverneur Marble consists mainly of calcite and dolomite marbles, with subordinate calcsilicates, quartzites, metapelite and quartz-feldspar gneisses. An evaporite environment is suggested by anhydrite which comprises local, mappable units, halite in veins, tourmaline rich units, lapis lazuli, and stromatolites.

The Gouverneur Marble is overlain by migmatitic Pleasant Lake Gneiss consisting of biotite-hornblende-quartz-feldspar gneiss with thin interbeds of marble, quartzite, and rusty gneisses.

2.2.2 <u>Lake George Group</u>. The layered sequence within the Adirondack Highlands is referred to as the Lake George Group (Figs. 2,3). Its basal member is a thin, discontinuous sillimanite-biotite-quartz-feldspar gneiss known as the Eagle Lake Gneiss (formerly the Older Paragneiss of Walton and deWaard (1963)). This unit is interpreted as a metamorphosed regolith above the Piseco Group and is correlated with similar lithologies at the base of the Baldface Hill Gneiss.

The Paradox Lake Formation overlies the Eagle Lake Gneiss and consists of calcitic and dolomitic marble interlayered with calcsilicates and quartzite. It is correlated with the Cedar River Formation of the central Adirondacks (McLelland and Isachsen, 1980), the Gouverneur Marble of the Northwest Lowlands, and the Sacandaga Formation of the southern Adirondacks (McLelland and Isachsen, 1980).

Above the Paradox Lake Formation, within the eastern Adirondacks, there occurs a migmatitic biotite-quartz-plagioclase with variable garnet, sillimanite, and graphite. This unit, known as the Treadway Mt. Formation, thins westward and is not recognized in the central Adirondacks.

Overlying the Treadway Mt. Formation are streaky green and pink granitic gneisses with metasedimentary interlayers. These are referred to as the Lake Durant Formation (deWaard, 1974) and are correlated with the Pleasant Lake Gneiss of the Northwest Lowlands.

The Spring Hill Pond Formation structurally overlies the Lake

Durant Fm. in the eastern Adirondacks. It consists of graphitic garnet-sillimanite-quartz-feldspar gneiss, quartzite, and marble. Local concentrations of graphite were mined in the past (Alling, 1917). This unit correlates with the Blue Mt. Lake Formation of the central Adirondacks (McLelland and Isachsen, 1980), with the Tomany Mt. Formation of the Southern Adirondacks (McLelland and Isachsen, 1980) and possibly with the Upper Marble of the Northwest Lowlands. If the latter is correct a major isoclinal axis must pass through the Lake Durant Formation making the Springhill Pond and Paradox Lake Formations the same.

Structurally above the Springhill Pond Formation is the Thunderbolt Mt. Formation (Walton and deWaard, 1963) consisting of charnockitic and granitic gneisses with metasedimentary interlayers. This unit correlates with the Little Moose Mt. Formation (McLelland and Isachsen (1980), in the central Adirondacks and with the Rooster Hill Formation in the southern Adirondacks (McLelland and Isachsen, 1980).

Overlying the Rooster Hill Formation in the southern Adirondacks are a sequence of formations (McLelland and Isachsen, 1980) not exposed to the north, although future research may result in correlations around as yet unrecognized folds and/or tectonic slides. From lower to upper, these units are the Peck Lake Formation consisting of migmatic garnet-biotite-quartz feldspar gneisses with sporadic sillimanite; an overlying thick section of interlayered garnetiferous leucogneiss, quartzite, and amphibolite known as the Green Lake Formation; thence thick quartzofeldspathic gneisses of the Canada Lake Charnockite and finally quartzites and garnetiferous quartzites of the Irving Pond Formation. These quartzites may correlate with similar lithologies within the Springhill Pond Formation, but this is as yet undemonstrated.

2.3 Metamorphosed Intrusive Rocks

The intrusive nature of these rocks is variously reflected by their textures, cross-cutting relationships, and the presence of rotated xenoliths.

2.3.1 Metanorthosites. Metamorphosed anorthositic rocks underly much of the central and northern Adirondacks (Figs. 1,2) and range from coarse-grained andesine anorthosite to anorthositic gabbro together with late iron-rich differentiates. The major anorthosite massifs in the Adirondacks are characterized by strong negative gravity (Simmons, 1964) and aeromagnetic anomalies (Gilbert and Zietz, 1968). Therefore strong negative gravity and aeromagnetic anomalies east of Lake George (Fig. 2) suggest a shallowly buried anorthosite mass there. This is supported by the presence of small bodies found by Hills (1964) and by recent mapping.

Interpretation of gravity anomalies associated with the Marcy massif suggests that the most plausible shape of the body is a 3-5 km thick slab with at least two deep extensions that may be feeder pipes (Simmons, 1964) or diapiric extensions. Other anorthosites in the Adirondacks may have similar sheet or slab-like shapes. However,

Morse (1969, 1982) has emphasized that these gravity interpretations depend critically on small differences in assumed average density. Therefore, other shapes are not ruled out.

The anorthosite masses are clearly intrusive into the layered sequences of the Oswegatchie and Lake George Groups (Isachsen and others, 1973). Oxygen isotope data of Valley and O'Neil (1982; Valley, 1985, this volume) from calcsilicates in contact with anorthosite at Willsboro and Cascade Slide (Fig. 5) appear to constrain emplacement to depths not greater than 10 km and probably less than 7 km. It remains to be seen if similar low $\delta^{18}O$ values will be found for the main massifs, especially the Marcy, because iron-rich inverted pigeonite in associated charnockite has been interpreted as indicative of minimum crystallization pressures of 8 kb (Ollila and others, 1984).

The composition of the parent magma of Adirondack anorthosites has long been debated and the arguments are summarized in Isachsen (1969). More recently Emslie (1978) has proposed that anorthosite evolves from gabbroic magmas that become ponded and crystallize olivine and aluminous pyroxenes at the crust-mantle interface and then rise into the crust as high-alumina tholeiites. These differentiate to plagioclase charged crystal mushes which are emplaced as leuconoritic and anorthositic magmas. Morse (1982) suggests that high-alumina gabbro may differentiate to anorthositic liquids by metastable supersaturation of plagioclase during slow ascent. Both models infer emplacement within anorogenic settings. Wiebe (1978) has presented evidence for the existence of anorthositic liquids in the Nain Province. These magmas have compositions similar to the gabbroic anorthosites that Buddington (1939) believed to be parental to Adirondack anorthosites. Presently known rare earth and other trace element data indicate a mantle source for the Marcy anorthosite, similar to that for MORB's (Ashwal and Wooden, 1983).

2.3.2 <u>Mangeritic and Quartz Syenitic Gneisses</u>. Mangerite and quartz-syenitic gneisses tend to be localized at borders of the anorthosite massifs (Fig. 2) where they intrude the anorthosite and contain xenoliths and xenocrysts derived from it. They have been variously interpreted as post-anorthosite intrusive (Buddington, 1939); differentiates from a granodioritic magma (deWaard, 1968); and as contact anatectitic melts (Isachsen, 1969; Isachsen and others, 1973; Ashwal, 1978; Emslie, 1978); or as deep crustal anatectites that left anorthositic restites behind (DeVore, 1979). Field relationships (Buddington, 1967) and trace element patterns (Ashwal and Wooden, 1983; Basu and Pettingill, 1983) appear to rule out models involving comagmatic relationships with the anorthosites.

2.3.3 <u>Hornblende Granitic Gneiss</u>. Pink to gray, medium grained hornblende-biotite granitic gneiss is common in the Adirondacks, particularly in the western Highlands (Fig. 2). A megacrystic variety in the Lowlands was named the Hermon granite gneiss by Buddington (1939). Local cross-cutting relationships and angular relict euhedral phenocrysts provide strong evidence for the intrusive origin of these rocks. They have been omitted from Fig. 4 which emphasizes stratigraphic units.

Numerous sheets of equigranular and megacrystic pink granitic gneiss occur throughout the Adirondacks. Many of these are probably intrusive, but clearly cross-cutting relationships are very rare.

2.3.4 <u>Olivine Metagabbro</u>. Figure 2 shows the location of the major bodies of olivine metagabbro in the Adirondacks. These tholeiite to olivine tholeiite intrusives appear to concentrate in the vicinity of the larger bodies of anorthosite (Fig. 2). This spatial association is not likely to be coincidental, and it is suggested that if Emslie's (1978) model of anorthosite genesis is applicable, these rocks may represent high level examples of tholeiitic magma that became ponded at the crust-mantle interface.

The olivine metagabbros exhibit a wide variety of coronas, the most common being those cored by olivine or oxides. Both hydrous and anhydrous reaction rims are common in a single specimen. These have been discussed by McLelland and Whitney (1977, 1980a, 1980b) and by Whitney and McLelland (1973, 1983). Detailed chemical investigations indicate that these coronas form in closed systems with limited transport of mobile species. Reactions written in aluminum-fixed frames of reference demonstrate that coronas within a single thin section interact with one another and form mass-balanced, closed systems. Spinel clouding and anorthite depletion of plagioclase are an integral part of corona formation.

3. ISOTOPIC AGE RELATIONSHIPS

Although a large number of isotopic ages have been determined within the Adirondacks, their interpretation is made difficult by the high metamorphic grade that affected the region. Many systems may have been reset during high grade metamorphism and closure may substantially post-date emplacement.

Silver (1969) obtained U/Pb ages on zircons from anorthositic rocks and charnockites in the Adirondack Highlands. The former yielded ages of 1074-1005±10 Ma, and based upon textural relationships of the rocks as well as their zircons, Silver interpreted these ages as minimum and probably metamorphic. U/Pb ages for charnockitic rocks yielded a concordia intercept age of 1113±10 Ma which Silver interpreted as the age of crystallization of the charnockites. Because of then presumed comagmatic relationships between the anorthositic and charnockitic rocks, he considered 1113±10 Ma to be the probable age of emplacement of the anorthosites.

Rb/Sr isochrons determined on charnockitic and granitic rocks exist for a number of localities in the Adirondacks. In the Highlands, these range from an age of 1174±14 Ma from the Snowy Mt. Dome (Hills and Isachsen, 1975) to 1075±20 Ma obtained from charnockitic rocks near the south end of Lake George (Hills and Gast, 1964). Bickford and Turner (1971) found ages of 1095±39 Ma and 1120±11 Ma for granitic gneiss of presumed anatectic origin near Brant Lake. Heath and Fairbairn (1969) obtained an age of 1033±31 Ma for mangeritic rocks of the Tupper-Saranac and Loon Lake Complexes.

The spread of Rb/Sr isotopic ages in the Highlands suggests that these ages date closure rather than emplacement. Similar conclusions have been reached by Grant and others (1984) for gneisses from the Northwest Lowlands which yield ages ranging from 1297±25 to 950 Ma. These ages are similar to the 1265±25 Ma reported by Grant and others (1981) for the basal Alexandria Bay Leucogranitic Gneiss and a zircon age of 1202±15 Ma obtained by Silver (1965) for quartzofeldspathic gneiss in the western Highlands.

Recently Sm/Nd methods have been applied to rocks of the anorthosite-charnockite suite by Basu and Pettingill (1983) and Ashwal and Wooden (1983). The former study yielded an isochron age of 1098±7 balanced by ratios from garnet. This strongly implies that the age determined is metamorphic. Ashwal and Wooden (1983) obtained a Sm/Nd isochron age of 1288±36 Ma from relatively pristine igneous rocks in a layered sequence from the southern part of the Marcy massif. Their isochrons from other rocks yielded younger ages down to 900 Ma and provide a 300 million year window for anorthosite crystallization. They consider that most of these ages are reset or date closure and that 1288±36 Ma represents the minimum age of emplacement of the anorthositic rocks.

If Ashwal and Wooden's (1983) conclusions are correct, and if Silver's 1113±10 Ma zircon age dates charnockite emplacement, then a very considerable time interval separates the charnockites and anorthosites. Also implied is a hiatus of 200-300 my between anorthosite emplacement and peak Grenvillian metamorphism at 1100-1000 Ma. Given this, the presence of rotated, foliated xenoliths in the anorthosites suggests a pre-Grenvillian orogeny. This matter is clearly of great significance and requires further clarification.

Although the interpretation of Adirondack isotopic ages is still open to debate, the older dates are consistent with the 1290±15 Ma zircon age obtained by Silver and Lumbers (1966) for Ontario paragneisses unconformably overlying the 1400-1500 Ma Algonquin batholith (Lumbers, 1979) and are related to intrusives dated at 1280 Ma (Rankin and others, 1983). The possibility is opened up that rocks of the Adirondacks accumulated 1400-1500 Ma ago and were intruded by anorthosites at shallow levels prior to the peak Grenvillian metamorphism 1100-1000 Ma ago.

4. MINERAL DEPOSITS OF THE ADIRONDACKS

The nature and origin of mineral deposits in the Adirondacks are integral parts of its Proterozoic evolution. Significant deposits are found in both the Adirondack Highlands and Lowlands. They represent not only considerable economic resources but also provide an important part of the data-base for geologic understanding of the region.

4.1 Metallic Deposits

4.1.1 Magmatic segregations of ilmenite-magnetite in massif anorthosites, pyroxene-granulite facies. Although several small

deposits accompany metagabbro, anorthositic rocks, and mafic syenitic gneiss in the Northwest Adirondacks (Leonard and Buddington, 1964), the largest deposits occur as sheets and lenses in anorthosites of the Marcy massif near Sanford Lake, N.Y. (Fig. 5). The ore minerals consist of both discrete ilmenite, and magnetite with exsolution lamellae of ilmenite. The ore occurs both as disseminated ore and in cross-cutting veins. Mineral chemistry (Kelly, 1979) and REE patterns (Ashwal, 1978) favor a late magmatic origin.

4.1.2 <u>Stratabound (Metasedimentary?) iron oxide deposits in gneisses at hornblende-granulite facies</u>. These non-titaniferous magnetite deposits occur principally within the leucogranitic Brant Lake and Lyon Mt. Leucogranitic Gneiss of the Highlands and the Baldface Hill Gneiss of the western Highlands.

The deposits consist of magnetite-rich layers and lenses that show a marked parallelism with the contacts and foliation of enclosing gneisses and have undergone isoclinal folding along with the rest of the sequence. The largest body, Benson Mines (Fig. 5), is located in the western Highlands. It is 11 km long from 6-60 meters wide and forms a tabular, hook shaped body on the east limb of a refolded isoclinal fold.

The stratiform and concordant configuration of the magnetite bodies, and their position near the base of the supracrustal section, suggest that they may represent metamorphosed iron-rich sediments.

4.1.3 <u>Stratabound (Metasedimentary?) sulfide deposits in marble-calcsilicate formations at amphibolite facies</u>. The most important of these are the stratabound sphalerite-pyrite-galena ores that occur in a multiply-folded sequence of anhydrite and dolomitic marbles in the Balmat-Edwards district of the Northwest Lowlands (Fig. 4).

Buddington and others (1969) and Whelan and DeLorraine (1984) have presented strong arguments for a sedimentary diagenetic origin followed by remobilization during deformation and regional metamorphism at sillimanite K-feldspar grade.

Other sulfide deposits in the Lowlands include pyrite-pyrrhotite which occur in rusty graphitic gneisses at sillimanite-muscovite grade. They exhibit a consistent stratigraphic position and appear to be of sedimentary-metamorphic derivation (Prucha, 1956).

4.2 Non-metallic Deposits

4.2.1 <u>Reconstituted metamorphic rocks</u>. <u>Wollastonite</u> deposits are found in close proximity to anorthositic rocks and the intrusive Diana Complex (Fig. 4). The largest and best known deposit occurs near Willsboro, N.Y. where wollastonite is developed in a tabular inclusion or infold of calcsilicate rock in metanorthosite. The ore contains approximately 65% wollastonite as well as grossularitic garnet and calcic pyroxene. This inferred metasomatic segregation may have been facilitated by the influx of meteoric waters as suggested by the extremely low $\delta^{18}O$ found in the wollastonite-clinopyroxene-garnet assemblages by Valley and O'Neil (1982).

Buddington (1939), Putman (1958), DeRudder (1962), and Valley (1985, this volume) consider the Willsboro wollastonite to be a variably metamorphosed skarn. The absence of quartz or calcite suggests either a perfect original stoichiometric mix or, more likely, that metasomatism played the major role in wollastonite formation. Valley and Essene (1980) have demonstrated the importance of compositional variations in metamorphic fluids in determining the occurrence of wollastonite and other calcsilicate minerals (see Section V, Metamorphism).

Talc-tremolite deposits of the Balmat-Edwards district (Fig. 4) are the largest of their type in the world. The deposits, which are conformably interlayered with units in the Gouverneur Marble, exhibit a retrograde paragenetic succession (diopside-forsterite-tremolite-anthophyllite-serpentine-talc). Engel and Engel (1962) interpreted the deposits as due to the retrograde metamorphism of siliceous magnesite-dolomite beds with varying degrees of metasomatism.

Garnet deposits occur at Gore Mt. (Fig. 5) and are unique for the size of garnet porphyroblasts (average diameter = 30 cm; largest garnet found approx. 100 cm). The garnets occur in a coarse amphibolite which formed from a typical olivine metagabbro, portions of which are still present in the walls of the open-pit mine (Bartholome, 1956). The large size of the garnets may be the result of limited nucleation sites together with local, abnormally high PH_2O at granulite facies grade (Luther, 1976). The water may have been introduced along a large adjacent fault during metamorphism.

4.2.2 Pegmatite Dikes. Straight walled and clearly discordant pegmatite occur in the Adirondacks and have been commercially exploited for feldspar. Most are found near the southern and eastern margin and have been described by Tan (1966). Mineral assemblages in these pegmatites indicate a decreasing thermal gradient from 640-700°C to 525-600°C from the east-central to the southern Adirondacks. These temperatures are in agreement with those of Bohlen and others (1980) and suggest that the pegmatites were intruded near the time of peak metamorphism but after deformation had ceased. The inferred date for this emplacement based on a Rb/Sr age on muscovite in a pegmatite from the southern Adirondacks is 930±40 my (B. Giletti, unpub. on specimen no. 2112 in New York State Museum Collection).

5. STRUCTURAL GEOLOGY

The Adirondacks are characterized by regional ductile deformation resulting in multiple, large scale folding (Figs. 4,5) manifested by at least five fold sets distinguishable on the basis of style and orientation. The two earliest fold sets are recumbent isoclines that may be accompanied by tectonic slides and thrusts. Second generation isoclines form large, nappe-like structures. These are followed by more open, upright folds, the first set of which is commonly coaxial with earlier isoclinal folds. A strong planar and linear fabric is de-

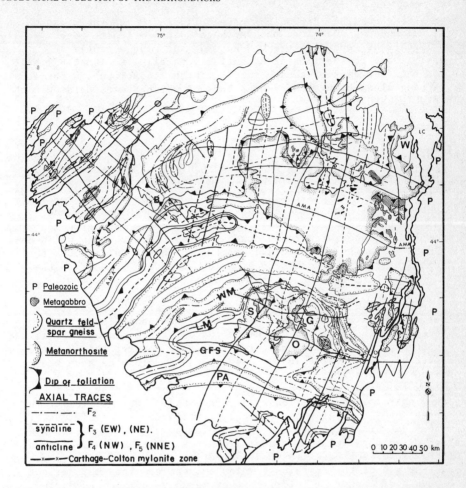

Figure 5. Axial trace map for the Adirondack Highlands and Lowlands.
AMA: Arab Mt. anticline; C: Canada Lake syncline; G: Gore Mt.; LM:
Little Moose Mt. syncline; O: Oregon Dome; PA: Piseco anticline; S:
Snowy Mt. Dome; W: Willsboro; WM: Wakeley Mt. nappe; P: Paleozoic; LC:
Lake Champlain; LG: Lake George; B: Benson Mines; GFS: Glens Falls
syncline

veloped throughout the Adirondacks especially in the southern and
central Adirondacks and along the Carthage-Colton mylonite zone.
Ribbon lineations generally trend E-W to NW and plunge gently to
moderately about the horizontal essentially parallel to isoclinal fold
axes and the tube axes of sheath folds. These lineations are particularly strong along high strain zones some of which are associated with
demonstrable thrusting.

5.1 Folds

Although each of the five fold sets is distinct, nothing is known about
the time intervals separating them, and they may partly or wholly,
represent phases of an orogenic continuum. Nonetheless, it is convenient to label these fold sets as F_1-F_5 and to discuss them accordingly.
Representatives of each set are present in both the Highlands and
Lowlands and have been tentatively correlated across the Carthage-
Colton mylonite zone (Wiener and others, 1984). Figure 4 shows in detail geological units and fold axes of the Northwest Adirondacks that
are typical of the Adirondacks as a whole. Figure 5 summarizes the
axial trace pattern for the entire Adirondacks.

5.1.1 F_1-Folds. Within the Highlands F_1 folds are recognizable only
as minor, intrafolial isoclines. Within the Lowlands Wiener (1983)
and Wiener and others (1984) postulate the existence of major F_1 folds
based upon apparent stratigraphic repetitions. However this hypothesis
remains to be demonstrated and no unequivocal F_1 closures have been
recognized at map scale.

Minor folds of the F_1 generation are characterized by the fact
that they only fold compositional layering (S_0). Regional foliation
(S_1) appears to be axial planar to the tightly appressed, similar F_1
folds that exhibit attenuated, disrupted limbs. Their axes lie
parallel or subparallel to regional lineation. Wiener (1983) assigns
a northwest over southeast movement direction to these folds.

The origin of the F_1 folds remains enigmatic. They may reflect
the presence of larger F_1 folds, or they may have originated during
an early thrusting episode, for example.

5.1.2 F_2-Folds. The earliest mappable folds consist of unusually
large isoclinal structures whose overall attitude is recumbent or
reclined (Figs. 4,5). They fold both S_0 and S_1 and exhibit a strong
axial planar foliation of their own (S_2). It is generally impossible
to distinguish S_1 from S_2 except in the axial region of F_2 folds.
Both foliations are formed by compositional layering and aligned
minerals generally oriented parallel or subparallel to compositional
layering.

Within the southern and central Highlands F_2 axes form a broad
E-W arc (Fig. 5) and plunge gently at 10-15° about the horizontal.
Strong E-W mineral and ribbon lineations parallel these axes. The
largest F_2 structures are the Wakeley Mt. nappe, the Little Moose Mt.
syncline, and the Canada Lake syncline (Fig. 5). The latter two
folds are interpreted as portions of a single, extremely large iso-

clinal syncline whose axial trace forms a loop, the closures of which lie beneath Paleozoic cover on the eastern and western flanks of the Adirondacks (Wiener and others, 1984). The amplitude of this fold may exceed 70 km., and prior to erosion it was overlain by the Wakeley Mt. nappe. It may be that low angle detachment surfaces developed along the lower limb of the Wakeley Mt. nappe; but, to date, mapping has failed to reveal any discontinuities of this nature.

In the western Adirondack Highlands, a number of F_2 axial traces tend to swing into parallelism with the Carthage-Colton mylonite zone as it is approached from the east (Fig. 5). This parallelism is poorly understood but is not believed to be coincidental. Portions of the western Highlands are currently being mapped by the New York State Geological Survey, and this investigation should help to resolve the matter.

Within the Adirondack Lowlands F_2 folds are isoclinal and exhibit axial traces trending NNE-NE (Figs. 4,5). Their axes trend N-S to E-W indicating a moderately reclined nature. Mineral lineations are oriented subparallel to these axes as are ribbon lineations along the Carthage-Colton mylonite zone. Wiener (1983) assigned a southeastward directed tectonic transport direction to these folds, but the case is uncertain, and west to northwest tectonic transport remains possible.

In both the Highlands and Lowlands F_2 folds exhibit a geometry in which relatively incompetent lithologies are greatly thickened in hinge areas and thinned on fold limbs, while competent units such as quartzites and charnockites retain approximately constant thickness throughout.

5.1.3 F_3-Folds. These are generally open and upright structures, although within the Lowlands some are relatively tight (Figs. 4,5). Axial planar foliation to these folds is generally poorly developed.

Within the southern and central Adirondacks F_3 folds of unusually large dimensions dominate outcrop patterns. The largest of these are the Piseco anticline, the Glens Falls syncline, and the Gloversville syncline (Fig. 5). Like the largest F_2 structures, these folds have axial traces exceeding 100 km. They are coaxial with F_2 folds; a relationship considered to be of critical importance in understanding the tectonic evolution of the region.

As with F_2 folds, the axial traces of F_3 structures swing parallel to the Carthage-Colton mylonite zone (Fig. 5). At the same place coaxiality between F_2 and F_3 disappears.

5.1.4 F_4-Folds. These open, upright, NW trending folds are best developed in the Northwest Lowlands and die out eastward in the Highlands (Fig. 5).

Interference between F_3 and F_4 folds in the Lowlands has produced a number of dome and basin structures. Alexandria Bay Leucogranitic Gneiss is exposed in the cores of fourteen of these domical culminations (Figs. 4,5). Buddington (1939) originally interpreted a number of these as phacoliths. However, they are clearly not intrusive, but rather structural, in origin (Wiener and others, 1984; Foose and Carl, 1977).

5.1.5 F_5-Folds. Fifth generation folds (Fig. 5) trend NNE and are upright and open although they exhibit marked tightening in the northeastern Adirondacks. Folds of this set die out to the northwest and are only sparsely represented in the Lowlands. Within the central and eastern Adirondacks, the interference of F_5-folds with earlier sets give rise to classic basin and dome patterns such as seen along the Piseco Anticline (Fig. 5). Indeed, the entire southern and central Adirondacks may be considered to be an erosionally breached F_5 dome centered on the Oregon and Snowy Mt. domes (Fig. 5). The same applies to the multidomical Marcy massif to the north (Fig. 5).

5.2 Low Angle Ductile Faults

Several low angle ductile faults, believed to be thrusts, have been mapped within the Northwest Lowlands (e.g., Wiener and others, 1984) and in the eastern Adirondacks (Berry, 1961)(Fig. 5). In addition to these, current work by Isachsen has disclosed asymmetric fabric indicative of simple shear along the Carthage-Colton mylonite zone (Geraghty and others, 1981). A growing body of information strongly suggests that other low angle detachment surfaces exist but have eluded detection because they are oriented essentially parallel to layering. The existence of these structures is suspected because of the widespread occurrence of ribbon lineations, sheath folds, mylonitic textures, and gneissic layering or lamination, all of which are characteristically associated with rotational strain accompanying thrusting (e.g. Davidson and others, 1982). Regional coaxiality between early fold sets, and their parallelism to ribbon lineation, provides further evidence for rotational strain (Cobbold and Quinquis, 1980).

Within the Northwest Lowlands Foose and Brown (1976) postulate a folded thrust, or tectonic slide, between the Lower Marble and the Major Paragneiss in the area of Reservoir Hill (Fig. 4). Displacement along this structure is in question (Wiener and others, 1984), but lithologic units are locally truncated, and the proposed fault is associated with marked thinning on the lower limb of an F_2 fold. Similar NNE faults occur throughout the Lowlands and are summarized by Brown (1983) who points out that they transect early fold structures but are accompanied by oriented high grade metamorphic minerals such as sillimanite. They are believed to have formed during F_2 folding and in some places, to have been reactivated at later times (Brown, 1983; Foose and Brown, 1976). DeLorraine and Dill (1982) have mapped early folded thrusts in the underground workings of the #4 mine of the Balmat zinc deposit (Fig. 4). These appear to accompany early isoclinal folding.

The northeast-trending Carthage-Colton mylonite zone (Figs. 4, 5) has been studied and described by Geraghty and others (1981) and by Wiener (1983) both of whom summarize earlier investigations. The zone, which separates the Adirondack Highlands from the Lowlands, constitutes a major structural element in the southeastern portion of the Grenville Province. It is exposed over a distance of 110 km and varies in width from a few meters to more than 5 km; the wider

portions being located in the south. Much of the zone is occupied by
igneous rocks of the Diana Complex. Wiener considers the zone to be
a faulted intrusive contact along which quartz syenites were emplaced
during F_2 folding. The entire zone dips to the northwest parallel to
regional foliation. Whitney (1983) and Baer (1976) consider the
Carthage-Colton zone to be continuous with the Chibougimou-Gatineau
Line (Fig. 7).

The zone is clearly one of high strain and is characterized by
ductile grain size reduction and recrystallization that occurred under
hornblende-granulite facies conditions (Geraghty and others, 1981).
A strong mylonitic foliation is present along with a pervasive down-
dip lineation that tends to parallel NNW to EW trending isoclinal
fold axes in at least the northern two thirds of the zone (Geraghty
and others, 1981). Wiener (1983) associates this fabric with F_2
folding and tectonic transport directed towards the southeast.
Geraghty and others (1981) suggest that the northern portion of the
zone may mark a southeastward directed fold-thrust nappe of unknown
displacement that developed during early recumbent folding, while in
the southern third of the zone strain and mylonitization were distributed over a wider belt marked by a broad northeast trending fold.

In the eastern Adirondacks Berry (1961) mapped several extensive
thrust faults of unknown displacement that occur along the eastern
shore of Lake George (Fig. 5). Berry's cross sections imply that the
thrusts are unfolded, but recent field investigations demonstrate
that they are folded by at least F_3 structures. The rocks comprising
the thrust sheets are mainly Piseco Group charnockitic-mangeritic
rocks containing occasional blue-grey xenocrysts of andesine. The
base of the sheets is commonly marked by lenses of anorthositic gabbro and garnetiferous noritic rocks. The lithologies as well as
overlying units exhibit grain size reduction, ribbon lineation, and
laminated layering, characteristic of mylonization. Down-dip isoclinal fold axes and lineation show a strong E-W preferred orientation. Mapping demonstrates that the thrust sheet truncates layering
in underlying metasediments. The truncations appear to effect an F_3
fold limb and, therefore, the thrust may have been locally overlapped
by F_3 folding. These thrusts suggest that similar structures may
exist throughout the area but have not yet been recognized because of
later deformation superimposed on detachment surfaces which formed
parallel, or sub-parallel, to layering.

5.3 Planar and Linear Fabric, and Sheath Folds

As indicated in earlier section most of the Adirondacks are characterized by strong tectonic foliation and lineation. These are well
developed along the Carthage-Colton mylonite zone and throughout
much of the southern and central Adirondacks. Particularly striking
are ribbon lineations developed in quartzofeldspathic lithologies.
McLelland (1984) has described this ribbon lineation and traced its
evolution from relatively undeformed megacrystic granitic gneiss to
ductiley deformed, grain size reduced ribbon gneiss. The ribbons
whose average dimensions are 40 cm x 1 cm x 2 mm, consist of aggre-

gates of quartz and K-feldspar with minor plagioclase, biotite, and hornblende. They are generally oriented EW-N70W parallel to F_2 and F_3 fold axes. Progressive deformation of initially subhedral K-feldspar megacrysts into aligned asymmetric augen, and thence into ribbons, indicates that the lineation lies approximately parallel to the direction of maximum finite elongation (McLelland, 1984). Parallelism of the lineation with early isoclinal fold axes, as well as asymmetry of feldspar tails, further suggests that the ribbons formed during widespread rotational strain with a dominant component of simple shear (McLelland, 1984). Orientation of feldspar tails with respect to foliation is generally consistent with a tentatively assigned east over west sense of shear.

Fabrics similar to those in deformed K-feldspar augen gneiss exist in many other rock types in the southern and central Adirondacks. The Sacandaga Fm. which directly overlies the Piseco Group in the Piseco Anticline is replete with laminar foliation and penetrative E-W lineations, especially quartz ribbons. The formation consists of alternating leucogneisses and mafic gneisses, all of which exhibit flaggy layering. They strongly resemble mylonitic "straight gneisses" that occur along the boundaries of the several thrust sheet domains of the Parry Sound, Ontario area (Fig. 7)(Davidson and others, 1982).

In view of the foregoing, it appears possible that much of the foliation and lineation of the central and southern Adirondacks, as well as the Carthage-Colton mylonite zone, are the result of processes of ductile grain size reduction that accompanied early folding and possible thrusting. The important role played by rotational strain is indicated by the E-W development of quartz ribbons and feldspar augen tails, as well as their parallelism to early fold axes. Numerous studies (Bryant and Reed, 1969; Quinquis and Cobbold, 1981) indicate that parallelism between lineation and fold axes is often due to fold axis rotation in response to rotational strain with a dominant component of simple shear which simultaneously results in lineations parallel to tectonic transport.

The development of sheath folds and the arcuation of fold axes have received much attention in recent years (Hansen, 1971; Ramsey, 1980; Quinquis and Cobbold, 1980; Henderson, 1981). These features, which are not always easy to recognize, arise when fold axes behave as passive markers during ductile differential flow. As demonstrated by Cobbold and Quinquis (1980) this behavior is satisfactorily accounted for in a regime where the bulk deformation is dominated by a progressive simple shear that rotates early formed fold axes towards the direction of maximum finite strain (X) that is marked by elongation lineations such as feldspar tails and quartz ribbons. Resultant sheath folds exhibit long axes parallel to maximum finite strain and have cross sections oriented at a high angle to this direction. Recent investigations (McLelland, 1984) in the eastern Adirondacks have revealed the existence of sheath folds in calcsilicate-bearing marbles. The long axes of these folds trend N70W to E-W and plunge gently about the horizontal. Strong mineral and ribbon lineations plunge down the sheaths parallel to their long axes. Elliptical cross sections have long axes that strike NNE. These sheath folds are considered to provide additional evidence for the presence of a strong E-W directed rotational strain that was active during the major (granulite facies) phase of ductile deformation in the region.

5.4 Neotectonics and Brittle Structures

Isachsen (1984) has summarized ongoing research and results related to the brittle structures and neotectonics of the Adirondacks. The salient results of these investigations are summarized below.

Several lines of evidence indicate that the NNE trending Adirondack dome is a young uplift that came into being during late Mesozoic or Tertiary time and is still rising. Such evidence includes: 1) drainage basin studies indicate that only approx. 18 million years are required to reduce the elevation of a mountain mass by nine-tenths, even allowing for isostatic compensation; 2) the drainage pattern is still largely consequent (radial), the streams being largely unadjusted to the great variation in erosional resistance of bedrock units. Isopach maps show no evidence of domical uplift of the region during Paleozoic time.

Possible Tertiary movement is suggested along the southeastern border fault system of the Adirondacks (Fig. 1) by the following: 1) the fault trace coincidences with the break in slope at the foot of a steep escarpment, and drag-folded weak shales of the down-thrown block still remain on the escarpment face; 2) both scarp relief and vertical displacement increase northward towards the more uplifted portion of the dome and thus appear to be dome-related; 3) carbonated springs issue along segments of the fault at Saratoga Springs, and such springs elsewhere in the world are generally restricted to areas of active tectonism or igneous activity.

Holocene movements in the region are suggested by low level but recurrent seismic activity and by releveling studies that suggest contemporary domical uplift. Releveling suggests uplift rates of 3.7 mm/year near the center of the dome and 2.2 mm/yr along the eastern margin of the Adirondacks. However, recent controversies have arisen concerning systematic errors in geodetic measurements leaving the uplift rates in doubt (Isachsen, 1984).

Brittle deformation, presumably associated with the domical uplift, occurs throughout the Adirondacks and is largely responsible for the NNE linear features that characterize the region. Those in the eastern Adirondacks are demonstrable high angle faults, but many others, especially near the center of the dome, appear to be steeply dipping "zero displacement crackle-zones". These zones, up to a few tens of meters in width, are interpreted as tensional features formed originally during late stages of unloading following the Grenville event and reactivated much later by crustal stretching at shallow levels over a rising elongate dome.

Low level, recurrent seismic activity within the Adirondack-Western Ontario province occurs along a 500 km long, 200 km wide northwesterly trending belt. Focal mechanism studies of 11 earthquakes in the belt, 8 of them in the Adirondacks, indicate thrust faulting along NNW to NW trending planes. The inferred direction of maximum compressive stress is largely uniform and trends ENE to WNW, nearly parallel to the calculated absolute plate motion of North America. No direct, genetic relationships have as yet been shown to exist between this seismic activity and Adirondack uplift and fracturing.

The causes of Adirondack doming are by necessity speculative but the dome is similar to others throughout the world that are the result of crustal expansion over thermal highs. The lack of volcanics and low heat flow of the Adirondacks indicate that any inferred thermal front lies at least 1-2 km below the surface. However, the high thermal conductivity at 20-25 km (Nekut and others, 1977) and the approx. 20 km reflector discovered by COCORP traverse (Brown and others, 1983) may be the result of small quantities of deep crustal intergranular melts caused by rising isotherms over a thermal high or deep intrusion.

6. METAMORPHISM

High grade regional metamorphism associated with the 1000-1100 Ma Grenville event has thoroughly effected the entire Adirondack region. Most of the Lowlands are characterized by sillimanite-almandine-muscovite assemblages of the upper amphibolite facies while the Highlands lie almost entirely within the hornblende granulite facies. Several isograds have been recognized within the western Adirondacks and are shown in Fig. 6.

The transition from amphibolite to granulite facies was studied by Engel and Engel (1953, 1960, 1962) and by Nielsen (1971) both of whom located an orthopyroxene isograd within amphibolite interlayers of the Major Paragneiss belt (Fig. 6). This isograd lies 5-15 km west of the Carthage-Colton mylonite zone and is oriented sub-parallel to it. The first appearance of orthopyroxene appears to be due to hornblende consuming reactions. Local occurrences of orthopyroxene west of this isograd suggest greater complexity than currently realized. Buddington (1963) also located an isograd for the first appearance of orthopyroxene in orthoamphibolites. This coincides with the disappearance of sphene from hornblende granitic rocks and is labeled "sphene" on Fig. 6. Bloomer (1969) mapped the short orthopyroxene isograd shown in the north-central portion of Fig. 6.

Engel and Engel (1960) mapped a K-feldspar-garnet isograd in metapelites that coincides with the orthopyroxene isograd (Fig. 6). The isograd appears to reflect the onset of the reaction muscovite + biotite + quartz = K-feldspar + garnet + H_2O, and is accompanied by an increase in migmatization probably due to the release of H_2O from muscovite and biotite.

DeWaard (1969, 1971) subdivided the granulite facies into several subfacies as a function of P_L, P_f, and T. The occurrence of a 10-15 km wide, northward terminating, wedge of cordierite bearing metapelites upgrade of the orthopyroxene isograd within the southeastern Lowlands (Fig. 6) places these rocks within the low pressure biotite-cordierite-almandine subfacies of the granulite facies (deWaard, 1966). Stoddard (1980) reports additional cordierite to the

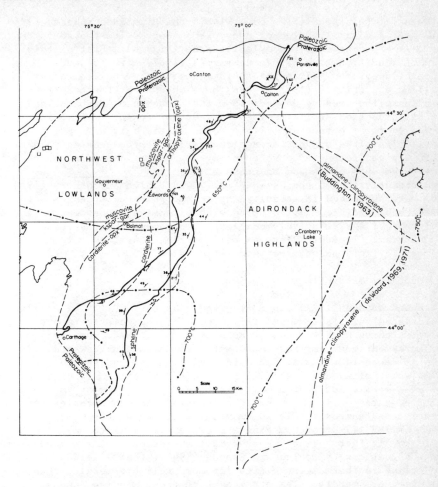

Figure 6. Map showing the Carthage-Colton mylonite zone (heavy outline); foliation; mapped isograds, and paleoisotherms. Individual isograds are discussed in the text. Isolated orthopyroxene occurrences (squares) are from Guzowski (1979, western cluster) and Foose (1974, eastern cluster). Isolated cordierite occurrences (X) are from Buddington (unpubl. guidebook) and Stoddard (1976, p. 67). Paleoisotherms from Bohlen and others (1980). Hachured line is boundary of metanorthosite to east (Figure from Wiener and others, 1984).

northeast of this zone (Fig. 6). Within the Highlands deWaard (1969, 1971) proposed an almandine-clinopyroxene isograd located 30-50 km east of the Carthage-Colton mylonite zone (Fig. 6). East of this isograd P,T conditions lie within deWaard's (1967) high pressure hornblende-clinopyroxene-almandine subfacies of the granulite facies. McLelland and Whitney (1977, 1980a) have shown that the occurrence of almandine-clinopyroxene depends upon the presence of Fe-oxide and give as a reaction Anorthite + (6 - α) Orthopyroxene + αFe-oxide + (α - 2) quartz = garnet and clinopyroxene where α is a function of the distribution of Fe and Mg between coexisting ferromagnesian phases. Quartz may be present as a reactant or a product, depending on the relative amounts of Fe and Mg present.

Recently Bohlen and Essene (1977) and Bohlen and others (1980, 1984) have succeeded in establishing metamorphic temperatures across the Adirondacks (Fig. 1). Coupled with extensive geobarometry, these results provide a relatively complete P,T framework for Adirondack metamorphism. These studies, as well as the nature and role of the metamorphic fluid phase, are summarized below.

6.1 Geothermometry

Bohlen and Essene (1977) and Bohlen and others (1980, 1984) utilized K-feldspar-plagioclase-magnetite-ilmenite geothermometry to obtain isotherms throughout the Adirondacks (Fig. 1). Their results are in good agreement with earlier thermometry summarized by deWaard (1967) and with determinations by Stoddard (1980), McLelland and Whitney (1977), and Valley and Essene (1980). More recently Metzger (1984) applied a variety of geothermometers (i.e. - two feldspar, magnetite-ilmenite, garnet-clinopyroxene, two pyroxene, and garnet-biotite) to rocks located approximately 20 km west of southern Lake George. Her result of $723°\pm23°C$ corroborates the extrapolated isotherms of Bohlen and Essene in that region. Furthermore she demonstrates, in agreement with observations by Bohlen and Essene (1980), that K-feldspar-plagioclase geothermometry yields far more reliable results than other geothermometers. The net result of these studies demonstrate (Fig. 1) that temperatures increase from approx. $600°\pm30°C$ in the Lowlands to approx. $800°\pm30°C$ at a point within the Marcy massif.

Bohlen and others (1980) have argued that the isotherms of Fig. 1 record temperatures at, or close to, peak metamorphic conditions. Whitney (1978) has suggested that these temperatures may record solidus temperatures of relatively dry granitic or quartz-dioritic melts. However, as pointed out by Bohlen and Essene (1980) similar temperatures were determined for anorthosites and amphibolites which could not have undergone partial melting at the given temperatures.

It is noteworthy that the isotherms exhibit a concentric arrangement about the Marcy anorthosite massif (Fig. 1). A smaller concentric configuration exists for the $700°C$ isotherm near the southern end of the Diana Complex which includes the Carthage anorthosite. This spatial distribution may be due to a thermal influx from the anorthosite themselves. However, the isotherms appear to be unaffected by regional folding, whereas the anorthosites are deformed by folds at least as early as F_2. Thus it may be that the anorthosites intruded prior to most deformation in the area. If this turns out to be the case, the present configuration of the isotherms may be the result of upward flexing of low-dipping isotherms by late diapiric uprise of the relatively less dense anorthositic bodies (Whitney, 1983). This question is unresolved, but recent results of Ashwal and Wooden (1983) favor early emplacement of anorthosite (see section on isotopic age relationships). This interpretation is consistent with the suggested emplacement of the anorthosites during a pre-Grenvillian rifting event (Whitney, 1983). However, very early syntectonic intrusion and/or retarded cooling of the anorthosites beneath thrust sheets cannot be ruled out, and the possibility remains that their heat contributed to the thermal regime of the Grenvillian metamorphism.

6.2 Geobarometry

Assignment of pressures in the Adirondacks are based upon seven independent geobarometers including garnet-rutile-aluminosilicate-ilmenite-quartz (Bohlen and others, 1983a,b); ferrosilite-fayalite-quartz (Bohlen and Boettcher, 1981; Jaffe and others, 1978); sphalerite-pyrrhotite-pyrite (Brown and others, 1978); fayalite-anorthite-garnet and ferrosilite-anorthite-garnet-quartz (Bohlen and others, 1983); akermanite-monticellite-wollastonite (Valley and Essene, 1980b); anorthite-grossular-aluminosilicate-quartz (Newton and Haselton (1981). Less precise, but nonetheless restrictive, mineral occurrences such as sillimanite, kyanite (Boone, 1978), biotite-cordierite-almandine (deWaard, 1966; Stoddard, 1980), and grossular-quartz (Valley and Essene, 1980b) are in good agreement with these geobarometers and indicate pressures of 6-6.5 kb within the Lowlands increasing to 7-8 kb in the Highlands (Bohlen, 1983; Newton, 1983). The uncertainty in these pressure estimates is on the order of ±0.5 kb. Bohlen (1983) gives a plot of approximately 50 pressure determinations, and although these are not contoured, the highest pressures appear to center on the Marcy massif giving a pattern similar to that shown by the 150-therms of Fig. 1.

6.3 Composition and Role of Fluid Phase

Buddington (1963) and deWaard (1969) emphasized the importance of the fluid phase in Adirondack metamorphism. More recently Valley and Essene (1977, 1980a,b) have examined a number of calcsilicate reactions in Adirondack marbles and have shown that variations in CO_2 pressure have had a pronounced effect on mineral assemblages. These authors have attributed variability in CO_2 pressure to (1) influxes of H_2O under conditions of $P_{load} = P_{fluid} = (PCO_2 + PH_2O)$, or (2) conditions under which $P_{fluid} \ll P_{load}$ and CO_2 activity was buffered by solid phases only. It is difficult to distinguish between these alternatives on the basis of petrology alone. Valley and O'Neil (1982, 1984) have utilized C-O-H stable isotopic analysis, in conjunction with petrologic considerations, to obtain reasonably quantitative estimates on the composition, homogeneity, and pervasiveness of the fluid phase in the Adirondacks. Their results indicate fluid/rock ratios were generally low but variable; that preserved, pre-metamorphic values indicate channelized fluid flow and preclude a pervasive, homogenizing fluid phase; and that XCO_2/XH_2O was highly variable.

As an example of fluid phase variability, Valley and others (1983) describe calcsilicate assemblages separated by only 1 km in the central Adirondacks where P,T conditions are close to 710°C and 7 kb. In one locality the reaction tremolite = enstatite + diopside + quartz + H_2O has buffered PH_2O at 0.9 kb. At the other locality the reaction calcite + quartz = wollastonite + CO_2 has buffered PCO_2 at 1.1 kb. If it is assumed that $P_{load} = PCO_2 + PH_2O$, these results indicate a marked variability, and sharp gradients, in fluid composition over a relatively short distance.

Valley and Essene (1980b) and Valley and O'Neil (1984) describe and analyze the assemblage akermanite + monticellite + wollastonite from a marble xenolith at Cascade Slide in anorthosite of the Marcy massif (Fig. 1). The stability of these and other phases fixes P,T at 7.4±1 kb and 750±30°C and indicates that local metamorphic fluids were unusually poor in CO_2. The presence of the assemblage calcite + quartz 100 m away shows that CO_2 fugacity varied by a factor of exceeding 60 over short distances. $\delta^{18}O$ values of 7-10 near the margin of the xenolith indicates some local exchange between calcite and anorthosite, but values rise to 26.1 at the xenolith 15 m away thus precluding the presence of a pervasive, homogenizing fluid phase. The low fugacity of CO_2 in these and other assemblages apparently is the result of $P_{fluid} \ll P_{load}$ (Valley and others, this volume).

Valley and O'Neil (1982, 1984) have utilized oxygen isotopes to study wollastonite deposits occurring in a 15 km-long belt in the northeastern Adirondacks. Along this belt the wollastonite has unusually low values of 3.1 to -1.3 within 125 m of adjacent anorthosite. These values reflect depletions whose magnitudes are explicable only by exchange with heated meteoric water. Valley and O'Neil (1982, 1984) propose that the wollastonite deposits formed as metasomatic skarns adjacent to anorthosite. Relatively shallow intrusion depths of 7-10 km are thus necessary such that hydrothermal/meteoric waters could circulate and facilitate wollastonite formation as well as cause low $\delta^{18}O$ values. Subsequently, the anorthosite and wollastonite skarn were buried and metamorphosed to granulite facies. The preservation of the steep $\delta^{18}O$ gradients and low $\delta^{18}O$ values in the skarn demonstrates that no pervasive fluid phases were present during peak regional metamorphism.

The oxygen isotope studies of Valley and O'Neil strongly imply that granulite facies fluids of the Adirondacks were highly variable in composition; were channelized rather than pervasive; and were accompanied by steep compositional gradients over short distances. These conclusions do not support pervasive CO_2-flooding as prime cause for anhydrous granulite facies metamorphism in the Adirondacks (Newton and others, 1980).

Valley (1985, this volume) has presented evidence for a pre-Grenvillian high temperature event associated with the emplacement of

anorthosite which may have occurred at relatively shallow depths
(Valley and O'Neil, 1982). Wollastonite bearing skarns formed, and
were devolatized, during this emplacement. Subsequently, these skarns
remained isolated from H_2O or CO_2 fluids and, therefore, underwent
vapor absent metamorphism ($P_V \ll P_L$) during the 1000-1100 Ma Grenvillian
events.

6.4 Timing of Metamorphism

On the basis of U/Pb isotope systems in a large variety of rock types,
Silver (1969) concluded that granulite facies metamorphism took place
between 1000 and 1100 Ma. Subsequent determinations discussed in
section 3 are in general agreement with this assignment. Field relationships throughout the Adirondacks indicate that high-grade metamorphism was essentially syntectonic but the unfolded nature of the isograds indicate that peak metamorphism outlasted folding (Wiener and
others, 1984). Anatexis terminated before second phase folding and in
the Lowlands was associated with the release of H_2O during muscovite
breakdown. Fourth and fifth phase folds postdate peak metamorphism but
F_5 appears to overlap locally developed retrograde metamorphism in the
eastern Adirondacks. Textural and mineralogical evidence for these conclusions is summarized in Wiener and others (1984).

7. DISCUSSION

Fig. 5 shows the axial trace map for the entire Adirondacks. Within
the northern Highlands most of these traces have been taken from the
literature and from foliation data shown on the New York State Geological map (Fisher and others, 1973) and must be regarded as speculative.
Nonetheless, the overall pattern is well established and clearly demonstrates the Adirondacks to be a multiply-deformed region of high metamorphic grade. Much of the ductile deformation overlaps the 1000-1100
Ma peak metamorphism which appears to have occurred at temperatures of
$650°-800°C$ and pressures of 6.5-8 kb (Whitney and McLelland, 1973;
Bohlen and Essene, 1977; Bohlen and others, 1984).

The M-discontinuity in the Adirondacks lies 32-35 km beneath the
surface (Katz, 1955), and metamorphic assemblages now exposed at the
surface indicate an original overburden of 20-25 km. This implies a
crustal thickness of 50-60 km during the 1.1-1.0 Ma Grenvillian Orogeny; essentially a double continental thickness similar to that observed in Tibet (e.g., Vine and Matthews, 1963; Dewey and Burke, 1973);
the Turkish-Iranian Plateau (Sengor and Kidd, 1979); and the Andean
Altiplano (James, 1971). Similar crustal thicknesses appear to have
accompanied Grenvillian metamorphism in other portions of the Grenville Province (Bourne, 1978; Davidson and Morgan, 1980).

The double crustal thickness of the Adirondacks requires a tectonic mechanism of regional dimensions. By analogy with the afore-
mentioned regions of double crustal thickness, it seems warranted to
postulate a convergent plate interaction as the driving force for
crustal thickening. Since no good candidates for a suture appear to

exist within the Grenville Province itself, McLelland and Isachsen (1980) suggested that the NNE trending N.Y. - Alabama Lineament of King and Zeitz (1978) may be a paleomagnetic expression of the plate margin. McLelland and Isachsen (1980) postulated that subduction along this margin resulted in the sequential underthrusting of 200-300 km wide crustal segments resulting in a double crustal thickness. Subsequent differential uplift of hot, ductile lower crust with internal density contrasts led to vertical gravity tectonics culminating in fold nappes similar to those produced by Ramberg (1967) in centrifuged scale-model experiments. It was further hypothesized that the apparent non-alignment of Grenville Province foliation patterns (Tectonic Map of Canada, 1968) could be explained by this mechanism. However, the pattern may be due to multiple folding (Wynne-Edwards, 1967, 1969).

Hamilton (1981) has presented a tectonic model in which the Adirondacks and Grenville Province are interpreted as an obliquely exposed section through a magmatic arc dominated by crustal extension and extensively invaded by magmas which cause the heating and anatexis associated with the Grenvillian metamorphism.

Whitney (1983) has adopted a three-stage plate tectonic model for Adirondack evolution. Anorthosites were intruded at shallow levels during an initial stage of intracontinental rifting. This was followed by convergence and intracontinental subduction (see also Seyfert, 1980) accompanied by folding, thrusting, and Grenvillian metamorphism. The third phase was marked by diapiric rise of the less dense anorthosite massifs causing deflection of metamorphic isotherms. An Andean arc with back-arc spreading followed by collision is compatible with this model (Whitney, 1983).

We suggest that significant clues to reconstructing the tectonic history of the Adirondacks are provided by the ribbon lineations discussed in section 5. These lineations are found in a large number of localities throughout the central and southern Adirondacks and are well developed along the Carthage-Colton mylonite zone. Their general E-W to NW down-dip orientation indicates that the maximum elongation direction of the finite strain ellipsoid maintained a relatively consistent regional orientation during the ductile deformation associated with their formation. Parallelism between these elongation lineations and early, isoclinal fold axes further suggests that rotational ductile strain was responsible for lineations, rotation of early fold axes, and sheath folds. The most satisfactory mechanism for providing this rotational strain is by the emplacement of large nappes, tectonic slices, and thrust sheets.

This interpretation allows much of Adirondack structure to be explained as stages in a tectonic continuum. A NNE trending, continent-continent margin southeast of the Adirondacks with eastward directed subduction would result in NNE-trending buckle folds and westward directed overthrusts together with complementary eastward underthrusting. Continued rotational strain would lead to tectonic thickening and at depth to grain size reduction, recrystallization, and ribbon lineations. Early folds would become tightly overturned to the west and begin to arcuate until fold axes became passive markers rotated into parallelism with elongation lineations. The E-W alignment of F_1 and F_2 folds may be explained by such rotation. Within the central and southern Adirondacks F_3 folds may have formed in response to convergent flow or resultant compressional strain oriented at high angles to the E-W to NW direction of tectonic transport. NNE trending F_5 folds may represent late pulses on the NNE trending plate margin. The Carthage-Colton mylonite zone lies at a high angle to the proposed direction of tectonic transport and appears to represent the focusing of this strain into a zone of unusually high strain. Passive markers in an asymmetric fabric indicate an E over W sense of movement (Isachsen, in preparation).

Although this model provides a cohesive tectonic framework for much of the Adirondacks, it does not offer a complete explanation for the orientation of F_3 and F_4 folds in the Lowlands nor for the observation (Fig. 5) that fold axes swing parallel to the Carthage-Colton mylonite zone as it is approached from the east. The latter configuration is not believed to be coincidental, and its explanation will be important in any complete understanding of Adirondack evolution.

The apparent W to NW-directed tectonic transport in the Adirondacks, and associated NNE-NE high strain zones, appear to be part of a much larger tectonic pattern that extends northwestward to the Grenville Tectonic Front itself (Fig. 7). The Carthage-Colton mylonite zone appears to be continuous with the boundary between the Central Granulite Terrain and the Central Metasedimentary Belt (Fig. 7). Within the Central Metasedimentary Belt, the Elzevir Terrain is separated from the Frontenac-Adirondack Terrain by a large NE trending zone of high strain (Brock and Moore, 1983). The Elzevir Terrain and the Bancroft Terrain are separated by another high strain zone that trends NE (Fig. 7). The boundary between the Bancroft Terrain and the Ontario Gneiss Segment is also a zone of intense ductile deformation and straightening (Thivierge and others, 1984). Strong SE plunging lineation and kinematic indicators indicate overthrusting to the NW. Within the Central Metasedimentary Belt there occur a number of NNE trending high strain zones (Fig. 7), and these may be the result of NW directed tectonic transport.

Within the Ontario Gneiss Segment near Parry Sound, Davidson and Morgan (1980) and Davidson and others (1982) have mapped several domains that are separated from one another by tectonized zones exhibiting widespread ductile deformation, grain size reduction, straight gneisses, ribbon lineations, sheath folds and other assorted manifestations of high strain mylonitization (Fig. 7). Kinematic indicators clearly demonstrate a northwest direction of overthrusting associated with each domain. SE-plunging, stretching lineations lie parallel to this sense of movement (Thivierge and others, 1984). Comparable features, with the same sense of relative movement are found along the Grenville Tectonic Front although this major feature strikes at an angle to the high strain zones within the Grenville Province itself.

The regional picture that emerges from this study is one of intense tectonic transport along approximately E-W to NW directed lines during the Grenville Orogeny. This tectonic flow regime appears to extend for almost 600 km from the Grenville Front Tectonic Zone in western Ontario to the eastern edge of the Adirondacks. Although

Figure 7. Generalized map showing the principal subdivisions of the Grenville Province in southern Ontario. Domains: A - Ahmic; BD - Britt; GHD - Go Home; MRD - Moon River; PSD - Parry Sound; RD - Rousseau; SD - Seguin. The CCZ and GCL appear to be continuous beneath Paleozoic cover and separate the Central Metasedimentary Belt on the west from the Central Granulite Facies Terrain on the east. Compiled from Davidson and others (1982) and Brock and Moore (1983).

uncertainties exist, observations to date indicate movement may have been towards the NW over the entire region. The tectonic style was characterized by the formation of early fold nappes whose axes and associated lineations are generally oriented EW to NW. Accompanying these were thrusts and tectonic slides which trend NS to NE and are displaced to the NW. These structures are thought to have produced the crustal thickening that occurred during the Grenville Orogeny, presumably driven by continent-continent collision along a suture lying southeast of the Adirondacks beneath Paleozoic cover.

8. ACKNOWLEDGEMENTS

Much of our current knowledge of Adirondack geology stems from earlier research by A.F. Buddington, M. Walton, and D. deWaard. Our own research has received generous support from the U.S. Nuclear Regulatory Commission (NUREG/CR-1865 to YWI) and the National Science Foundation (NSF EAR 77-23425 and NSF EAR 80-25373 to JMM).

9. REFERENCES

Alling, H.L., 1917. 'The Adirondack graphite deposits': New York State Museum Bulletin, no. **199**, 150 p.

Anderson, S.L., Burke, K.C., Kidd, W.S.F., Putman, G.W., 1983. 'Mylonite zones, thrusts, and tectonics of the Grenville in the Adirondacks' (abs): Geol. Soc. Amer., Abstracts with Programs, v. **15**, no. 3, p. 123.

Ashwal, L.D., 1978. Petrogenesis of massif-type anorthosites: Crystallization history and liquid line of descent of the Adirondack and Morin Complexes (Ph.D. thesis): Princeton, New Jersey, Princeton University, 136 p.

Ashwal, L.D. and Wooden, J., 1983. 'Sr and Nd isotope geochronology, geologic history, and origin of Adirondack anorthosite': Geochimica et Cosmochimica Acta, v. **47**, no. 11, p. 1875-1887.

Baer, A.J., 1976. 'The Grenville Province in Paleohelikian times: A possible model of evolution': Royal Society of London Philosophical Trans. Ser. A, v. **280**, p. 499-515.

Baer, A.J., and others, 1974. 'Grenville geology and plate tectonics': Geoscience Canada, v. **1**, no. 3, p. 54-60.

Bartholome, P.M., 1956. Structural and petrological studies in Hamilton County, New York (Ph.D. thesis): Princeton, New Jersey, Princeton University, 113 p.

Basu, A.R. and Pettingill, H.S., 1983. 'Origin and age of Adirondack anorthosites re-evaluated with Nd isotopes': Geology, v. **11**, p. 514-518.

Berry, R.H., 1961. Precambrian geology of the Putnam-Whitehall area, New York (Ph.D. thesis): New Haven, Connecticut, Yale University.

Bickford, M.E., and Turner, B.B., 1971. 'Age and probable anatectic origin of the Brant Lake Gneiss, southeastern Adirondacks, New York': Geol. Soc. of Amer. Bulletin, v. **82**, no. 8, p. 2333-2342.

Bloomer, R.O., 1969. 'Garnets in various rock types in the Canton area in northern New York': New York Academy of Science Annals v. **172**, p. 31-48.

Bohlen, S.R. and Boettcher, A.L., 1981. 'Experimental investigations and geological applications of orthopyroxene geobarometry': Am. Mineral., v. **66**, p. 951-964.

Bohlen, S.R. and Boettcher, A.L., 1982. 'The quartz-coesite transformation: A precise determination of the effects of other components': Jour. Geophys. Res., v. 87, p. 7073-7078.

Bohlen, S.R., Boettcher, A.L., and Wall, V.J., 1981. 'Experimental calibration of two geobarometers for garnet-bearing assemblages' (abs.): Geol. Soc. Amer., Abstracts with Programs, v. 13, p. 412.

Bohlen, S.R. and Essene, E.J., 1977a. 'Feldspar and oxide thermometry in the Adirondack Highlands': Contributions to Mineralogy and Petrology, v. 62, p. 153-169.

Bohlen, S.R. and Essene, E.J., 1977b. 'Errors in applying olivine-quartz-orthopyroxene barometry': American Geophysical Union Transactions, v. 58, p. 1242.

Bohlen, S.R. and Essene, E.J., 1978. 'The significance of metamorphic fluorite in the Adirondacks': Geochimica et Cosmochimica Acta, v. 42, p. 1669-1678.

Bohlen, S.R. and Essene, E.J., 1979. 'A critical evolution of two-pyroxene thermometry in Adirondack granulites': Lithos, v. 12, p. 335-345.

Bohlen, S.R. and Essene, E.J., 1980. 'Evaluation of coexisting garnet-biotite, garnet clinopyroxenes, and other Mg-Fe exchange thermometers in Adirondack granulites': Geol. Soc. of Amer. Bulletin, v. 91, p. 685-719.

Bohlen, S.R., Essene, E.J., Boettcher, A.L., 1980. 'Reinvestigation and application of olivine-quartz-orthopyroxene barometry': Earth and Planetary Sci. Lett., v. 47, p. 1-10.

Bohlen, S.R., Essene, E.J., and Hoffman, K.S., 1980. 'Update on feldspar and oxide thermometry in the Adirondack Mountains, NY': Geol. Soc. of Amer. Bulletin, v. 91, p. 110-113.

Bohlen, S.R., Essene, E.J., Valley, J.W., 1985. 'Review of geothermometry and geobarometry in the Adirondacks (in press)': Jour. Petrology.

Bohlen, S.R., Wall, V.J., Boettcher, A.L., 1983a. 'Experimental investigations and geological applications of equilibria in the system $FeO-TiO_2-Al_2O_3-SiO_2-H_2O$': Am. Mineral., v. 68, p. 1049-1058.

Bohlen, S.R., Wall, V.J., and Boettcher, A., 1983b. 'Experimental investigation and application of garnet granulite equilibria': Contributions to Mineralogy and Petrology, v. 83, p. 52-61.

Bohlen, S.R., Wall, V.J., and Boettcher, A.L., 1983b. 'Geobarometry in granulites', in, Saxena, S.K. (ed.), Advances in Physical Geochemistry, v. **3**, p. 141-171.

Boone, G.M., 1978. 'Kyanite in Adirondack Highlands sillimanite rich gneiss and P-T estimates of metamorphism': Geol. Soc. Amer., Abstracts with Programs, v. **10**, p. 34.

Bourne, J.H., 1978. 'Metamorphism in the eastern and southwestern portions of the Grenville Province', in, Metamorphism in the Canadian Shield, Geological Survey of Canada Paper **78-10**, p. 315-328.

Brock, B.S., 1980. 'Stark Complex (Dexter Lake area): Petrology, chemistry, structure, and relation to other green rock complexes and layered gneisses, northern Adirondacks, New York': Geol. Soc. Amer. Bulletin, Part II, v. **91**, no. 2, p. 433-504.

Brock, B.S. and Moore, J., 1983. 'Chronology, chemistry and tectonics of igneous rocks in terranes of the Grenville Province, Canada' (abs): Geol. Soc. Amer., Abstracts with Programs.

Brown, C.E., 1983. 'Mineralization, mining, and mineral resources in the Beaver Creek area of the Grenville Lowlands in St. Lawrence County, N.Y.': Geol. Survey Prof. Paper **1279**, p. 1-21.

Brown, J.S., 1936. 'Structure and primary mineralization of the zinc mine at Balmat, New York': Economic Geology, v. **31**, no. 3, p. 233-258.

Brown, J.S., and Engel, A.E.J., 1956. 'Revision of Grenville stratigraphy and structure in the Balmat-Edwards district, northwest Adirondacks, New York': Geol. Soc. Amer. Bulletin, v. **67**, p. 1599-1622.

Brown, L., Ando, C., Klemperer, S., Oliver, J., Kaufman, S., Czuchra, B., Walsh, T., and Isachsen, Y., 1983. 'Adirondack-Appalachian crustal structure: The COCORP northeast structure': Geol. Soc. Amer. Bulletin, v. **94**, p. 1173-1184.

Brown, P.E., Essene, E.J., and Kelly, W.C., 1978. 'Sphalerite geobarometry in the Balmat-Edwards district, N.Y.': Am. Mineral., v. **63**, p. 250-257.

Bryant, B. and Reed, J., 1969. 'Significance of minor folds near major thrust faults in the southern Appalachians and British and Norwegian Caledonides': Geol. Mag., v. 106, p. 412-419.

Buddington, A.F., 1939. 'Adirondack igneous rocks and their metamorphism': Geol. Soc. Amer. Memoir **7**, 354 p.

Buddington, A.F., 1963. 'Isograds and the role of H_2O in metamorphic facies of orthogneisses of the northwest Adirondack area, New York': Geol. Soc. of Amer. Bulletin, v. 74, no. 9, p. 1155-1182.

Buddington, A.F., and Leonard, F., 1962. 'Regional geology of the St. Lawrence County magnetite district, northwest Adirondacks, New York': U.S. Geol. Survey Prof. Paper 376, 145 p.

Carl, J.D., 1978. 'Sampling of a migmatite: New chemical data and possible parent material for the major paragneiss, Northwest Adirondacks, New York': Geol. Soc. of Amer., Abstracts with Programs, Northeastern Section, v. 10, no. 2, p. 36.

Carl, J.D. and Van Diver, B.B., 1975. 'Precambrian alaskite bodies as ash-flow tuffs, northwest Adirondacks, New York': Geol. Soc. Amer. Bulletin, v. 86, no. 12, p. 1691-1707.

Cobbold, P.R. and Quinquis, H., 1980. 'Development of sheath folds in shear regimes': Jour. Structural Geology, v. 2, p. 119-126.

Davidson, A., and Morgan, W.C., 1980. 'Preliminary notes on the geology east of Georgian Bay, Grenville Structural Province, Ontario in, Current Research, Part A, Geological Survey of Canada, Paper 81-1A, p. 291-298.

Davidson, A., Culshaw, N.G., Nadeau, L., 1982. 'A tectono-metamorphic framework for part of the Grenville Province, Parry Sound Region, Ontario in, Current Research, Part A, Geological Survey of Canada, Paper 82-1A, p. 173-190.

deLorraine, W., 1979. Geology of the Fowler orebody, Balmat #4 Mine, Northwest Adirondacks, N.Y. (unpubl. M.S. Thesis): Amherst, Massachusetts, University of Massachusetts, 159 p.

deLorraine, W., and Dill, D., 1982. 'Structure, stratigraphic controls and genesis of the Balmat zinc deposits, Northwest Adirondacks, New York', in, Precambrian Sulphide Deposits, Hutchinson, R.W. (ed.), Geol. Assoc. Canada, Special Paper 25, p. 571-596.

deRudder, R.D., 1962. Mineralogy, petrology, and genesis of the Willsboro, N.Y. wollastonite deposits (Ph.D. Thesis), Indiana University.

deWaard, D., 1969. 'Facies series and P-T conditions of metamorphism in the Adirondack Mountains': Amsterdam, Koninklijke Nederlandse Akademie van Wetenschappen Proceedings, ser. B, v. 72, no. 2, p. 127-131.

deWaard, D., 1971. 'Threefold division of the granulite facies in the Adirondack Mountains': Kristlanlinikum, v. 7, p. 85-93.

Emslie, R.F., 1978. 'Anorthosite massifs, rapakivi granites and late Precambrian rifting of North America': Precambrian Research, v. 7, no. 1, p. 61-98.

Engel, A.E. and Engel, C.G., 1953. 'Grenville series in northwest Adirondack Mountains, New York: Part I: General features of the Grenville Series': Geol. Soc. Amer. Bulletin, v. 64, p. 1013-1048.

Engel, A.E. and Engel, C.G., 1960. 'Progressive metamorphism and granitization of the major paragneiss, Northwest Adirondack Mts., N.Y.': Geol. Soc. Amer. Bulletin, v. 71, p. 1-58.

Engel, A.E. and Engel, C.G., 1962. 'Progressive metamorphism of amphibolite, Northwest Adirondack Mountains, New York': Geol. Soc. Amer., **Buddington Volume**, p. 37-82.

Fisher, D.W., Isachsen, Y.W., Rickard, L.V., 1971. 'Geologic map of New York': N.Y. State Mus. and Science Service Map and Chart Series, no. 15, scale 1:250,000.

Fleuty, M.J., 1964. 'The description of folds': Geol. Assoc. Proc., v. 75, Pt. 4, p. 461-489.

Foose, M.P., 1974. The structure, stratigraphy and metamorphic history of the Bigelow area, Northwest Adirondacks, New York (Ph.D. Thesis): New Jersey, Princeton University.

Foose, M.P., and Brown, C.E., 1976. 'A preliminary synthesis of structural, stratigraphic and magnetic data from part of the northwest Adirondacks, New York': U.S. Geol. Survey Open-File Report **76-281**, 21 p.

Foose, M.P., and Carl, J.D., 1977. 'Setting of alaskite bodies in the Northwest Adirondacks': Geology, v. 5, no. 2, p. 77-81.

Foose, M.P., Mose, D.G., Nagel, M.S., and Tunsory, A., 1981. 'The Rb-Sr ages and structural and stratigraphic relationships of Precambrian granitic rocks in the Northwest Adirondacks, N.Y.': Geol. Soc. Amer., Abstracts with Programs, Northeastern Section, v. 13, no. 3, p. 133.

Geraghty, E.P., 1978. Structure, stratigraphy, and petrology of part of the Blue Mountain 15' quadrangle, central Adirondack Mountains, New York (Ph.D. Thesis): Syracuse, N.Y., Syracuse University, 186 p.

Geraghty, E.P., Isachsen, Y.W., and Wright, S.F., 1981. 'Extent and character of the Carthage-Colton mylonite zone, Northwest Adirondacks, N.Y.': Nuclear Regulatory Commission, Final Report, **NUREG/CR-1865**.

Grant, N.K., Carl, J.D., LePak, R.J., Hickman, M.H., 1984. 'A 350-Ma crustal history in the Adirondack Lowlands' (abs): Geol. Soc. Amer., Abstracts with Programs, v. 16, no. 1, p. 19.

Grant, N.K., Maher, T.M., and Lepaki, R.J., 1981. 'The age and origin of leucogneisses in the Adirondack Lowlands, New York': Geol. Soc. Amer., Abstracts with Programs, Annual Meeting, v. 13, no. 7, p. 463.

Hamilton, W.S., 1981. 'Crustal evolution by arc magmatism': Phil. Trans. Roy. Soc. London, Ser. A, v. 301, p. 279-291.

Hansen, E., 1971. Strain Facies. Springer-Verlag, Berlin, 207 p.

Heath, S.A., and Fairbairn, H.W., 1969. '^{87}Sr/^{86}Sr ratios in anorthosites and some associated rocks', in, Origin of Anorthosite and Related rocks (ed. Y.W. Isachsen): New York Museum and Science Service Memoir 18, p. 99-110.

Henderson, J., 1981. 'Structural analysis of sheath folds with horizontal X-axes, northeast Canada': Jour. Structural Geology, v. 3, p. 203-210.

Henderson, J., 1983. 'Structure and metamorphism of the Aphebian, Penhayn Group and its Archean basement complex in the Lyon Inlet district, Mellville Peninsula, District of Franklin': Geol. Survey of Canada Bulletin 324, p. 1-50.

Hills, A. and Isachsen, Y., 1975. 'Rb-Sr isochron data for mangeritic rocks from the Snowy Mt. massif, Adirondack Highlands' (abs.): Geol. Soc. Amer., Abstracts with Programs, v. 7, p. 73.

Hills, F.A., 1964. The Precambrian geology of the Glens Falls and Fort Ann quadrangles, southeastern Adirondack Mountains, New York (1:24,000) (Ph.D. Thesis): Princeton, New Jersey, Princeton University.

Hills, F.A., and Gast, P.W., 1964. 'Age of pyroxene-hornblende granite gneiss of the eastern Adirondacks by the rubidium-strontium whole-rock method': Geol. Soc. Amer. Bulletin, v. 75, p. 759-766.

Isachsen, Y.W. (ed.), 1969. 'Origin of anorthosite and related rocks': New York State Museum and Science Service Memoir 18, 466 p.

Isachsen, Y.W., 1984. 'Structural and tectonic studies in New York State': Nuclear Regulatory Commission NUREG/CR-3178, 74 p.

Isachsen, Y.W., and Landing, W., 1983. 'First Proterozoic stromatolites from the Adirondacks' (abs.): Geol. Soc. Amer., Abstracts with Programs, v. 15, no. 6, p. 601.

James, D.E., 1971. 'Plate tectonic model for the evolution of the central Andes': Geol. Soc. Amer., v. 82, p. 3325-3346.

Katz, S., 1955. 'Seismic study of crustal structure in Pennsylvania and New York': Bulletin of Seismological Society of America, p. 303-325.

King, E.R., and Zeitz, I., 1978. 'The New York-Alabama Lineament: Geophysical evidence for a major crustal break beneath the Appalachian basin': Geology, v. 6, p. 312-318.

Lewis, J.R., 1969. Structure and stratigraphy of the Rossie Complex, northwest Adirondacks, New York (Ph.D. Thesis): Syracuse, N.Y., Syracuse University, 141 p.

Lumbers, S.B., 1979. 'The Grenville Province of Ontario', in Morey, G.B. (ed.), Fieldtrip Guidebook for the Archean and Proterozoic stratigraphy of the Great Lakes area: Minn. Geol. Soc., Guidebook 13, p. 1-6.

Luther, F., 1976. Petrological evolution of the garnet deposit at Gore Mt., Warren County, New York (unpubl. Ph.D. Thesis): Lehigh University, 218 p.

Maher, T.H., Lepak, R.J., and Grant, N.K., 1981. 'Rb-Sr ages, crustal prehistory and stratigraphic sequence: Leucogneisses and marbles of the Adirondack Lowlands, New York': Geol. Soc. Amer., Abstracts with Programs, v. 13, no. 3, p. 144.

McLelland, J., 1984. 'Origin of ribbon lineation within the southern Adirondacks, U.S.A.': Jour. Structural Geology, v. 6, p. 147-157.

McLelland, J., and Isachsen, Y.W., 1980. 'Structural synthesis of the southern and central Adirondacks: A model for the Adirondacks as whole and plate tectonics interpretations': Geol. Soc. Amer. Bulletin, Part II, v. 91, p. 208-292.

McLelland, J., and Whitney, P.R., 1977. 'The origin of garnet in the anorthosite-charnockite suite of the Adirondacks': Contributions to Mineralogy and Petrology, v. 60, no. 2, p. 161-181.

McLelland, J., and Whitney, P.R., 1980a. 'A generalized garnet-forming reaction for metaigneous rocks in the Adirondacks': Contributions to Mineralogy and Petrology, v. 72, p. 111-122.

McLelland, J., and Whitney, P.R., 1980b. 'Compositional controls on spinel clouding and garnet formation in plagioclase of olivine metagabbros, Adirondack Mountains, New York': Contributions to Mineralogy and Petrology, v. 73, p. 243-251.

McLelland, J., Geraghty, E., and Boone, G., 1978. 'The structural framework and petrology of the southern Adirondacks': New York State Geological Association 50th Annual Meeting Guidebook, p. 58-103.

Metzger, E., 1984. Structure, lithologic succession, and petrology of the Stony Creek area, Warren Co., Adirondacks, N.Y. (unpubl. Ph.D. Thesis): Syracuse, N.Y., Syracuse University, 242 p.

Morse, S.A., 1969. 'Layered intrusions and anorthosite genesis, in, Y.W. Isachsen (ed.), Origin of anorthosite and related rocks: New York State Museum and Science Service Memoir **18**, Albany, N.Y.

Morse, S.A., 1982. 'A partisan view of Proterozoic anorthosites': Am. Mineral., v. **67**, p. 1087-1100.

Nekut,, A., Connery, J.E.P., Kuckes, A.F., 1977. 'Deep crustal electrical conductivity; evidence for water in the lower crust': Geophys. Research Letters, v. **4**, p. 239-242.

Newton, R.C., 1983. 'Geobarometry of high grade metamorphic rocks': Am. Jour. Sci., v. **283-A**, p. 1-28.

Newton, R.C. and Haselton, H.T., 1981. 'Thermodynamics of the garnet-plagioclase-Al_2SiO_5-quartz geobarometer', in, Newton, R.C., Navrotsky, A., Wood, B.J. (eds.), Thermodynamics of Minerals and Melts, p. 129-145, Springer-Verlag, N.Y.

Newton, R.C., Smith, J.V. and Windley, B.F., 1980. 'Carbonic metamorphism, granulites, and crustal growth': Nature, v. **288**, p. 174-196.

Nielsen, P.A., 1971. Metamorphism and metamorphic isograds in the Northwest Lowlands of the Adirondacks (unpubl. Master's Thesis): Binghamton, N.Y., State Univ. at Binghamton, 59 p.

Olilla, P., Jaffe, H., and Jaffe, E., 1984. 'Iron rich inverted pigeonite: Evidence for the deep emplacement of the Adirondack anorthosite massif' (abs.): Geol. Soc. Amer., Abstracts with Programs, v. **16**, no. 1, p. 54.

Postel, A.W., 1952. 'Geology of the Clinton County magnetite district, New York': U.S. Geol. Survey Professional Paper **237**, 88 p.

Prucha, J.J., 1956. 'Pyrite deposits of St. Lawrence and Jefferson Counties, N.Y.': New York State Museum and Science Service Bulletin **357**, 87 p.

Putman, G.W., 1958. *Geology of some wollastonite deposits in the eastern Adirondacks, N.Y.* (unpubl. Master's thesis): Pennsylvania State University, 105 p.

Ramberg, H., 1967. *Gravity, deformation, and the earth's crust as studied by centrifuged models*, New York, Academic Press, 214 p.

Ramsey, J.G., 1980. 'Shear zone geometry: A review': *Jour. Structural Geology*, v. **2**, p. 83-99.

Rankin, D.W., Stern, T.W., McLelland, J., Zartman, R.E., and Odom, A.L., 1983. 'Correlation chart for Precambrian rocks of the eastern United States': *U.S. Geol. Survey, Professional Paper* **1241-E**, p. 1-18.

Sengor, A.M.C., and Kidd, W.S.F., 1979. 'The post-collisional tectonics of the Turkish-Iranian Plateau, and a comparison with Tibet': *Tectonophysics*, v. **55**, p. 361-377.

Seyfert, C.R., 1980. 'Paleomagnetic evidence of a middle Proterozoic collision between North America and Gondwanaland as a cause for metamorphism and deformation in the Adirondacks': *Geol. Soc. Amer. Bulletin, Parts I and II*, v. **91**, p. 118-120, 816-842.

Silver, L.T., 1965. 'U-Pb isotopic data in zircons of the Grenville series of the Adirondack Mountains, New York': *American Geophysical Union Transactions*, v. **46**, p. 164.

Silver, L.T., 1969. 'A geochronologic investigation of the anorthosite complex, Adirondack Mountains, New York', in, Isachsen, Y.W. (ed.), *Origin of anorthosites and related rocks*: New York State Museum and Science Service Memoir **18**, p. 233-252.

Silver, L.T., and Lumbers, S.B., 1966. 'Geochronologic studies in the Bancroft-Madoc area of the Grenville Province, Ontario' (abs): *Geol. Soc. Amer. Spec. Publ.*, v. **87**, p. 156.

Simmons, G., 1964. 'Gravity survey and geological interpretation, northern New York': *Geol. Soc. Amer. Bulletin*, v. **75**, p. 81-98.

Stoddard, E.F., 1976. *Granulite facies metamorphism in the Colton-Rainbow Falls area, northwest Adirondacks, New York* (Ph.D. Thesis): Los Angeles, California, University of California, 271 p.

Tan, L.P., 1966. 'Major pegmatite deposits of New York State': *New York State Museum and Science Service Bulletin*, no. 408, 138 p.

Thivierge, R., Brock, B., Hanmer, S., 1984. 'Structural relationships at the northwestern margin of the Central Metasedimentary Belt, Grenville Province, Ontario' (abs.): Geol. Assoc. Canada Meetings, London, Ontario.

Turner, B., 1980. 'Polyphase Precambrian deformation and stratigraphic relations, central to southeastern Adirondack Mountains, New York': Geol. Soc. Amer. Bulletin **91**, Part II, p. 293-325.

Valley, J.W., 1985. 'Polymetamorphism in the Adirondacks: Granulite facies wollastonite' (this volume).

Valley, J.W., and Essene, E.J., 1977. 'Regional metamorphic wollastonite in the Adirondacks' (abs.): Geol. Soc. Amer., Abstracts with Programs, v. **9**, no. 3, p. 325-327.

Valley, J.W. and Essene, E.J., 1980a. 'Calcsilicate reactions in Adirondack marbles: The role of fluids and solid solutions, Parts I and II': Geol. Soc. Amer. Bulletin **91**, p. 114-117, 720-815.

Valley, J.W. and Essene, E.J., 1980b. 'Akermanite in the Cascade Slide Xenolith and its significance for regional metamorphism in the Adirondacks': Contributions to Mineralogy and Petrology v. **74**, p. 143-152.

Valley, J.W., McLelland, J.M., Essene, E., and Lamb, W., 1983. 'Metamorphic fluids in the deep crust: Evidence from the Adirondacks': Nature, v. **301**, p. 226-228.

Valley, J.W. and O'Neil, J.R., 1982. 'Oxygen isotope evidence for shallow emplacement of Adirondack anorthosite': Nature, v. **300**, p. 497-500.

Valley, J.W. and O'Neil, J.R., 1984. 'Fluid heterogeneity during granulite facies metamorphism in the Adirondacks: Stable isotope evidence': Contributions to Mineralogy and Petrology, v. **85**, p. 158-173.

Vine, F., and Matthews, D., 1963. 'Magnetic anomalies over mid-ocean ridges': Nature, v. **199**, p. 947-949.

Walton, M.S., and deWaard, D., 1963. 'Orogenic evolution of the Precambrian in the Adirondack Highlands, a new synthesis': Koninklijke Nederlandse Akademie van Wetenschappen Proceedings, ser. B, v. **66**, no. 3, p. 98-106.

Whelan, J.F., Rye, R.O., deLorraine, W., 1984. 'The Balmat-Edwards Zn-Pb deposits; synsedimentary ore from Mississippi Valley-type fluids': Econ. Geol. and Bull. Soc. Econ. Geol., v. **79**, no. 2, p. 239-265.

Whitney, P.R., 1983. 'A three-stage model for the tectonic history of the Adirondack region, New York': Northeastern Geology, v. 5, p. 61-72.

Whitney, P.R., 1978. 'The significance of garnet "isograds" in granulite facies rocks of the Adirondacks', in, Metamorphism of the Canadian Shield, Geol. Soc. Can. Paper 78-10, p. 357-366.

Whitney, P.R., and McLelland, J., 1973. 'Origin of coronas in meta-gabbros of the Adirondack Mts., New York': Contributions to Mineralogy and Petrology, v. 39, p. 81-98.

Wiener, R.W., 1977. 'Timing of emplacement and deformation of the Diana Complex along the Adirondack Highlands-Northwest Lowlands boundary': Geol. Soc. Amer., Abstracts with Programs, v. 9, no. 3, p. 329-330.

Wiener, R.W., 1981. Stratigraphy, structural geology and petrology of bedrock along the Adirondack Highlands-Northwest Lowlands boundary near Harrisville, N.Y. (Ph.D. Thesis): Amherst, Massachusetts, University of Massachusetts, 195 p.

Wiener, R.W., 1983. 'Adirondack Highlands-Northwest Lowlands "boundary": A multiply folded intrusive contact with fold-associated mylonization': Geol. Soc. Amer. Bulletin, v. 94, no. 9, p. 1081-1108.

Wiener, R.W., McLelland, J.M., Isachsen, Y.W., and Hall, L., 1984. 'Stratigraphy and structural geology of the Adirondack Mts., New York: Review and synthesis', in, The Grenville Event in the Appalachians, Bartholomew, M.J. (ed.): Geol. Soc. Amer. Special Paper 194, p. 1-55.

Wynne-Edwards, H.R., 1967. 'Westport map area, Ontario, with special emphasis on the Precambrian rocks': Geol. Survey of Canada Memoir 346, 142 p.

Wynne-Edwards, H.R., 1969. 'Tectonic overprinting in the Grenville Province', in Wynne-Edwards, H.R., ed., Age relations in high-grade metamorphic terrains: Geol. Assoc. of Canada Special Paper 5, p. 5-16.

Zeitz, I., and Gilbert, F.P., 1981. 'Aeromagnetic map of New York, Adirondack Mountains sheet': U.S. Geol. Survey Geophysical Investigation Map GP-938 (sheet 5 of 5), scale 1:250,000.

POLYMETAMORPHISM IN THE ADIRONDACKS: WOLLASTONITE AT CONTACTS OF
SHALLOWLY INTRUDED ANORTHOSITE

John W. Valley

Department of Geology and Geophysics
The University of Wisconsin, Madison, WI 53706, USA

ABSTRACT

Two distinct metamorphic events can be resolved in the N.E. Adirondacks through study of phase equilibria and stable isotope geochemistry at occurrences of wollastonite. Low values of $\delta^{18}O$ (to -1.3) at the wollastonite mines near Willsboro indicate that large amounts of heated meteoric water were involved in the formation of skarns at anorthosite contacts. This process is only consistent with magmatic intrusion and crystallization of anorthosite at relatively shallow depths (<10 km). Likewise, the formation of wollastonite, monticellite, and akermanite in a marble xenolith, surrounded by anorthosite at Cascade Slide, can best be explained by low pressure decarbonation reactions. Alternate theories of high pressure genesis for these minerals require large amounts of fluid infiltration to dilute CO_2. Sharp gradients in $\delta^{18}O$ across this xenolith prove that H_2O could not have been such a diluent and no evidence exists for the presence of large amounts of any other fluid component. This evidence of shallow contact metamorphism is in marked contrast to the depths indicated by mineral barometers for the ensuing granulite facies metamorphism (23-26 km). These results, in conjunction with recent Sm-Nd geochronology, support shallow (<10 km) anorthosite intrusion and contact metamorphism at ~1.3 by. followed by deep granulite facies metamorphism at 1.1-1.0 by.

INTRODUCTION

Wollastonite is a common product of high temperature, low pressure contact metamorphism of siliceous limestones and it is classically regarded as an index mineral of the sanidinite facies. Experimental results support this view for conditions of high f_{CO_2}, but also show wollastonite as more stable than calcite + quartz at greatly reduced temperatures for conditions with CO_2-poor fluids (1). Thus, an ambiguity arises and wollastonite is also stable at higher pressures under amphibolite and granulite facies conditions at low f_{CO_2}.

At the higher pressures of regional metamorphism, CO_2-poor fluids are necessary to stabilize wollastonite at T<800°C. The common association of wollastonite and low pressure contact aureoles reflects the CO_2-rich fluids that are generated by prograde metamorphism of limestones. Occurrences of regionally metamorphosed wollastonite must therefore represent more complex fluid conditions than those of simple volatilization. In particular, low f_{CO_2} could result from fluid-absent metamorphism or from H_2O infiltration. Thus, wollastonite occurrences provide important clues as to the compositions and migration histories of fluids in the deep crust.

In the Adirondacks, numerous wollastonite occurrences are found in metasediments associated with anorthosite or syenite. These rocks were deformed and recrystallized during granulite facies metamorphism raising the question of genesis; did wollastonite originally form in a low pressure contact aureole (moderate to high f_{CO_2}) that was subsequently overprinted by granulite facies metamorphism (fluid-absent) or was wollastonite formed during the granulite facies event (H_2O infiltration)? Related questions include: did anorthosite intrude at shallow or deep levels in the crust and did crystallization of anorthosite provide heat causing the granulite event?

Oxygen isotopic ratios are sensitive to fluid movements, especially at limestone contacts with igneous rocks where pre-metamorphic isotopic differences are large. In the Adirondacks oxygen isotopic compositions from detailed traverses across contacts in conjunction with phase equilibria, indicate an early low pressure (P<3kbar) contact metamorphism, locally involving large amounts of heated meteoric water, followed by high pressure (7-8kbar), fluid-absent granulite facies metamorphism. This evidence of polymetamorphism provides an important constraint for tectonic models of the Adirondacks and the Grenville Province as a whole.

ANORTHOSITE INTRUSION AND METAMORPHISM

Recent workers in the Adirondacks have tended towards accepting one of two broad tectonic models; "polymetamorphism" or "isobaric-cooling", as shown in Figure 1 (2-11). Certain aspects of the pressure-temperature-time relations for the younger event in these models, Grenville metamorphism, are well-known, but earlier events, particularly the intrusion of anorthosite, are controversial, as are many of the details.

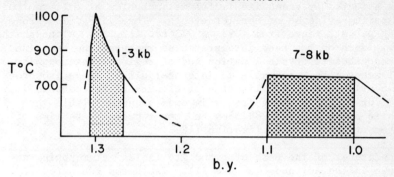
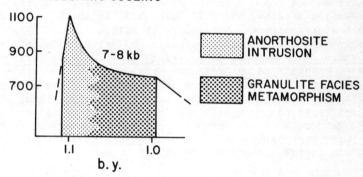

Figures 1a,b. Approximate pressure, temperature, and time relations for magmatic anorthosite intrusion and granulite facies metamorphism in the Adirondacks. Figure 1a shows a polymetamorphic tectonic model where anorthosite intrusion is a separate, shallow event followed by regional metamorphism. Figure 1b shows an isobaric cooling model where anorthosite intrusion is deep and where isobaric crystallization and cooling of anorthosite provides heat for granulite facies metamorphism.

All radiogenic age-dating systems in the Adirondacks have been strongly affected by granulite facies metamorphism at 1.1. to 1.0 by., obscuring evidence of earlier events. Three Sm-Nd isochrons, including garnet, range from 1098 to 973 my. (12,13). Although the 1098±7my. isochron has been interpreted as representing both the time of anorthosite crystallization and of regional metamorphism (13), it seems more reasonable to interpret all of these isochrons as representing the crystallization or closure of metamorphic garnet. One Sm-Nd isochron from a layered sequence within the anorthosite suite yields 1288±36my and may represent the time of primary anorthosite crystallization (12).

Temperatures of the peak of granulite facies metamorphism are well-known in the Adirondacks (see 11). Isotherms contoured from reintegrated feldspar pairs at 56 localities and reintegrated Fe-Ti oxides at 11 localities reveal a "bulls-eye" pattern nearly concentric with the main Anorthosite Massif (fig. 2). Estimated temperatures exceed 775°C near the core of the massif, decrease to 725° near the margins, and fall below 625° in the NW Lowlands. These temperature estimates are generally within ±15° of those from other systems including calcite-dolomite and garnet-clinopyroxene and this concordance extends to a variety of rock types including anorthosite, granitic to gabbroic gneisses, marbles, and semi-pelitic gneisses. Such concordance indicates that these are peak metamorphic temperatures rather than retrograde temperatures which would be scattered due to variations in blocking temperature. Calculated temperatures are further supported by metamorphic grade which is upper amphibolite facies in the NW Lowlands (600-700°C) and granulite facies elsewhere (700-800°C).

Pressures of granulite facies metamorphism are likewise well-known from application of 8 geobarometers at 52 localities (fig. 2). Values range from 7.0±0.4 kbar in the NW Lowlands to 7.6±0.3 kbar in the east-central Adirondacks (11).

The magmatic temperatures and pressures of anorthosite intrusion are less well known than those of metamorphism. The magmatic temperature has been estimated at 1100±100°C from reintegration of orthopyroxene-clinopyroxene pairs (14) and at present there are no estimates of magmatic pressures for anorthosite based on mineral equilibria. As will be discussed later, oxygen isotope ratios in wollastonite skarn suggest that anorthosite intrusion was shallow (7).

POLYMETAMORPHISM IN THE ADIRONDACKS 221

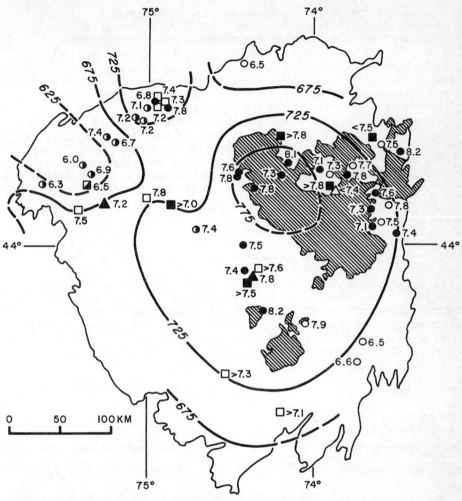

☐ ILMENITE-SILLIMANITE-QUARTZ-GARNET-RUTILE
■ FERROSILITE-FAYALITE-QUARTZ
◨ SPHALERITE-PYRRHOTITE-PYRITE
○ FAYALITE-ANORTHITE-GARNET
● FERROSILITE-ANORTHITE-GARNET-QUARTZ
◐ ANORTHITE-GARNET-SILLIMANITE-QUARTZ
▲ KYANITE-SILLIMANITE
△ AKERMANITE

Figure 2. Peak metamorphic values of temperature and pressure in the Adirondack Mountains, New York, U.S.A. Isotherms are contoured from estimates of feldspar and Fe-Ti oxide thermometry and pressures are estimated from eight independent geobarometers, (after Bohlen, Valley and Essene (1985), reference 11). Anorthosite is shaded.

It has been proposed, based on optical reintegration, that the compositions of finely exsolved Fe-rich orthopyroxenes from syenitic gneisses represent inverted pigeonites and indicate a minimum pressure of 8 kbar for magmatic crystallization and that Adirondack anorthosites were likewise emplaced deeply (15). However, the following alternate interpretations are consistent with shallow anorthosite intrusion and must also be considered: equilibration or crystallization of pigeonite during granulite facies metamorphism at 7-8 kbar, magmatic crystallization of impure, natural pigeonite at 3 kbar, or deep emplacement of syenite that is unrelated to shallowly emplaced anorthosites. Evaluation of these possibilities will require direct analysis by electron microprobe to confirm pyroxene compositions and high pressure experimentation to accurately determine the stability of natural Fe-rich pigeonites that deviate from the ideal CaO-MgO-FeO system (16; Lindsley, pers. comm. 1984).

Garnet coronas, armoring orthopyroxene or Fe-Ti oxides, are a prominent feature in many Adirondack anorthosites and gabbros. The persistence of these coronas results from slow diffusion rates due to low water activity during granulite facies metamorphism. Their formation is consistent with polymetamorphism (fig. 1a) with prograde garnet growth due to increasing P and T of granulite metamorphism. However some workers have attributed coronas to isobaric-cooling (fig 1b) with retrograde garnet growth due to decreasing T at constant P after igneous intrusion (3,9,11).

WOLLASTONITE

Localities of wollastonite and of calcite + quartz are shown in Figure 3. These occurrences range from coarse-grained wollastonite + clinopyroxene + garnet skarn to siliceous marbles that may contain only minor wollastonite. The detailed mineralogy, equilibria, and locations of these rocks have been previously described (6,17,18).

The wollastonite + clinopyroxene + garnet association is seen at many localities associated with anorthosite in the NE Adirondacks. Most notably it is found in 3 deposits near Willsboro, two of which have been economically developed and contain over 10^7 tons of $CaSiO_3$ (localities 1 and 2 in figure 3). At the Willsboro deposits, typical ore contains 75% wollastonite with quartz and calcite being totally absent. It is clear that such large amounts of a high variance mineral assemblage formed through metasomatism and that great quantities of fluid were locally involved.

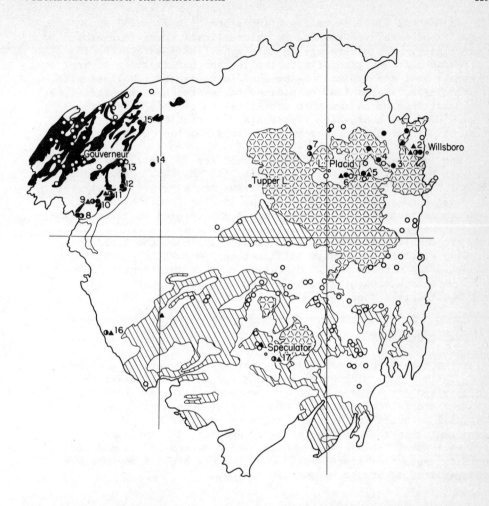

Figure 3. Generalized geologic map of the Adirondacks, N.Y. Major rock units are shown as marble (black), anorthosite (triangles) and metasediments (diagonal lines). Unpatterned areas are principally orthogneisses and the Diana Complex (in the NW near localities 10 and 11) is outlined. Sample localities are keyed by mineral assemblage: wollastonite (solid circle), calcite + quartz (open circle), wollastonite + calcite + quartz (half-filled circle), and grossular-rich garnet + quartz (triangle). Wollastonite in the NE and central Adirondacks is always near anorthosite. Wollastonite in the NW and west-central is near syenite. All Wollastonite localities have experienced granulite facies metamorphism.

At Cascade Slide (locality 6 in figure 3) a 30 x 200 m. xenolith of marble surrounded by calc-silicate skarn is enclosed by anorthosite. The mineralogy in this xenolith is unique for the Adirondacks or any granulite facies terrane and includes several minerals that are typical of the sanidinite facies: wollastonite, monticellite, akermanite, cuspidine, harkerite, and wilkeite (4,6). The conditions of metamorphic equilibration of this body were 750±30°C, 7.4±1 kbar which represents the granulite event, consistent with either polymetamorphism or isobaric-cooling (6).

One important aspect of these localities is their proximity to anorthosite (in the NE) and syenite (in the NW). Many occur directly at contacts or within xenoliths (refs. 2,4,6,7,10,17,18, 19), although a few localities are up to 1 km from the nearest igneous contact (in 3 dimensions). If interpreted in terms of the polymetamorphic model this relationship indicates preservation of the pregranulite facies contact aureole, while in the isobaric cooling model large amounts infiltrating H_2O must be postulated in order to decrease f_{CO_2}. Migration of this infiltrating H_2O must have been concentrated along contacts in order to explain the spatial relationship.

MINERAL EQUILIBRIA

The metamorphism of siliceous marbles is generally portrayed by a T-X(H_2O-CO_2) diagram at constant pressure. T-X diagrams are well-suited for studies of contact metamorphism where pressure changes are seldom important. They may also be appropriate for regional metamorphism if the effects of variable X_{CO_2} are more important than those of variable P. However, the inherent assumptions of T-X diagrams should always be evaluated, and for certain types of regional metamorphism, they are not appropriate and may lead to errors of interpretation.

By assuming constant P_{Total} for a T-X diagram it is assumed that $P_{H_2O} + P_{CO_2} = P_{Total}$. This assumption is never totally correct, even if the metamorphic fluid is truly represented by the C-O-H system, as some finite fugacity of CH_4, CO, and H_2 is always present. For many metamorphic environments the total partial pressure of these three species is less than 1% of P_{Total} and they may safely be neglected. However, at low f_{O_2}, particularly in the presence of graphite, CH_4 can become the dominant fluid species (20,21). During granulite facies metamorphism a second assumption, that of fluid saturation, is controversial. If granulites form by fluid-absent metamorphism due to removal of H_2O-rich fluids by melting (22), rather than by CO_2-flooding (23), then $P_{H_2O} + P_{CO_2}$ may be much less than P_{Total}. Likewise, many high temperature igneous rocks and their contact aureoles crystallize with low H_2O contents. Subsequent granulite facies metamorphism of such rocks, in the absence of CO_2-infiltration, may also be fluid-absent.

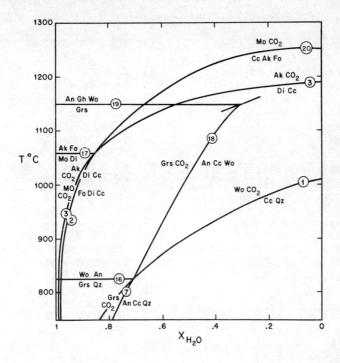

Figure 4. T-X (H_2O-CO_2) diagram at 8 kbar showing reactions that restrict the stability of wollastonite, grossular, monticellite, and akermanite (see Table 1). Conditions of fluid-absent granulite facies metamorphism cannot be properly modelled by this diagram because of the inherent assumption $P_{H_2O} + P_{CO_2} = P_{Total}$.

Figure 4 shows reactions that restrict the stability of wollastonite, grossular, monticellite and akermanite as a function of T-X at 8 kbar. Table 1 lists abbreviations and further identifies these reactions. Reaction locations were calculated using the program EQUILI (Wall and Essene, unpbd) and experimentally reversed equilibrium points as described previously (6,17).

Table 1. Mineral abbreviations and reactions.

ABBREVIATIONS			REACTIONS	
Ak	akermanite	$Ca_2MgSi_2O_7$	1)	$Cc + Qz = Wo + CO_2$
An	anorthite	$CaAl_2Si_2O_8$	2)	$Fo + Di + 2Cc = 3Mo + 2CO_2$
Cc	calcite	$CaCO_3$	3)	$Cc + Di = Ak + CO_2$
Clt	clintonite	$CaMg_2Al_4SiO_{10}(OH)_2$	4)	$3Tr + 5Cc =$
Di	diopside	$CaMgSi_2O_6$		$2Fo + 11Di + 5CO_2 + 3H_2O$
En	enstatite	$MgSiO_3$	5)	$Tr + 3Cc + 2Qz =$
Fo	forsterite	Mg_2SiO_4		$5Di + 3CO_2 + H_2O$
Gh	gehlenite	$Ca_2Al_2SiO_7$	6)	$Ru + Cc + Qz = Spn + CO_2$
Grs	grossular	$Ca_3Al_2Si_3O_{12}$	7)	$An + 2Cc + Qz = Grs + 2CO_2$
Kfs	K-feldspar	$KAlSi_3O_8$	8)	$Spn + Cc = Pv + Wo + CO_2$
Lm	lime	CaO	9)	$Fo + Cc = Mo + Per + CO_2$
Mo	monticellite	$CaMgSiO_4$	10)	$Grs + Cc = Gh + 2Wo + CO_2$
Per	periclase	MgO	11)	$3Clt + Fo + CO_2 =$
Ph	phlogopite	$KMg_3AlSi_3O_{10}(OH)_2$		$2Di + 6Spl + Cc + 3H_2O$
Pt	portlandite	$Ca(OH)_2$	12)	$Tr = 2Di + 3En + Qz + H_2O$
Pv	perovskite	$CaTiO_3$	13)	$Pt + CO_2 = Cc + H_2O$
Qz	quartz	SiO_2	14)	$Cc = Lm + CO_2$
Ru	rutile	TiO_2	15)	$Ph + 3Qz = 3En + Kfs + H_2O$
Spl	spinel	$MgAl_2O_4$	16)	$Grs + Qz = 2Wo + An$
Spn	sphene	$CaTiSiO_5$	17)	$3Mo + Di = 2Ak + Fo$
Tr	tremolite	$Ca_2Mg_5Si_8O_{22}(OH)_2$	18)	$An + Cc + Wo = Grs + CO_2$
Wo	wollastonite	$CaSiO_3$	19)	$2Grs = An + Gh + 3Wo$
			20)	$Cc + Ak + Fo = 3Mo + CO_2$

Temperatures near the anorthosite contact are likely to have been 800-900°C at the time of magmatic crystallization. At 8 kbar figure 4 shows that wollastonite is only stable at $X_{H_2O}>0.5$ while monticellite and akermanite require $X_{H_2O}>0.97$. If much higher temperatures of 1100°C are assumed, equal to the anorthositic magma, then monticellite still requires $X_{H_2O}>0.8$. Thus the isobaric-cooling model requires that large amounts of H_2O infiltrated all wollastonite and monticellite-bearing rocks to dilute the pure CO_2 fluid produced by reactions (1), (2) and (3). The quantity of H_2O necessary can be calculated from modal analysis, X_{H_2O}, and reaction stoichiometry. For instance consideration of reaction progress for reaction (2) shows that the prominent rock at Cascade Slide which is 15-25% monticellite and 75-85% calcite would have required infiltrating water equivalent to 1.13 times the weight of the rock (H_2O/rock = 2.25 if ratioed by atoms of oxygen) at 780°C, $X_{H_2O} = 0.982$.

In the polymetamorphic model, the low initial pressure of 1-3 kbar enhances the formation of wollastonite, grossular, monticellite and akermanite at 800-900°C in CO_2-rich conditions because all volatilization reaction temperatures in figure 4 are greatly reduced with decreased pressure. It is an important contrast that this model does not require infiltration of H_2O, unlike isobaric cooling at 7-8 kbar.

During the subsequent granulite facies metamorphism (P = 7-8 kbar) any fluids that were present must have been CO_2-poor regardless of which model one considers. This may mean that a H_2O-rich fluid was present. However, in a polymetamorphic scenario the rocks are already decarbonated and it is also possible that the granulite facies metamorphism was fluid-absent. The possibility of fluid-absent metamorphism cannot be portrayed in figure 4.

Figure 5 is calculated with P and T constant at 8 kbar, 780°C. Reactions are located by calculation from experimental reversals and thermochemical data (21). The axes may be related to $X_{H_2O}-CO_2$ by the relation $f_i = \gamma_i P_i = X_i \gamma_i P_T$ where γ_{H_2O} = 1.2 and γ_{CO_2} = 14. All reactions plot as straight lines in figure 5 with slope controlled only by the stoichiometry of H_2O and CO_2. The dashed curve (BVL, binary vapor line) solves the equation 8 kbar = $P_{H_2O} + P_{CO_2}$ and represents a horizontal line across figure 4 at 780°C. Outside of the BVL (upper right) fluid pressure exceeds lithostatic pressure. As rocks have little strength at these P-T such over-pressures cannot be sustained for geologically significant periods of time and fluids would escape upwards within the crust. The area outside of the BVL is thus inaccessible to geological processes while the area inside of the BVL is of the most interest for the study of granulites. Inside of the BVL represents conditions where $P_{H_2O} + P_{CO_2} < P_{Total}$ regardless of whether this condition is due to fluid-absent metamorphism, high f_{CH_4} or non-C-O-H fluid components (N,S,Cl,F). While small amounts of non-C-O-H fluids are surely present during metamorphism, there is no evidence that either the compositions or the quantities required by the following discussion are ever attained during regional metamorphism.

In figure 5 it can be seen that wollastonite indicates $\log f_{CO_2} < 4.3$ which corresponds to $X_{CO_2} < 0.2$, $X_{H_2O} > 0.8$ if conditions were on the BVL or fluid-absent conditions at lower X_{H_2O}. Unfortunately few reactions have been identified in wollastonite-bearing samples that involve H_2O. Such reactions would cross the vertical reactions in figure 5 and would provide important limits in conjunction with the many decarbonation reactions.

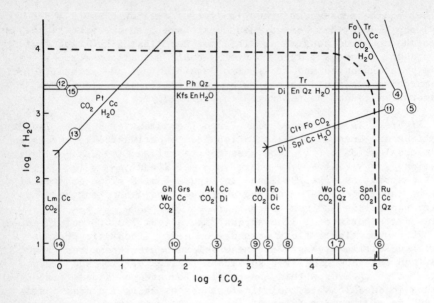

Figure 5. Plot of log f_{H_2O} vs. log f_{CO_2} at 780°C, 8 kbar. In the marble xenolith at Cascade Slide mineral assemblages record equilibration during granulite facies metamorphism, show gradients of over a factor of 60/120m in f_{CO_2} and 10/100m in f_{H_2O}, and indicate fluid-absent metamorphism.

One rock unit in the Cascade Slide Xenolith contains the assemblage monticellite + forsterite + diopside + calcite + spinel ± vesuvianite ± wilkeite ± sphalerite (4,6). This assemblage is buffered along reaction (2) and below reaction (11). Adjustment of reactions (2) and (11) to correct for the effects of solid solution does not significantly change their locations because quantitative analysis shows these minerals to be nearly stoichiometric (sample LP77-201, ref. 6). While most of the reactions in figure 5 are believed to be accurately placed ±0.2 log unit, reaction (11) is less certain as its calculation involved estimation of ΔG_f (clintonite) from experimental reversals of chlorite-clintonite reactions (24), but the accuracy is still believed to be better than ±0.5 log unit. Even if a large error of 0.5 log unit in f_{H_2O} or fCO_2 is assumed fluid-absent metamorphism is indicated for this sample unless the fluid was comprised of over 90% species other than CO_2 or H_2O.

Figure 5 also shows that fluids were locally buffered in the rock units at Cascade Slide. Akermanite + monticellite + wollastonite assemblages restrict f_{CO_2} to the left of reaction (3) (log f_{CO_2} < 2.5, sample LP77-210-10) while 30 m. away rocks are buffered along reaction (2) and 100 m. away assemblages including calcite + quartz indicate values to the right of reaction (1) (log f_{CO_2} > 4.3, sample LP201-3). Thus f_{CO_2} varied by over a factor of 60 within less than 100 m. Values of f_{H_2O} are likewise indicated to vary by over a factor of 10 between a sample containing tremolite (above reaction (12)) and those buffered to below the intersection of reactions (2) and (11). These fluid gradients further support the conclusion that granulite facies metamorphism was fluid-absent for these rocks as large amounts of a pervasively infiltrating fluid would exhaust the buffer capacity of these rocks and homogenize fluid compositions.

OXYGEN ISOTOPES

Oxygen isotopic ratios provide the first clear evidence supporting the polymetamorphic model over isobaric cooling. These ratios are particularly sensitive to movements of H_2O-CO_2 fluids across anorthosite-marble contacts because of large premetamorphic differences in $\delta^{18}O$. In the Adirondacks marbles have the highest values of $\delta^{18}O$ of any rock type, averaging 19.0 for 54 granulite facies calcites (7,10). Adirondack anorthosites have slightly enriched values, relative to anorthosite elsewhere, and average $\delta^{18}O=9.7$ for plagioclase (10,25). Values of wollastonite range from 21.1‰ in marble to -1.3‰ in skarn (Table 2).

Values of $\delta^{18}O$ for calcite vary greatly about the mean from 12.3 to 26.1. Small variations of ±1-2‰ may be due to differences in metamorphic temperature and amounts of volatilization, but calculations show that this will not be a larger effect. The scatter seen in Adirondack marbles must therefore be due to differences in premetamorphic composition or fluid infiltration (10). On average differences in premetamorphic composition must be the dominant cause of sample variability as 37 unmetamorphosed limestones of Grenville age show a similar range from 11.3 to 26.2‰ (average = 20.7‰, ref. 26). This view is supported by the general correlation of low $\delta^{18}O$ with lower modal percentage of carbonate minerals which suggests variable amounts of sediment mixing, $\delta^{18}O$-rich carbonates with $\delta^{18}O$-poor silicates (10).

Table 2. Values of $\delta^{18}O$ for Adirondack wollastonites

Sample	$\delta^{18}O$	Locations in Figure 3	
AUS 78-4	-1.2	3	skarn, ore-rock, Lewis Mine
81-AUS-1	+1.2	1	skarn, ore-rock, Deerhead Deposit
GOV 25-4	15.4	11	marble
GOV 76	8.8	11	skarn, ore-rock, Valentine Deposit, Harrisville Mine
GOV 77-300-1	21.1	15	marble
GOV 77-307-10	12.5	11	marble, 4 m. from syenite contact
GOV 77-308-1	15.6	11	marble
GOV 78-14	16.2	10	marble
81-GOV-1	13.2	11	marble
LP 77-211-2	15.6	5	calc-silicate
LP 77-212-11	10.4	7	Mt. Pisgah
LP 77-213-17	10.8	7	marble
PL 1197	7.5	16	calc-silicate
V 3-11-461	4.3	10	skarn, ore-rock, Valentine Deposit, Harrisville Mine
W 50-1	0.0	1	skarn, ore-rock Willsboro Mine
W 50-161	0.9	1	skarn, ore-rock Willsboro Mine
W 50-163	-1.3	1	skarn, ore-rock Willsboro Mine
W 50-181	3.5	1	skarn, ore-rock Willsboro Mine
W 50-183	1.0	1	skarn, ore-rock Willsboro Mine
W 50-186	7.0	1	skarn, ore-rock Willsboro Mine
W 81-3-88	2.9	1	skarn, ore-rock Willsboro Mine
W 81-3-91	2.5	1	skarn, ore-rock Willsboro Mine
W 81-7-1	3.1	1	skarn, ore-rock Willsboro Mine
82-W-1	-0.7	1	skarn, ore-rock Willsboro Mine

In addition to the above arguments which suggest that, *on average*, premetamorphic differences are important for controlling $\delta^{18}O$ values in marbles, it is often possible to show that *an individual sample* has preserved its premetamorphic value (10). This is the case because large amounts of fluid infiltration and isotopic exchange will have an homogenizing effect, smoothing out $\delta^{18}O$ gradients and eliminating extreme values. In the Adirondacks three situations are recognized that, based on oxygen isotope data alone, indicate preservation of premetamorphic values: areas of sharp gradients in $\delta^{18}O$, samples with very high $\delta^{18}O$ calcite (25 to 27.2), and samples with extremely low $\delta^{18}O$ wollastonites (-1.3 to 3.1). In some instances mineral equilibria can be found that are restrictive and these results further support the view that fluid dominated processes were not important during granulite facies metamorphism of many rocks.

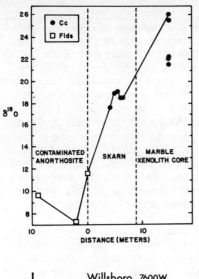

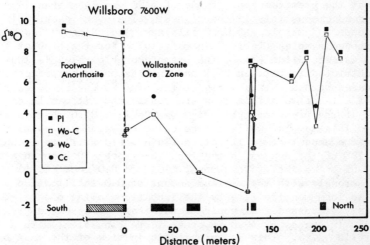

Figures 6a,b. Values of $\delta^{18}O$ for samples from across contacts of anorthosite and wollastonite skarn. Figure 6a traverses the marble xenolith at Cascade Slide showing a sharp gradient in $\delta^{18}O$ (16‰/15 m.) and preservation of high $\delta^{18}O$ calcite in the core. Figure 6b traverses the wollastonite deposit at Willsboro, showing a sharp gradient in $\delta^{18}O$ (6‰/1 m.) and low $\delta^{18}O$ wollastonite (to -1.3) near the anorthosite contact. These gradients and extreme values formed before granulite facies metamorphism and show that these rocks have not experienced sufficient amounts of pervasive fluid infiltration to homogenize them.

Cascade Slide Xenolith

Figure 6a shows $\delta^{18}O$ values from a traverse across the xenolith at Cascade Slide from contaminated, carbonate-free anorthosite across garnet + clinopyroxene ± wollastonite skarn that is 3-20% carbonate and into carbonate-rich marble in the xenolith core. In the anorthosite values average 9.5‰, similar to other Adirondack anorthosite, and some increase in $\delta^{18}O$ is seen towards the skarn which has higher values of 17.6 to 19.3. In the core of the xenolith calcite in monticellite marble has $\delta^{18}O$ = 26.1. It is clear that the intermediate $\delta^{18}O$ values in skarn represent mixing between marble and anorthosite on the scale of 1-10m, but the mechanism and timing of this exchange is unknown. Of more importance is the high $\delta^{18}O$ marble from the core which is surrounded by lower $\delta^{18}O$ anorthosite. Any fluid infiltration of this marble during metamorphism would have decreased its $\delta^{18}O$.

The amount of fluid that infiltrated a high $\delta^{18}O$ sample can be estimated from knowledge of its premetamorphic isotopic composition and the conditions of metamorphic volatilization. It is almost certain that the premetamorphic value of $\delta^{18}O$ was less than 28.1 in the Adirondacks as this is the highest value ever measured in 215 limestones and marbles of Grenville age, including 69 low grade or unmetamorphosed samples (Table 2 in ref. 10). For the highest $\delta^{18}O$ calcite at Cascade Slide (26.1) the premetamorphic value is thus tightly restricted to between 26.1 and 28.1 as no known process could have increased such an already high value of $\delta^{18}O$ during metamorphism. When the effects of volatilization are considered, it can be calculated that if the premetamorphic $\delta^{18}O$ was 26.1 to 27.0 then no fluid infiltrated this sample, while if the original value was as high as 28.1, then the amount of infiltrating fluid would still represent less than 10% of the oxygen in the rock (H_2O/rock = 0.1 on an atomic basis or 0.05 by weight). This upper limit to H_2O/rock (<0.1) contrasts strongly with the minimum amount of H_2O infiltration necessary to form monticellite by the isobaric-cooling model (⩾2.25) as previously discussed. It would be necessary to assume a premetamorphic $\delta^{18}O$ in excess of 40.0 in order to be consistent with a H_2O/rock ratio ⩾2.25. This is unrealistic in view of the vast body of data indicating that $\delta^{18}O$ was below 28.1.

As previously mentioned the petrologic limit to H_2O/rock of ⩾2.25 only holds for the isobaric cooling model. If the Cascade Slide Xenolith was first metamorphosed at low P=1-3 kbar then the temperatures (and X_{H_2O}) of reactions (1),(2) and (3) are reduced and H_2O infiltration is not required. Thus the isotope data in conjunction with the mineral equilibria are not consistent with isobaric cooling and support polymetamorphism and shallow intrusion of anorthosite.

Willsboro Skarn Belt

Figure 6b shows $\delta^{18}O$ values from a traverse across the wollastonite deposit at Willsboro (7,10). The traverse extends from Whiteface facies gabbroic anorthosite in the footwall to the south, across 125 m. of wollastonite + clinopyroxene + garnet skarn that shows extreme ^{18}O depletion and is interlayered with amphibolite which may represent metasomatized anorthosite sheets. Mixed rocks including wollastonite skarn, amphibolite and granitic gneiss occur in the hanging wall to the north. The bars along the bottom of figure 6b are coded to rock type; diagonal lines represent anorthosite and solid black represents major intersection of wollastonite ore-rock.

The values of $\delta^{18}O$ in wollastonite from the ore-zone include the lowest that have yet been reported from a granulite facies terrane (-1.3 to 3.1). These low $\delta^{18}O$'s represent a depletion of 20‰ relative to average Adirondack carbonates. Metamorphic decarbonation has undoubtedly caused a small amount of this depletion, but could not account for more than 1/3 of this 20‰ anomaly. Exchange with heated meteoric water during skarn formation best explains these results. It is well-documented that meteoric waters circulate in aureoles of shallow plutons where rocks are fractured, but it is impossible that waters under hydrostatic pressure could infiltrate to the depths of granulite facies metamorphism where any fluids would be at lithostatic pressure (~ 3 $P_{hydrostatic}$). These results thus indicate that skarn formation proceeded at relatively shallow depths (<10km, ref. 7). This is only possible under polymetamorphism.

ADIRONDACK TECTONICS

The results of mineral equilibria and stable isotope studies of wollastonite occurrences at anorthosite contacts in the NE Adirondacks support a polymetamorphic model as shown in figure 1a. In this model anorthosite intrusion and crystallization is indicated to be an early and tectonically distinct event as supported by one 1288 my. Sm-Nd isochron (12). Anorthosite intrusion may have been at various crustal levels throughout the Grenville Province, but in the NE Adirondacks it was shallow, <10 km. Skarn formation, involving large amounts of heated meteoric water, accompanied magmatic crystallization. Some marbles, which can now be recognized by average or high $\delta^{18}O$ values, reacted at the anorthosite contacts to form calc-silicate minerals including wollastonite, grossular, monticellite, and akermanite. In contrast to the skarns little or no meteoric water infiltrated these rocks. A complex deformation history has been superimposed on these contacts and granulite facies metamorphism has overprinted all rocks.

The conditions of the granulite facies metamorphism were deep burial (7-8 kbar, 23-26 km) at 700-800°C. Granulite fluid conditions were highly variable; f_{H_2O} was generally low, wollastonite rocks were low in f_{CO_2}, and many rocks were not fluid saturated indicating fluid-absent metamorphism. There is no clear evidence supporting pervasive CO_2 or H_2O infiltration during regional metamorphism though some rocks may have locally been infiltrated.

Other, sometimes controversial conclusions can be made if the polymetamorphic model is assumed. Garnet coronas are indicated to grow in response to increasing P and T during regional metamorphism and are not retrograde. The circular pattern of metamorphic isotherms results from domical uplift after regional metamorphism. The heat of magmatic crystallization of anorthosite was largely dissipated by convective fluid flow though some heat may have fused granitic rocks. The crystallization of anorthosite was not the heat source for regional metamorphism and other as yet undated metaigneous rocks may be candidates.

It is likely that many of the features of this model apply to the Grenville Province as a whole. In particular, if anorthosite intrusion was a separate tectonic event from regional metamorphism in the Adirondacks then models of anorthosite genesis during crustal rifting and spreading followed by regional metamorphism during continent-continent collision are supported (5,11,27,28).

Support from the National Science Foundation (EAR 75-22388, 78-22568, 81-21214, 83-11772) and The Gas Research Institute (5083-260-0852) is appreciated.

REFERENCES

(1) Greenwood, H.J. (1967). Wollastonite: stability in H_2O-CO_2 mixtures and occurrence in a contact-metamorphic aureole near Salmo, British Columbia, Canada. Am. Mineral. 52, 1669-1680.
(2) Isachsen, Y.W., McLelland, J., Whitney, P.R. (1975). Anorthosite contact relations in the Adirondacks and their implications for geological history. Geol. Soc. Am. Abstr. w. Prog. 7, 78-79.
(3) Whitney, P.R. (1978). The significance of garnet "isograds" in granulite facies rocks of the Adirondacks. Geol. Surv. Can. Pap. 78-10, 357-366.
(4) Tracy, R.J., Jaffe, H.W., Robinson, P. (1978). Monticellite marble at Cascade Mt., Adirondack Mountains, N.Y. Am. Mineral. 63, 991-999.
(5) McLelland, J., Isachsen, Y. (1980). Structural synthesis of the southern and central Adirondacks: A model for the Adirondacks as a whole and plate-tectonic interpretations. Geol. Soc. Am. Bull. 91, 68-72, 208-292.
(6) Valley, J.W., Essene, E.J. (1980). Akermanite in the Cascade Slide Xenolith, and its significance for regional metamorphism in the Adirondacks. Contr. Min. Pet. 74, 143-152.
(7) Valley, J.W., O'Neil, J.R. (1982). Oxygen isotope evidence for shallow emplacement of Adirondack anorthosite. Nature 300, 497-500.
(8) Whitney, P.R. (1983). A three-stage model for the tectonic history of the Adirondack Region, N.Y. New England Geol. 5, 61-72.
(9) Whitney, P.R., McLelland, J.M. (1983). Origin of biotite-hornblende-garnet coronas between oxides and plagioclase in olivine metagabbros, Adirondack Region, N.Y. Contr. Min. Pet. 82, 34-41.
(10) Valley, J.W., O'Neil, J.R. (1984). Fluid heterogeneity during granulite facies metamorphism in the Adirondacks: stable isotope evidence. Contr. Min. Pet. 85, 158-173.
(11) Bohlen, S.R., Valley, J.W., Essene, E.J. (1985). Metamorphism in the Adirondacks. I. Pressure and temperature. J. Petrol., in press.
(12) Ashwal, L.D., Wooden, J.L. (1983). Sr and Nd isotope geochronology, geologic history, and origin of the Adirondack Anorthosite. Geochim. Cosmochim. Acta 47, 1875-1885.
(13) Basu, A.R., Pettingill, H.S. (1983). Origin and age of Adirondack anorthosites reevaluated with Nd isotopes. Geology 11, 514-518.

(14) Bohlen, S.R., Essene, E.J. (1978). Igneous pyroxenes from metamorphosed anorthosite massifs. Contr. Min. Pet. 65, 433-442.
(15) Ollila, P.W., Jaffe, H.W. and Jaffe, E.B. (1984). Iron-rich inverted pigeonite: evidence for the deep emplacement of the Adirondack Anorthosite Massif. Geol. Soc. Am. Abstr. w. Prog., 54.
(16) Lindsley, D.H. (1983). Pyroxene thermometry. Am. Mineral. 68, 477-493.
(17) Valley, J.W., Essene, E.J. (1980). Calc-silicate reactions in Adirondack marble: the role of fluids and solid solution. Geol. Soc. Am. Bull. 91, 114-117, 720-815.
(18) Putman, G.W. (1958). Geology of some wollastonite deposits in the Eastern Adirondacks, New York. MS thesis, Penn. State Univ., 105 p.
(19) Buddington, A.F. (1939). Adirondack igneous rocks and their metamorphism. Geol. Soc. Am. Mem. 7, 354 p.
(20) French, B.M. (1966). Some geological implications of equilibrium between graphite and a C-O-H gas phase at high temperatures and pressures. Rev. in Geophys. 4, 223-253.
(21) Valley, J.W., Peterson, E.U., Essene, E.J., Bowman, J.R. (1982). Fluorphlogopite and fluortremolite in Adirondack marbles and calculated C-O-H-F fluid compositions. Am. Mineral. 67, 545-557.
(22) Fyfe, W.S. (1973). The granulite facies, partial melting, and the Archean crust. Phil. Trans. Roy. Soc. Lond. 273A, 457-461.
(23) Newton, R.C., Smith, J.V., Windley, B.F. (1980). Carbonic metamorphism, granulites, and crustal growth. Nature 288, 45-50.
(24) Hoschek, G. (1976). Zur Stabilität von Clintonite im system $CaO-MgO-Al_2O_3-SiO_2-H_2O-CO_2$. Fort. der Min., 39-40.
(25) Taylor, H.P. (1969). Oxygen isotope studies of anorthosites with particular reference to bodies in the Adirondack Mts., N.Y. N.Y. State and Sci. Ser. Mem. 18, 111-134.
(26) Veizer, J., Hoefs, J. (1976). The nature of $^{18}O/^{16}O$ and $^{13}C/^{12}C$ secular trends in sedimentary carbonate rocks. Geochim. Cosmochim. Acta 40, 1387-1395.
(27) Emslie, R.F. (1978). Anorthosite massifs, rapikivi granites, and late Proterozoic rifting of North America. Precamb. Res. 7, 61-98.
(28) Morse, S.A. (1982). A partisan review of Proterozoic anorthosites. Am. Mineral. 67, 1087-1100.

Pb-ISOTOPIC STUDIES OF PROTEROZOIC IGNEOUS ROCKS, WEST GREENLAND, WITH IMPLICATIONS ON THE EVOLUTION OF THE GREENLAND SHIELD

Feiko Kalsbeek (1) and Paul N. Taylor (2)

(1) The Geological Survey of Greenland (GGU), Øster Voldgade 10, 1350 Copenhagen K, Denmark
(2) Department of Geology and Mineralogy, University of Oxford, Parks Road, Oxford OX1 3PR, U.K.

ABSTRACT. Archaean rocks have been recognised over most of Greenland, but it is not clear whether they once formed one coherent continental unit or whether they represent a number of originally dispersed smaller 'continents'. Over much of Greenland the Archaean rocks were strongly affected by Proterozoic deformation and metamorphism, and Proterozoic intrusive complexes occur locally. Pb-isotope studies of the latter, in combination with Rb-Sr isotopic information, are used to see whether or not the rocks have been contaminated by components derived from Archaean crustal material at depth. Where no contamination is detectable this may indicate the presence of ancient plate margins.

1. INTRODUCTION

The Archaean craton of South Greenland (1) consists largely of variable gneisses and migmatites (ca. 85%) and amphibolites (ca. 15%) with local important occurrences of anorthositic rocks. Most of the rocks have ages between ca. 3000 Ma and ca. 2600 Ma but in the Godthåbsfjord area, east of Godthåb (Fig.1), early Archaean rocks (ca. 3800 - 3600 Ma) also occur.
On most regional and global map compilations the Archaean craton is shown as indicated on Fig.1, and it is not generally realised that Archaean rocks in Greenland have in fact a much wider distribution. The Nagssugtoqidian mobile belt, north of the craton (Fig.1) consists in part of Archaean rocks, deformed and metamorphosed during the Nagssugtoqidian (Hudsonian) orogeny, 1900 - 1600 Ma ago, and in the Rinkian mobile belt to the north, an Archaean crystalline basement underlies a sequence of presumably early Proterozoic supracrustal rocks which, together with the basement, were strongly deformed and metamorphosed during the Hudsonian orogeny. Also in the Caledonian fold belt in East Greenland, remnants of Archaean rocks have locally been recognised. Fig.1 gives a survey of the localities where reliable isotopic evidence of the presence of Archaean rocks north of the Archaean craton sensu stricto has been found (2,3).

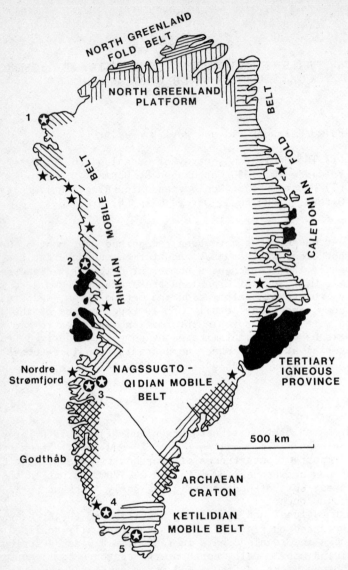

Figure 1. Sketch of the main geological divisions in Greenland with localities mentioned in the text numbered. Filled stars indicate the presence of proven Archaean rocks outside the Archaean craton of South Greenland (2,3); open stars indicate Proterozoic intrusive rocks (West Greenland only). 1: Etah meta-igneous complex; 2: Prøven granite/charnockite; 3: Granitic (s.l.) gneisses in the Nagssugtoqidian mobile belt; 4: Ketilidian granites in the border zone of the Archaean craton; 5: Ketilidian granites in the central part of the Ketilidian mobile belt.

Proterozoic intrusive rocks appear to have a more restricted distribution in West Greenland. In the northernmost part of the Rinkian mobile belt, the Etah meta-igneous complex (locality 1, Fig.1) has yielded a Rb-Sr whole-rock isochron age of ca. 1850 Ma (4). The Prøven granite/charnockite in the Rinkian mobile belt (loc.2, Fig.1) has an age of ca. 1860 Ma (Rb-Sr whole-rock isochron (2)). For some gneisses in the Nagssugtoqidian mobile belt (loc.3, Fig.1) ages of ca. 2200 - 1850 Ma have been obtained (Pb-Pb whole-rock isochrons (5) and U-Pb zircon data (6)). In these cases the Proterozoic intrusive rocks are spatially associated with Archaean rocks but their mutual relationships are not always clear.

In the Ketilidian mobile belt, south of the Archaean craton, Proterozoic granites occupy vast areas (loc.5, Fig.1). Ages between ca. 1850 and ca. 1750 Ma have been obtained (Rb-Sr whole-rock and U-Pb zircon data (7, 8)). Ketilidian granites also occur in the southernmost part of the Archaean craton (loc.4, Fig.1) It was earlier believed that the Ketilidian mobile belt contained an Archaean basement, but the isotopic evidence of Van Breemen et al.(7) has not confirmed this.

In view of the wide distribution of Archaean rocks throughout most of Greenland, the question arises whether these, together with rocks of similar age in Canada, Scotland and Norway, once formed a continuous continental unit, or whether they represent a number of originally dispersed 'micro-continents'. Orogenic activity was wide-spread over most of Greenland during the Proterozoic, and such activity is nowadays most commonly related to plate margins rather than within-plate settings. Still, over most of Greenland no independent evidence of the presence of plate margins has yet been described, and several aspects of the main tectonic evolution of the Greenland shield are as yet not well understood.

In order to gain some insight into these problems we have studied the whole-rock lead isotope systems in several Proterozoic intrusive rocks in Greenland to see which of these have been contaminated with components derived from Archaean rocks at depth. Pb-isotopes, especially in combination with Sr-isotopic information, are particularly useful in this respect because deep crustal rocks are often markedly depleted in U relative to Pb. As a consequence the increase in $^{207}Pb/^{204}Pb$ and $^{206}Pb/^{204}Pb$ ratios by the decay of uranium is here much slower than in the mantle or in the upper crust. Rocks contaminated with such 'retarded' lead, derived from the deep older crust, will have low initial Pb-isotopic ratios compared to pristine mantle-derived magma of the same age, and this is reflected in the low first stage model μ values obtained for such rocks ($\mu_1 = {}^{238}U/{}^{204}Pb$ for the model source from which the magma was derived). For non-contaminated igneous rocks normal mantle μ values are found (generally in the order of 7.5 - 8 for Archaean rocks, increasing to ca. 8 for Proterozoic rocks (9, 10)), and contaminated rocks often have model μ_1 values significantly lower than this (e.g. 11). Three examples of this approach will be shortly reviewed and it will be shown that the information obtained can be used in deciphering the main lines in the structure of the Greenland shield.

2. GRANITES FROM THE KETILIDIAN MOBILE BELT

We have obtained Pb-Pb whole-rock isochrons for a number of Ketilidian granites from South Greenland (12). The ages found vary from ca. 1700 Ma to ca. 1850 Ma and are in good agreement with the Rb-Sr isochron ages obtained by Van Breemen et al.(7). The Pb-Pb isochrons for three granite bodies emplaced in the southern border zone of the Archaean craton (the Kærne granite, Quiartorfik granite, and Storø granite, respectively KG, QG and SG in Fig.2) define much lower model μ_1 values than those obtained for granites from the central part of the mobile belt (Fig.2; μ_1 values can be read off on the diagram from the point of intersection of the individual isochrons with the geochron line). Simultaneously, the Ketilidian granites from the border zone of the Archaean craton have significantly higher initial $^{87}Sr/^{86}Sr$ (Sr_i) ratios than those from the mobile belt proper (12, 7). These data clearly indicate that the granites in the Archaean border zone contain large proportions of Archaean crustal material – not surprising in view of their mode of occurrence.

More significantly, the granites from the central part of the Ketilidian mobile belt show no evidence of having been contaminated with Pb and Sr derived from much older rocks at depth, and this indicates that most of the Ketilidian mobile belt is not underlain by Archaean crust. The same conclusion was reached by Van Breemen et al.(7) on the basis of Sr-isotopic evidence. It may be concluded that the Ketilidian mobile belt consists almost exclusively of newly-formed Proterozoic crust, and that the southern border of the Archaean craton in Proterozoic times was a continental plate margin.

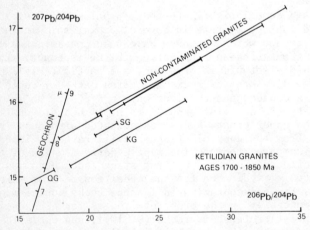

Figure 2. Synopsis of Pb-Pb isochrons (individual data points not shown) obtained for Ketilidian granites in South Greenland. KG, QG and SG are the Kærne, Quiartorfik and Storø granites from the border zone of the Archaean craton (loc.4 in Fig.1). The other isochrons represent granites from the central part of the Ketilidian mobile belt (loc.5 in Fig.1). Full information on the data is given in ref.(12).

3. DOLERITE DYKES FROM THE SOUTHERN PART OF THE ARCHAEAN CRATON

The southern part of the Archaean craton contains major swarms of often very well preserved dolerite dykes. These dykes are important chronological markers: they were intruded after the end of the Archaean tectono-metamorphic activity at ca. 2500 Ma and before the onset of the Ketilidian deformation and metamorphism to the south at ca. 1850 Ma (1, 13, 14).

We have obtained Rb-Sr and Pb-Pb whole-rock isotope data on two major dykes, the Naujat dyke and the Torssut dyke (15). The Rb-Sr isochrons (together with data obtained from a third dyke for which we have no Pb-isotopic information) gave a pooled weighted mean age of 2130 ± 65 Ma, the ages obtained for the individual dykes being the same within error, but the initial $^{87}Sr/^{86}Sr$ ratio obtained for the Torssut dyke (0.70277 ± 0.00012) was significantly higher than that of the Naujat dyke (0.70155 ± 0.00018). Pb-isotope measurements were carried out to test whether this difference in Sr_i values could be explained by crustal contamination. The samples from the Naujat dyke (Sr_i 0.70155) scattered in the Pb-Pb isochron diagram (not shown) about a 2050 Ma line-of-best-fit. Model μ_1 values, calculated for individual samples assuming an age of 2150 Ma, range from 7.41 - 7.96, with a mean value of 7.68.

The samples from the Torssut dyke do not define an isochron but form a cluster at a $^{206}Pb/^{204}Pb$ ratio of ca. 15 (Fig.3). Model μ_1 values

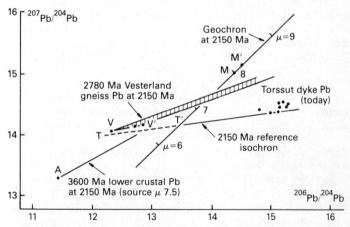

Figure 3. Pb-isotope evolution diagram showing the isotopic composition of Pb in the Torssut dyke today (dots) and 2150 Ma ago (field between T and T') compared to that of the surrounding gneisses 2150 Ma ago (Vesterland gneisses, hatched area, data from (16)) and early Archaean gneisses, presumed to be present at depth beneath the Torssut dyke, 2150 Ma ago (line extending from A). Pristine mantle-derived magmas may be expected to have Pb-isotopic compositions on or to the right of the 2150 Ma geochron between M and M'. Mixing of Pb of this composition with contaminant lead from the Vesterland or comparable gneisses cannot explain the isotopic composition of the lead in the Torssut dyke - see the text.

calculated assuming an age of 2150 Ma range from 6.57 to 6.87, with a mean value of 6.71. These very low model μ_1 values indicate that the Torssut dyke contains a large proportion of lead derived from Archaean rocks at depth. For the Naujat dyke such contamination, if present at all, was much less extensive than for the Torssut dyke.

Attempts to model the Pb-isotopic composition of the Torssut dyke samples by comparison with that of the surrounding gneisses gave an unexpected result: at the time of its formation, the Torssut dyke had significantly less radiogenic Pb-isotopic compositions than any of the gneisses we have analysed from the southern part of the Archaean craton (15, 16). This implies that, even if all the lead in the dyke had been introduced into the magma as a contaminant, this lead could not have been derived from late Archaean gneisses like those now exposed in the area. In the more probable case that the Torssut dyke also contains a proportion of lead derived from the mantle at 2150 Ma the problem is even more manifest since such mantle-derived lead is much more radiogenic than that in the analysed gneisses (Fig.3). Obviously, the contaminant lead in the Torssut dyke must have been much less radiogenic than that in the late Archaean gneisses.

No suitable source for such very unradiogenic contaminant lead has yet been identified in the area, but we think that at depth early Archaean rocks, comparable to the early Archaean Amîtsoq gneisses (ca. 3600 Ma) from the Godthåbsfjord area, may be present. Such rocks have very unradiogenic Pb-isotopic compositions (17) and would be an adequate source for the contaminant lead in the Torssut dyke.

The suggestion of the presence of early Archaean, Amîtsoq-type, gneisses at depth in the southern part of the craton is supported by Pb-isotope data on sample suites from the late Archaean gneisses (16). These rocks have given ages between ca. 3100 and 2800 Ma, and the isochrons define much lower model μ_1 values (7.20 - 7.25) than would be expected for non-contaminated late Archaean intrusive rocks. A similar phenomenon has been described from the Godthåbsfjord area, where early and late Archaean rocks occur side by side, and where the late Archaean rocks often have low model μ_1 values because of contamination at depth with lead derived from the early Archaean Amîtsoq gneisses (11).

The evidence suggests, then, that apart from the Amîtsoq 'continent' in the Godthåbsfjord area, at least one other early Archaean crustal block may be present at depth. These two blocks appear to be separated by pristine late Archaean crust, but the isotopic evidence from the intervening areas is as yet too scanty for this to be certain.

4. GNEISSES FROM THE NAGSSUGTOQIDIAN MOBILE BELT

The Nagssugtoqidian mobile belt is among the least well known areas in West Greenland. Detailed mapping is mainly restricted to the border zone near the Archaean craton to the south, and to an area in the Nordre Strømfjord region (for details and references see (18)). Isotopic data are scarce. Since Ramberg's first description of the Nagssugtoqidian mobile belt (19) it has been generally believed that most of it consists of deformed Archaean gneisses (e.g. 20), but there is as yet very little isotopic evidence to substantiate this. The mobile belt is characterised by a number of ENE trending shear zones, in which there is a strong parallelism for all structural elements, alternating with areas in which older, more open, fold structures are preserved.

Recent Pb-Pb and Rb-Sr isotope work (5, 21) has confirmed the presence of at least some Archaean rocks (a granite body in the coastal area, for example, has given an age of ca. 2600 Ma (5)), but it was also found that part of the gneisses that make up the Nagssugtoqidian mobile belt yield Proterozoic isochron ages.

It would appear that these gneisses were formed from Proterozoic intrusive granitic (s.l.) rocks, and there is no evidence to show that these rocks had been contaminated with Archaean crustal material at depth: our Pb-Pb isochrons (ages 1930 $\pm$ 170 Ma and 2210 $\pm$ 160 Ma) define model μ_1 values of 7.96 and 8.10, respectively, which is normal for non-contaminated Proterozoic igneous rocks. Hickman and Glassley (21) have obtained well defined Proterozoic Rb-Sr whole-rock isochrons for a variety of gneisses and granitic rocks at a locality in the western part of Nordre Strømfjord. The over-all (site-mean) age for a number of sublocalities was 1933 $\pm$ 60 Ma, Sr_i 0.7031 $\pm$ 0.0008. We would interpret these data as indicating a Proterozoic age for these rocks, and formation from a virtually non-contaminated parent, but Hickman and Glassley (21) prefer an origin by profound chemical changes of Archaean precursors of the rocks during the Proterozoic shear zone formation. Zircon U-Pb analyses on three samples from our collections by R.T. Pidgeon (6) have yielded an age of ca. 1850 Ma, and provided no evidence of the presence of older (inherited) zircons.

The occurrence of non-contaminated granitic gneisses in the central part of the Nagssugtoqidian mobile belt, if confirmed by further isotopic evidence, may indicate the presence of a suture between two originally separated Archaean continental plates. A geochemical study of (probably early Proterozoic) supracrustal rocks from the Nordre Strømfjord area (22) has shown that some of these are of volcanic origin and have a distinct calc-alkaline chemical signature. At present such rocks are characteristically found at 'destructive' (ocean/continent) plate boundaries, and if this was also the case during the Proterozoic, it would support the presence of an ancient plate margin in this area.

5. CONCLUSIONS

The crystalline shield of Greenland consists largely of Archaean gneisses, locally alternating with gneissified Proterozoic intrusions. In the field it is commonly impossible to differentiate between these gneisses of very different age. Lead isotope studies, especially in combination with Rb-Sr isotope data, are a very powerful tool, not only to give chronological information, but also to study the sub-surface structure of the Greenland shield. It is possible to differentiate between Proterozoic igneous rocks that have intruded through Archaean continental crust and Proterozoic rocks that have developed in non-continental settings. It can be shown, for example, that the Ketilidian mobile belt in South Greenland consists of pristine Proterozoic rocks, and that the Archaean 'continent' of Greenland therefore ends at the border with the Ketilidian mobile belt. There is (as yet incomplete) evidence that the Nagssugtoqidian mobile belt contains pristine Proterozoic granitic (s.l.) rocks, which suggests the presence here of a suture between once separated Archaean continental blocks. The presence of such sutures, if confirmed by further evidence, would indicate that the Archaean rocks in the North Atlantic areas originally did not form a coherent continent. Within the Archaean craton there is evidence for the occurrence of at least two kernels of early Archaean material, separated by pristine late Archaean crust, also here suggesting the presence of originally separated continental masses.

ACKNOWLEDGEMENTS

The geochronological facilities at Oxford, where the isotopic analyses have been carried out, are funded by a grant from the Natural Environment Research Council to Dr. S. Moorbath. F.K. acknowledges reception of a grant from the Danish Natural Science Research Council which enabled him to participate in the NATO Advanced Study Institute 1984 in South Norway. Publication of this paper was permitted by the Director of the Geological Survey of Greenland.

REFERENCES

1. Bridgwater, D., Keto, L., McGregor, V.R. and Myers, J.S. 1976. Archaean gneiss complex in Greenland, in: Geology of Greenland, ed. by Escher, A. and Watt, W.S., Grønlands Geologiske Undersøgelse, Copenhagen, pp. 18 - 75.
2. Kalsbeek, F. 1981, Precambrian Res. 14, pp. 203 - 219.
3. Kalsbeek, F., Rapp. Grønlands geol. Unders. (in press).
4. Larsen, O. Personal communication, 1984.
5. Kalsbeek, F., Taylor, P.N. and Henriksen, N., Can. J. Earth Sci. 21, pp. 1126 - 1131.
6. Pidgeon, R.T. Personal communication, 1983.
7. Van Breemen, O., Aftalion, M. and Allaart, J.H. 1974, Bull. geol. Soc. Amer. 85, pp. 403 - 412.
8. Gulson, B.L. and Krogh, T.E. 1975, Geochim. cosmochim. Acta 39, pp. 65 - 82.
9. Moorbath, S. and Taylor, P.N. 1981. Isotopic evidence for continental growth in the Precambrian, in: Precambrian plate tectonics, ed. by Kröner, A., Elsevier, Amsterdam, pp. 419 - 525.
10. Oversby, V.M. 1974, Nature 248, pp. 132 - 133.
11. Taylor, P.N., Moorbath, S., Goodwin, R. and Petrykowski, A.C. 1980, Geochim. cosmochim. Acta 44, pp. 1437 - 1453.
12. Kalsbeek, F. and Taylor, P.N., Earth planet. Sci. Lett. (in press).
13. Bondesen, E. and Henriksen, N. 1965, Bull. Grønlands geol. Unders. 52 (also Meddr Grønland 179/2) 42 p.
14. Allaart, J.H. 1976. Ketilidian mobile belt in South Greenland, in: Geology of Greenland, ed. by Escher, A. and Watt, W.S., Grønlands Geologiske Undersøgelse, Copenhagen, pp. 120 - 151.

15. Kalsbeek, F. and Taylor, P.N., Contr. Min. Petr. (in press).
16. Taylor, P.N. and Kalsbeek, F. Unpublished data.
17. Black, L.P., Gale, N.H., Moorbath, S., Pankhurst, R.J. and McGregor, V.R. 1971, Earth planet. Sci. Lett. 12, pp. 245 - 259.
18. Korstgård, J.A. (ed.) 1979. Nagssugtoqidian geology. Rapp. Grønlands geol. Unders. 89, 146 p.
19. Ramberg, H. 1948, Meddr Dansk geol. Foren. 11, pp. 312 - 327.
20. Watterson, J. 1978, Nature 273, pp. 636 - 640.
21. Hickman, M.H. and Glassley, W.E. 1984, Contr. Min. Petr. 87, pp. 265 - 281.
22. Rehkopff, A. 1984, Rapp. Grønlands geol. Unders. 117, 26 pp.

CORRELATIONS BETWEEN THE GRENVILLE PROVINCE AND SVECONORWEGIAN OROGENIC BELT - IMPLICATIONS FOR PROTEROZOIC EVOLUTION OF THE SOUTHERN MARGINS OF THE CANADIAN AND BALTIC SHIELDS.

Charles F. Gower
Department of Mines and Energy,
Newfoundland and Labrador
P.O. Box 4750, St. John's,
Newfoundland, Canada, A1C 5T7

ABSTRACT. Adopting the hypothesis that the Grenville and Sveconorwegian fronts were part of the same tectonic margin during Lower-Middle Proterozoic times, and that the Baltic Shield rotated clockwise relative to the Canadian Shield prior to, and (or) during the Grenville Orogeny, the following model is proposed. The Archean cratons extending from Wyoming, U.S.A. to central Europe formed a linear protocontinent, on the peripheries of which continental margin sediments accumulated. Subsequently the southern Canadian and Baltic Shields became the site of a long-lived Lower to Middle Proterozoic continental margin convergent plate boundary. Rotation of the Baltic Shield relative to the Canadian Shield initiated an intermittent Middle Proterozoic rifting regime and eventually caused progressive collision along the Canadian-Baltic Shield mutual southern boundaries, resulting in interior thrusting and regional structural trends oblique to orogen margins. A complimentary tensional regime farther north eventually triggered Late Proterozoic rifting that presaged the early Phanerozoic Iapetus Ocean.

1. INTRODUCTION

Following initial proposals for Sveconorwegian-Grenvillian correlation (1,2) and a compilation of data on Proterozoic rocks across the southern Canadian Shield and the Baltic Shield (3), evidence has steadily accumulated demonstrating that the Grenville Province and Sveconorwegian Orogenic Belt were formerly part of the same orogen. Recognition of this correlation should, in turn, permit clarification of relationships between pre-Grenvillian orogenic belts in both regions. Unfortunately, partly because most continental reconstructions between the Canadian and Baltic Shields result in positioning the Grenville Front at almost 90° to the Sveconorwegian Front, two alternative models have emerged and hence introduce complications with respect to earlier Proterozoic reconstructions. In one model the Grenville and Sveconorwegian Fronts are interpreted as formerly part of the same tectonic margin of a continuous orogen, which was closed to the north (2,4). The other model considers the two fronts as opposing tectonic margins with the continuation of the orogen northward (5,6,7).

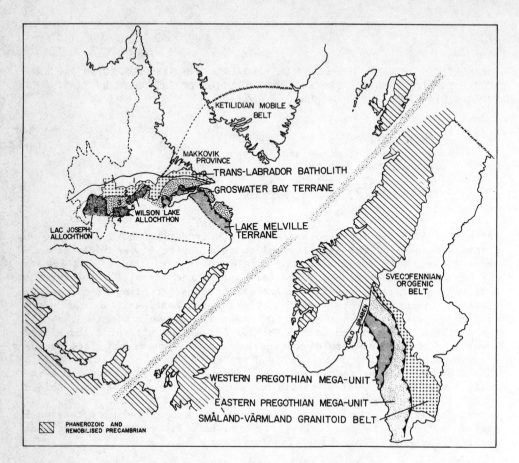

Figure 1. Correlations between lithotectonic terranes in the Grenville Province of eastern Labrador and Sveconorwegian Orogenic Belt in southern Sweden. Modified after (8) to indicate correlative regions in central and western Labrador. Allochthon names and numbers (of lithotectonic blocks) refer to nomenclature used elsewhere in this volume (9,10).

Recently, Gower and Owen (8) subdivided part of the Grenville Province of eastern Labrador into three major lithotectonic terranes (Trans-Labrador batholith, Groswater Bay terrane and Lake Melville terrane). They presented lithological and structural descriptions and geological histories of each region and, from a review of Scandinavian geological literature concluded that each of these lithotectonic terranes has a direct counterpart in southern Sweden (Figure 1). These correlations were interpreted to mean that both regions formed part of the same tectonic margin during Middle Proterozoic times.

This paper briefly reviews the correlations established by Gower and Owen (with additional illustrations to emphasize particular points), and uses their model to explore some broader aspects of Proterozoic evolution between the Canadian and Baltic Shields.

This exploration is done in conjunction with constraints suggested from paleomagnetic studies, from which three major conclusions have emerged (11,12,13,14,15,16,17,18). The first conclusion is that apparent polar wandering paths for the Canadian and Baltic Shields display similar trends for the period 1800-1200 Ma. This denies the possibility of extensive motion (excluding near in-situ rotation) of the Canadian Shield relative to the Baltic Shield during this period. The second conclusion is that there was a major clockwise rotation (up to 90°) of the Baltic Shield relative to the Canadian Shield during the same period. Thirdly, the apparent polar wandering path for northwest Scotland follows that for the Canadian Shield until Late Proterozoic times when it also became compatible with that for the Baltic Shield (19,15).

2. CORRELATIONS BETWEEN PROTEROZOIC ROCKS OF EASTERN LABRADOR AND SOUTHERN SWEDEN.

The major correlations identified by Gower and Owen (8) are summarized below. The supporting references are quoted by Gower and Owen and are omitted here.

2.1 Makkovik Province - Svecofennian Orogenic Belt

(i) Both the Makkovik Province (and its Greenland extension, the Ketilidian Mobile Belt) and the Svecofennian Orogenic Belt comprise Lower Proterozoic (ca. 2100-1700 Ma) tholeiitic mafic volcanic rocks and turbidite sediments, overlain by calc-alkaline felsic volcanic rocks and intruded by syntectonic granitoid plutons.

(ii) Structural trends in the Makkovik Province and the Svecofennian Orogenic Belt are truncated obliquely (in the same angular sense, Figure 2) by granitoid rocks of the Trans-Labrador batholith and Småland-Värmland granitoid belt respectively. The granitoid belts contain evidence of those trends however.

(iii) A suite of mafic intrusions were emplaced at ca. 950 Ma parallel to the Grenville and Sveconorwegian fronts in Labrador and Sweden respectively.

2.2 Trans-Labrador batholith - Småland-Värmland granitoid belt

(i) The Trans-Labrador batholith and Småland-Värmland granitoid belt were both emplaced at ca. 1650-1600 Ma but include earlier granitoid intrusions with Makkovikian or Svecofennian trends.

(ii) Both granitoid belts separate Lower Proterozoic rocks from lithologies affected by the Grenvillian - Sveconorwegian Orogenies.

(iii) Both granitoid belts (and hence the Grenvillian and Sveconorwegian Fronts - see below) coincide with major negative Bouguer gravity anomalies (Figure 3).

(iv) Both granitoid belts are characterized by granodiorite, granite, monzodiorite, monzonite and late (ca. 1625 Ma) syenitic fractionates.

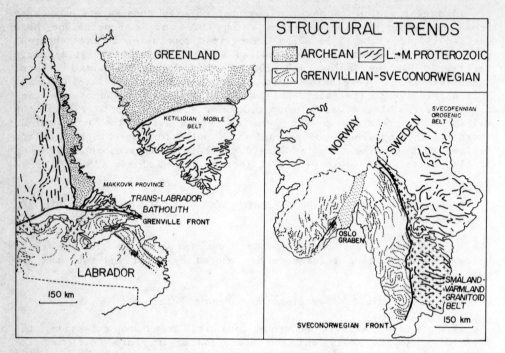

Figure 2. Structural Provinces and structural trends and their relationships to the Trans-Labrador batholith and Småland-Värmland granitoid belt.

(v) High crustal emplacement levels are inferred for the Trans-Labrador batholith and Småland-Värmland granitoid belts with transitional contacts between intrusive rocks and inferred co-genetic felsic volcanic rocks.

(vi) Scattered remnants of layered gabbro-diorite-monzodiorite intrusions characterize both granitoid belts. These intrusions appear to pre-date, but to be genetically associated with, the youngest (ca. 1625 Ma) granitoid plutons.

(vii) Both granitoid belts are unconformably overlain by continental arkoses, conglomerates and sedimentary breccias of Middle Proterozic age.

2.3. Grenville-Sveconorwegian front

(i) The Grenville-Sveconorwegian Front, defined as the northern (eastern) limit of widespread Grenvillian (Sveconorwegian) deformation, passes obliquely through the Småland-Värmland granitoid belt such that the southern end of the Småland-Värmland granitoid belt is entirely outside the Sveconorwegian Orogenic Belt and the western end of the Trans-Labrador batholith entirely within the Grenville Province (Figure 2).

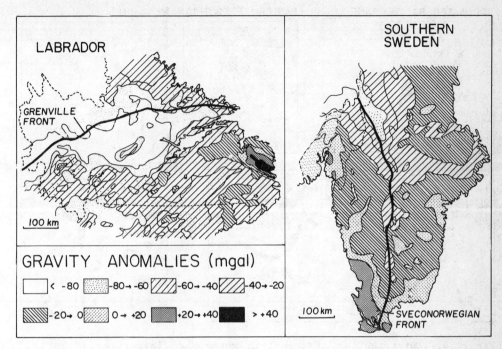

Figure 3. Bouguer gravity anomaly maps for southern Labrador and southern Sweden illustrating negative Bouguer anomalies associated with the Grenville and Sveconorwegian Fronts.

(ii) This line appears to mark the limit of Grenvillian-Sveconorwegian K-Ar age dates and also the northern (eastern) limit of coronitic metagabbro (see below).

2.4 Groswater Bay terrane - Eastern Pregothian mega-unit.

(i) The Groswater Bay terrane and Eastern Pregothian mega-unit are of comparable width (40-60 km) and bounded by major thrust zones, especially on their southern (western) flanks.
(ii) Both regions are characterized by orthogneiss and deformed granitoid plutons subsidiary supracrustal rocks.
(iii) An undated gneissic basement to the earliest (Lower or early Middle Proterozoic) supracrustal rocks has been inferred for both regions.
(iv) The (Lower or early Middle Proterozoic) pelitic supracrustal rocks are typically kyanite-bearing in both regions.
(v) Gneissic granitoid rocks in the Eastern Pregothian mega-unit have been dated at ca. 1770 Ma but several pulses of intrusion are inferred. Within the Groswater Bay terrane an age range of 1680-1610 Ma on gneissic rocks has been obtained.
(vi) Both regions are characterized by abundant mafic intrusions with coronitic textures (hyperites). Partial distribution is shown in Figure 4. An age of 1550 Ma has been obtained for one hyperite in

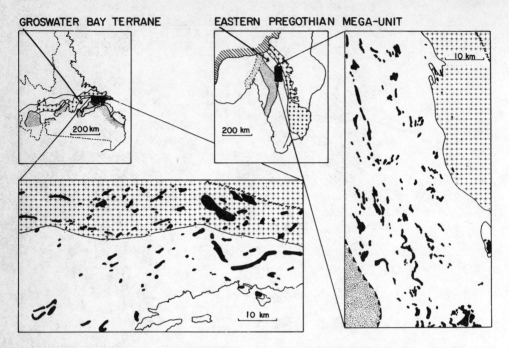

Figure 4. Distribution of coronitic metagabbro (hyperite) in part of the Groswater Bay terrane and Eastern Pregothian mega-unit.

southern Sweden but it has been suggested that two unrelated suites have been grouped together. In eastern Labrador major pulses of mafic intrusion (with coronitic textures) occurred at ca. 1600 Ma and ca. 1400 Ma.

2.5 Lake Melville terrane - Western Pregothian and Åmål Mega-units

(i) The Lake Melville terrane and Western Pregothian/Åmål mega-units are both characterized by a higher proportion of metasedimentary gneiss than in the Groswater Bay terrane and Eastern Pregothian mega-unit.

(ii) In both regions the metasedimentary gneiss is interpreted to be of Lower or early Middle Proterozoic age.

(iii) Both regions are characterized by several periods of granitoid pluton emplacement. In the Åmål mega-unit these ages have been grouped as pulses at 1700-1650 Ma, 1430-1370 Ma and 1230-1200 Ma. In the Lake Melville terrane geochronological studies are at a preliminary stage, but two granitoid units have been reliably dated (U-Pb zircon, T.E. Krogh, personal communication, 1984) at 1657 Ma and 1273 Ma.

(iv) In both regions Rb-Sr systematics have been disturbed. The granitoid unit in the Lake Melville terrane giving a 1273 Ma zircon age has yielded a Rb-Sr whole rock age of 1085 Ma (C. Brooks, personal communication, 1984).

(v) Layered ultramafic-mafic intrusions are a feature of the 'hanging wall' thrust front region in the Lake Melville terrane. Similar bodies have been described in an analogous structural setting (adjacent to the Mylonite Zone) in southern Sweden.

(vi) Granulite facies rocks are found in the 'hanging wall' thrust front region in both terranes.

3. PRE-GRENVILLIAN OROGENIC BELTS

The previously outlined correlations are taken as justification for interpreting eastern Labrador and southern Sweden as the same tectonic margin during Middle Proterozoic times. Recent discussion (9,20) has drawn comparisons between the Lower and Middle Proterozoic orogens in Labrador with similar aged orogens in the central U.S.A. Extending these interpretations, a generalized reconstruction of Lower Proterozoic orogens along the southern Canadian and Baltic shields is presented in Figure 5a.

In Figure 5a the Baltic Shield has been positioned to achieve a compromise between previously suggested pre-Grenvillian paleomagnetic positions (15,17) and the distribution of Archean cratons and Lower Proterozoic mobile belts in North America and northern Europe. In this configuration the Archean cratons form a belt of discontinuous blocks on a global scale. They may thus be interpreted as either a single linear Archean protocontinent extending from Wyoming, U.S.A. to central Europe (18), or a series of microcontinents swept together prior to, or during, Lower Proterozoic orogenesis. Continental margin sediments are envisaged as having formed around the peripheries of the protocontinent during early Lower Proterozoic times.

Following earlier models (21,22,23) a continental margin convergent plate boundary is envisaged as the tectonic setting for Lower Proterozoic orogenesis along the southern margin of the arcuate Archean protocontinent. The 1900-1700 Ma tectonothermal activity includes orogenies variously referred to (west to east), Penokean (U.S.A.), Makkovikian/ Ketilidian (Labrador), Ketilidian (Greenland) and Svecofennian or Svecokarelian (Scandinavia). Van Schmus and Bickford (22) have pointed out that, for the western end of this belt, orogenic activity becomes progressively younger moving outward from the Archean cratons. They revive the name Mazatzal Orogeny (24) for the ca. 1680-1610 Ma orogenic activity. The term Labradorian Orogeny has been introduced (28) for similar aged tectonothermal activity in Labrador and the name Ghost-Gothian Orogeny used (6) for temporally equivalent activity in Scandinavia.

South of rocks grouped with the Mazatzal Orogeny in midcontinental U.S.A. is a granite-rhyolite terrane dated between 1480-1380 Ma. Rocks of comparable lithology and age extend eastward into the Grenville Province of southern Ontario. Similar aged rocks are found in an analogous tectonic position (i.e. marginal to the 1680-1610 Ma orogenic belts) in Scandinavia and referred to as the Western Orogeny (6).

The region in which comparable rocks might be expected to occur in Labrador is largely unmapped and undated and entirely within the Grenville Province. Because the southern margin of 1680-1610 Ma magmatism and the extent of Grenvillian intrusive activity have yet to be defined

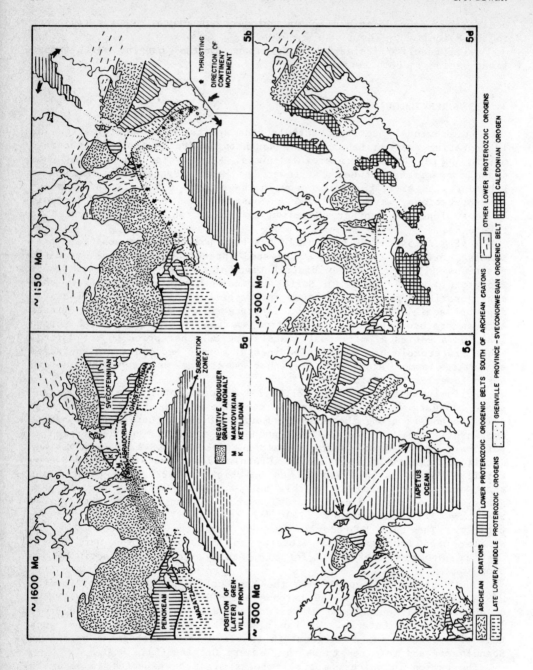

in Labrador, it is difficult to assess the potential extent of any analogous 1480-1380 acidic terrane in Labrador. Nevertheless, manifestations of Middle Proterozoic felsic magmatism are present in Labrador, and include the felsic volcanic rocks of the Bruce River and Blueberry Lake Groups (1530 Ma and 1540 Ma respectively), and the Otter Lake granite (1496 Ma). This magmatism is somewhat older than the 1480-1380 Ma acidic magmatism in midcontinental U.S.A. and is farther north with respect to 1680-1610 Ma orogenic activity. Nevertheless, in that it reflects younger magmatism it still invites comparison (albeit diachronous) with the 1430-1380 plutons in central U.S.A., especially those north of the main granite-rhyolite terrane.

4. THE GRENVILLIAN OROGEN

In Figure 5b the Baltic Shield is envisaged as having rotated clockwise, through about 30-40°, and is now positioned in accordance with the post-Grenvillian orientation suggested from paleomagnetic data (17). This rotation may have involved complete early Middle Proterozoic separation (from 1500 Ma until 1200 Ma) of the Baltic and Canadian Shields (15), but it seems rather coincidental that the Canadian and Baltic Shields would then return to their (more-or-less) former positions after 300 Ma. Therefore, an approximately in-situ reorientation is envisaged here. During this time intermittent tensional and compressional regimes operated, and have been previously suggested to have existed in Labrador (24,25,26).

The model depicted in Figure 5b has several implications with respect to Grenvillian orogenesis. The rotation reduces and perhaps eventually closes the inferred oceanic area originally situated south of the Lower Proterozoic orogens, and, in effect, generates a partial continent-continent collision. It also means that, what was the same tectonic margin of an orogen prior to Grenvillian events eventually, in a sense, becomes the opposite tectonic margin during the Grenvillian Orogeny. If limited separations between the Canadian and Baltic Shields occurred, their rejoining could account for Grenvillian age dates north of the Grenville and Sveconorwegian fronts in east Greenland and northern Scandinavia.

Collisional models have been proposed many times for the Grenvillian Orogeny but have foundered for lack of convincing evidence of a suture, or the involvement of other continental landmasses. In this model there

Figure 5. Reconstructions between the Canadian and Baltic Shields a) ca. 1600 Ma. Note that negative Bouguer gravity anomaly coincides with regions of inferred major crustal overlap. b) ca. 1150 Ma. Baltic Shield has rotated clockwise initiating a compressive regime in the Grenville Province. c) ca. 500 Ma. Tensional regime initiated earlier between east Greenland and the northern Baltic Shield eventually developed with Iapetus Ocean. d) ca. 300 Ma. Subsequent closure of Iapetus Ocean during Caledonian Orogeny, finally linking northwest Scotland with the remainder of Great Britian and creating 2000 km 'mismatch' of Grenville and Sveconorwegian Fronts.

is no need to appeal to other continental landmasses and the nature of any suture would be rather different from that for 'face-on' continental collisions.

The rotation depicted in Figure 5b would establish a compressional regime in the region now represented by the Grenville Province, and a tensional region in the Archean and Lower Proterozoic orogenic belts farther northeast. The compressional regime finds its release in thrusting which now defines the margins of the Grenville Province, and boundaries of various lithotectonic terranes within it. Also anticipated are regional structural trends oblique to the regional trend of the Grenvillian and Sveconorwegian orogens. The sense of structural trend obliquity would be opposite in the Grenville Province to that in the Sveconorwegian Orogenic Belt, as is in fact the case (7,4,6). Furthermore the model also predicts major dextral strike-slip faults in Labrador and similar sinistral faults in Scandinavia, as the interior Grenville-Sveconorwegian orogen is compressed with respect to its margin (cf. flextural-slip folds). A right-lateral sense of displacement is evident in major southeast-trending faults in eastern Labrador (Figure 2) and in Scandinavia left-lateral movement is particularly evident along the Bamble shear zone.

The region of tension north of the Grenville Province ultimately becomes a zone of crustal rupture which gradually extends south. Rifting was clearly active prior to 850 Ma in northernmost Norway and northwest Scotland (which, according to this model were in close proximity at the time). A late Precambrian age for continental red-beds (Double Mer Formation) in Labrador (27), and the sparagmites of southern Sweden and Norway suggests rifting occurred somewhat later farther south. Thus, the crustal separation, which began as a consequence of Middle Proterozoic rotation, becomes the forerunner of Iapetus Ocean.

5. EARLY PHANEROZOIC CONTINENTAL RELATIONSHIPS

Figures 5c and 5d complete the model up to the Caledonian Orogeny and illustrate the manner in which northwest Scotland finally becomes affixed to the remainder of the British Isles. It is simply the result of a 2000 km 'mismatch' of continents when Iapetus Ocean finally closed, as schematic opening and closing vectors on Figure 5c attempt to illustrate.

6. ACKNOWLEDGEMENTS

This paper was critically reviewed by G. Nunn and is published with the permission of the Director, Mineral Development Division, Newfoundland Department of Mines and Energy.

7. REFERENCES

1. Wynne-Edwards, H.R. and Hasan, Z.U., 1970. American Jour. of Sci., **268**, pp. 289-308.
2. Bridgwater, D. and Windley, B.F., 1973. Symposium on granites, gneisses and related rocks. (L.A. Lister, ed.), Geol. Soc. of South Africa, Special Publication No. 3, pp. 307-317.
3. Falkum, R. and Pedersen, J.S., 1980. Geologische Rundschau, **69**, pp. 622-647.
4. Max, M.D., 1979. Geology, **7**, pp. 76-78.
5. Berthelsen, A., 1980. International Geological Congress, Colloquium C6. Paris 1980, pp. 5-21.
6. Baer, A.J., 1981. Precambrian Plate Tectonics (A. Kroner, ed.) Elsevier, Amsterdam, pp. 353-385.
7. Gower, C.F. and Owen, V., 1984. Can. Jour. of Earth Sci., **21**, pp. 678-693.
8. Thomas, A., Nunn, G.A.G. and Wardle, R.J., this volume.
9. Rivers, T. and Nunn, G.A.G., this volume.
10. Kratz, K.O., Gerling, E.K. and Lobach-Zhuchenko, S.B., 1968. Can. Jour. of Earth Sci., **5**, pp. 657-660.
11. Neuvonen, K.J., 1974. Bull. of Geol. Soc. of Finland, **50**, pp. 31-37.
12. Ueno, H. Irving, E. and McNutt, R.H., 1975. Can. Jour. of Earth Sci., **12**, pp. 209-226.
13. Poorter, R.P.E, 1976. Physics of Earth and Plan. Int., **12**, pp. 241-247.
14. Poorter, R.P.E., 1981. Precambrian Plate Tectonics (A. Kroner, ed.) Elsevier, Amsterdam, pp. 599-622.
15. Patchett, P.J. and Bylund, G., 1977. Earth and Planet. Sci. Lett., **35**, pp. 92-104.
16. Patchett, P.J., Bylund, G. and Upton, B.G.J., 1978. Earth and Planet. Sci. Lett., **40**, pp. 341-364.
17. Piper, J.D.A., 1982. Earth and Planet. Sci. Lett., **59**, pp. 61-69.
18. Beckmann, G.E.J., 1976. Jour. Geol. Soc. London, **132**, pp. 45-59.
19. Wardle, R.J., Rivers, C.T., Gower, C.F., Nunn, G.A.G., and Thomas, A. (in Press). New perspectives on the Grenville Problem. Geol. Assoc. Can. Special paper.
20. Van Schmus, W.R., 1976. Phil. Trans. Royal Soc. of London, Series A, **280**, pp. 605-628.
21. Van Schmus, W.R. and Bickford, M.E., 1981. Precambrian Plate Tectonics (A. Kroner, ed.) Elsevier, Amsterdam, pp. 261-296.
22. Baragar, W.R.A. and Scoates, R.F.J., 1981. Precambrian Plate Tectonics (A. Kroner, ed.) Elsevier, Amsterdam, pp. 297-330.
22. Wilson, E.D., 1939. Geol. Soc. America, Bull., **50**, pp. 1113-1164.
23. Nunn, G.A.G., Noel, N. and Culshaw, N.G., (M.J. Murray, J.G. Whelan and R.V. Gibbons, eds.), Report 84-1, pp. 30-41.
24. Emslie, R.F., 1978. Precambrian Research, **7**, pp. 61-98.
25. Ryan,B., in press. Newfoundland Dept. Mines and Energy, Memoir 3.
26. Gower, C.F. and Ryan, A.B., in press. New perspectives on the Grenville Problem. Geol. Assoc. Can. Special Paper.
27. Gower, C.F., Erdmer, P. and Wardle R.J., in prep.

THE EVOLUTION OF THE SOUTH NORWEGIAN PROTEROZOIC AS REVEALED BY THE MAJOR AND MEGA-TECTONICS OF THE KONGSBERG AND BAMBLE SECTORS

I. C. STARMER

Department of Geology, University College London,
Gower Street, London WC1E 6BT, U. K.

ABSTRACT   Detailed surveys of the Kongsberg and Bamble Sectors have shown that both underwent early, major folding, with amphibolite-granulite facies metamorphism (probably associated with CO_2 flushing and possible mantle degassing) during the Kongsbergian Orogeny (~1600-1500 Ma). A series of ductile shearing episodes mylonitised rocks, particularly at the junctions with the Telemark block, until folding of regional extent occurred again during the second phase of the Sveconorwegian Orogeny (~1100-1000 Ma) which reached greenschist to amphibolite facies grade. The sequence of tectonic events is punctuated by intrusive and metamorphic episodes which can be dated relatively and absolutely. The Kongsberg-Bamble belt probably formed by deposition in a marginal basin of the Svecofennian craton and the Proterozoic Supercontinent. Early shelf deposits were followed by deeper water sediments with intermediate and basic volcanics. Some overlap into the Telemark block is preserved, in places. The major and mega structures suggest a more stable, thicker Telemark block surrounded by a Kongsberg-Bamble mobile belt which has an early antiformal axis still recognisable in the oldest metasediments. The different structural styles and ductile shear directions explain the apparently anomalous NE-SW orientation of the Bamble Sector and the preservation of remnant Kongsbergian ages in certain domains. The more intensive Sveconorwegian deformation and metamorphism in the Telemark Sector is also explained.

1. INTRODUCTION

The Proterozoic of South Norway, lying to the west of the Oslo Region rocks, consists of three sectors. The central Telemark Sector, covering most of South Norway, comprises the Telemark Basement Gneiss Complex, overlain, in parts, by the Telemark Supracrustal Suite. It is bordered to the southeast, along the Skagerrak coast, by the NE-SW Bamble Sector, extending for 150 Km and reaching a width of 30 Km. To the east, it is separated from the Oslo Region rocks by the N-S Kongsberg Sector which reaches a width of 30 Km. The three divisions

are now separated by late faulting, the so-called 'friction breccia' of
A. Bugge (1928), which has now been shown to follow earlier mylonitised
margins of both the Kongsberg and Bamble Sectors (Starmer 1977). The
distinction is most graphically illustrated in the Bamble Sector (Fig 2)
where the 'friction breccia' actually cuts out the earlier mylonitised
margin. The mylonite zones developed largely from variable quartz-
plagioclase-biotite-hornblende gneisses and had remarkable longevity:
the overprints of successive ductile shear episodes can be separated,
particularly because of their effect on adjacent intrusives. Three
main episodes are recognised in the mylonite zones with another affecting
the adjacent Telemark granitic gneiss: two minor shearing events had
only localised effects.

   The Telemark Basement Gneiss Complex consists mainly of
migmatised gneisses, granite gneisses and late granites: it abuts the
Kongsberg and Bamble Sectors which contain a great diversity of rock
types and a much lower proportion of granitic material.

   Southern Norway is traditionally described as part of the
Sveconorwegian (Grenvillian) Province, but there is now evidence of an
earlier orogeny in the Bamble Sector (Starmer 1969a & b, 1972a; Field
and Raheim 1979) and in the Kongsberg Sector (Starmer 1977; Jacobsen
and Heier 1978). The earlier orogeny ($\sim$ 1600-1500 Ma) is best
described as the 'Kongsbergian' after Oftedahl (1980) and the later
event ($\sim$ 1300-1000 Ma) is called the 'Sveconorwegian Orogeny' or
'Regeneration'.

   The general features of the Kongsberg and Bamble Sectors will
first be described. Preliminary studies by the author (Starmer 1977)
have shown that their geological histories were similar and they will
be considered in terms of what is now suggested to be one integral
'Kongsberg-Bamble mobile belt'. Its margin with the central Telemark
block was a continuously re-activated lineament of ductile shearing
and later faulting.

2. GEOLOGY OF THE KONGSBERG AND BAMBLE SECTORS

2.1 Major Features of the Kongsberg Sector

<u>2.1.1 General.</u> The Kongsberg Sector is overlain, in the southeast, by
Cambro-Ordovician sediments of the Oslo Region. In the west and northwest, it is faulted, against the Telemark Basement Gneiss Complex
(TBGC) of the Telemark Sector, by the 'friction breccia'. In the
north, it passes into the TBGC.

   The various parts of the Kongsberg Sector have been discussed
in detail previously and recently a model has been proposed for the
structure and evolution of the whole region (Starmer 1977, 1979, 1980,
1981, 1985). The Kongsberg Sector (Fig. 1) can be divided into three
natural divisions: the Western Kongsberg Complex (WKC), the Modum
Complex (MC) and the Eastern Kongsberg Complex (EKC). These are
exposed in early major folds, with the distinctive Modum Complex (MC)
in a N-S antiform flanked by synforms of the WKC and EKC. These first
order major structures are complicated by smaller (second and third
order) major folds and minor folds.

EVOLUTION OF THE KONGSBERG AND BAMBLE SECTORS

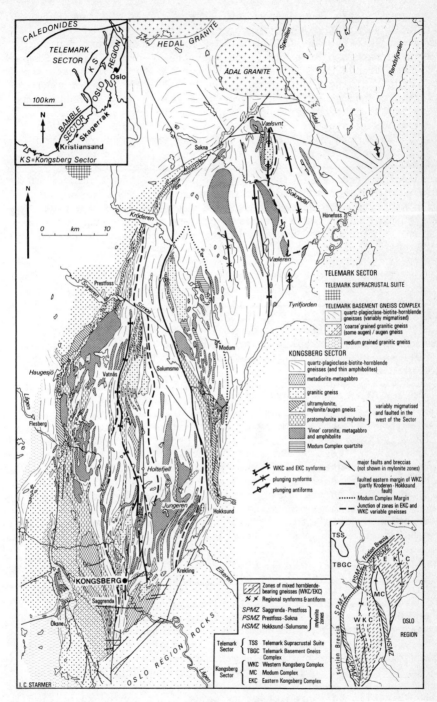

Figure 1. The geology of the Kongsberg Sector.

2.1.2 Major Faults. Four major fault zones affect the Kongsberg Sector. These are the 'friction breccia' zone curving along its western margin (with the TBGC), the N-S Kröderen-Hokksund fault at the eastern margin of the WKC (with both the MC and EKC), the Simoa fault which trends ESE-WNW between the two faults above, and a zone of E-W step faulting in the north around Ådal. These major faults are linked, exposing different structural levels and, with numerous smaller faults, show a series of displacements from Precambrian to post-Permian times causing extensions into the Oslo Region rocks.

The 'friction breccia' downthrows the Kongsberg Sector relative to the Telemark Sector, to the west. It is a curved zone of normal faulting which, in the south, has a downthrow of many hundreds of metres: as it turns north eastwards, the displacement gradually diminishes and finally dissipates in splay faults south of the Ådal granite. It may be associated with the Oslofjord fault (Ramberg 1976) which has a downthrow of 3 Km in the south and 1 Km in the north. (A hinged effect of similar magnitude would cause no significant re-orientation of earlier structures in the Kongsberg Sector).

The Kröderen-Hokksund fault trends approximately N-S, following a mylonite zone N of Hokksund and then truncating it further N. To the north of the Simoa fault, it underwent at least one downthrow of the western (WKC) side, but further S, displacements were smaller.

The transverse Simoa fault links the friction breccia with the Kröderen-Hokksund fault: it transects the WKC with a downthrow (probably as large as 1 Km) to the north, causing exposure of much higher structural levels.

In the north of the EKC, between Hönefoss and the Ådal granite, several major E-W faults (Fig. 1) are associated with a mosaic of smaller dislocations. The major faults are normal, with a dip and downthrow to the south, forming a zone of step faulting, exposing successively lower structural levels of the EKC northwards. The zone is limited in the west by the friction breccia, along which the eastward downthrow gradually diminishes northwards, partly cancelling displacement in an earlier protomylonite which also dissipated northwards. Therefore in the north there is no significant difference between the variable gneisses (of the EKC and TBGC) on either side of the 'friction breccia' (Starmer 1981). Moreover the step faulting causes EKC gneisses to pass northwards into similar rocks, continuous with those of the TBGC, in a series of open major and minor Sveconorwegian folds. Dips of rocks are generally shallower (15-45°) than throughout the rest of the Kongsberg Sector. Further N, these structures persist in migmatised gneisses and amphibolites around Randsfjorden and Sperillen. All these rocks and structures are typical of the TBGC further W. This area therefore seems to preserve a downward passage of the EKC into the TBGC, the protoliths of the variable gneisses having been deposited by an overlap of the Kongsberg Sector supracrustals into the proto-Telemark block.

2.1.3 Major Mylonite Zones. Throughout the Kongsberg Sector, there are numerous small zones of ductile shearing (e.g. around the Vatnås granite, Fig. 1), but three major mylonite zones can be identified.

These are the Saggrenda-Prestfoss mylonite zone (*SPMZ*) and the Prestfoss
-Sokna mylonite zone (*PSMZ*), at the western margin of the Kongsberg
Sector, together with the Hokksund-Solumsmo mylonite zone (*HSMZ*) at the
western margin of the Modum Complex.

The Saggrenda-Prestfoss mylonite zone (*SPMZ*) forms the western
margin of the Kongsberg Sector in the deeper levels exposed S of the
Simoa fault. Protomylonites trend N from the Oslo Region rocks to
Flesberg (dipping 65-80°E) and are intruded by the Telemark granitic
gneiss. From Flesberg, an ultramylonite thins N to Haugesjö (with a dip
decreasing from 60 to 30°E). At Haugesjö, it bends NE as a result of an
early curvature and break (before emplacement of the Telemark granitic
gneiss) with the later development of a Sveconorwegian basin structure.
It trends NE (with dips from 30-60°SE concordant to structures in the
Kongsberg Sector), to another basin at Prestfoss, where it is cut out
to the north by a combination of the Simoa fault and the steeper,
discordant Prestfoss-Sokna mylonite zone (*PSMZ*).

The Prestfoss-Sokna mylonite zone (*PSMZ*) gradually thinned
north eastwards, with less intense shearing. It continued to dissipate
NE of Sokna, splaying into a number of bands, later folded and broken
by Sveconorwegian deformation. South of Prestfoss, it continued as
protomylonites along the eastern side of the *SPMZ*.

The Hokksund-Solumsmo mylonite zone (*HSMZ*) formed at the
western margin of the Modum Complex when this was thrust obliquely upwards to the northwest, against the WKC. 15 Km N of Hokksund, it gradually dissipated in more micaceous protomylonites and these have been
truncated further N by the Kröderen-Hokksund fault.

2.1.4. Major Structure. The first order major folds, exposing the MC
in a central antiform and the EKC and WKC in flanking synforms, were
early structures developed before the intrusion of Kongsbergian granites
and gabbro-diorites. Most major folds plunged ~30-40°N and dome and
basin structures were not formed, except in the extreme north (E of Sokna)
where a dome resulted from later effects around a granite intrusion and
particularly from Sveconorwegian deformations. The latter (discussed
in sections 2.1.2 and 2.1.3) produced major folds at the junctions with
the Telemark Sector, in the north of the EKC and in the *SPMZ* and *PSMZ*.
Elsewhere, Sveconorwegian deformation caused only gentle flexuring of
the major structure in the Kongsberg Sector and some minor folding.

The Telemark Basement Gneiss Complex (TBGC) has markedly
shallower dips (20-70°E) than the Kongsberg Sector where dips of lithobands and foliations vary from 60° to vertical. In the south, the
friction breccia, with its easterly downthrow of many hundred metres,
cuts the outcrop of the coarse Telemark granitic gneiss. East of the
fault zone, the granitic gneiss has ubiquitous augen and a constant
planar structure with very gently Sveconorwegian flexuring. West of
the fault, the deeper structural levels have sporadic augen developments
and a more intensive Sveconorwegian folding on N-S axes with some cross
folding.

2.1.5. Lithologies: metasupracrustals and intrusives. In the Kongsberg
Sector, the major structure and the ductile upthrust of the Modum
Complex (MC) indicate that the original sediments of the latter were

overlain by the EKC and WKC deposits. The WKC is downthrown relative to
the EKC by the late Kröderen-Hokksund fault and, whilst the two
complexes may be partly lateral equivalents, some higher stratigraphic
levels may be represented in the WKC.

The Modum Complex (MC), contains variable quartz-biotite-
plagioclase gneisses (± almandine, hornblende, cordierite, sillimanite,
graphite and rare orthopyroxene) with thick units of pure quartzite
associated with sillimanite-rich gneisses and quartzites which develop
quartz-sillimanite 'nodules' (lensoids) Cordierite-phlogopite schists
have also formed. Concordant pyrite-pyrrhotite layers ('fahlbånds')
contain more chalcopyrite and graphite than similar rocks in the EKC and
WKC: their Co mineralisation is related to later, basic intrusions
(Gammon 1966). Granitic and granodioritic gneisses (± sillimanite) are
extensive. Basic ('Vinor') intrusions were altered to coronites,
metagabbros and amphibolites in the Sveconorwegian Orogeny which also
scapolitised them and formed some orthoamphibole-cordierite rocks, large
granitic and quartz-plagioclase pegmatites and extensive albitites.
(Some orthoamphibole-cordierite rocks seem to represent original supracrustal layers). Modum Complex rocks have been likened to those of the
Bamble Sector (A. Bugge 1937, Jösang 1966). It has also been noted
(Starmer 1980) that the variable gneisses are almost exclusively metasediments, in contrast to the WKC, but that they, and all the other
lithologies, including the more unusual types, are typical of parts of
the Bamble Sector.

The Western Kongsberg Complex (WKC) and Eastern Kongsberg
Complex (EKC) contain variable quartz-plagioclase-biotite-hornblende
gneisses which have a zonation related to the early, first and second
order major structures. A zone of hornblende-poor or hornblende-free
rocks occurs in the west of the WKC and in a complementary position
around second order structures in the east of the EKC. Elsewhere, the
rocks are a mixed series of hornblende-bearing gneisses. Throughout
the WKC and EKC, concordant basic layers of 'hornblende gneiss'
(hornblende-plagioclase-quartz-biotite rocks) are developed. All types
of variable gneiss develop pyroxene (instead of biotite) in the WKC
around the granite N of Kongsberg and in the EKC around its axial trace.
Sulphide-bearing 'fahlbånds' form concordant layers throughout the WKC
and EKC.

Chemically, the variable gneisses of the WKC and EKC have
characteristics of dacitic-andesitic volcanics and of greywackes. Some
very thin quartz-rich layers suggest sedimentary origins, whereas
gradations with the 'hornblende gneisses' suggest volcanic origins,
since the latter represent basic effusives (± intrusives). The whole
sequence seems to represent a mixed volcanic-sedimentary series. This
would be further supported by an exhalative origin for the sulphidic
fahlbånds, although Gammon (1966) has suggested they were sapropelic
muds.

During the early deformation, which produced the first order
major folds, high viscosity contrasts around the large quartzite units
of the MC caused the formation of major and minor N-S buckle folds.
The low viscosity contrasts of gneisses in the WKC and EKC synforms,
caused them to react by flow folding induced by heterogeneous simple

shear, probably resulting from tightening in underlying synforms of MC rocks.

After the early folding, the Kongsberg Sector underwent a number of ductile shearing episodes punctuated by intrusive events.

Kongsbergian granites intruded near the axial zones of the WKC, MC and EKC. In the WKC, two large bodies occur N and SE of Kongsberg with a zone of migmatites intruding protomylonites W of the town. The northern body is charnockitic-enderbitic in its northern part and adjacent gneisses and xenoliths of newly intruded metabasites developed pyroxenes. A protomylonite zone developed around its western margin. (Further N, the Vatnås granite is a later intrusion). Kongsbergian granites also intruded near the axial zones in the north of the MC and EKC. (in the MC, later migmatisation which notably produced sillimanitic granitic gneisses (± 'nodules'), was contemporaneous with the intrusion of the Vatnås granite in the WKC).

Large gabbro-diorite-tonalite bodies intruded the west of the WKC (Fig. 1) with smaller sheets emplaced across its entire width. Irregular, in situ differentiation was complicated by ductile shearing and Kongsbergian metamorphism.

A later series of basic 'Vinor' bodies intruded in three distinct periods. 'Early phase' leucocratic gabbros were enclosed by the coarse Telemark granite, which was intruded by 'main phase' stocks and sheets (now coronites), themselves cut by coarse metagabbro dykes. The coronites show a differentiation from troctolite to olivine gabbro, olivine norite, gabbro and olivine ferrogabbro, often within single bodies. A 'late phase' of hypabyssal, olivine-free dolerites formed innumerable sheets, now amphibolitised.

The coarse Telemark granite has already been mentioned. It transgressed into the Kongsberg Sector, truncating the westernmost metadiorite body and intruding the mylonitised margin which had previously been bent at Haugesjö. Megacrysts (reaching 3 cm size) later formed augen. The medium grained Telemark granite intruded after the 'late phase' Vinor sheets and the deformation of augen textures in the coarse granite.

The Hedal and Ådal granites (the 'Flå granite') were post-tectonic intrusions in the north, where EKC rocks had been interfolded with the TBGC.

2.2 Major Features of the Bamble Sector

2.2.1 General. The Bamble Sector (Fig. 2) is overlain in the northeast by Cambro-Ordovician sediments of the Oslo Region. In the northwest, it is faulted against the Telemark Basement Gneiss Complex (TBGC) by the 'friction breccia' or 'Porsgrunn-Kristiansand fault' of Morton et al (1970).

Data for the Bamble Sector are taken from previous studies (Starmer 1967, 1969a, 1969b, 1972a, 1972b, 1976, 1978) and from a survey of the whole Sector(Starmer, ms) with an accompanying colour map(scale 1:100,000) of which a preliminary copy (1981) is lodged with Norges Geologiske Undersökelse.

The Bamble Sector has a NE-SW regional grain which is somewhat

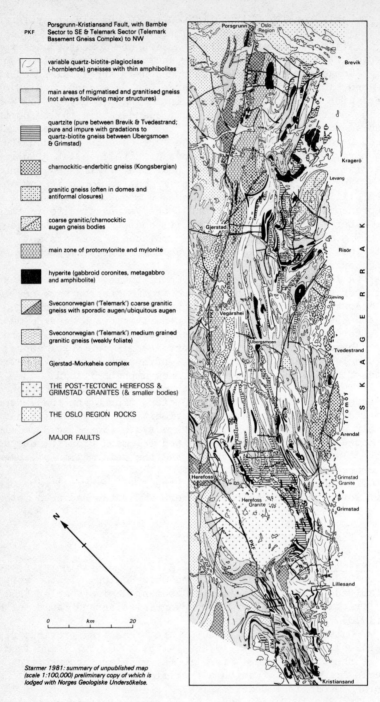

Figure 2. The geology of the Bamble Sector.

arcuate and convex towards the Telemark Sector and becomes more N-S in the southeast. This results largely from the mega-tectonics of this region. The Bamble Sector has a more complex major structure than the Kongsberg Sector, but the oldest supracrustals are suggested to show the same general relationship with remnants of a first order antiformal axis preserved in quartzites and sillimanite-rich rocks (± nodules).

Granulite facies charnockitic and enderbitic rocks occur along the coast from Risör to Arendal, but the author's research group (Milne 1981, Milne & Starmer 1982) have now found a second, inland zone of enderbitic gneisses, between Kragerö and Gjerstad. These have now been dated (Field et al, this vol.) and seem to have formed at the same time as the coastal lithologies. In contrast to the Kongsberg Sector, these Kongsbergian granulite facies rocks and penecontemporaneous granitic bodies have complex major and minor folds and interference patterns.

Migmatisation and granitisation are significantly more widespread than in the Kongsberg Sector and resulted from a series of events.

2.2.2 Major Faults. The Porsgrunn-Kristiansand fault is not a continuation of the 'friction breccia' at the Kongsberg Sector margin and the latter term therefore causes some confusion. It is a zone of late, brittle faults which cuts out an earlier mylonitised junction in lake Nelaug and forms the Bamble Sector margin SW from here to Kristiansand. Northeast of Nelaug, the fault zone has downthrown the Bamble Sector mylonites which it follows to Gjerstad, thence transgressing N into the Telemark granitic gneiss. As at the Kongsberg Sector margin, there were a series of displacements from Precambrian to post-Permian times, with later dislocations extending into the Oslo Region rocks. The overall effect was a downthrow of the southeastern (Bamble) side calculated to be in excess of 0.5 Km (Smithson 1963) and possibly more than 2 Km (Ramberg and Smithson 1975). It has exposed levels in the Bamble Sector which were much shallower than the adjacent Telemark Sector rocks during the Sveconorwegian Orogeny. The zone consists of faults (often forming a mosaic) dipping 45°SE to vertical, with both oblique and normal displacements. It is accompanied by a number of sub-parallel (NE-SW) faults, particularly the Risör-Nidelva fault and the Arendal-Grimstad (Tromöysund) fault, both of which are similarly arcuate. Major transverse (NW-SE) faults also occur and cause a mosaic dislocation of the Bamble Sector. Most faults had a tensional, normal or oblique displacement, but a few (including the Tromöysund fault) show earlier strike-slip movements which were dextral along NE-SW fractures and sinistral along NW-SE lines (Starmer 1967, 1972a).

2.2.3 Major Mylonite Zones. Unlike the Kongsberg Sector, major mylonites and protomylonites are only developed at the junction with the Telemark Sector. This ductile shear zone trends NE-SW from Porsgrunn to Nelaug, but 10 Km N of Gjerstad it turns N-S for a short distance, with a series of anastomosing mylonites surrounding protomylonite lenses (up to 1 Km across).

2.2.4 Major structure. Major structures comprise first, second and third order folds domes and basins which are generally elongated NE-SW and

overturned NW, but are not totally homoaxial. This is because they
result from both folding and ductile shearing. Lithobands and
foliations have a dominant dip of 55-75°SE, sub-parallel to the axial
planes of most major folds. As in the Kongsberg region, the Telemark
Sector rocks have shallower dips (of 20-60°SE).

Remnants of a first order antiformal axis, of Kongsbergian
age, are preserved in the 'Coastal Quartzite Complex' (between Brevik
and Tvedestrand) and the 'Nidelva Quartzite Complex' (between
Ubergsmoen and Grimstad, around the eastern side of the Herefoss
Granite). Large units of quartzite and sillimanite-rich rock
($\pm$ 'nodules') form the same association as in the Modum Complex (MC)
antiform of the Kongsberg Sector However, in the Bamble Sector,
numerous second order major folds and fold interference structures
(principally domes and basins) developed around the axial planar trace
of the first order antiform which closes southwestward in the 'Coastal
Quartzite Complex' just W of Risör (Starmer 1976) but reappears on
the northeastern side of the Herefoss Granite in the 'Nidelva Quartzite
Complex' where it has been bent. This bend and the separation of
the major outcrops of the two Quartzite Complexes result from folding
modified by dextral strike slip shearing. Between the major outcrops,
isolated quartzite layers are deformed by asymmetric folds with steep
axes. These commonly reflect a dextral strike slip, but, in some zones,
result from a sinistral lag. The latter effect is seen on a larger
scale around the Levang granitic dome (which contains second order
domes and basins): here, the second order domes and basins of the
surrounding 'Coastal Quartzite Complex' were transposed around the
Levang granite, bending its remobilised southwestern apophysy. This
occurred after emplacement of the main granite and its satellite
bodies, but before intrusion of gabbroid rocks ('hyperites') and the
ensuing phase of the Sveconorwegian Regeneration (Starmer 1978). In
the sillimanite-rich rocks, the orientations of 'nodules' can be
related to the tectonics (Starmer, in prep.).

The two Quartzite Complexes are surrounded by a sequence of
variable quartz-biotite-plagioclase-hornblende gneisses which show some
systematic zonation (see section 2.2.5) and may be exposed in remnants
of the complementary first order synforms, as in the Kongsberg Sector.
The Quartzite Complexes underwent early buckle folding whereas the
surrounding variable gneisses deformed by flow folding. The gneisses
have less well-defined early structures, greatly modified by later events
and migmatisation. The early major and minor structures are only
preserved well in the two Quartzite Complexes, in the Levang granitic
dome and in the two charnockitic-enderbitic complexes.

The main outcrop of the coastal charnockitic-enderbitic rocks
is an elongate, NE-SW lens between Tvedestrand and Arendal, where
orthogneisses were mixed with migmatised supracrustals, but were
confined largely within a pre-existing (first order synform) structure:
around Arendal, quartzites recur on the southeastern side of the
orthogneisses (and within their outcrop). In detail, the major outcrop
pattern (Fig. 1) results from fold interferences modifying some
discordant relationships. It is further complicated by later
migmatisation, which becomes more intensive towards the margin of the

lens. In certain zones, minor fold interference patterns are developed, but Types 1, 2 and 3 are often juxtaposed within a few metres, since the interference results from a number of fold episodes. Some domes and basins were produced by differential flattening of tight to isoclinal NE-SW folds, but most formed from refolding of these structures by later NE-SW dextral strike-slip. The minor NE-SW folds (λ=a few cm to few m) have subhorizontal axes and are overturned NW: they commonly form the long axes of minor domes and basins, affecting charnockitic-enderbitic orthogneisses and supracrustals with intercalated basic layers and some early intrusive metabasites (pyribolites). Many intrusive pyribolites and leucosomes show only the dextral (and local sinistral) shear with its associated boudinage and steep axis folds, which further complicated pre-existing structures. Concordant, discordant and irregular amphibolites (with rare pyribolitic patches) belong to the later 'hyperite' intrusions and are commonly unfolded although they have a contact parallel foliation.

The enderbitic gneisses outcropping inland between Kragerö and Gjerstad, seem to result from an intrusion into a planar, partly mylonitised zone in the variable gneisses, probably within a first order synform. The enderbites include a few septae of mylonite in the northwest and a few layers of gneiss in the southeast. Isolated pyribolite layers also occur. The rocks are now exposed in a major dextral strike slip fold, with a transposed northwestern limb, its separation being accentuated by later retrogressions.

The Sveconorwegian Orogeny produced very little major folding in the Bamble Sector. Minor structures are common, forming open buckle folds (which are occasionally tighter), angular folds and kink bands: overturned folds and drag folds sporadically have thrust limbs and some conjugate shear zones have developed. Amphibolitised margins of newly intruded gabbroid coronites (the 'hyperites') have developed contact parallel foliations with some minor folding. A few open major flexures are Sveconorwegian in age, but SE from Grimstad and Herefoss to Kristiansand, major folds occur, particularly forming a quartzite dome immediately S of the Herefoss Granite. At this time, the Telemark Sector rocks were at much lower crustal levels and were deformed by major Sveconorwegian folds which are particularly obvious in the coarse granitic gneiss and its augen developments (see section 2.3).

2.2.5. Lithologies: metasupracrustals, intrusives and migmatites. The 'Coastal Quartzite Complex' (between Brevik and Tvedestrand) consists largely of major units of pure quartzite and of sillimanite rich gneisses, schists and quartzites (± 'nodules'). It is identical to the Modum Complex and, like the latter, was extensively affected by Sveconorwegian scapolitisation and albitite development. The 'Nidelva Quartzite Complex' more commonly has gradations of the quartzites with quartz-biotite(-plagioclase) gneisses and was not affected by this Sveconorwegian activity. The other lithologies in both Quartzite Complexes are similar to those in the Modum Complex. Variable quartz-biotite-plagioclase gneisses may contain almandine, hornblende, cordierite, sillimanite, graphite and rare pyroxenes. Concordant pyrite-pyrrhotite rich layers often carry chalcopyrite and graphite. Granitic rocks (± sillimanite) are extensive, and numerous large granitic and quartz-plagioclase pegmatites occur. Orthoamphibole-cordierite rocks are common in the Coastal Quartzite Complex and represent original supracrustals and Sveconorwegian metasomatic rocks.

As in the Kongsberg Sector, the metasedimentary Quartzite Complexes are overlain by a sequence of variable gneisses for which mixed volcanic-sedimentary origins have been proposed (Morton 1971, Beeson 1978). Sulphidic layers with graphite and concordant early amphibolites (effusives ±intrusives) occur throughout.

The variable gneisses are quartz-biotite-plagioclase rocks (± hornblende) with some sillimanite in certain layers. Despite extensive migmatisation, certain zonations can be seen. Around the southwestward closure of the 'Coastal Quartzite Complex', W of Risör (Starmer 1976, 1978) actinolite bearing gneisses pass NW into a zone of 'mixed gneisses' and then into more monotonous rocks showing an increase in hornblende content towards the Porsgrunn-Kristiansand fault. (The 'mixed gneisses' consist of alternating quartzites, biotite-plagioclase-quartz gneiss (± hornblende), biotite schists, amphibolite and some granitic gneiss in bands from a few cm to 2 m wide). The zonation probably reflects an early, first order synform, complementary to the antiformal axis of the 'Coastal Quartzite Complex'. Effects of the early major structures upon disposition of lithologies are also seen on the islands and mainland coast, between Lillesand and Kristiansand, where a distinctive medium to coarse plagioclase-hornblende-quartz-biotite gneiss is of very constant composition over a wide area and probably represents intermediate metavolcanics.

Marbles, skarns and calc-silicate rocks have minor developments throughout the Bamble Sector, but are particularly concentrated between Risör and Lillesand in an arcuate belt around, and within the coastal charnockitic-enderbitic gneisses. Individual horizons are commonly thin (< 1 m wide) but several normally occur together. They are complexly folded on both a major and minor scale, with ductile flowage and boudinage forming pods and lenses. The skarns show an early development of garnet-pyroxene types and a later (post-Sveconorwegian folding) growth of epidote bearing varieties.

The supracrustal lithologies (above) have been intruded by a variety of rocks and extensively migmatised at various times.

The two developments of Kongsbergian charnockitic-enderbitic gneisses consist of low to intermediate pressure granulites with orthopyroxene-plagioclase stable and sillimanite, almandine, cordierite and rare orthopyroxene in the enclosed variable gneisses. The coastal lens contains orthogneisses mixed with layers of variable gneiss, quartzite, graphitic schist, marble, skarn, pyribolite and their migmatised remnants, but it is cut by several generations of later, basic intrusives The orthogneisses are mainly enderbitic (plagioclase-hypersthene/ferrohypersthene-quartz rocks ± diopside/augite) on Tromöy, but mainly charnockitic (plagioclase- K feldspar-hypersthene/ferrohypersthene-quartz-hornblende-biotite rocks ± diopside/augite) on the mainland. The latter have more widespread migmatisation and retrogression (hydration) which commonly changes the characteristic green-brown colour to pink, regardless of whether it is accompanied by K-feldspathisation. Patches and layers of partly retrogressed charnockitic gneiss occur NW of the main outcrop and NE of it, where they range from charnockitic to enderbitic and quartz-hypersthene dioritic compositions, with relict orthopyroxenes in adjacent variable gneisses (Starmer 1972b).

There is thus some systematic lithological variation within the coastal charnockitic-enderbitic gneisses. The Tromöy rocks are strongly LILE depleted (Moine et al 1972, Field et al 1980). Moine et al (1972) considered the same characteristics occurred in the mainland rocks around Arendal, with a passage northwards into non-depleted rocks towards Tvedestrand. Field et al (1980) considered all the mainland rocks to be non-depleted and recognised a series of NE-SW striking zones(zones A to D of Field et al, this volume): Zone D (Tromöy) with LILE and REE depleted "charnockitic" rocks, passed NW into Zone C (mainland, Arendal to Tvedestrand) with non LILE depleted charnockitic rocks, and further NW to Zone B (NW of the charnockitic-enderbitic outcrop) with orthopyroxene in metabasites but not in acid-intermediate rocks, and finally NW into Zone A of amphibolite facies lithologies.

The enderbitic gneisses between Kragerö and Gjerstad are plagioclase-hypersthene-quartz rocks ±hornblende, clinopyroxene with retrograde hornblende and biotite). They have similar (tonalitic-trondhjemitic) compositions to the Tromöy enderbites and preliminary studies(Starmer, unpubl) suggest they are similarly depleted in LILE, REE, Ba and Zr and enriched in Na. They grade outwards into amphibolite facies lithologies as a result of original metamorphism and subsequent retrogressions.

Genetically, the coastal charnockitic-enderbitic complex has been considered to result from igneous protoliths which underwent a chemical fractionation produced by metasomatism during metamorphism (Cooper & Field 1977) or primary, igneous differentiation during high grade metamorphism (Field et al 1980). Whilst some lithologies have igneous protoliths, there are numerous migmatised remnants of earlier rocks and it is suggested that the chemistry of the whole complex was modified during granulite facies metamorphism. Many of the enclosed variable gneisses show depletions in K, Rb, Sr and Ba and enrichments in Na, Sc and Y. Field & Clough (1976) and Clough & Field (1980) have shown that enclosed pyribolites ("metabasites") are depleted in K, Rb, Sr, Ba, Zr and enriched in Na in their "Zone D" (Tromöy). This was considered to result from metasomatism(Field & Clough 1976) or primary crystallisation under high grade conditions (Clough & Field 1980). In the Bamble Sector, only the two areas of charnockitic-enderbitic gneisses show these peculiar chemical features and contain tonalitic-trondhjemitic (-granitic) suites of this age. The chemical depletions and enhancements may therefore be due to granulite facies metamorphism soon after their emplacement. Possibly (as suggested by Starmer 1972b) intrusion and migmatisation occurred under amphibolite and hornblende granulite facies conditions with granulite facies metamorphism very soon afterwards. (Amphibolitic enclaves in charnockitic rocks developed orthopyroxene-rich margins). The zones A to D cut major structures which have also deformed the charnockitic-enderbitic gneisses. The zonation results partly from granulite facies metamorphism before this deformation, partly from similar metamorphism afterwards and partly from retrograde effects which decreased southeastwards. The retrogression in the transition zone (Zones B & C) is patchy and orthopyroxene isograds of different authors (eg. Bugge 1940, Moine et al 1972, Andreae 1974, Field & Clough 1976) vary in position. Some of the zone boundaries are tectonic: the C-D junction is the Tromöysund fault and the A-B junction is partly a fault.

Kongsbergian granites were emplaced particularly in major first and second order domes of the Coastal Quartzite Complex. The large Levang granite is a migmatised dome containing coarse granitic gneiss and migmatised remnants of variable gneisses, quartzites and amphibolites (metabasites) which outline second order domes and basins within it. It may be the upper levels above a more homogeneous granitic gneiss body. Smaller, satellite bodies of granitic gneiss and migmatite formed in second order quartzite domes and some have cores of more homogeneous granitic gneiss (Starmer 1978).

The Kongsbergian granites were later reworked and many bodies in the Bamble Sector are composite Kongsbergian-Sveconorwegian masses, as are the extensive migmatites and their major granitic layers. Medium grained granitic gneisses are both Kongsbergian and Sveconorwegian. Coarse granitic gneisses (± porphyroblasts) are also of both ages, although bands in migmatite zones are dominantly Sveconorwegian. Some homogeneous, medium grained granitic gneisses are late Sveconorwegian and clearly derived from partial melting of earlier granitic rocks, hydrated, feldspathised charnockitic gneisses and possibly some variable gneisses. Some of these (eg. between Lillesand and Kristiansand) vary from strongly foliate to non-foliate, enclosing early amphibolites, but being cut by late amphibolite sheets and contrasting with the medium grained Telemark granite at Kongsberg.

The series of basic intrusions in the Bamble Sector have already been mentioned. Apart from concordant basic layers (effusives ± intrusives) intercalated in the supracrustals, there were two main periods of basic intrusive activity, with several distinct phases in each. The first series of 'metabasites' were affected by the Kongsbergian metamorphism and deformation, whilst the second series of 'hyperites' were affected only by the Sveconorwegian events. The first series now form thin sheets of amphibolite (and pyribolite in the granulite facies areas) with early phase bodies involved in dome and basin structures of the Levang 'granite' and the charnockitic-enderbitic complexes, but the majority resulting from later phase injections which underwent only dextral strike slip shearing and folding. The second series of 'hyperites' can be divided into an 'early phase' a 'main phase' and a 'late phase'. 'Early phase hyperites' consist of amphibolites and meta-leucocratic gabbros (gabbros, leucocratic gabbros anorthositic gabbros and diorites). Rarely, pyribolitic patches formed in the anhydrous granulite facies areas. They occur across the Bamble Sector as sheets, but are similar to contemporaneous, larger masses associated with granitic-charnockitic augen bodies (discussed below). Like the latter, they have alkaline affinities. The 'main phase hyperites' occur in large stocks in the northeast, around the Coastal Quartzite Complex, but elsewhere are found mainly in large and small sheets, some of which partially follow earlier folds. These gabbroid rocks formed coronites, which were metamorphosed to metagabbro and amphibolite, but show a differentiation from troctolite to olivine gabbro olivine norite, gabbro and norite, with several varieties present in larger bodies. Peridotites and pyroxenites are rare. The 'late phase hyperites' are now thin amphibolite sheets which were apparently olivine-free dolerites.

It is suggested that five genetically related bodies of granitic and charnockitic augen (and megacrystic) gneiss occur in large masses at Gjerstad, Vegårshei and Ubergsmoen, with a large mass and smaller northwestern satellite at Gjeving. Across the Bamble Sector they form elongate NE-SW bodies within a NW-SE zone which is displaced dextrally into the Telemark Sector at Gjerstad. All are clearly intrusive, with remnant porphyritic igneous textures and small apophyses. They were partly confined by existing structures and the Gjeving bodies were emplaced in elongate domes. At the southwestern end of the Vegårshei body (at Hovdefjell) charnockitic apophyses intruded existing mylonites, but were subsequently mylonitised themselves and underwent later (brittle) cataclasis associated with the adjacent Porsgrunn-Kristiansand fault. The Gjerstad body discordantly cuts structures in the Telemark Sector. It ranges in composition from granitic to quartz monzonitic and charnockitic and cuts the genetically related Gjerstad-Morkeheia Complex which comprises iron and alkali enriched metagabbros, metadiorites and metagranodiorites of anorthositic affinities (Milne & Starmer 1982). Metagabbroic and metadioritic rocks are also associated with the Vegårshei and Gjeving bodies and in the latter are extensively granitised by the granitic augen gneisses (Starmer 1972b).

The granitic-charnockitic bodies have megacrysts and augen reaching 6 cm in size. The least-tectonised zones sometimes show megacryst alignments and rare sigmoidal patterns suggestive of passive, dextral strike-slip during consolidation. Tectonised zones show a later, pure shear followed by simple shear related to thrusting and mylonitisation. The pure shear involved a flattening in the regional foliation, producing discoid augen (with no directional elongation) in zones up to 10 m wide, but sometimes much more extensive. The subsequent thrusting (also seen in adjacent gneisses and protomylonites) produced elongated augen by progressive simple (rotational) shear of megacrysts. This occurred in zones a few m wide, but became ubiquitous towards the northwestern margin of the Ubergsmoen and Vegårshei bodies and the southeastern margin of the Gjerstad body (i.e. towards the marginal mylonite zone of the Bamble Sector). Within the mylonite zone, the augen of the Vegårshei and Gjerstad bodies were often destroyed by mylonitisation. The extreme NW margin of the Vegårshei body, where it is outside the mylonite zone (Fig. 2), shows only pure shear without superimposed thrusting. This is also the case throughout the Gjeving bodies, which show only flattening of augen, since they are many Km distant from the mylonite zone. The related basic-intermediate rocks show the same tectonisation as adjacent augen gneisses.

Sveconorwegian quartz-plagioclase pegmatites and leucosomes usually carry the mafic minerals of their host rocks. They were both syn- and post- tectonic. Granitic pegmatites were usually post-tectonic and more commonly formed large, discrete bodies.

The large Herefoss and Grimstad Granites (and several smaller

bodies) were post-tectonic intrusions although the Herefoss Granite
was dissected by later movements along the Porsgrunn-Kristiansand
fault. Just south of the Herefoss Granite, slightly earlier granites
and quartz diorites intruded a quartzite dome.

2.2.6 Major Zonation of Metamorphic Facies. In the Bamble Sector, the
Kongsbergian Orogeny produced two zones of granulite and hornblende
granulite facies rocks surrounding the developments of charnockitic-
enderbitic orthogneisses between Risör and Arendal and between Kragerö
and Gjerstad. These zones passed outwards into upper amphibolite facies
assemblages which were developed over the rest of the Bamble Sector.
The transitional zones (see section 2.2.5) result from original
metamorphisms, before and after deformation, and from retrograde effects.
The latter ware largely of Sveconorwegian age and were intensive in
the Telemark Sector: they had a limited effect in the higher level
Bamble Sector, although in the northwest, they were largely responsible
for some of the NE-SW isograds located by Touret (1968) and Moine et al
(1972).

2.3 The Telemark Sector Adjacent to the Kongsberg and Bamble Sectors

The relationships of the Kongsberg Sector with the adjacent Telemark
Basement Gneiss Complex (TBGC) have already been considered. The
coarse and medium grained Sveconorwegian granitic gneisses (and the
variable quartz-plagioclase-biotite gneisses) of the TBGC can be
correlated (Fig. 3) from the Kongsberg Sector margin (where they have
a N-S grain) to the northern end of the Bamble Sector margin (where they
have a NE-SW grain). The coarse granitic gneiss has ubiquitous augen
along the western side of the Kongsberg Sector and the Oslo Region:
further west, it has sporadic augen developments. This spatial
relationship curves round to a NE-SW orientation adjacent to the northern
end of the Bamble Sector, as a result of major Sveconorwegian fold
interferences deforming the foliation and augen textures. Around
Frier (as in the Kongsberg area) thin amphibolites cut the coarse
granitic gneiss, but earlier amphibolite lenses and schlieren are
enclosed and granitised. In places (eg. N of Kragerö), stocks of
'main phase' hyperite coronite cut the coarse granitic gneisses. The
coarse granitic gneiss was later intruded by the medium grained granite,
which is often weakly foliate or massive. The consistent time relationships of granitic rocks at the Kongsberg Sector margin and the northern
end of the Bamble Sector do not necessarily apply to similar rocks
throughout the Telemark Sector.

Pre-Sveconorwegian events are also recorded. The oldest rocks
are variable quartz-biotite-plagioclase (-hornblende) gneisses, similar
to the metasediments and metavolcanics of the Kongsberg and Bamble
Sectors; although sillimanite is absent, muscovite is much commoner
and migmatisation is more extensive. Marbles occur in a number of
places (eg. at Tveit, 15 Km N of Kristiansand - Falkum 1966). An
early migmatite complex of variable gneisses, granodioritic gneisses,
granitic gneisses and augen gneisses (with intruded basic sheets)
underwent mid-amphibolite facies metamorphism. Accompanying deformation
produced fold interferences which were generally more open structures

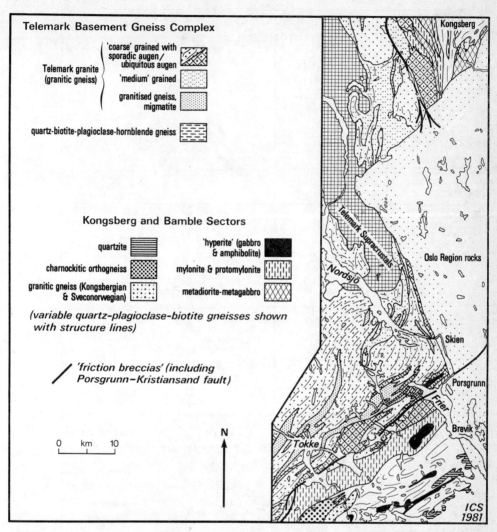

Figure 3. The region between the Kongsberg and Bamble Sectors

than in the Kongsberg and Bamble Sectors. Some later granitic gneisses and augen gneisses formed before being cut discordantly by the Gjerstad -Morkeheia Complex and the Gjerstad augen gneiss body, which are themselves cut by Sveconorwegian amphibolites, granitic sheets and pegmatites.

The Telemark Supracrustal Suite (Fig. 3) was deposited on the Telemark Basement Gneiss Complex, but the latter was subsequently rejuvenated, particularly during the Sveconorwegian Orogeny. Whether some rocks in the Basement Gneiss Complex represent metamorphosed and migmatised remnants of the Supracrustal Suite is a problem which remains unresolved.

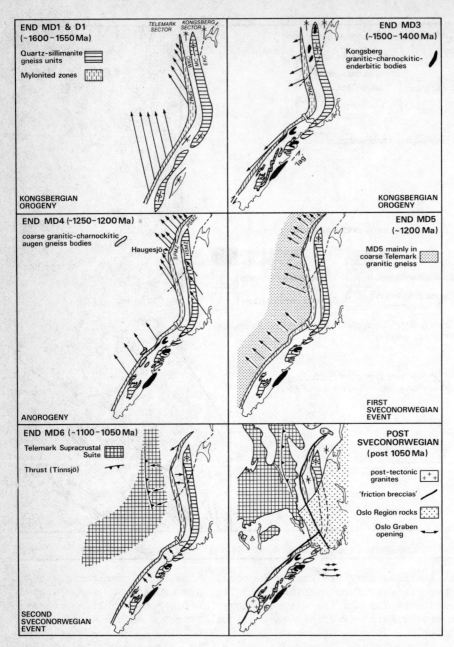

Figure 4. The evolution of the Kongsberg-Bamble mobile belt through the main mylonitisation (MD) periods. (Arrow lengths vary in some mylonite zones to show variations in the amount of displacement).

2.4 The Evolution of the Kongsberg-Bamble Mobile Belt.

The evolution is summarised in Table 1, and stages in the development of the Kongsberg-Bamble mobile belt are shown on Figure 4. Radiometric data (recalculated to $\lambda^{87}Rb = 1.42 \times 10^{-11}$. yr^{-1}) are taken from Jacobsen & Heier (1978) unless otherwise stated. Movement directions during ductile shearing were derived from mineral lineations, porphyroclast rotations, augen elongations, microfolding and drag folding.
The deposition of the original supracrustals (now quartzites and variable gneisses) seems to have occurred in a N-S trough, which was a marginal basin of an older craton of Svecofennian (Hudsonian) age. The subsequent tectonics reflect a more stable, proto-Telemark block (thicker crust) to the north and west of the Kongsberg-Bamble mobile belt, with ductile shearing concentrated along the junction. In the Kongsberg Sector, shears dissipated in the north, with a decrease in north-westward upthrusting: later faulting was also hinged in the north, although movements were in the opposite sense. The whole region was near the margin of the Proterozoic Supercontinent, the motions of which produced internal shear zones.

The oldest quartzites and sillimanite-rich rocks, exposed in the Modum Complex (MC) antiform and its continuation in the 'Coastal' and 'Nidelva Quartzite Complexes' of the Bamble Sector, represent miogeosynclinal, shelf sediments. These were overlain by (?eugeosynclinal) deeper water sediments and intermediate volcanics, now represented by the variable gneisses exposed in the EKC and WKC synforms and in the Bamble Sector. Throughout the sequence, sulphidic layers and basic effusives (± intrusives) occurred. In the north of the EKC, these rocks overlapped into the Telemark Basement Gneiss Complex and a more widespread deposition in the latter may account for observed Kongsbergian ages within it.
The first deformation (D1) involved E-W compressions and produced first, second and third order major, buckle folds in the main Modum Complex-Bamble Quartzite Complex axis (where viscosity contrasts were high) and flow folding in the variable gneisses of the EKC, WKC and Bamble Sector. A regional S1 foliation developed and was axial planar to these F1 major and minor (often isoclinal) folds.
Late D1 deformation and the first major mylonitisation (MD1) decoupled the Kongsberg-Bamble belt from the Telemark block by overthrusting. The Kongsberg Sector underwent an oblique NW upthrust with displacement increasing southwards, giving a widening mylonitised junction (the *SPMZ*) and a small protomylonite zone to the east, which was migmatised by Kongsbergian granites. Movement lineations pitched 50-70°S on planes dipping steeply E and are still preserved in the WKC in protomylonite, ultramylonite and screens in the margins of a metadiorite body. F1 folds were transposed and S1 foliations were deformed into intrafolial folds. The finite stresses correspond with D1 which produced major folds plunging ~30-40°N. Further east, in the MC, the F1 buckle folds were transposed and in the EKC, first and second order major folds developed (plunging 30-40°N) with minor folds causing transposition and intrafolial isoclines of the S1 foliation. The Kongsberg Sector now consisted of a central MC antiform flanked by the EKC and WKC synforms, with no major domes and basins.

The Bamble Sector was thrust obliquely onto the Telemark block and was bent, as was the main quartzite axis, to a more NE-SW orientation. The mylonitised front now dips ~60°SE with movement directions pitching 60-75°SW. To the south east, a shear foliation was accompanied by isoclines which were often intrafolial. Further southeast, first and second order major folds became non-cylindroidal with axial culminations eventually forming elongate domes and basins. Most major and minor folds were overturned NW. Near the southwestward closure of the 'Coastal Quartzite Complex', axial culminations were probably formed by inhomogenous compressive strain along the length of the axial plane (Starmer 1976), but further NE, in its interior, observable cross folds (Starmer 1978) may have been produced by a biaxial compressive stress field resulting from the oblique overthrusting. Hopgood (1980) has shown how progressive deformation, in a single stress field, can be accompanied by axial curvature due to frictional or viscosity differences and subparallel axes can be superimposed on penecontemporaneous curved axes.

In the Bamble Sector, these MDI fabrics (dipping 55° to 75° SE) and the decoupling interface with the Telemark block (dipping 60°SE) posthumously controlled later deformations, particularly dextral shearing (MD3) which developed folds with parallel axial planes. The MDI deformation phase initially produced the wedge of Bamble Sector rocks shown by gravity studies (eg. Smithson 1963), but this also reflects modifications resulting from later deformations.

Kongsbergian acidic intrusions (~1550-1520 Ma) resulted from anatexis following the crustal thickening of D1-MD1. In the Kongsberg, Bamble and Telemark Sectors, slightly earlier basic intrusions are represented by enclaves in the acidic rocks (eg. the Kongsberg granites, the Levang granitic dome and the charnockitic-enderbitic gneisses), but the main basic activity occurred later (~1500 Ma). In the Kongsberg Sector, granitic bodies intruded near the axial zones of the WKC, MC and EKC, forming migmatites in MD1 protomylonites, just west of Kongsberg. They appear to have been water-undersaturated magmas with charnockitic-enderbitic rocks developed near the WKC first order synform axis N of Kongsberg and granulite facies assemblages in adjacent gneisses and xenoliths. Depletions in HREE, Y, Rb, Th, U and erratic enrichments in Ba, Sr, Zr occur within charnockitic-enderbitic rocks and their xenoliths (Craig, unpubl): they have been suggested to reflect flushing by CO_2, possibly during mantle degassing, soon after intrusion (Starmer 1985). Granulite facies gneisses were also developed further east in the WKC (at Holtefjell) around a second order synform, sporadically in the MC, and around the first order synform axial zone of the EKC.

In the Bamble Sector, granitic bodies (particularly the Levang 'granite' and its satellites) intruded domes in the first order quartzite antiform. The central parts of the Levang dome have yielded ages ~1580 Ma (O'Nions & Baadsgaard 1971). Charnockitic-enderbitic rocks were emplaced in the first order synforms and like the Levang dome, retained some supracrustal gneiss layers. Enderbitic rocks near Gjerstad were elongated in planar (partly mylonitised) structures. The charnockitic-enderbitic rocks on the coast, around Arendal, were largely confined by an elongate major structure. Their granulite facies metamorphism occurred soon after intrusion, ~1540 Ma (Field & Raheim 1979) and, as at Kongsberg, affected the surrounding gneisses (Starmer 1972b). Touret (1971) has shown, from fluid inclusion studies, that transitions from amphibolite to granulite facies domains were caused by increased P_{CO_2}, possibly linked with mantle degassing, the metamorphic conditions being 700-800°C and 6-8 kb. This process would explain the LILE depletions and the development of granulite facies rocks only around the charnockitic-enderbitic orthogneisses.

A minor phase of ductile shearing (MD2) is seen locally in the Kongsberg Sector where an oblique SW upthrust produced foliation and narrow mylonites in the granites of the MC and WKC, before they were cut by dioritic bodies. Some MF2 folds of MS1 mylonites are preserved in xenoliths within metadiorite.

In the Bamble Sector, NW-SE compressions increased its rotation relative to the Kongsberg Sector. New shear folds and isoclines formed, NE-SW folds were tightened and differentially flattened and interference patterns developed. Effects were particularly marked in the newly emplaced granitic and charnockitic-enderbitic rocks, which developed tight to isoclinal NE-SW folds, overturned NW and deforming early metabasites, with formation of some domes and basins: mylonitised zones developed in the northwest of the Sector, but were minor compared with the folding.

Basic and basic-intermediate intrusions were emplaced ~1500 Ma. In the Kongsberg Sector, large bodies underwent in situ differentiation to gabbro-diorite-tonalite rocks, with a trondhjemitic trend. Smaller sheets were also emplaced. These rocks intruded the Kongsbergian granites and were metamorphosed under static, upper amphibolite facies conditions before ductile shearing affected them, in places. These metamorphic conditions were operative during intrusion, since the bodies underwent a plagioclase-hornblende fractionation under high P_{total} and P_{H_2O}, forming some cumulates.

In the Bamble Sector and in the Telemark Sector, basic sheets ('metabasites') intruded and are now amphibolites and pyribolites. They seem to have had calc-alkaline affinities and iron enrichment trends.

The second major period of ductile shearing (MD3) occurred when the Kongsberg and Bamble Sectors were virtually in their present relative position. Mylonitisation, at their junctions with the Telemark block, was produced by a southwestward, dextral strike slip of the Bamble Sector (with a slight upthrust) and an oblique SW upthrust of the Kongsberg Sector.

Near the western margin of the Kongsberg Sector, gneisses and metagabbro-metadiorite bodies were mylonitised. In the WKC, protomylonites developed around the western margin of the large Kongsbergian granites (which were also sheared internally) and in several small zones, notably at the site of the later Vatnås granite. Movement directions pitched 70°N on planes dipping steeply E.

In the Bamble Sector, the effect of dextral strike slip movements is seen over its entire exposed width. Mylonitisation at its NW margin is still preserved in a number of places, notably at Flakvarp,

Dates (Ma) & Source	Conditions, Metamorphic facies		EVENTS:	Kongsberg Sector	Bamble Sector
?1700 -1600	1		Deposition	SUPRACRUSTALS: Shelf sediments overlain by deeper water sediments & intermediate volcanics. Sulphidic layers & basic effusives (± intrusives) throughout. Overlap in TBGC.	
			SINKING OF REGION		
				D1(E-W COMPRESSION): F1(N-S) buckle in major quartzite units, flow in variable gneisses. Regional foliation.	
			MD1, LATE D1:	WKC thrust obliquely NW against TBGC, increasingly to S (in SPMZ) MC transposition of F1 EKC major & minor F1	Thrust obliquely N onto TBGC & bent towards NE-SW. Wedge cross-section initiated. Elongate major domes & basins
~1550 ~1520	1 3	K O R N O G N S G B E E Y R G I A N	Upper Amphibolite -granulite	INTRUSION: Granites (± charnockite-enderbites) near axial zones WKC (1), MC, EKC	Granites (e.g. Levang (2)) near quartzite axial zone. Charnockite-enderbites at Arendal (3) & Gjerstad
(~1580)	2			METAMORPHISM: Local granulite facies (?mantle degassing)	
			MD2:	Oblique upthrust SW. Foliation & mylonite in granites. Folds of MD1 fabrics	NW-SE compression, increased bend, tightening of folds, new folds & foliation. Same thrust NW over TBGC.
~1500	1		Upper Amphibolite (local granulite in Bamble)	INTRUSION: gabbro-diorite-tonalite (1)	gabbro-dolerite
				STATIC METAMORPHISM	
			MD3:	WKC thrust obliquely SW against TBGC & internally. Mylonites in granites	Dextral shear, with slight upthrust, against TBGC. (Folds mylonites, enderbites & transposes quartzites. Folds interfere).
~1390 -1340	3 1		(?Uprise, deposition TSS, sinking?)	INTRUSION: Vatnås granite (1)	Granite sheets (3)
<1340	1			INTRUSION: 'Early phase' Vinor gabbros, leucogabbros(1) (Anorthositic Affinities?)	Granitic-charnockitic (augen) bodies (3) ± gabbro-diorite-granodiorite. 'Early phase' hyperites.
1275-1235	3				?Passive dextral slip: augen & streaking in above bodies
		S O V R E O C G O E N N Y O R W.	Mid-Amphibolite	MD4: Internal dislocation. Oblique upthrust NW. Western margin: SPMZ broken & bent NE & cut out by PSMZ. MC-WKC margin: HSMZ formed	Thrust NW against TBGC: augen in some granitic charnockitic bodies
~1200	1			INTRUSION: Coarse Telemark granite (also in Kongsberg & Bamble Sectors)	
				MD5: (Augen in coarse Telemark granite) Thrust WNW & NW. PSMZ enhanced	Thrust NW. Minor isoclines
~1200	1		Tension Mid-Amphibolite	INTRUSION: Gabbroid stocks & sheets in depth (now coronites) 'Main phase' Vinor (Kongsberg(1) & hyperite (Bamble) bodies	
				HIGH PT IN DEPTH: Corona growth, intrusion gabbroid dykes, mobilisation of gneisses (WKC), some amphibolitisation	
				EPEIROGENIC UPRISE OF REGION	
				(?Deposition TSS?)	
			Sinking	INTRUSION: Hypabyssal, olivine-free dolerites 'Late phase' Vinor (Kongsberg) and hyperite (Bamble) sheets	
(~1100?)	1	S O V R E O C G O E N N Y O R W E G I A N	Lower (-mid) Amphibolite	MD6: Local shears & drag folds WKC thrust W in W & E in E	Local shears & folds Thrust NW
				D6: F6 open, concentric, buckle folds: various axes, domes & basins Major in TBGC & Kongsberg Sector margin (SPMZ, PSMZ & N of EKC), minor in TBGC, Kongsberg & Bamble Sectors. Major, gentle flexures WKC	
~1050	1 3 4			STATIC METAMORPHIC CLIMAX: Almandines overprint MD6 & D6	
				INTRUSION: Medium grained Telemark granite (also in Kongsberg (1) & Bamble (3) Sectors). Pegmatites (4), aplites. Reworking older granites	
(~1000?)			Epidote Amphibolite	OVERPRINT: Epidote-muscovite-biotite fabrics	Pegmatites continue (4)
			Greenschist	FAULTING: Probable initiation of 'friction breccias'(including Porsgrunn-Kristiansand fault)	
970-900	1 5		UPRISE	INTRUSION: Post-tectonic granites Hedal & Ådal (Flå) (5)	Herefoss & Grimstad (1)
				FAULTING: Continues to post-Permian times INTRUSIONS & VEINS in Permian	

Table 1 (opposite) Events, with deformation (D), mylonitisation (MD) and folding (F) episodes numbered consecutively. WKC, MC and EKC are the Kongsberg Sector's divisions and SPMZ, PSMZ and HSMZ its major mylonite zones: TBGC is the Telemark Basement Gneiss Complex and TSS the Telemark Supracrustal Suite. Correlated radiometric data are from:
1 - Jacobsen and Heier (1978); 2 - O'Nions and Baadsgaard (1971);
3 - Field and Råheim (1979, 1981, 1983 - see text); 4 - Broch (1964);
5 - Killeen and Heier (1975).

near Porsgrunn. Movement directions pitch 10-15°NE on planes dipping 55-70°SE. Just SE of the mylonite zone, the enderbitic body at Gjerstad developed a major dextral shear fold and asymmetric minor folds with axes plunging 50° to 160-180° and axial planes dipping 55-70° SE. The northwestern limb of the major fold was transposed in the mylonite zone, the intervening area containing comminuted pyroxenes with later amphibole coronas showing the transposition may have been effected, in part, by a later sinistral lag which also bent the adjacent mylonite zone to a local N-S orientation. Away from the mylonite zone, the Levang granite dome was tightened and its western apophysy was bent as the surrounding quartzite domes and basins were transposed around it by sinistral lag (Starmer 1978). Their separation from the 'Nidelva Quartzite Complex' (originally due to the folding) was enhanced, with isolated, intervening quartzites developing dextral drag folds. The axial planar trace of the 'Nidelva Quartzite Complex' was bent (NE of the Herefoss Granite) and some minor sheath folds developed. Throughout the Bamble Sector, folds, domes and basins were formed or accentuated, with some zones showing local, sinistral lags. In the coastal charnockitic-enderbitic rocks (where conditions were still those of the granulite or hornblende-granulite facies), new basic intrusions and leucosomes developed only the folding of this strike slip phase, which deformed earlier masses into complex interference patterns with their host rocks: axial planes of the new folds dip 55-60° SE with axes and lineations plunging 45-75° towards 170-215°.

A period of anorogeny followed. In the Kongsberg Sector, the Vatnås granite intruded (~1340 Ma) cutting an MD3 mylonite zone, although it was somewhat sheared by MD4 movememts. Migmatites also formed particularly in the antiformal Modum Complex. 'Early phase' Vinor gabbro and leucocratic gabbro stocks and sheets (with anorthositic affinities) were emplaced, partly in the metadiorites: they were sheared by MD4 and subsequently veined and enclosed by the coarse Telemark granite.

In the Bamble Sector, near Tvedestrand, granitic layers intruded ~1390 Ma (Field & Råheim 1981). Some granitic gneisses (±augen) in the Telemark Sector seem to be of the same age.

A number of elongate bodies of coarse megacrystic granite-charnockite (now augen gneiss) were emplaced. They were associated with alkali and iron-enriched gabbros, diorites and granodiorites (of anorthositic affinities) in the Gjerstad-Morkeheia Complex (Milne & Starmer 1982) and at Vegårshei, and with alkaline gabbroic bodies ('early phase' hyperites) at Gjeving (Starmer 1972b). Some of the bodies intruded mylonites, causing thermal metamorphism, but they were subsequently sheared during MD4. Smalley et al (1983) considered the Gjerstad augen gneiss body to have intruded between ~1350 and ~1250 Ma, the latter (Rb-Sr whole rock) age representing cooling after high T deformation and corona growth and probably equivalent to MD4 deformation in the present study. SE of the Gjerstad body, dioritic sheets, showing less extreme iron enrichment, intruded the mylonite zone between the separated enderbitic units, but were mylonitised themselves by MD4 and subsequently cut by Sveconorwegian 'hyperites', granitic veins and pegmatites as were the Gjerstad-Morkeheia Complex and the augen gneiss bodies. One of the latter was altered to an adinole by a 'hyperite' gabbro, just north of Gjeving (Starmer 1972b). The augen gneiss bodies were elongated NE-SW, possibly in tensional features developed by passive strike-slip movements: they occur across the Bamble Sector in a NW-SE zone which is dextrally displaced into the Telemark Sector, at Gjerstad. There may have been a passive dextral strike-slip at the Bamble-Telemark junction, before they were totally consolidated. This would explain the alignment of some megacrysts and possibly the elongate major apophyses at the northeastern end of the Vegårshei body. The third major mylonitisation (MD4) at ~1250 Ma, marked the end of the anorogenic period and resulted from a new (Sveconorwegian) stress field which thrust the Kongsberg and Bamble Sectors NW, causing internal dislocation of the former. Again, the upthrust dissipated northwards at the Kongsberg Sector margin and there was some clockwise rotation of the mobile belt relative to the Telemark block.

Before the coarse Telemark granite intruded, the Kongsberg Sector frontal thrust (the *SPMZ*) was broken and bent at Haugesjö, from where it trended NE (at ~35°). At Prestfoss, further NE, it was truncated (together with the Vatnås protomylonite zone) by the *PSMZ*. The latter thinned and shallowed NE, finally splaying and dissipating NE of Sokna, where it was affected by Sveconorwegian folding. The western margin of the Modum Complex was thrust obliquely upwards to the north west, developing the *HSMZ*. Movement directions pitch 60-70°S on planes dipping steeply E (in the *SPMZ*) or W (in the *HSMZ*), although in the extreme south they pitch 3o-35°S: where the western margin trends NE-SW (in the *SPMZ* and *PSMZ* north of Haugesjö) they pitch 80°SW on planes dipping moderately SE (with MD1 and MD3 directions now pitching 80-90°SW and 45-50°NE, respectively).

In the Bamble Sector marginal mylonite zone, there was thrusting obliquely N to NNW on planes dipping ~60°SE, although the same lineations (plunging ~50° towards 140-160°) were imposed on other planes (eg. those striking N-S, near Gjerstad). In the megacrystic granitic-charnockitic bodies, which were not totally consolidated, augen developed by pure shear and subsequent simple shear (see section 2.2.5). The latter reflected an oblique northwestward upthrust with movement senses, on steep SE dipping foliations, pitching 50° to 80°SW in the Gjerstad and Ubergsmoen bodies and 70-90°SW in the Vegårshei body. Pyroxenes were granulated without retrogression, but developed later amphibole coronas (Milne & Starmer 1982). Similarly, there was little hydration in the granulite and hornblende granulite facies areas and metabasites retained pyroxenes: elsewhere in the Bamble Sector, amphibolite facies conditions prevailed.

The intrusion of the coarse Telemark granite (~1200 Ma) resulted from anatexis following the crustal thickening of MD4. It was emplaced in the Telemark Basement Gneiss Complex but intruded upwards into overlying Kongsberg and Bamble Sector rocks. This occurred after the MD4 bending of the Kongsberg Sector margin and xenoliths of the PSMZ were incorporated. The granite crystallised slowly in depth and developed feldspar megacrysts (later often forming augen).

The fourth major period of ductile shearing (MD5) formed augen and mylonitic laminae in the coarse Telemark granite and adjacent granitised mylonites of the Kongsberg and Bamble Sectors. There was a progressive simple shear, with little flattening across foliae, and some microfolding of quartz ribbons.

In the south of the Kongsberg Sector, where its margin was N-S, there was an oblique WNW thrust with ubiquitous augen developed only in areas downthrown E of the friction breccia, indicating that maximum shearing occurred in the upper levels where the granite had intruded Kongsberg Sector rocks. In the north, where the margin was NE-SW, there was an up-dip thrust NW and the PSMZ was enhanced along the granite margin, dissipating where it disappears in the northeast.

At the Bamble Sector margin, there was an up-dip thrust NW in the granite and isoclines developed in the mylonite zone, with axial planes having a shallow dip NE. Similar structures deformed amphibolites in the granite. The latter became mylonitised, in places.

The MD4 thrust directions at the Kongsberg Sector margin converge with those at the Bamble Sector margin (Fig. 4). This results from a slight relative rotation of the latter during the subsequent Sveconorwegian folding (D6) which produced major structures in the granitic gneiss of the hinge zone (Fig. 3).

A second period of anorogeny was marked by deep level intrusions of gabbroid stocks and sheets (now coronites). These 'main phase' Vinor intrusions (Kongsberg) and 'main phase' hyperites(Bamble) cut the coarse Telemark granitic gneiss and its MD5 fabrics at the margins of the Bamble and Kongsberg Sectors, but may be contemporaneous with the large Evje-Iveland intrusions further SW in the Telemark Sector (and with the 'Moss globuliths' of Berthelsen (1969) in Østfold). Chemically (Starmer 1985), they are transitional alkali basalt-tholeiite-high alumina types, often considered typical of continental thinning without rifting. Dominantly, they have mildly alkaline affinities with alkali and Fe enrichment trends. Several pulses were intruded from a large underlying mass of alkali-olivine basic magma(Starmer 1969b). They underwent contamination and fractionation during crystallisation under high P and moderate T.

High PT conditions and successive magma injections prolonged the subsolidus cooling of the gabbroids which developed coronas.The stocks were often cut by coarse metagabbro dykes in the WKC and by coarse coronite dykes in the Bamble Sector (Starmer 1978, 1979). Later amphibolitisation with the formation of contact parallel foliations was accompanied, in the WKC and MC, by mobilisation of adjacent gneisses and back-veining.

Epeirogenic uprise of the whole region occurred before the intrusion of hypabyssal sheets of olivine-free dolerite (now amphibolitised). These formed the 'late phase' hyperites (Bamble) and Vinor intrusions

(Kongsberg) and were finer grained than the 'main phase' bodies, being intruded over a much wider area in the Bamble, Kongsberg and Telemark Sectors. They are probably equivalent to the Kattsund-Koster dykes of Hageskov and Pedersen (1981) in the Oslofjord area.

In the Bamble Sector near Gjerstad, a much later metagabbro dyke trends discordantly ENE-WSW for 7 Km. It was post-tectonic with chilled margins and may be equivalent to the basic dykes intruded ~950-850 Ma in SW Norway and S Sweden.

The Telemark Supracrustal Suite consists of shallow water sediments and volcanics, deposited on a Telemark Basement Gneiss Complex. They are cut by amphibolitised sheets similar to the 'late phase' Vinor intrusions and may have been deposited after uprise, but before the latter were intruded during tensional sinking. It is possible that some (or all) of them were deposited during the earlier anorogenic period, since Nilsen (1982) found the lowermost Rjukan group had a strong E-W transposed foliation and metamorphism (thought to be of 16-1400 Ma age) whilst the overlying Bandak group showed only the later, Sveconorwegian metamorphism and N-S folding (~1050-900 Ma). Some workers consider that the volcanics may be syn-orogenic.

The second phase of the Sveconorwegian Orogeny involved local ductile thrusting (MD6), buckle folding (D6) and lower to mid-amphibolite facies metamorphism which caused further amphibolitisations of the basic intrusives. It was a period of extensive fluid activity with formation of granites, pegmatites and ubiquitous leucosomes.

Local MD6 thrusting produced narrow mylonite zones (particularly in the coarse Telemark granitic gneiss and the amphibolitised margins of 'main phase' stocks) and minor drag folds (particularly in 'late phase' amphibolite sheets). In the Kongsberg Sector, the WKC was thrust outwards to the east and west, where its margins were N-S, and underwent strike slip and oblique shear where they were NE-SW. In the Bamble Sector, folding occurred particularly between gabbroid stocks, but small thrusts and conjugate shears were more widespread. In the Telemark Supracrustal Suite, near Tinnsjö (30 Km W of Kongsberg) a westward thrust probably occurred at this time.

D6 major and minor buckle folds (F6) with some crossfolds were partly contemporaneous and partly later than MD6. Major folds only developed in the deeper levels of the Telemark Sector, upthrown W and NW of the 'friction breccia' and at the Kongsberg Sector margin (in the north of the EKC and in the SPMZ and PSMZ). Elsewhere, only gentle major flexuring occurred in the higher levels, now downthrown and forming the entire Bamble and Kongsberg Sectors and marginal parts of the Telemark Sector. Significantly, where the friction breccia displacement disappears in the north of the Kongsberg Sector, Sveconorwegian major folding becomes prominent as the EKC grades into the Telemark Sector. Minor flexural slip folds, again with cross folds producing domes and basins, are ubiquitous, but more intensively developed in the Telemark Sector. The structures usually have steep axial planes with various axial directions, dominantly N-S, but also commonly E-W in the Bamble Sector.

The static metamorphic climax (just before 1060 Ma) occurred after most of the F6 folding with almandines overprinting MD6 and D6 structures and fabrics, particularly in amphibolites. This corresponds with low grade metamorphism recorded in the Telemark Supracrustal Suite by Kleppe and Råheim (1979).

The intrusion of the medium grained Telemark granite (~1060 Ma) occurred just after the metamorphic climax and the crustal thickening of MD6-D6. It was emplaced in the Telemark block at the margins with the Kongsberg Sector and the north of the Bamble Sector: it contains biotite and muscovite in contrast to the earlier coarse Telemark granites which contain hornblende, almandine and biotite and are more strongly foliated. Thinner sheets and veins, together with some small bodies of homogeneous, poorly foliate granitic gneiss intruded the Kongsberg and Bamble Sectors. Field & Råheim (1979) recorded an age of ~1060 Ma for granitic sheets in the Arendal area. Pegmatites, aplites and homogeneous bodies were partly derived from older granites and were partly juvenile: they particularly intruded the Bamble Sector and the antiformal Modum Complex which were at higher levels than the presently exposed Telemark Sector rocks, where pegmatites formed much smaller bodies. Pegmatites and leucosomes formed over an extended period, particularly in the Bamble Sector, and may contain early garnet, scapolite and kyanite (the latter formed by break down of sillimanite and cordierite) with later epidotes and chlorites. They solidified after most of the buckle folding, but some were deformed by late, open structures developed under epidote-amphibolite facies conditions.

An epidote-amphibolite facies overprint produced epidote-muscovite-biotite fabrics (± actinolite) in all rocks of the Kongsberg and Telemark Sectors and in parts of the Bamble Sector, where a few open, minor folds and conjugate shears developed. It is possibly equivalent to the Fjordzone deformation (~1015 Ma) of Hageskov and Pedersen (1981). Pegmatite intrusion continued for some time in the Bamble Sector.

Post-tectonic granites (intruded ~970-900 Ma) formed the Herefoss and Grimstad bodies across the junction of the Bamble and Telemark Sectors and the Ådal and Hedal (Flå) granites in the north of the Kongsberg Sector, where it passed into the Telemark block. The Herefoss granite was dissected by later movements on the Porsgrunn-Kristiansand ('friction breccia') fault, but the Ådal granite straddled the northern hinged termination of the friction breccia and underwent only minor fracturing.

Late faulting, initially under greenschist facies conditions, may have begun before the post-tectonic granites were intruded. Sporadic movements continued until post-Permian times. The 'friction breccia' faulting disconnected the Kongsberg Sector from the Bamble Sector and the mylonite front of the latter was truncated by the Porsgrunn-Kristiansand fault. The friction breccia was hinged in the north of the Kongsberg Sector and the arcuate Porsgrunn-Kristiansand fault zone of the Bamble Sector was a mosaic of tensional, normal and oblique faults: displacements may have slightly accentuated the bend of the Kongsberg-Bamble belt, since they were associated with the formation of the Oslo (-Skagerrak) Graben which had a greater extension in the south (Ramberg 1976). The early dextral NE-SW and sinistral NW-SE strike slip faults in the Bamble Sector may be reflections of earlier Graben tectonics.

2.5 General Conclusions on the Kongsberg-Bamble Mobile Belt

In recent years, a number of different Cordilleran and Himalayan plate-tectonic models have been suggested to apply to the Proterozoic of this region (eg. Torske 1977, Falkum & Petersen 1980, Berthelsen 1980, Smalley et al 1984). The present work does not lead to a justifiable assessment of these models. The presently exposed Kongsberg-Bamble mobile belt evolved at deep crustal levels, after the earliest recorded events which might represent some form of marginal basin that was subsequently thrust outwards to east and west. The post-Kongsbergian events seem to reflect processes in continental crust, with intrusion of a series of voluminous granites, differentiated alkaline complexes and gabbroid bodies (with continental affinities).

The rotation of the Bamble segment relative to the Kongsberg Sector, and other sectors further E (Fig. 4) occurred largely in the early (Kongsbergian) stages of MD1 and MD2, but was accentuated later, particularly during the (Sveconorwegian) MD5 and MD6 stages, but also during the (Phanerozoic) 'friction breccia' faulting, associated with the opening of the Oslo Graben. The continuity of the Kongsberg-Bamble belt had been retained during the Sveconorwegian Orogeny, with albitites and scapolitisation in the Modum Complex antiform and the Coastal Quartzite Complex, but not in the Nidelva Quartzite Complex which had been transposed relative to them during MD3. The larger hyperite stocks were also emplaced in the Kongsberg Sector and around the Coastal Quartzite Complex. The 'friction breccia' faulting effectively disconnected the Kongsberg Sector from the Bamble Sector. Some relative rotation of the Bamble Segment may be associated with late Precambrian (post 1200 Ma) rotations of the Fennoscandian Shield relative to the Laurentian Shield, suggested by palaeomagnetic studies (eg. Patchett & Bylund 1977, Patchett et al 1978, Stearn & Piper 1984).

Anatectic magmas were emplaced after identifiable periods of crustal thickening. Associated migmatisation was more extensive in the Bamble Sector, since it had a wedge cross-section overlying the Telemark block, contrasting with the steep junction of the Kongsberg Sector.

The primary crust forming event in the Kongsberg, Bamble and Telemark Sectors, was the Kongsbergian Orogeny. The Sveconorwegian Orogeny was a major event in the Telemark Sector, but had a limited effect, particularly in terms of deformation, in the Kongsberg and Bamble Sectors, which were then at higher crustal levels.

The Kongsbergian Orogeny formed most major fold structures in the Kongsberg and Bamble Sectors and the metamorphism (associated with CO_2 flushing and possible mantle degassing) produced localised granulite facies domains with outward transitions to the more widespread amphibolite facies rocks. The main granulite facies developments were in the D1-MD1 first order, major synforms, namely those of the EKC and WKC in the Kongsberg Sector and at Gjerstad and Arendal in the Bamble Sector.

The Sveconorwegian Orogeny formed minor buckle folds in the Kongsberg and Bamble Sectors but usually only produced major folds at their margins with the Telemark Sector. Metamorphism, at mid-amphibolite facies grade, amphibolitised the 'Vinor' and 'hyperite' intrusions and produced some axial planar biotite growths (± hornblende) in the

hinges of minor folds. New isograds in the northwest of the Bamble Sector were related to metamorphism in the Telemark Sector (then at lower levels) but southeastwards towards the coast, the thickening of the Bamble wedge allowed certain domains to retain earlier structures and metamorphic features. These domains were the large Quartzite Complexes, the centre of (upper levels of) the large Levang dome and the charnockitic-enderbitic complexes. In the widest part of the Bamble Sector, the coastal charnockitic-enderbitic complex, around Arendal, retained Kongsbergian ages and complex major and minor structures since it was insulated from the Telemark block beneath by several Km of underlying, major structures probably involving a first order synform of large quartzite units. It is significant that Field & Raheim (1981) recorded that some of the charnockitic rocks were affected only by the intrusion of "post-tectonic" granitic sheets at ~1390 and ~1060 Ma. Kongsbergian ages are more widely preserved in the Kongsberg Sector, because of its steeper junction with the Telemark block.
     The Sveconorwegian orogeny occurred in two distinct phases, separated by anorogenic basic intrusions associated with attempted rifting. This correlates with the data of Weis and Demaiffe(1983) from Rogaland, where they suggested two Sveconorwegian events occurred at 1300-1200 and ~1000 Ma.

ACKNOWLEDGEMENTS
Thanks are due to Janet Baker for financial assistance to reproduce the diagrams and for the original cartography, to Ray Crundwell for photography, and to Teresa Anthony and Elizabeth Murray for preparation of the original and final manuscript, respectively.

REFERENCES
Andreae, M.O. 1974. 'Chemical and stable isotope composition of the high grade metamorphic rocks from the Arendal area, Southern Norway'. Contrib. Mineral Petrol. 47, 299-316.
Beeson, R. 1978. 'The geochemistry of meta-igneous rocks from the amphibolite facies terrain of South Norway'. Nor. geol. Tidsskr. 58, 1-16.
Berthelsen, A. 1969. 'Globulith. A new type of intrusive structure, exemplified by metabasic bodies in the Moss area, SE Norway'. Nor. geol. Unders 266, 70-85.
Berthelsen, A. 1980.'Towards a palinspastic tectonic analysis of the Baltic Shield.' 26th Int. Geol. Congr. Paris, Colloq. 108, 5-21.
Broch, O.A. 1964. 'Age determination of Norwegian minerals up to March 1964'. Nor. geol. Unders. 228, 84-113.
Bugge, A. 1928. 'En forkastning i det Syd-Norske Grunnfjell'. Nor. geol. Unders. 130, 124 pp.
Bugge, A. 1937. 'Flesberg og Eiker'. Nor. geol. Unders. 143, 118 pp.
Bugge, J.A.W. 1940. 'Geological and petrographical investigations in the Arendal district'. Nor. geol. Tidsskr. 20, 171-209.
Clough, P.W.L. & Field, D. 1980. 'Chemical variation in metabasites from a Proterozoic amphibolite-granulite transition zone, South Norway'. Contrib. Mineral. Petrol. 73, 277-286.
Cooper, D.C. & Field, D. 1977. 'The chemistry of Proterozoic low-potash,

high-iron, charnockitic gneisses from Tromoy, South Norway'. Earth Planet Sci. Lett. 35, 105-115.

Falkum, T. 1966. 'The complex of metasediments and migmatites at Tveit, Kristiansand'. Nor. geol. Tidsskr. 46, 85-110.

Falkum. T. & Petersen, J.S. 1980. 'The Sveconorwegian orogenic belt, a case of late Proterozoic plate-collision'. Geol. Rundsch. 69, 622-647.

Field, D. & Clough, P.W.L. 1976. 'K/Rb ratios and metasomatism from a Precambrian amphibolite-granulite transition zone'. Journ. Geol. Soc. 132, 277-288.

Field, D, Drury, S.A. & Cooper, D.C. 1980. 'Rare-earth and LILE element fractionation in high grade charnockitic gneisses, South Norway'. Lithos 13, 281-289.

Field, D. & Råheim, A. 1979. 'Rb-Sr total rock isotope studies on Precambrian charnockitic gneisses from South Norway: evidence for isochron resetting during a low-grade metamorphic deformational event'. Earth planet. Sci. Lett. 45, 32-44.

Field, D. & Råheim, A. 1981. 'Age relationships in the Proterozoic high grade gneiss regions of southern Norway'. Precambrian Res. 14, 261-275.

Field, D. & Råheim, A. 1983. 'Age relationships in the Proterozoic high grade gneiss regions of southern Norway'. Reply to "Discussion and comment" by Weiss and Demaiffe. Precambrian Res. 22, 157-161.

Gammon, J.B. 1966. 'Fahlbånds in the Precambrian of southern Norway'. Econ. Geol. 61, 174-188.

Hageskov, D. & Pedersen, S. 1981. 'Rb/Sr whole rock age determinations from the western part of the Östfold basement complex, SE Norway'. Bull. geol. Soc. Denmark 29, 119-128.

Hopgood, A.M. 1980. 'Polyphase fold analysis of gneisses and migmatites'. Trans. Roy. Soc. Edinb.: Earth Sciences, 71, 55-68.

Jacobsen, S.B. & Heier, K.S. 1978. 'Rb-Sr isotope systematics in the Kongsberg Sector, South Norway'. Lithos. 11, 257-276.

Jösang, O. 1966. 'Geologiske og petrografiske undersökelser i Modumfeltet'. Nor. geol. Unders. 235, 148 pp.

Killeen, P.G. & Heier, K.S. 1975. 'Th, U, K and heat production measurements in ten Precambrian granites of the Telemark area, Norway'. Nor. geol. Unders. 319, 59-83.

Kleppe, A. & Raheim, A. 1979. 'Rb/Sr investigations on Precambrian rocks in the central part of Telemark'. Proc. ECOG VI Lillehammer 1979, 53 (abstract)

Milne, K.P. 1981. 'The structure, petrology and geochemistry of rocks around a Proterozoic mylonite zone in the Gjerstad region, South Norway'. PhD Thesis Univ. of London.

Milne, K.P. & Starmer, I.C. 1982. 'Extreme differentiation in the Proterozoic Gjerstad-Morkeheia Complex of South Norway'. Contrib. Mineral. Petrol. 79, 381-393.

Moine, B., de la Roche, H. & Touret, J. 1972. 'Structures géochimiques et zonéographie métamorphique dans le Précambrien catazonal du Sud de la Norvège'. Sci. de la Terre 17, 131-164.

Morton, R.D. 1971. 'Geological investigations in the Bamble Sector of the Fennoscandian shield, S. Norway. No. 2 Metasediments and metapyroclastics (?) within the Precambrian metamorphic suite of the S. Norwegian Skaergaard'. Nor. geol. Tidsskr. 51, 63-83.

Morton, R.D., Batey, R.H. and O'Nions, R.K. 1970. Geological investigations in the Bamble Sector of the Fennoscandian Shield in South Norway I: The geology of eastern Bamble'. Nor. geol. Unders. 263, 1-72.

Nilsen, K.S. 1982. 'Mineraliseringer i Vest-Telemark og deres genetiske forhold til bergartene og tektoniske problemer'. Norsk. Geol. Forenings VIII Landsmøte. Geolognytt 17, 40-41 (Abstract).

Oftedahl, O. 1980. 'Geology of Norway'. Nor. geol. Unders. 356, 3-115.

O'Nions, R.K. & Baadsgaard, H. 1971. 'A radiometric study of polymetamorphism in the Bamble Region, Norway'. Contrib. Mineral. Petrol. 34, 1-21.

Patchett, P.J. & Bylund, G. 1977. 'Age of the Grenville belt magnetisation: Rb-Sr and palaeomagnetic evidence from Swedish dolerites'. Earth Planet. Sci.Lett. 35, 92-104.

Patchett, P.J., Bylund, G. & Upton, B.G.J. 1978. 'Palaeomagnetism and the Grenville Orogeny: new Rb-Sr ages from dolerites in Canada and Greenland'. Earth. Planet. Sci. Lett. 40, 349-364.

Ramberg, I.B. 1976. 'Gravity interpretation of the Oslo Graben and associated igneous rocks'. Nor. geol. Unders. 325 (Bull. 38) 194 pp.

Ramberg, I.B. & Smithson, S.B. 1975. 'Geophysical Interpretation of crustal structure along the southeastern coast of Norway and Skagerrak'. Geol. Soc. Am. Bull. 86, 769-774.

Smalley, P.C., Field, D.& Raheim, A. 1983. 'Resetting of Rb-Sr whole rock isochrons during Sveconorwegian low grade events in the Gjerstad augen gneiss, Telemark, Southern Norway'. Isotop. Geosci. 1, 269-282

Smalley, P.C., Field, D.& Raheim, A. 1984. 'Geochronology of the Telemark Sector of Southern Norway and development of the SW Scandinavian crust'. Terra Cognita Spec. Issue, ECOG VIII, 9(Abstract, B5).

Smithson, S.B. 1963. 'Granite studies I: a gravity investigation of two Precambrian granites in South Norway'. Nor geol. Unders. 214B,51-140.

Starmer, I.C. 1967. 'The geology of the Risør area, South Norway'. PhD Thesis, Univ. of Nottm.

Starmer, I.C. 1969a. 'The migmatite complex of the Risør area'. Nor. geol. Tidsskr. 49, 33-56.

Starmer, I.C. 1969b. 'Basic plutonic intrusions of the Risör-Söndeled area,South Norway'. Nor. geol. Tidsskr. 49, 403-431.

Starmer, I.C. 1972a. 'The Sveconorwegian Regeneration and earlier orogenic events in the Bamble Series, South Norway'. Nor. geol. Unders. 277, 37-52.

Starmer, I.C. 1972b. 'Polyphase metamorphism in the granulite facies terrain of the Risör area, South Norway'. Nor. geol. Tidsskr. 52, 43-71.

Starmer, I.C. 1976. 'The early major structure and petrology of rocks in the Bamble Series, Söndeled-Sandnesfjord, Aust-Agder'. Nor. geol. Unders. 327, 77-97.

Starmer, I.C. 1977. 'The geology and evolution of the southwestern part of the Kongsberg Series'. Nor. geol. Tidsskr. 57, 1-22.

Starmer, I.C. 1978. 'The major tectonics of the Bamble Series between Söndeledfjord and Kilsfjord (Aust-Agder and Telemark)'. Nor. Geol. Unders. 338, 37-58.

Starmer, I.C. 1979. 'The Kongsberg Series margin and its major bend in the Flesberg area. Numedal, Buskerud'. Nor. geol. Unders. 351, 99-120.

Starmer, I.C. 1980. 'A Proterozoic mylonite zone in the Kongsberg Series north of Hokksund, south central Norway'. Nor. geol. Tidsskr. 60 189-193.

Starmer, I.C. 1981. 'The northern Kongsberg Series and its western margin' Nor. geol. Unders. 370, 25-44.

Starmer, I.C. 1985. 'The geology of the Kongsberg district and the evolution of the entire Kongsberg Sector, South Norway'. Nor. geol. Unders. (in Press).

Stearn, J.E.F. & Piper, J.D.A. 1984. 'Palaeomagnetism of the Sveconorwegian mobile belt of the Fennoscandian Shield'. Precambrian Res. 23, 201-246.

Torske, T. 1977. 'The South Norway Precambrian region - a Proterozoic Cordilleran - type orogenic segment'. Nor. geol. Tidsskr. 57, 97-120.

Touret, J. 1968. 'The Precambrian Metamorphic Rocks around the Lake Vegår (Aust-Agder, Southern Norway)'. Nor. Geol. Unders. 257, 45 pp.

Touret, J. 1971. 'Le faciès granulites en Norvège méridionale: II, les inclusions fluides'. Lithos 4, 423-436.

Weis, D. & Demaiffe, D. 1983. 'Age relationships in the Proterozoic high-grade gneiss regions of Southern Norway: Discussions and Comment'. Precambrian Res. 22, 149-155.

TECTONIC ENVIRONMENT AND AGE RELATIONSHIPS OF THE TELEMARK SUPRA-
CRUSTALS, SOUTHERN NORWAY

T.S. Brewer and D. Field
Department of Geology
University of Nottingham
Nottingham NG7 2RD
UK

ABSTRACT. Two contrasting models have been proposed for the formation
of the Proterozoic Telemark supracrustals: (a) within continental rift
zones [9] and (b) at a destructive plate margin [10,11]. There has
been no previous attempt to constrain these models using geochemical
data. From our large data base of major and trace elements (including
REE) it is evident that the two relatively low-grade metamorphic events
(i.e. regional greenschist facies and a late thermal event) have caused
considerable chemical disturbance and the major elements are invalid
as discriminators. However, for the dominant metabasalts,
incompatible element plots (spidergrams) produce a consistent type of
pattern which is extremely similar to those of modern basalts erupted
at destructive plate margins. The Telemark supracrustals probably
formed in this type of environment during the Mid-Proterozoic. All
of the lithologies have suffered major disturbances of the Rb-Sr
isotopic systems. Newly presented Rb-Sr dates of c.1000 Ma do not
record eruptive events, but the late thermal event.

1. INTRODUCTION

The Telemark supracrustals comprise the only low-grade metamorphic
suite(s) in southern Norway. Their age, structural and metamorphic
history and tectonic environment of formation remain problematical.
The stratigraphical nomenclature (Rjukan, Seljord and Bandak groups)
was established in central Telemark [1] (Fig.1), and has been followed
in a broader lithostratigraphical correlation attempt across southern
Norway [2]. Meta-volcanics are absent from the Seljord group, but
dominate the Rjukan and Bandak sequences.
    According to Dons [3] the rocks of the Rjukan group (lowest unit)
have undergone two folding episodes, those of the Seljord group three
episodes, and the Bandak group (uppermost unit) up to four episodes.
Dons [1,3] also recorded that the three groups are separated from each
other by angular unconformities, which he interpreted as representing
two orogenic events separating the three internally conformable units.

The only previous geochemical studies have been extremely limited in terms of both sample size/distribution and elements analysed [4.5]. There are no published trace element data.

Rb-Sr isotopic studies have singularly failed to date the age of eruption, and the high degree of scatter on Nicolaysen diagrams is believed to be evidence of significant disturbance during metamorphism [6,7]. Reference isochron 'dates' of c.1050Ma have been interpreted as representing the time of metamorphism, assumed in one case as high-grade upper amphibolite facies [6] and, in the other, as low-grade greenschist facies [7]. Menuge's Sm-Nd age of 1190±37 for the Rjukan group [8] was interpreted by him as dating the extrusive event. An attempt to date the Bandak group by the same method was unsuccessful.

Despite the uncertainties concerning field relationships and structural and metamorphic history, and the lack of any supporting chemical data, two quite contrasting models have been proposed for the formation of the Telemark supracrustals: (a) within continental rift zones [9] and (b) at a destructive plate margin [10,11]. We have attempted to constrain these models by a combination of field, petrographic and geochemical studies. We have established a large geochemical data base covering all of the outcrops of the Telemark supracrustals. Herein we present some of the critical field and chemical data for the largest, and stratigraphically most complete, region - central Telemark (Figure 1).

2. FIELD RELATIONSHIPS

There are four important aspects to be considered in any attempted tectonic reconstruction: (a) the supracrustal/gneiss relationships, (b) the lithologies of the supracrustals, (c) their distribution patterns and (d) the metamorphic/structural history.

(a) Supracrustal/gneiss contacts

In sub-area Valldal (Figure 1), in the vicinity of Nyastøl bru (map reference 810350, Sauda 1:250,000) there is unequivocal evidence that the volcanics rest unconformably on a pre-existing (1.5 - 1.6 Ga old?) gneissic basement [12], but within Central Telemark we have no evidence for primary contacts and none have been reported. The contacts are either (i) steeply dipping, where they are related to the late-post kinematic granite intrusions, or (ii) reworked:

(i) Although large rafts of supposed supracrustals have been reported further to the south [1,13] the southern boundary of the main outcrop occurs in the region of Kviteseid-Nissedal (Figure 1). The generally shallow dip of the supracrustals increases towards this margin until it is sub-vertical near the contact. Falkum & Petersen [9] believe that this type of contact is reflecting marginal deposition at the edge of a graben, and that the contact is fault bounded. However, these margins are also distinctive by virtue of the presence of numerous late-kinematic granites (i.e. Rb/Sr ages cluster around 900 Ma [6,14, 15, 16]), and our observations are that it is these which were

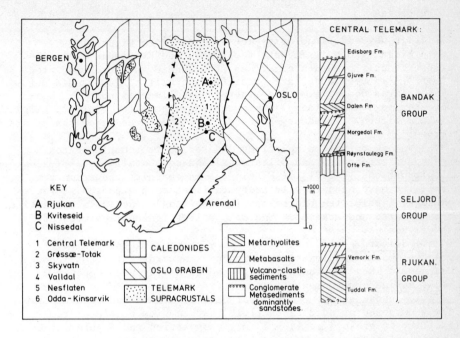

Figure 1. The distribution of low-grade supracrustal outcrops in Southern Norway. Inset shows the detailed stratigraphy of the Central Telemark region.

responsible for the locally exaggerated dips in the supracrustals. Forceful intrusion is supported by the presence of large rafts of supracrustals within these late granites.
(ii) There is an excellent example of a reworked contact in the Rjukan area (Figure 1), where the base of the Tuddal formation consists of flow-banded metarhyolites which are in contact with a weakly foliated biotite granite. Near the contact, the grain size in the metarhyolites increases and the flow banding has been enhanced by secondary, metasomatic growth of alkali feldspar and biotite. At the contact itself there is a 5-10m wide zone in which the two rock types are practically indistinguishable. Whatever the nature of the original contact (cover/basement?) it has now been obscured because the contact zone has acted as a major conduit for metasomatic fluids which were probably related to the large number of post-kinematic, undeformed pegmatites in this vicinity.

(b) Lithologies

The stratigraphy and lithologies are summarised in Figure 1.

(i) <u>Rjukan group.</u> In both the Rjukan and Kviteseid areas the lowermost Tuddal formation comprises flow-banded and autobrecciated metatuffs. In the Tuddal area itself, spherulitic flow-banded metarhyolitic lavas dominate. The mineral assemblage is quartz-orthoclase-plagioclase (An_{12}) ± microcline, muscovite, biotite, calcite, sphene and zircon. Phenocrysts of each of the three major phases are preserved.

The Vemork formation comprises metasediments (conglomerates, cross-bedded sandstones and siltstones) and metabasalts in approximately equal proportions [c.f.21]. Within the metabasalts the primary mineralogies have been totally recrystallised to a <u>greenschist facies</u> assemblage of chlorite-epidote-plagioclase (An_8) ± actinolite, biotite, calcite, quartz and opaques. The only relict igneous features are blasto-amygdales.

(ii) <u>Seljord group.</u> This entirely metasedimantary unit is composed of basal conglomerates which pass upwards into cross-bedded sandstones. They have been interpreted as having been deposited in a shallow, <u>marine</u> environment [17].

(iii) <u>Bandak group.</u> The basal Ofte formation has been described as being composed of acid lavas [1,18] but in view of the extreme variability in grain size (1 - 12mm), composition and angularity of the principal minerals, (quartz, feldspars) and exotic basaltic fragments, it is more likely to be a volcanogenic sediment.

The Røynstanlegg formation is another sedimentary deposit of probable marine origin [17]. Near Ofte it rests unconformably on rocks of that formation and elsewhere on rocks of the Rjukan group.

The Morgedal formation is predominantly metabasaltic with local intercalations of cross-bedded sandstones and conglomerates. In the metabasites the only relict features are blasto-amygdaloidal textures. The assemblages are the same as in the Vemork formation with the important exception that hornblende porphyroblasts are sometimes developed. It may be this development of hornblende that has led to the misconception that these rocks have undergone amphibolite facies metamorphism on a regional scale [6]. They have not - the regional assemblage is greenschist facies, with the hornblende resulting from a later thermal imprint in the vicinities of the post kinematic granites. The crystals are poikiloblastic and randomly overgrow the regional fabric and greenschist mineralogy. To our knowledge this is the first time that this important distinction has been recorded - unlike the basement gneisses, the supracrustals have <u>not</u> undergone any regional amphibolite facies metamorphism.

The Gjuve formation is similar to the Morgedal formation, whilst the uppermost Eidsborg formation is again totally sedimentary of inferred shallow margin origin [17].

(c) Distribution pattern

The metavolcanics are strongly bimodal, especially with respect to SiO_2

(Figure 2). The more basic of the lithologies have suffered extensive chloritisation and this may have involved some losses of SiO_2, perhaps diluting these rocks by 1-2%. This, however, cannot explain the compositional gap of some 15% (52-67%). There is an absence of intermediate compositions such as in some modern island arcs, e.g. Tonga-Kermadec [19]. In modern tectonic environments, andesites are usually indicative of subduction-related magmatism, but basalt is the dominant lithology in immature arcs [20], and where an arc extends into continental crust the initial outflow is rhyolitic (e.g. New Zealand [21, 22, 23]) and later volcanic activity is basaltic and andesitic (e.g. Chilean/Peruvian Andes [19]). The lack of andesitic compositions in Telemark is real but the possibilities that some intermediate rocks were once extruded but have since been eroded or have never been exposed cannot formally be discounted.

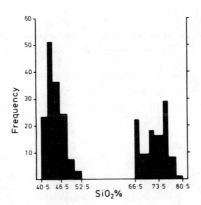

Figure 2. SiO_2% histogram for the central Telemark metavolcanics.

(d) Structural history

Previous studies have suggested a complex history with major unconformities between the three groups and with each group recording a different number of fold phases [3] - see introduction. Our investigations have yielded a quite different interpretation. All groups exhibit only a single, prominent, penetrative cleavage (S1). This is developed on a regional scale, is related to N-S trending folds and contains the regional greenschist facies mineralogy. Later folds occur only in the vicinities of the post-kinematic granites. Style and orientation vary. In the Morgedal formation the thermally developed hornblende porphyroblasts are clearly related to this folding. They overprint the regional (S1) fabric, cross-cut undeformed quartz and granite veins which themselves relate to the late granites, are randomly oriented, and are spatially related to the granites: They developed in thermal aureoles and the locally developed second deformation was caused by the intrusions. There were no other regionally developed deformational episodes. Consequently different stratigraphical units do not exhibit different structural histories and so the between-group unconformities recognised by Dons [1, 3] cannot represent major orogenic episodes. They are best interpreted as sedimentary features - as continuously deposited pile (others have been recognised within the groups); post-depositionally, the whole sequence suffered a single regional metamorphism/deformation, never above greenschist facies.

3. GEOCHEMISTRY

Previous geochemical studies have employed only major elements and the data sets are small [4,5]. The considerable variation in the major element compositions of the basalts is illustrated in Figure 3. For instance, the spread of $K_2O + Na_2O$ against SiO_2 precludes even discrimination into a calc-alkaline or tholeiitic grouping. The extreme compositional range (Table 1) can occur even within a single outcrop, and we shall provide details elsewhere how the spread in alkalis and some other elements has been caused by differential element mobility during the greenschist facies metamorphic event. The major element chemistry gives no reliable guide to the eruptive environment; neither do trace element discrimination diagrams employing elements such as Zr, Y and TiO_2 (Figure 4); for instance, in the Zr v Zr/Y plot consideration of the Gjuve rocks alone would suggest a MORB-affinity, whereas all but two of the Vemork rocks plot in the within-plate field and the majority of the Morgedal rocks fall in neither. 'Spidergram' plots are more profitable. Despite the evidence that elements such as Rb, Sr and Ba have suffered some disturbance [12 and unpubl.data] a consistent pattern is produced on all scales, and within all of the basaltic formations (Figure 5). This is one of variable LILE (large ion lithophile element) enrichment, accompanied by a negative Nb anomaly and a positive Ce anomaly. With the possible exception of a few samples close to the lower bounding profile for the Gjuve formation (Figure 5), the patterns are internally consistent with an overall shape similar to those of modern basalts generated at destructive plate margins [27, 28, 29] (Figure 5). The higher concentrations of K, Rb and Ba in some of the Telemark rocks (see upper bounding profiles, Figure 5) must in part be due to the differential mobility induced during the greenschist metamorphism, but this cannot account in full for the scale of the enrichment in Rb and Ba of up to ten times the normalised values for e.g. Sr. It can be inferred that the primary chemistries must have been LILE enriched and Nb-deficient, consistent with partial melting under hydrous conditions when the LILE are transported in the aqueous phases derived

Table 1. Range of chemical compositions within the metabasaltic formations of central Telemark, indicating major elements in wt%, trace elements in p.p.m. All Fe calculated as Fe_2O_3.

	VEMORK n = 26	MORGEDAL n = 78	GJUVE n = 44
SiO_2	42.15 - 52.29	40.94 - 49.30	40.59 - 47.93
Al_2O_3	13.64 - 16.85	13.24 - 17.03	13.74 - 18.13
TiO_2	1.68 - 2.67	1.64 - 4.63	1.16 - 2.06
Fe_2O_3	11.37 - 17.10	11.66 - 21.55	10.39 - 14.84
MgO	6.20 - 8.69	2.31 - 8.76	6.43 - 12.64
CaO	2.36 - 9.02	3.42 - 12.86	3.06 - 10.01
Na_2O	1.70 - 4.34	1.21 - 5.95	1.40 - 4.59
K_2O	0.24 - 2.43	0.20 - 2.19	* - 2.75
MnO	0.16 - 0.31	0.13 - 0.20	0.09 - 0.30
P_2O_5	0.28 - 0.44	0.20 - 0.82	0.09 - 0.29
Ba	76 - 791	98 - 964	* - 753
Ce	20 - 56	11 - 90	* - 31
Co	* - 60	41 - 85	48 - 77
Cr	4 - 437	33 - 167	7 - 196
Cu	6 - 217	2 - 401	3 - 186
La	* - 23	1 - 30	* - 8
Nb	16 - 21	2 - 11	* - 5
Ni	12 - 105	56 - 135	95 - 238
Pb	* - 36	* - 23	* - 10
Rb	6 - 67	7 - 115	* - 70
Sr	98 - 405	129 - 653	92 - 328
Th	* - 6	* - 12	* - 7
U	*	* - 4	*
V	157 - 258	125 - 297	87 - 241
Y	31 - 53	32 - 77	22 - 40
Zn	121 - 495	72 - 236	84 - 149
Zr	151 - 198	136 - 357	78 - 158

* indicates element below detection limit

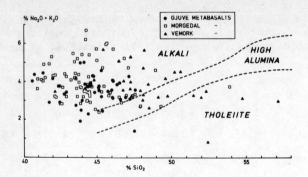

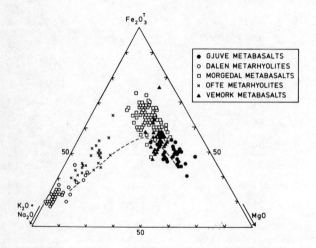

Figure 3. Major element discrimination diagrams; fields in SiO_2 v $(Na_2O + K_2O)$ from [24] and the calc-alkaline division in the AFM diagram from [25].

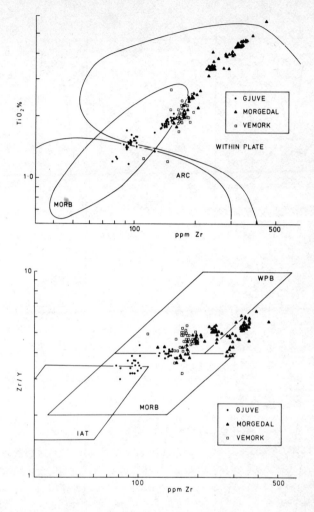

Figure 4. Zr v Zr/Y and TiO_2%. Fields from [26].

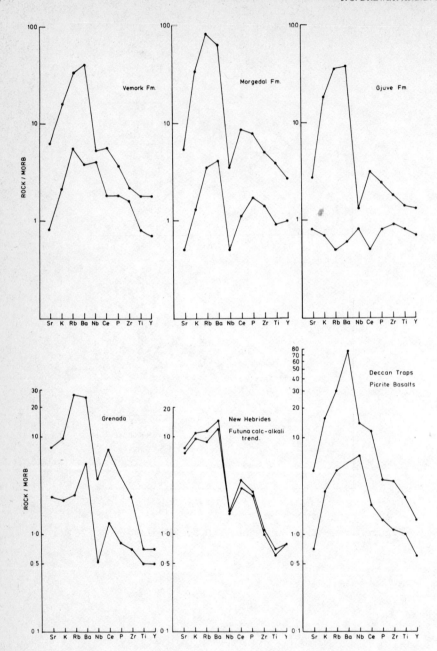

Figure 5. Comparison of central Telemark metabasalts with basalts of known tectonic affinity. For clarity only the maximum and minimum bounding surfaces are drawn. Normalising values from [26]. Modern fields from [27, 28, 29].

from the subducting slab, whilst elements with high ionic potential (e.g. Nb) remain in the residual phases [30, 33]. They most probably formed at a destruction plate margin.

4. ISOTOPIC EVIDENCE

Neither of the two previous Rb-Sr isotopic studies [6,7] were able to date conclusively either the eruptive or metamorphic ages. In each case both the eruptive 'ages' (c. 1570 Ma [6], 1600-1400 Ma [7] and the metamorphic 'ages' (c. 1067 Ma [6], 1050 Ma [7] were obtained from "reference isochrons" over which there was no geological control. Although the ages are remarkably similar, the metamorphic ages are interpreted as representing a high grade [6] and low grade [7] event. The fact that no actual data points were used in the construction of these reference isochrons invalidates their geological significance. We have sampled each formation, with sample sets taken from within individual outcrops. The data are presented in Figure 6. The most salient features are that the data points are scattered for each formation and that there is a lack of homogenisation at the outcrop scale. This we interpret as being due to metamorphic effects, both regional and thermal.

During the regional greenschist metamorphism muscovite, chlorite, biotite, epidote and calcite were all newly developed phases and hornblende and secondary biotite both formed during the later thermal overprint. This group of minerals can variably accommodate any unstabilised Ca and Sr (i.e. epidote, calcite [34, 35]) or K and Rb (i.e. muscovite, hornblende, biotite [36]). Chlorite will exclude both groups [35]. The isotopic composition of any one sample is therefore governed by the fact that the mobilised material (e.g. calcite, epidote veinlets) will not have had the same composition as its host rocks. After the redistribution Rb and/or Sr may be enriched or depleted, within a sample, and in the case of Sr enrichment the isotopic effect will be dependant upon the source of the introduced Sr.

It may be significant that the only three samples from the Morgedal formation which contain the thermal hornblende and biotite (a) were collected from within a few metres of each other and (b) define an age of 966 ± 99 Ma (MSWD = 1.1) which overlaps with the ages obtained from the late granites themselves [6, 14, 15, 16]. The remaining data points are more probably reflecting disturbance introduced during the earlier greenschist metamorphism.

So far there has been only one Sm-Nd study [8], with an isochron age of 1190 ± 37Ma (MSWD = 0.6) for the Rjukan group being interpreted by Menuge [8] as the age of eruption. No geologically meaningful age was obtained for the Bandak group. Menuge's samples used to obtain the isochron age included both metabasalts and metarhyolites. Using Menuge's data [8] we find that the T_{DM} model age calculations [37] for all of the acid rocks in this area produce very similar results - Kviteseid gneisses: 1430 - 1400 Ma, Rjukan metarhyolites 1460-1400 Ma, Dalen metarhyolites 1450 Ma; late granites 1400-1390 Ma. These data suggest that they were all derived from a common pre-existing

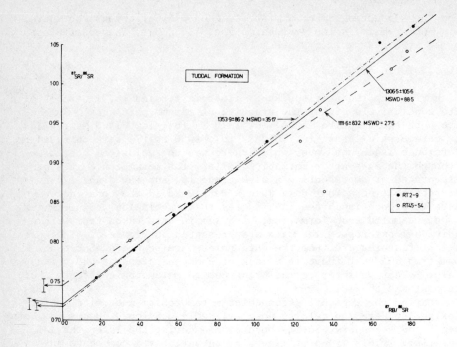

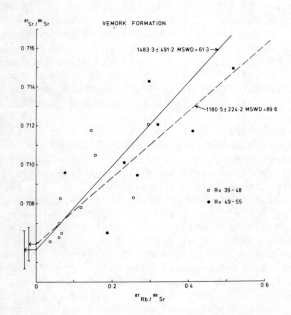

Figure 6. Nicolaysen diagrams for the major formations of the Central Telemark region.

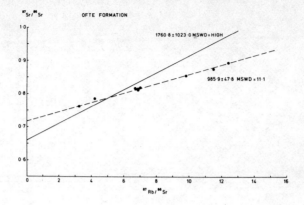

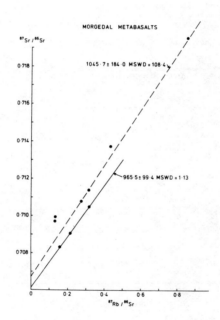

Figure 6. (continued)

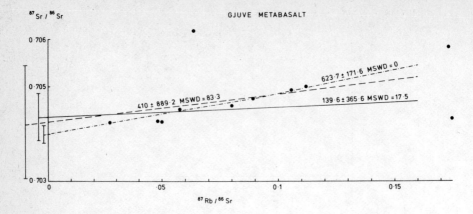
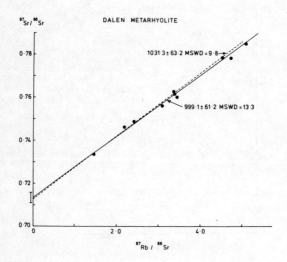

Figure 6. (continued)

crustal source c. 1.5 - 1.4 Ga old. T_{DM} ages for the metabasalts, however, are (a) scattered and (b) generally much younger (range 1600-1140 Ma). The scatter may have been produced either by crustal comtamination or by preferential REE mobility during metamorphism. These data, together with the lack of intermediate compositions between 52 - 67%, suggest that the basic and acid components of the Rjukan group may well have had different sources, probably with different initial $^{143}Nd/^{144}Nd$ ratios. If so, the fact that they plot on the same 1190 ± 37 Ma isochron may well be reflecting a secondary homogenisation rather than the primary eruption, perhaps due to the greenschist metamorphism. Here it is pertinent to point out that the mineral epidote is a known repository for REE [34, 35, 36] and that the Rjukan metabasalts contain variable amounts of this secondary mineral , often replacive and in the form of veinlets. In the light of these data we would argue that Menuge's [8] interpretation of the isochron data should be treated with some caution, and that at this stage, the possibility of the isochron age reflecting the regional metamorphism be regarded as a viable alternative hypothesis.

5. CONCLUSIONS

Although it is now clear that the supracrustals were deposited in a more or less continuous succession and that they have suffered a single regional greenschist facies metamorphism, with only a locally developed higher-grade thermal overprint, neither the age of eruption nor the ages of the metamorphisms have yet been firmly established.
Geochemical data indicate that magma genesis occurred at a destructive plate margin but differential element mobility during subsequent metamorphism does not allow a formal distinction to be made between an island arc or a continental margin setting. Indirect evidence provided by (a) the abundance of acid volcanics in the oldest (Rjukan) group, (b) the cyclic nature of the volcanism in the upper part of the Bandak group - which is extremely similar to the stratigraphy in the Chilean and Peruvian Andes [19] - and (c) the inferred marine nature of the intercalcalated sediments [17] all point to a continental margin environment similar to the Cordilleran type [39]. Here, as in Telemark, the acid volcanics represent (partial?) melts of pre-existing crust. This model is in accord with palaeomagnetic reconstruction, in which southern Norway is part of the marginal zone to a Proterozoic Supercontinent [40, 41, 42]. Neither the age of eruption nor the precise orientation and position of the subduction zone can yet be deduced from studies of the Telemark supracrustals.

ACKNOWLEDGEMENTS

TSB is in receipt of a NERC research studentship. Drs P.K.Harvey and B.P.Atkin are thanked for assistance with XRF determinations. Isotopic analyses were undertaken at Oxford University, where Dr P.N.Taylor provided assistance. Technical assistance was provided by
Mrs J Wilkinson and Mr D.Jones.

REFERENCES

[1] Dons, J.A. (1960). in 'Geology of Norway' (O.Holtedal, ed.) Nor.Geol.Unders.,**208**, pp.49-58.
[2] Sigmond, E.M.O. (1978). Nor.Geol.Unders.,**341**, pp. 1-94.
[3] Dons, J.A. (1972). Sci.de la Terre,**17**, pp. 23-29.
[4] Hasan, Z. (1971). Nor.Geol.Tidsskr,**51**, pp. 287-310.
[5] Moine, B. and Ploquin, A. (1972). Sci. de la Terre, **17**, pp. 47-80.
[6] Priem, H.N.A., Boelrijk, N.A.I.M., Hebeda, E.H., Verduemen, E.A.Th, and Verschure, R.H. (1973), Nor.Geol.Unders,**289**, pp.35-53.
[7] Kleppe, A.V. (1980). unpublished hovedoppgave Oslo Univ. 157pp.
[8] Menuge, J.F. (1983). unpublished PhD thesis, Cambridge Univ. 257pp.
[9] Falkum, T. and Petersen, J.S. (1980). Geol.Rundschau , **69** , pp. 522-647.
[10] Torske, T. (1977). Nor.Geol.Tidsskr.,**57**, pp.97-120.
[11] Berthelsen, A.)1980). 26th Int.Geol.Cong., Paris., Colloq. , pp. 5-21.
[12] Brewer, T.S. and Field, D. (1981). Abst. 1st Workshop European Geotraverse (Northern Segment), Copenhagen.
[13] Stout, J.H.)1972). J.Petrology,**13**, pp.99-145.
[14] Ploquin, A. (1980). Sci. de la Terre,**38**, pp. 1-389.
[15] Killeen, P.B. and Heier, K.S. (1975). Mat.-Naturv.Kl.Str.Ny serie, **35** , pp. 1-5.
[16] Venugopal, D.V. (1970). Nor.Geol.Tidsskr.,**50**, pp.257-260.
[17] Singh, I.B. (1969). Nor.Geol.Tidsskr. **49** , pp. 1-13.
[18] Bugge, C. (1931). Nor.Geol.Tidsskr.**12**, pp.149-170.
[19] Carmichael, I.S.E., Turner, F.J. and Verhoogen, J. (1974), 'Igneous Petrology', McGraw-Hill Inc. 759pp.
[20] Baker, P.E. (1o68). Bull.Volc.,**32**, pp. 198-206.
[21] Ewart, A. (1967). N.Z.J.Geol.Phy.,**10**, pp. 182-197.
[22] Ewart, A.)1968). N.Z.J.Geol.Geophys.,**11** , pp.478-545.
[23] Ewart, A. and Stipp, J.J. (1968). Geochim.Cosmochim.Acta., **32**, pp.697-735.
[24] Kuno, H. (1966). Bull.Volc.,**29**, pp.195-222.
[25] Irvine, T.N. and Barager, W.R.A. (1971). Can.J.Earth Sci. **8**, pp.523-548.
[26] Pearce, J.A. (1980). in 'Ophiolites', Proceedings International Symposium, Cyprus 1979. (A.Panayiotou ed.) pp.261-272.
[27] Krichnamurthy, P. and Cox, K.G. (1977). Contrib.Mineral.Petrol. **62**, pp. 53-75.
[28] Dupy, C., Dostal, J., Marchebt, G., Bougault, H., Joron, J-L., and Treuil, M. (1982). Earth Planet. Sci.Lett.,**60**, pp.207-225.
[29] Shimizu, N. and Arculus, R.J. (1975). Contrib.Mineral Petrol. **50**, pp.231-240.
[30] Best, M.G.)1975). J.Petrol. **16**, pp.212-236.
[31] Saunders, A.D. and Tarney, J. (1979). Geochim. Cosmochim. Acta., **43**, pp.555-572.
[32] Wood, D.A., Joron, J-L. and Treuil, M. (1979). Earth Planet Sci. Lett., **45** , pp.326-336.

[33] Pearce, J.A. (1983). in 'Continental Basalts and Mantle Xenoliths' (C.J.Hawkesworth and M.J.Norry, eds), Shiva, pp.230-249.
[34] Exley, R.S. (1982). J.Geophy.Res., **87**, pp.6547-6557.
[35] Dickin, A.P. and Jones, N.W. (1983). Contrib.Mineral.Petrol. **82**, pp.147-153.
[36] Deer, W.A., Howie, R.A. and Zussman, J. (1978). 'An Introduction to the Rock Forming Minerals', Longman, pp.1-528.
[37] De Paolo, D.J. (1981). Nature, **291**, pp.193-196.
[38] Dollase, W.A. (1971). Am.Mineral., **56**, pp.447-464.
[39] Thorpe, R.S. (1982), in 'Andesites' (R.S.Thrope, ed). J.Wiley and Sons, pp.1-7.
[40] Piper, J.D.A. (1982). Earth Planet. Sci.Lett, **59**, pp.61-89.
[41] Piper, J.D.A. (1983). Geophys. J.R.Astr.Soc. **74**, pp.163-197.
[42] Stearn, J.E.F. and Piper, J.D.A. (1984) Precambrian Res. **23**, pp.201-246.

GEOTECTONIC EVOLUTION OF SOUTHERN SCANDINAVIA IN LIGHT OF A LATE-PROTEROZOIC PLATE-COLLISION

T. Falkum
Geological Institute
Aarhus University
DK-8000 Århus C
Denmark

ABSTRACT. The major crustal accretion episode in central and southern Scandinavia took place during the Svecokarelian orogeny (2100-1700 Ma) and the extensive postkinematic plutonism and volcanism explains the long duration.
The subsequent interorogenic period developed in a prevailing tensional regime, producing a basin and range type of crustal segment. Basin subsidence with extensive bimodal volcanism of rhyolitic and basaltic compositions, interlayered with shallow-water- and continental-type sediments evolved into aulacogen-like structures (the Telemark supracrustals).
Extensive intrusions of granitoid and charnockitic plutons dominated large regions of southern Norway and southwestern Sweden during this period, with local deformation and metamorphism.
The Sveconorwegian orogeny (1200-850 Ma) was heralded by basic and intermediate intrusions, probably in response to tensional stresses created during the initiation of lithospheric subduction in connection with a period of crustal thinning. If the age of the Rjukan rhyolites is accepted as being 1200 Ma, and not 1500 Ma, the Telemark supracrustal trough may accordingly represent the earliest stages of the Sveconorwegian orogenic event, developed in a back-arc position in response to a N-S running plate collision zone to the west.
After 1150 Ma a compressional regime caused several periods of deep-seated plastic deformations and high-grade regional metamorphism. Several episodes of massive invasions of granitoid magmas can be separated by the number of deformations they have suffered.
The prevailing structures define an asymmetrical N-S running orogenic belt, and the intrusions suggest a unilateral zonation with an intrusion dominated western core-zone in Vest-Agder and Rogaland, a central zone with a considerable amount of reworked rocks in Aust-Agder, Telemark, Bamble and Østfold, which was gradually forced eastwards into a marginal zone of mega-tectonic imbricate structures of mega-units, separated by thrust zones and ultimately bordered by the Sveconorwegian Frontal Thrust Zone (SNFTZ) in SW Sweden equivalent to the Grenville Front in Canada. Upthrusting of large blocks and possibly crustal wedging into the mantle characterize this marginal zone of the Sveconorwegian orogeny.

Oblique to this trend the Bamble linear belt forms a major shear-zone, repeatedly reactivated, ultimately controlling the site and shape of the Permian Oslo Rift.
Major postkinematic batholiths were emplaced in all zones, although the most pronounced is the Agder-Rogaland batholith belt. 800 Ma dyke-swarms indicate cratonic-like conditions and suggest rapid post-orogenic uplift and erosion in a stabilized crustal segment.

INTRODUCTION

The southwesternmost part of the Fennoscandian peninsula (the western part of the Baltic Shield) has been referred to as the southern tip of Norway (Barth, 1960). It consists of a Proterozoic basement segment formed and/or reworked during the Sveconorwegian Orogeny (1200-900 Ma). The exposed section contains medium to high grade metamorphic rocks of igneous and supracrustal origin, penetrated by deep-level igneous plutons from small plugs to large batholiths.
The region is classical in the history of geological research, from an early start more than one and a half century ago. Esmark (1823) introduced the name norite for a dark rock from the island of Hidra, and Kolderup (1897, 1904) later introduced rock-names like farsundite and birkremite. A dozen new minerals have also been described from this region, of which malacon, blomstrandin and thortveitite are the best known. Of interest is also that Vogt (1887) described the Farsund charnockite as a 'quartz-norite' and that Holland (1900) later refers to this description claiming it to be identical to his charnockite from the Indian type-locality (Holland, 1893).
The Rogaland igneous province with the norite-anorthosite-charnockite association became a popular target for many geologists (Kolderup, 1896, 1914, Barth, 1933, Michot & Michot, 1969, and Duchesne & Demaiffe (1978), and has remained a major project for Belgian petrologists, slightly overlapping with the Utrecht geologists (Tobi, 1965, Hermans et al., 1975) who have mainly concentrated their efforts on unravelling the metamorphic history within the enveloping metamorphic sequence (see this book).
The prevailing view up to 1960 was that most rocks in this region formed by metamorphic and particularly by metasomatic processes. This contribution will emphasize the importance of magmatic processes combined with subsequent tectonic activity as the main controlling factors for the formation of the diverse rock types in southern Norway. Thus, it demonstrates the tremendous change in the concept of Proterozoic evolution, and comparison with modern geotectonics shows that the tectono-magmatic evolution in the late Proterozoic has much in common with later orogenic and cratogenic tectonics.

The lithostructural units

The mapping of the south Norwegian migmatites has become possible by establishing so-called litho-structural units or formations. It mainly implies that some characteristic rock units have been used as a kind of

marker horizons, which have been followed step by step through the
revealed structures. These rocks are normally profoundly migmatized
and therefore apparently altered when going from the limb of a fold to
the hinge zone. Visually, and also petrographically it is in many cases
difficult to reconcile the two rock types although step by step mapping
clearly demonstrates a rather continuous change of the rock unit along
its strike.

In some cases the foliation and layering may bend reasonably regularly around the fold hinges, although other hinges are dominated by
both small-scale and large-scale folds with axial surfaces parallel to
the axial surface of the megascopic fold. Thus the local strike and dip
measurements are totally misleading as regards to revealing the shape
of the macro-structures.

Another complication is the mineral facies change from amphibolite
to granulite facies, sometimes across the lithological boundaries. With
pure petrographical mapping in such an area, many structures will not
be revealed.

The most useful litho-structural unit is the banded gneiss, in
which most structural elements have been preserved. The interveening
granitic gneisses are more homogeneous rocks with only a faint foliation preserved, and in the most leucocratic parts even a foliation is
hardly seen. Consequently in many cases they have been taken for granites. In order to reveal the geological evolution of this complex
which has undergone deep-seated plastic and polyphase deformation, it
is absolutely necessary to recognize the structural and metamorphic
overprint on the igneous rocks.

Structurally the migmatitic banded gneisses are the oldest litho-
structural unit, followed by the homogeneous granitic gneisses and the
synkinematic megacryst gneisses. The late-kinematic plutons may still
preserve discordant contacts revealing their true origin, as also the
postkinematic plutons do. In the following each unit will be described
separately.

The migmatitic banded gneisses

This litho-structural unit includes a great diversity of rock-types,
ranging from regularly banded or layered gneisses with alternating
layers of felsic and mafic gneisses and amphibolites/pyribolites,
through all types of migmatites to heterogeneous grey to pink gneisses.
Amphibolites dominate in the west and appear to be mostly ortho-amphi-
bolites (Falkum, 1969 a, b), while micaceous rocks are more abundant
to the east showing more paragneissic features (Falkum 1966 a, b).

Included in the banded gneiss unit are several rock-types which
can be interpreted as metasediments, such as marbles, quartzites and
aluminous garnet-sillimanite-cordierite-biotite schists and gneisses.
These supracrustal rocks are supposed to have been deposited before
the onset of the Sveconorwegian orogeny around 1200 Ma.

The intense deformation and anatexis have migmatized the banded
gneisses to varying degrees, so all types of migmatites occur, even to
the extent that the term banded or layered has to be abandoned in order
to describe these rocks properly. Furthermore they have been dissected

by mafic and/or felsic dykes and sills in several episodes, adding
complications to the observed pattern of the rocks from this unit.
Thus this unit contains the largest variety of rock-types spanning
over the largest time interval of geological evolution as they contain
the oldest rocks dated to around 1650 Ma.

The heterogeneous gneisses

Large areas, especially in the east, are dominated by quartzo-feldspatic
gneisses, often with irregular layers and lenses of slightly more
micaceous rocks, giving a heterogeneous migmatitic appearance to this
unit. Most age determinations of these rocks indicate that they are
pre-Sveconorwegian in age, and constitute a basement for the Telemark
supracrustals.

The intricate appearance of the heterogeneous gneisses is certainly partly due to orogenic overprinting during the Sveconorwegian
orogeny, although this overprinting may be rather weak in many areas.
They have been referred to as the Setesdal gneisses (Falkum & Petersen,
1980) and are thought to be an important element for understanding
the earlier evolution of southern Norway.

The homogeneous granitic gneisses

In the southwestern sector (Agder-Rogaland) one of the most common
rock units contains several types of medium to fine grained, leucocratic gneisses of granitic to granodioritic composition. Biotite is
the main mafic mineral, normally present in less than 10%. In the most
leucocratic parts, the biotite often tends to be fine grained, making
it difficult to readily see the foliation. Hence the earlier tendency
to classify these gneisses as granites. However, most of them appear as
relatively thin sheets of enormous lateral extensions, deformed and
folded by at least four phases of deformation (Falkum, 1966 b, 1976 a).
The metamorphic grade ranges from amphibolite facies to granulite
facies.

The majority of these granitoid gneisses appears to be older than
the youngest generation of megacryst gneisses. However, some ages
around 1000 Ma are found, and if these are intrusion ages, they apparently intruded quite late during the period of deformation and metamorphism. The majority of ages cluster around 1020 to 1050 Ma (Rb/Sr-
ages), and ages up to 1200 Ma have been recorded (Pasteels & Michot,
1975), suggesting that the intrusion of the homogeneous granitoid
plutons occurred throughout the Sveconorwegian orogeny.

They have a low U- and Th-content, and also low initial Sr-ratios,
indicating derivation from Rb-poor source material and short crustal
residence time before their final emplacement. Although these units
normally occur as conformable sheets, their homogeneous appearance,
the angular and often rotated inclusions of mafic gneisses and amphibolites along their borders, together with their minimum melting composition, suggest an intrusive origin. In the central part of the Mandal
map sheet (Falkum, 1982) they constitute more than two-thirds of the
complex.

The synkinematic megacryst gneisses

A suite of porphyritic granites containing alkali feldspar megacrysts with a large size variation from a few cm and up to 25 cm, intruded the banded gneiss/granitic gneiss complex around 1020-1050 Ma ago. They were all deformed by the third phase of deformation and recrystallized under amphibolite facies condition in the central Agder region, while granulite facies grade was reached farther west. During this episode they deformed into very conspicuous augen gneisses, although some zones, particularly along the borders were deformed so the megacrysts became elongated ovoids. Further deformation resulted in a streaky gneiss in which no trace of the original megacryst texture can be found.

Despite the thorough refolding suffered by the megacryst granites, resulting in entirely recrystallized augen gneissic texture or even a streaky gneiss appearance, some of them can still be demonstrated as being discordant on a large scale, leaving little doubt as to the intrusive character of most of these gneisses.

The late kinematic plutons

Several plutons were forcefully emplaced into the older gneiss complex already deformed and metamorphosed during the third phase of deformation. These plutons have partly discordant, partly concordant contacts, clearly showing their intrusive nature, and subsequently to their emplacement they were deformed and recrystallized during the fourth phase of deformation. Emplaced as alkali feldspar porphyritic plutons of granitic composition, they are in places deformed to augen gneisses (Falkum, 1976 a).

A Rb/Sr age from the Homme granite yields 990 Ma (±14 Ma) which is considered to be the age of intrusion. The fourth phase of deformation is the last deformation with extensive recrystallization (amphibolite facies). The Holum granite is interpreted as the oldest postkinematic pluton and with a Rb/Sr age of 980 ± 34 Ma (Wilson et al., 1977) it limits the last metamorphism between 1000 and 980 Ma ago or at least within 1010 and 950 Ma.

The postkinematic intrusive plutons

Numerous postkinematic plutons were forcefully emplaced into the south Norwegian basement between 980 and 850 Ma ago. Cross-cutting border relations are common and even contact-metamorphic aureoles occur (Petersen, 1977, Falkum et al., 1979). Several are large batholiths while others are smaller stocklike plutons. Although they occur all over southern Norway, a certain concentration is found in a 20-30 km broad N-S running zone in the central part of the Mandal map. This Agder-Rogaland batholith belt coincides with the belt of homogeneous granitoid gneisses, previously termed the Flekkefjord gneisses (Falkum & Petersen, 1980).

The postkinematic plutons are mainly biotite and/or hornblende granites of calc-alcaline affinities, while another suite of plutons are the anorthosite-norite-mangerite-charnockite intrusions of South Rogaland Igneous Complex (SRIC).

During this period the region was cratonized and extensive dyking accompanied rapid uplift of South Norway. Several of the roughly E-W trending dykes give ages around 800 Ma.

THE PRE-SVECONORWEGIAN EVOLUTION

Rocks belonging to the Svecokarelian orogeny and post-tectonic intrusions have ages older than 1650-1700 Ma, and rocks of this age occur only in SE Norway and W Sweden. Thus, the Svecokarelian basement is apparently preserved in several mega-tectonic units in the eastern part of the Sveconorwegian orogenic belt. In the central Telemark and Kongsberg region, the oldest gneisses give Rb/Sr-ages around 1650-1500 Ma. This activity has been assigned to an orogenic event, the "Kongsbergian Orogeny" (Oftedahl, 1980).

It is, however, a matter of discussion what these ages mean. Firstly, are they true ages or have the systems suffered later influence opening the isotope systems? Next, are they ages of intrusions or extrusions in a cratonal period, or are they defining a deformational and regional metamorphic episode? These questions have not yet been satisfactorily answered, although the concept of an orogeny between 1650 and 1500 Ma ago seems to gain supporters.

The extent of the Svecokarelian basement in south central Norway around 1500-1600 Ma ago is, as yet unknown. It seems, however, that the Telemark gneisses (Dons, 1960) became rather stabilized as a block during this period, while the surrounding Kongsberg-Bamble zone was deformed into a linear belt still including old basement blocks more than 1600 Ma old (Falkum & Petersen, 1980).

The period from around 1450 Ma ago and until the onset of the Sveconorwegian orogeny 1200 Ma ago, or slightly earlier, seems to have been an inter-orogenic period with mainly tensional tectonics and volcanism. Most of southern Norway was probably sialic crust with intracontinental basin and range tectonics, with depressions allowing the sea to invade a broad zone along the present coast line.

The situation before the onset of the Sveconorwegian orogeny is the existence of a pre-Sveconorwegian basement consisting of rocks from at least one, possibly two orogenic periods, separated by interorogenic rocks formed in a tectonic regime with tensional tectonics.

SVECONORWEGIAN OROGENIC EVOLUTION

There is a remarkable lack of dates between 1350 Ma age and 1200 Ma in southern Norway. Probably it reflects a quiet period before the first recordable signs of the onset of a new orogeny. This is heralded by numerous intrusions around 1200 Ma ago, mostly as dykes, plugs and stocks, but also batholith-sized acid intrusions were emplaced. Gabbros and hyperites were emplaced in great numbers.

They cut across old structures and a foliation which is supposed to be pre-Sveconorwegian in age, and they are deformed during the subsequent orogenic movements. Thus the first Sveconorwegian deformational phase (F_1) formed a general foliation with mesoscopic intrafolial folds displaying axial plane foliation. These isoclinal folds are found within the migmatitic banded gneiss units, and are always conformable with the lithological layering. This deformation transposed all earlier structures into a parallel, penetrative foliation, so that the F_1-phase in Vest-Agder and Rogaland controls the direction of all earlier structures. Farther eastwards, older structures have obviously retained their own direction, although these directions may have been considerably rotated.

It has not yet been possible to date this early event, but it may be as old as 1200 Ma, or possibly even slightly older. This deformation is supposed to be connected with the earliest movements during the initiation phase of the Sveconorwegian orogeny. If the palaeomagnetic data, suggesting that Fennoscandia has been rotated 90° in relation to Laurentia (Poorter, 1981) is correct, a large part of this rotation most probably has taken place between 1300 and 1200 Ma ago. The F_1-deformation could then have been a result of this movement. Furthermore, it is here suggested that the opening of the Telemark basin also occurred during this phase, as a combination of rotational movements and initiation of an eastwards subducting slab somewhere to the west of Telemark. This subduction zone may well have migrated westwards during the next 200 Ma.

This subduction transposed the Agder-Rogaland banded gneiss complex eastwards against the Telemark block. Reaction from this block caused it to split up with a basin depression along the incipient rifting zone similar to the aulacogen structures between the Angara and Aldan Blocks in Siberia (Salop & Scheinmann, 1969). Bimodal volcanism accompanied this stage (Rjukan group). Later sagging with high flanks was responsible for the sedimentation of thick quartzites and conglomerates (Seljord group) and later more rift-like tectonics resulted in acid and basic lavas, arkosic sandstones, schists and a few marble units (Bandak group).

The reason for the failure to develop into a true rift was the continued eastwards moving rock-complex, which ultimately began to close the Telemark basin probably not later than 1150 Ma. All formations were folded along megatectonic folds with N-S trending fold axes, probably concurrently with the third phase of deformation in Agder-Rogaland.

After the F_1-deformation, several large mafic intrusives were emplaced into the banded gneisses. Later, huge amounts of acid magma pervasively invaded the banded gneiss units, particularly to the west of Kristiansand (Fig. 1). This stage could be equivalent to the incipient stage of Telemark with rhyolitic and basaltic volcanism (Rjukan group).

In Agder-Rogaland the mafic intrusives and granitic batholiths were deformed into large-scale isoclinal folds and metamorphosed to pyribolites and the homogeneous granitic gneisses. These F_2-folds are isoclinal with the fold axis plunging moderately to the east. As the axial plane strikes N-S and also dips moderately to the east, these folds become so-called reclined folds.

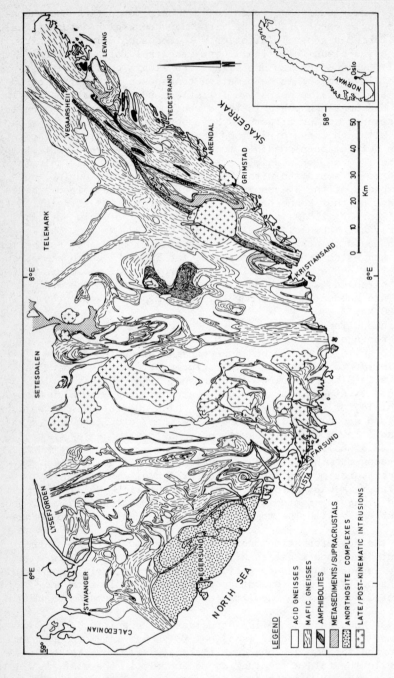

Fig. 1. Geological sketch map of the southern tip of Norway showing the main lithological units. (From Falkum & Petersen, 1980).

The F_2-phase probably deformed the Rjukan group and possibly also the Seljord group. After this deformation and high-grade metamorphism, several large masses of porphyritic granites of batholitic dimensions intruded the homogeneous gneiss/banded gneiss complex. The porphyritic granites comprise the Feda augen gneiss with alkali feldspar megacrysts up to 25 cm in length. These feldspars were deformed into augen shape by the F_3-deformation and concomitant high-grade metamorphism which probably took place in the time interval between 1050 Ma (1080?) and 1000 Ma ago.

The F_2- and F_3-phase deformations took place under very deep catazonal conditions with plastic rheid-type folding. The style of folding is nearly identical for both phases, and isoclinal folds on all scales are refolded by the isoclinal F_3-folds, which normally have N-S trending fold axes and moderately eastward-dipping axial surfaces.

The F_3-phase of deformation was also succeeded by the intrusion of a few coarse grained, porphyritic granites which retain partly discordant contact relationship although being deformed by the subsequent F_4-phase of deformation and metamorphosed under amphibolite facies conditions. These late-kinematic granites were emplaced by forceful intrusion, folding and tilting the country rocks (Falkum, 1976 a, b). The F_4-phase of deformation refolded the earlier isoclinal folds along N-S to NE-SW trending fold-axes, into several basin and dome structures. Most of the F_4-folds have steep axial surfaces and smaller amplitudes than the F_2- and F_3-folds, which explains the survival of the older folds.

The last phase of deformation (F_5) refolded the earlier structures into open to gentle folds with roughly E-W trending fold axes and steep axial planes. Mesoscopically these folds are ubiquitous, and can also be recognized in a large scale. This phase of deformation could possibly be connected to the intrusion of the late and/or postkinematic plutons.

The F_4-phase of deformation, however, concluded the regional metamorphism before the intrusion of the Holum granite (980 ± 34 Ma),(Wilson et al., 1977) and after the intrusion of the Homme granite (990 ± 16 Ma). Structurally it can be shown that the postkinematic intrusions, which were emplaced during the next 150 Ma, tilted the adjacent rock complex and changed the regional fabric which had been developed during the five recognized phases of deformation. Thus the postkinematic plutons stabilized the southern tip of Norway before extensive dyking around 800 Ma demonstrates the cratonal conditions with rapid upheaval of the Sveconorwegian fold belt.

A PRELIMINARY DYNAMIC MODEL

The Svecokarelian orogeny created a basement in southern Sweden and Norway. It has not yet been possible to find the westernmost boundary of this basement as it was intensely penetrated by stock- to batholith-sized intrusions and later severely reworked during the Sveconorwegian orogeny. As the intensity of this deformation and plutonic activity increases westwards, no certain Svecokarelian rocks have been found to the west of the Oslo Rift. The majority of pre-Sveconorwegian rocks in southern Norway are 1650 to 1350 Ma old. This Gothian or Kongsbergian episode may reflect a true orogenic event between 1600 Ma and 1500 Ma, although the possibility that it was an intracratonic magmatic event cannot yet be ruled out; possibly it was connected with intracratonic rifting and volcanism. Local deformation and metamorphism could occur under such circumstances.

Geotectonically the gneiss regions in Telemark and Hardangervidda evidently cratonized during this period and behaved like a semi-rigid block during the later tectonism. The Sveconorwegian orogeny initiated with eastward movements, probably induced from a N-S running subduction zone somewhere to the west of the present landmass of Norway. The prevailing E-W trend of Svecokarelian and later rocks were deformed into a N-S trend turning to northwest farther northwards (Fig. 2). Early in this process the central block was split into the Telemark and the Hardangervidda blocks, towards which the western gneisses were pushed.

The splitting created the Rjukan basin which was filled with rhyolitic and basaltic lavas and tuffs. Continued movements opened the basin even more, and the Seljord sandstones were deposited, followed by a deformation, which may be synchronous with the first deformation (D_1) farther west or the later D_2-phase of deformation. Subsequently the continued eastward movements opened the Telemark basin with rift-like volcanism and sedimentation in the Bandak group. Ultimately this aulacogen was closed during the next major phase of deformation, possibly equivalent to the D_3-phase in Agder-Rogaland.

The western segment was deformed eastwards against Telemark block, and the intracontinental ductile shear belt in Bamble formed as a response to these movements along the southern boundary of the Telemark block. The dextral movement indicates that the Agder-Rogaland segment moved northeastwards compared with Bamble region.

The Agder-Rogaland segment was catazonally deformed throughout the orogeny and the recognition of separate phases of deformation has been possible due to the interception by vast amounts of plutonic intrusions. It is mostly the intrusive events which have been dated, leaving a certain time interval for the intervening deformation and metamorphism.

Generally, two important conclusions emerge. The deformation and intrusive activity show a faint younging from east to west. It can most clearly be demonstrated that the volume of Sveconorwegian intrusions increase drastically towards west. These observations lead to recognition of a unilateral zonation of a roughly N-S to NW-SE running orogenic belt (Fig. 3).

The Agder-Rogaland segment is classified as the core zone of the Sveconorwegian orogeny (Falkum & Petersen, 1980) because it forms the site for a massive invasion of multiple synorogenic intrusions covering as much as 70-80% of the presently exposed part of the crust. Even if several of the intrusions are remobilized older crust or anatectic derivatives from deeper crust, it is considered that the core zone represents a true continental accretion zone with a considerable crustal addition.

Farther eastwards into the Aust-Agder-Telemark sector, the core zone merges into the central zone consisting of reworked pre-orogenic

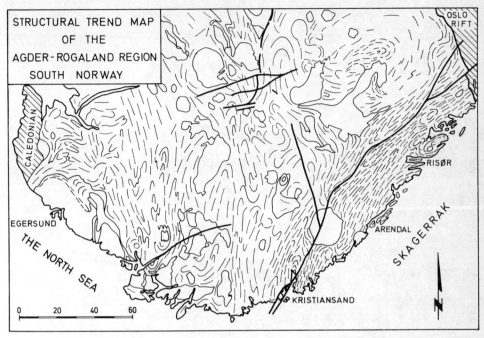

Fig. 2. A structural trend map showing the pattern of foliation in southern Norway. The blank areas are late- and postkinematic intrusions and the black lines are faults.

gneisses intimately mixed with numerous syn- and postkinematic intrusions. Sveconorwegian intrusions become less abundant farther eastwards into eastern Norway and south western Sweden, and the deformational and metamorphic overprinting also fade out towards the Sveconorwegian Frontal Thrust (SNFT) in central Sweden. Several shear zones are found from the Oslo region and all the way to the eastern border (SNFT). These shear zones run roughly N-S with a convex curvature to the east, comprising a sequence of megatectonic segments forming an imbricate structure imposed from the west. This sector is the marginal zone of the Sveconorwegian orogeny but is mainly composed of pre-orogenic rocks. It is suggested that it is a continuation of the Grenville Frontal Zone in Canada (Falkum & Petersen, 1980; Gower & Owen, 1984).

The unilateral model can be explained by eastwards movement on an orogenic scale. The massive invasion of magmas in the core zone combined with the catazonal deformation and high-grade metamorphism have the magmatectonic imprint of rock assemblages close to active subduction zones. The deformation has been controlled by the eastwards moving fold belt and its collision with stable blocks, creating shear zones around them. The curving of the fold belt is also a controlling factor of the deformation. Furthermore, taking the curving of the Grenville belt in Labrador into account, the supposed rotation of Scandinavia in relation to Laurentia may have been considerably less than 90°.

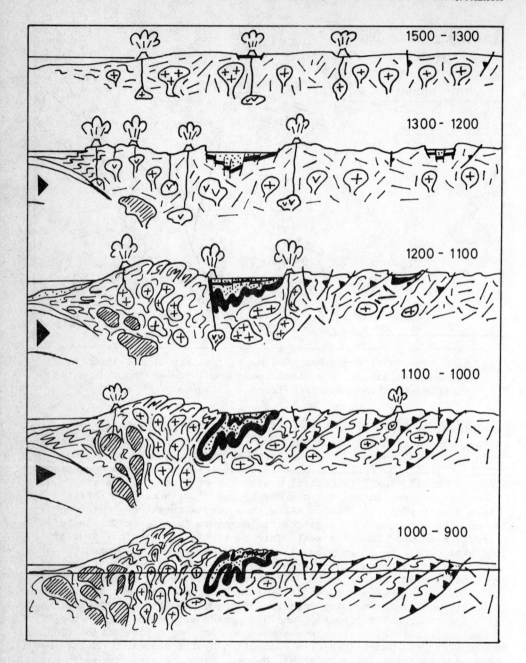

Fig. 3. Imaginative and schematic presentation of the Proterozoic development in southern Scandinavia after the Svecokarelian orogeny.

REFERENCES

Barth, T.F.W. 1933. The large pre-Cambrian intrusive bodies in the southern part of Norway. Intern. Geol. Congr. XVI, Rept. 1933, 297-309.
Barth, T.F.W. 1960. Precambrian of Southern Norway. In Holtedahl, O. (ed.), Geology of Norway. Nor. Geol. Unders. 208, 6-48.
Dons, J.A. 1960. Telemark supracrustals and associated rocks. In Holtedahl, O. (ed.), Geology of Norway. Nor. Geol. Unders. 208, 49-58.
Duchesne, J.C. & Demaiffe, D. 1978. Trace elements and anorthosite genesis. Earth Planet. Sci. Lett. 38, 249-272.
Esmark, J. 1823. Om noritformationen. Mag. f. Naturvidensk. 1, 205-215.
Falkum, T. 1966a. The complex of metasediments and migmatites at Tveit, Kristiansand. Geological investigations in the Precambrian of southern Norway. No. 1. Norsk Geol. Tidsskr. 46, 85-110.
Falkum, T. 1966b. Structural and petrological investigations of the Precambrian metamorphic and igneous charnockite and migmatite complex in the Flekkefjord area, Southern Norway. Norges Geol. Unders. 242, 19-25.
Falkum, T. 1969a. A deformed agglomerate in the Precambrian migmatites of southern Norway. Geological Investigations in the Precambrian of southern Norway. No. 3. Norsk Geol. Tidsskr. 49, 367-386.
Falkum, T. 1969b. Geology of the island Kinn in the Kristiansand skaergaard. Geological Investigations in the Precambrian of Southern Norway. No. 2. Norges Geol. Unders. 258, 185-227.
Falkum, T. 1976a. The structural geology of the Precambrian Homme granite and the enveloping banded gneisses in the Flekkefjord area, Vest-Agder, Southern Norway. Nor. Geol. Unders. 324, 79-101.
Falkum, T. 1976b. Some aspects of the geochemistry and petrology of the Precambrian Homme granite in the Flekkefjord area, southern Norway. Geol. Fören. Stockh. Förh. 98, 133-144.
Falkum, T. 1982. Geologisk kart over Norge, berggrunnskart MANDAL, 1-250.000. Norges geologiske undersøkelse.
Falkum, T. & Petersen, J.S. 1980. The Sveconorwegian orogenic belt, a case of Late-Proterozoic plate-collision. Geol. Rundsch. 69, 622-647.
Falkum, T., Wilson, J.R., Petersen, J.S. & Zimmermann, H.D. 1979. The intrusive granites of the Farsund area, south Norway. Their interrelations and relations with the Precambrian metamorphic envelope. Nor. Geol. Tidsskr. 59, 125-139.
Gower, C.F. & Owen, V. 1984. Pre-Grenvillian and Grenvillian lithotectonic regions in eastern Labrador - correlations with the Sveconorwegian Orogenic Belt in Sweden. Can. J. Earth Sci. 21, 678-693.
Hermans, G.A.E., Tobi, A.C., Poorter, R.P.E. & Maijer, C. 1975. The high-grade metamorphic Precambrian of the Sirdal-Ørsdal area, Rogaland/Vest-Agder, Southwest Norway. Nor. Geol. Unders. 318, 51-74.
Holland, T.H. 1893. The petrology of Job Charnock's tombstone. Jour. Asiatic Soc. Bengal, 62, 162-164.
Holland, T.H. 1900. The charnockite series, a group of Archean hyper-

sthenic rocks in peninsular India. Geol. Surv. India Mem. 28, 119-249.
Kolderup, C.F. 1897. Die Labradorfelse des westlichen Norwegens. I. Das Labradorfelsgebiet bei Ekersund und Soggendal. Bergens Mus. Aarbog 1896, 5, 222 pp.
Kolderup, C.F. 1904. Die Labradorfelse des westlichen Norwegens. II. Die Labradorfelse und die mit denselben verwandten Gesteine in dem Bergensgebiete. Bergens Mus. Aarbog 1903, 12, 129 pp.
Michot, J. & Michot, P. 1969. The problems of anorthosites: The South-Rogaland igneous complex, southwestern Norway. In: Isachsen, Y.W. (ed.), Origin of anorthosite and related rocks. New York State Mus. Sci. Serv. Mem. 18, 399-410.
Oftedahl, C. 1980. Geology of Norway. Nor. Geol. Unders. 356, 3-114.
Pasteels, P. & Michot, J. 1975. Geochronologic investigation of the metamorphic terrain of southwestern Norway. Nor. Geol. Tidsskr. 55, 111-134.
Poorter, R.P.E. 1981. Precambrian palaeomagnetism of Europe and the position of the Balto-Russian plate relative to Laurentia. In: Kröner, A. (ed.), Precambrian Plate Tectonics, Elsevier, 599-621.
Salop, L.I. & Scheinmann, Y.M. 1969. Tectonic history and structure of platforms and shields. Tectonophys. 7, 565-597.
Tobi, A.C. 1965. Fieldwork in the charnockitic Precambrian of Rogaland (SW Norway). Geologie en Mijnbouw 44, 208-217.
Vogt, J.H.L. 1887. Norske ertsforekomster. V. Titanjern-forekomstene i noritfeltet ved Ekersund-Soggendal. Archiv. f. Mathem. og Naturvidensk. 12, 1-101.
Wilson, J.R., Pedersen, S. Berthelsen, C.R. & Jakobsen, B.M. 1977. New light on the Precambrian Holum granite, south Norway. Nor. Geol. Tidsskr. 57, 347-360.

THE MANDAL - USTAOSET LINE, A NEWLY DISCOVERED MAJOR FAULT ZONE IN SOUTH NORWAY

Ellen M.O. Sigmond
Norges geologiske undersøkelse
P.O.Box 3006, 7001 Trondheim, Norway

ABSTRACT. Survey mapping and compilation in connection with the new 1:1 million geological bedrock map of Norway have revealed that the Precambrian rocks of South Norway can be divided into two blocks by a large tectonic line, the Mandal-Ustaoset Fault Zone. The location of such a tectonic line is indicated by the presence of ultracataclasites and mylonites along its central and northern parts; by a belt of post-tectonic granites west of the fault zone; by the distribution of the Telemark supra-crustals; and by the higher and more irregular magnetic field, and more numerous radiometric anomalies, west of the fault. It is uncertain whether the tectonic line represents a continent-continent collision suture, a collision which in this case must have taken place before the deposition of the Bandak Group of the Telemark supracrustals (pre-1200 Ma), or a mega-fault zone. The latest movements along the fault zone took place in its central and northern parts along low-angle faults, after intrusion of the post-tectonic granites (1000-800 Ma) and before deposition of the Cambrian shales.

GEOLOGICAL SETTING

The Precambrian rocks of central southern Norway can be divided into three provinces:
1) The volcanic and sedimentary rocks of the Telemark Suite;
2) The gneiss-migmatite-granite complex; and
3) The Kongsberg-Bamble area (Fig. 1) (these rocks are separated from the others by a major fault).

Younger granites (<1000 Ma) intrude the rocks of all three provinces (Fig. 1). The Telemark Suite consists of three groups - the Rjukan, Seljord and Bandak Groups - separated from one another by major unconformities. Whereas the youngest (Bandak) group is clearly deposited upon, and is younger than the gneiss-migmatite-granite basement (Naterstad 1973, Sigmond 1978), the oldest (Rjukan) group in places has a more diffuse transitional boundary with this same basement. The age of the supracrustals is somewhat uncertain, but it is generally accepted that rocks of the Bandak Group date from about 1000-1200 Ma B.P., and those of the Rjukan Group from about 1400-1600 Ma (Kleppe 1980). The surrounding and underlying gneisses and migmatites are considered to be part of a polymetamorphic terrain which carries evidence of at least two metamorphic episodes; one gneiss-migmatite forming event of about 1500-1600 Ma, and a younger (Sveconorwegian) event at about 1200-1000 Ma B.P. (O'Nions & Heier 1972, Versteeve 1975, Jacobsen & Heier 1978, Field & Råheim 1979, 1980). This latter episode resulted in places in an isotopic homogenization of the rocks. This is especially the case in the area around the Egersund anorthosite-charnockite complex where gneisses and migmatites generally have ages of 1200-1000 Ma, although a few (relict) ages of 1500 Ma have been reported (Versteeve 1975, Wielens et al. 1981, Priem & Verschure 1982).

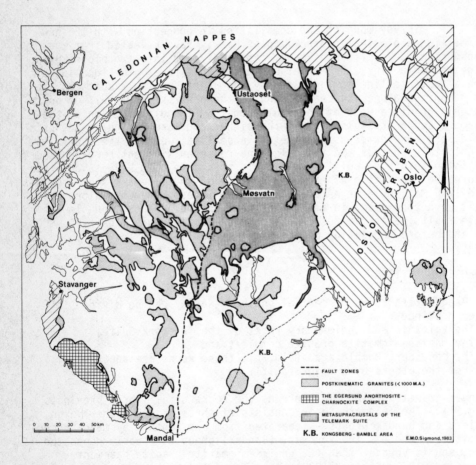

Figure 1. Relationship between the Telemark supra-crustals, the postkinematic granites (< 1000 Ma) and the Man-Us Fault Zone.

THE MANDAL-USTAOSET FAULT

The author's mapping and compilation in connection with the new geological bedrock map of Norway in a scale of 1:1000,000 (Sigmond et al. 1984) have revealed that the central part of southern Norway is divided into two blocks by a large tectonic line - the Mandal-Ustaoset Fault Zone (Sigmond in Berthelsen 1983, Sigmond 1984) - here called the Man-Us Fault.

Large-scale movements must have occurred along this line in at least two periods. The youngest movements occurred later than the intrusion of the post-kinematic granites (1000-800 Ma), but before the deposition of Cambrian shales. These granites are cataclastically deformed along the fault in the northern and central parts of the region.
According to the present author's interpretation of the Mandal 1:250000 geological map (Falkum 1982), the fault dies out about 70 km north of Mandal, ending up as a set of easterly dipping fault planes. Further south, there is no evidence in the field either of this younger brecciation or of an older shear zone (T. Falkum and R. Wilson, pers. comm. 1984). However, from the southern end of the above-mentioned faults an augen-gneiss forms a remarkably linear north-south unit, extending for about 70 km down to Mandal. Although the surrounding gneisses, augen-gneisses and migmatites on each side of this augen-gneiss show intricate fold patterns on the map, this particular augen-gneiss shows a much simpler structure. In the present author's opinion this metagranite is younger than the neighbouring gneisses and has intruded along the old zone of weakness, thus obliterating the traces of the oldest movements along the fault in this area.
This age relation is partly supported by available radiometric data; according to Pedersen (1981) one of the intricately folded augen-gneisses is ca. 1350 Ma old, while the linear augen-gneiss (metagranite) has a metamorphic age of around 1000 Ma (Wilson et al. 1977). Because of a greater uplift of the coastal areas relative to the central parts of South Norway, deeper parts of the crust have reached erosion levels in the south. At this crustal level the later fault movements may have resulted in plastic deformation and recrystallization, thus forming an augen-gneiss instead of brecciated granites. These younger movements are probably the result of a regional extensional stress, because the Telemark supracrustals east of the Man-Us line are downfaulted along easterly-dipping low angle fault planes. Structures ascribed to the older, and more regionally important, movements along this line are difficult to trace in the field. This is because the old fault zone has been intruded by post-tectonic granites, is also covered by the overlying rocks of the Bandak Group (Fig. 1) and here one can observe that the older movements have produced a 2.5-3 km broad belt consisting of zones of highly sheared gneisses with intervening zones of less deformed rocks. At Kaldhovd (25 km north of Møsvatn) basal conglomerates of the Bandak Group are unconformably overlying these sheared gneisses, which is a clear indication of a phase of ductile displacement older than the deposition of the Bandak conglomerates.

The presence of this major tectonic line is further revealed by a belt of late- and post-tectonic granites west of the fault and the distribution of the Telemark supracrustals (Fig. 1). Furthermore the pre-Bandak regional tectonic pattern west of the fault is different from that east of the fault (Sigmond et al. 1984). Aeromagnetic measurements show that the Man-Us line divides Southern Norway into a western area with an irregular magnetic pattern and high anomalies, and

Figure 2. Aeromagnetic residual map showing a generally higher and more variable magnetic field west of the fault than east of the fault. The map is based on the NGU aeromagnetic maps (1:250 000) Arendal, Mandal, Stavanger, Haugesund, Sauda, Skien, Oslo, Hamar, Odda and Bergen.

an eastern area with a generally very low magnetic content. The two blocks thus have different magnetic properties (Fig. 2).

Radiometric measurements in the southern part of the area (Fig. 3) show that the rocks west of the Man-Us Fault have a distinctly higher content of thorium and uranium than those east of the fault. [These measurements are made on fresh rocks in road-cuts and the points indicate localities with intensities greater than 100 impulses/sec]. The radiometric anomalies west of the fault occur not only in the postkinematic granites, but are in fact just as numerous in the surrounding rocks. According to Hysingjord (1976, 1977) the anomalies here are

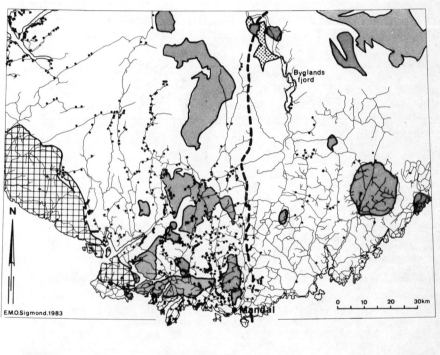

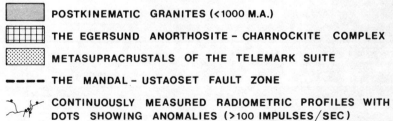

POSTKINEMATIC GRANITES (<1000 M.A.)

THE EGERSUND ANORTHOSITE - CHARNOCKITE COMPLEX

METASUPRACRUSTALS OF THE TELEMARK SUITE

――― THE MANDAL - USTAOSET FAULT ZONE

CONTINUOUSLY MEASURED RADIOMETRIC PROFILES WITH DOTS SHOWING ANOMALIES (>100 IMPULSES/SEC)

Figure 3. Radiometric measurements in South Norway.

associated with gneisses, granites and pegmatites, and some of the strongest anomalies derive from pegmatites. Thorkildsen (1975) noted that the measured anomalies east of the fault are all connected with the pegmatites. The distinctly higher content of radioactive minerals in the western block thus points to a fundamental difference between the migmatites and gneisses west and east of the zone. Such a difference has not yet been recognized on any geological map.

Looking at the distribution of various metals, copper and nickel occur more commonly both in the Telemark supracrustals and in the gneisses east of the fault zone than west of it (Fig. 4). Molybdenum on the other hand is more prevalent west of the Man-Us Fault (Fig. 5).

Figure 4. Copper and nickel occurrences in South Norway.

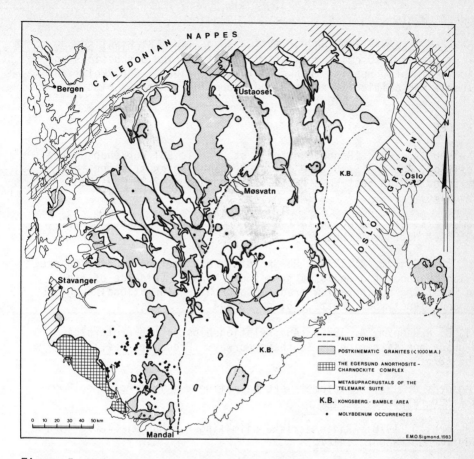

Figure 5. Molybdenum occurrences in South Norway.

CONCLUSIONS

The existence of a major tectonic zone from Mandal to Ustaoset dividing southern Norway into two large blocks is apparently well supported by geological and geophysical data. The old mylonite belt was formed before the deposition of the Bandak supracrustals (pre-1200 Ma), while the younger east-dipping normal faulting in the central and northern part took place after the intrusion of the post-tectonic granites (post-900 Ma). Whether the old ductile shear zone represents a continent-continent collision zone or merely large-scale fault movements remains to be determined in future studies.

ACKNOWLEDGEMENTS

The present work was carried out at the Norwegian Geological Survey, and I am grateful to colleagues at NGU for inspiring discussions and critical comments. Special thanks go to K.S. Heier, F.Chr. Wolff and D. Roberts for critically reading and improving the manuscript.

REFERENCES

Berthelsen, A. 1983: 'Report on the 1 st. Nordic EGT Workshop on the Composition, Age and Structure of the Precambrian crust along the South-Scandinavian East-West Profile.' Copenhagen, 7-28 May 1983 (polycopied).

Falkum, T. 1982: 'Geologisk kart over Norge, berggrunnskart Mandal 1:250 000.' Nor. geol. unders.

Field, D. & Råheim, A. 1979: 'Rb-Sr total rock isotope studies on precambrian charnockitic gneisses from south Norway: Evidence of isochron resetting during a low-grade metamorphic-deformational event.' Earth and Planet. Sci. Lett., 45, pp. 32-44.

Field, D. & Råheim, A. 1980: 'Age relationships in the Proterozoic high-grade gneiss region of Southern Norway.' Precambr. Res., 14, pp. 261-275.

Hysingjord, J. 1977: 'Radiometriske bilmålinger i Vest-Agder og Rogaland.' NGU-rapport 1416/15.

Hysingjord, J. 1978: 'Radiometriske bilmålinger i Mandal-Flekkefjord området.' NGU-rapport 1416/8.

Jacobsen, S.B. & Heier, K.S. 1978: 'Rb-Sr isotope systematics in metamorphic rocks, Kongsberg sector, South Norway.' Lithos 11, nr. 4, pp. 257-276.

Kleppe, A.V. 1980: 'Geologiske undersøkelser fra sentrale deler av Telemark med hovedvekt på geokronologi og kontaktrelasjonen mellom granittisk gneis og suprakrustalene.' Thesis Univ. Oslo.

Naterstad, J., Andresen, A. & Jorde, K. 1973: ' Tectonic succession of the Caledonian nappe front in the Haukelisaeter-Røldal area, Southwest Norway.' Nor. Geol. unders., 292, pp. 1-20.

O'Nions, R.K. & Heier, K.S. 1972: 'A reconnaissance Rb-Sr geochronological study of Kongsberg area, South Norway.' Nor. Geol. Tidsskr., 52, pp. 143-150.

Pedersen, S. 1981: 'Rb/Sr age determinations on late Proterozoic granitoids from the Evje area, South Norway.' Bull. geol. Soc. Denmark, vol. 29, pp. 129-143.

Priem, H.N.A. & Verschure, R.H. 1982: 'Review of the Isotope Geochronology of the High-Grade Metamorphic Precambrian of SW Norway.' Geol. Rundschau, 71, 1, pp. 81-84.

Sigmond, E.M.O. 1978: 'Beskrivelse til det berggrunnsgeologiske kartbladet Sauda 1:250 000.' Nor. geol. unders., 341, pp. 1-94.

Sigmond, E.M.O. 1984: 'En mega-forkastningssone i syd-Norge (abstract), in Meddelanden från Stockholms Universitets Geologiska Institution.' Nr. 255, Amands, S. & Schager, S. editors.

Sigmond, E.M.O., Gustavson, M. & Roberts, D. 1984: 'Berggrunnskart over Norge - M 1:1 million.' Nor. geol. unders.

Thorkildsen, C.D. 1975: 'Radiometriske bilmålinger på Sørlandet og i Telemark.' NGU-rapport 1389/9.

Versteeve, A.J. 1975: 'Isotope geochronology in the high-grade metamorphic Precambrian of SW Norway.' Nor. geol. unders., 318, pp. 1-50.

Wielens, J.B.W., Andriessen, P.A.M., Boelrijk, N.A.I.M., Hebeda, E.H., Priem, H.N.A., Verdurmen, E.A.Th. & Verschure, R.H. 1981: 'Isotope geochronology in the high-grade metamorphic Precambrian of Southwestern Norway: New Data and Reinterpretations.' Nor. geol. unders., 359, pp. 1-30.

Wilson, J.R., Pedersen, S., Berthelsen, C.R. & Jakobsen, B.M., 1977: 'New Light on the Precambrian Holum Granite South Norway.' Nor. Geol. Tidsskr., 57, pp. 347-360.

TERRANE DISPLACEMENT AND SVECONORVEGIAN ROTATION OF THE BALTIC SHIELD:
A WORKING HYPOTHESIS

Tore Torske
Institute of Biology and Geology,
University of Tromsø,
P.O. Box 3085,
N-9001 Tromsø,
Norway

ABSTRACT. A hypothetical, sinistral displacement of c. 500 km is postulated to have taken place between the South Norway Precambrian Region and the main Baltic Shield in Sveconorvegian times. This is an inference based on the integration into a plate tectonic model of the following two premises: 1) the concept of displaced terranes as an integral tectonic element of actualistic cordilleran-type continental margins should be taken into account; 2) previously published paleomagnetic reconstructions, indicating a c. 90° clockwise rotation of the Baltic Shield relative to the Laurentian Shield in the time interval 1200-1000Ma ago. Restoring the South Norway Precambrian Region to its hypothetical pre-displacement position would bring an inferred post-Svecofennian, cordilleran-type orogenic segment in South Norway into alignment with a coeval, fragmentary, cordilleran-type orogenic belt in Southwest Sweden.

INTRODUCTION

The Telemark-Agder-Rogaland block of the South Norway Precambrian Region is separated from the main part of the Baltic Shield by a number of major shear zones with intervening crustal blocks or mega-units (Fig. 1; Lindh 1974, Zeck & Malling 1976, Hageskov 1978, 1980a, b, Lundqvist 1979, Berthelsen 1980a, b, Falkum & Petersen 1980, Gorbatschev 1980, 1984, Lindh & Malmström 1980, Sigmond 1984, Skjernaa & Pedersen 1982).
  It has been tentatively interpreted in terms of a cordilleran-type orogenic segment (Torske 1976, 1977, Berthelsen 1980a, Falkum & Petersen 1980). This segment is characterized by a tripartite division into 1) an area in the west and south-west (Rogaland-Agder area) with sparsely distributed high-grade metamorphic remnants of inferred continental margin type pelitic, psammitic and calcareous sedimentary rocks, scattered among volumetrically predominant, younger, plutonic rocks. East of this follows 2) a volcano-plutonic belt including scattered occurrences of andesitic/dacitic metavolcanic and other metasupracrustal rocks (e.g., Hardanger: Torske 1977, 1982), surrounded and intruded

by granitoid plutons of a largely calc-alkaline batholithic belt.
To the east and north-east, this belt is followed by 3) an interior
province consisting of the less deformed and metamorphosed, more continental-type volcanic and sedimentary rocks of the Telemark Suite
(Dons 1960, 1972). Based on the lithological and petrological differences between the metasupracrustal rocks of these three belts an active
continental margin, overlying an eastward dipping subduction zone, is
inferred to have existed on the seaward side of the present SW coast of
Norway (Torske 1976, 1977, Berthelsen 1980a, Falkum & Petersen 1980).
Recent geochemical data appear, in a general way, to be consistent with
such a model (Brewer & Field 1985, Smalley & Field 1985).

The age of the Proterozoic, subduction-related metasupracrustal
rocks and orogenic activity in South Norway has been interpreted as
either a) Sveconorvegian (c. 1200-900 Ma ago) or b) pre-Sveconorvegian
(c. 1700-1200 Ma ago), followed by pervasive resetting of the isotopic
systems caused by Sveconorvegian intrusive and tectonothermal events.
These different age assignments of the South Norway cordilleran-type
activity reflect contrasting interpretations of geochronological data
from the region. One opinion holds that the widespread and predominant
radiometric dates within the 1200-900 Ma age bracket record rock-forming events during which most of the preserved subduction-related volcanic, sedimentary and plutonic rocks were emplaced (Falkum & Petersen
1980, Menuge 1985). Others maintain that while the widespread occurrence of pervasive Sveconorvegian deformation, metamorphism, and intrusive activity is not in dispute, a large part of the age data may represent tectonothermal resetting of radioactive clocks in older supracrustal and infracrustal rocks (Priem et al. 1973, Versteeve 1975,
Torske 1977, Smalley et al. 1983, Dahlgren 1984, Smalley & Field 1985,
Verschure 1985). Much more work is needed to decide this important
issue.

The South Norway Precambrian Region appears to represent only a
small fragment of a cordilleran-type orogen; its continuation beyond
the limits of South Norway is unknown. The purpose of this note is to
present evidence, largely circumstantial, indicating that the present
situation of the South Norway Precambrian Region may have been caused
by its severance, in Sveconorvegian times, from an older, more extensive orogenic belt.

TENTATIVE CORRELATIONS BETWEEN SOUTH NORWAY AND WEST SWEDEN

The interpretation of Sveconorvegian radiometric ages recorded from the
Telemark supracrustal rocks is controversial, as referred to above.
Accepting that they may be resetting ages, the inferred pre-Sveconorvegian rock units in the South Norway Precambrian Region may be seen in
conjunction with, and perhaps integrated with, the NNW-trending, post-
Svecofennian and pre-Sveconorvegian Småland-Värmland-Dala plutonic-
volcanic belt in western Sweden, bordering the 'Sveconorvegian Front'
or 'Protogine Zone' (Gorbatschev 1980); and which has been assigned to
a South-west Scandinavian orogen (Gorbatschev 1980, Gorbatschev & Welin
1980, Nyström 1982).

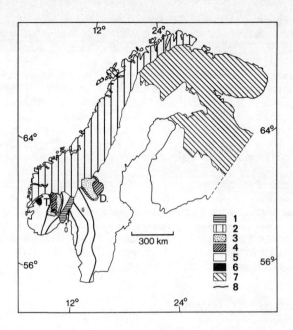

Fig. 1. Major geological divisions of the Baltic Shield, and some topical features of the South Scandinavian Precambrian. - D, Dala province; T, Telemark province; G, location of Gräsmark Fm.; O, Oslofjord Shear Zone. Legend: 1, Oslo Paleorift; 2, Caledonides (including the Sparagmite region); 3, Dala Sandstone and Seljord Group (Telemark); 4, Dala Volcanics and Rjukan Group (Telemark); 5, undifferentiated Proterozoic rocks; 6, Hardanger metasupracrustals; 7, region with Archean basement; 8, major Sveconorvegian shear zones in Southern Scandinavia.

Based on petrological and lithostratigraphical similarities, the largely volcanic Rjukan Group (Dons 1960, 1972) and the sedimentary, diabase-intruded Seljord Group (ibid.) in central Telemark have been tentatively correlated with the sub-Jotnian Dala Volcanics, dated at c. 1635 Ma (Welin & Lundqvist 1970; recalculated to λ 87Rb = 1.42×10^{-11}/yr), and with the overlying, Jotnian, diabase-intruded Dala Sandstone in western Sweden, respectively (Fig. 1; Priem et al. 1970, 1973, Torske 1977). The Telemark rocks are more deformed than their Swedish counterparts and have been metamorphosed under greenschist facies conditions (Dons 1960, 1972), while the rocks of the Dala sequence are largely unaltered, or show prehnite-pumpellyite to pumpellyite-actinolite facies metamorphism (Hjelmqvist 1966, 1973, Lundqvist 1968, Nyström 1980).

Based on its prominent greywacke and basalt protoliths (Lundqvist 1979, Park et al. 1979, Gorbatschev 1980, Samuelsson & Åhäll 1984), the Stora Le - Marstrand metasupracrustal sequence in the coastal areas of SW Sweden may be interpreted as a possible continental margin sequence (Lundqvist 1979, Gorbatschev 1980). This would indicate a tectonic setting similar to that suggested for the high-grade metasedimentary, pre-Sveconorvegian rocks in Southwest Norway (Demaiffe & Michot 1985), viz., the Faurefjell metasediments and the Gyadal and Hunnedal garnetiferous migmatites (Tobi 1965, Hermans et al. 1975, Tobi et al. 1985, Wilmart et al. 1984).

In the Swedish 'Eastern pre-Gothian Mega-unit' a deformed and metamorphosed supracrustal sequence, the Gräsmark Formation (Lundegårdh 1977, 1980) contains greywacke type metasediments and andesitic to dacitic, as well as rhyolitic, metavolcanic rocks. These rocks may be contemporaneous with the sub-Jotnian Dala volcanics (ibid.). The lithological composition of this sequence indicates that it may possibly have formed part of a calc-alkaline volcanic arc, near a continental margin.

The remnants of ?pre-Sveconorvegian, post-Svecofennian supracrustal rock sequences in South Norway and in Southwest Sweden may thus constitute a couple of two fragmentary cordilleran-type orogenic segments: in South Norway the three zones, marginal, central and interior (Torske 1976, 1977); in Sweden the marginal Stora Le - Marstrand sequence in the SW, the Gräsmark sequence farther NE, and the Dala sequence on the farther continentward side.

TERRANE DISPLACEMENT AND SHIELD ROTATION

After the cordilleran models for the South Norway Precambrian Region were proposed, the North American prototype cordillera -- and indeed, the entire circum-Pacific plate margins -- have been extensively reinterpreted in terms of large numbers of allochthonous, exotic terranes; transported and emplaced along the Pacific continental margins (e.g., Nur & Ben-Avraham 1982, Ben-Avraham & Nur 1983, Silver & Smith 1983, Schermer et al. 1984). These recent, major advances in the understanding of the archetype should be taken into account in re-appraisals of the cordilleran models of our region.

Cordilleran orogenic models presuppose a plate tectonic regime. A number of recent paleomagnetic studies have presented Precambrian continental reconstructions between the Laurentian and Baltic Shields involving a c. $90°$ clockwise, relative rotation of the latter with respect to the former within the Grenvillian/Sveconorvegian time span, c. 1200-1000 Ma ago (Fig. 2; Poorter 1976, Patchett et al. 1978, Piper 1979, 1980, 1982, Stearn & Piper 1984). Displacements like this, if true, require continental drift and, by implication, sea floor spreading. In this connection, it is of interest to note that evidence from the Borden Basin, Baffin Island (X, Fig. 2) indicates that incipient ocean opening, with subsequent closure took place between Greenland and the Canadian Shield in the time interval 1200-1000 Ma ago (Jackson & Iannelli 1984), that is, coevally with the Grenvillian/Sveconorvegian events in the Laurentian and Baltic Shields.

Among the geological features of the Laurentian-Baltic continental reconstruction, two stand out (Fig. 2): 1) the Grenville and Sveconorvegian fronts become closely aligned, and 2) the excentrically situated Archean, Kola nucleus of the Baltic Shield is juxtaposed with the Archean Craton of South Greenland.

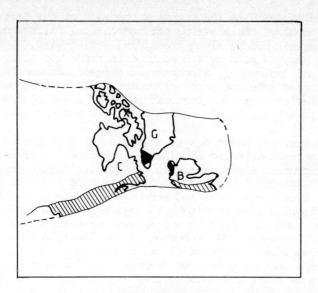

Fig. 2. Laurentian-Baltic part of Proterozoic Supercontinent reconstruction (redrawn and simplified after Piper 1982). - C, Canadian Shield; X, Borden Basin, Baffin Island; G, Greenland; B, Baltic Shield. Ruled area: Grenvillian-Sveconorvegian mobile belt. Black: Archean Craton in Greenland, Kola nucleus in the Baltic Shield.

For the present Baltic Shield, the former of these features has the following, important consequences: a) the Sveconorvegian belt and front should continue northwards, underneath the allochthonous, Scandinavian Caledonides (Berthelsen 1980b, Gorbatschev 1980); and b) the present western rim of the shield may have had a protracted, pre-Sveconorvegian history as part of the southern margin of the Proterozoic Supercontinent of Piper (1982), instead of having been continuous with the East Greenland continental crust, as envisaged in more conventional continental reconstructions (Davies & Windley 1976, Zwart & Dornsiepen 1978).

Sedimentary sequences deposited during early, passive periods in the history of this continental margin could have been affected by tectonothermal events during active periods involving subduction, and accompanied by volcanism, plutonism, metamorphism and orogenic deformation. Remnants of such sedimentary sequences may be represented by the high-grade metasupracrustal gneisses in the envelopes and septa of the plutons of the younger, Rogaland Igneous Complex. A generally similar plate tectonic situation could be ascribed to the Stora Le - Marstrand belt in SW Sweden. Alternatively, both these metasupracrustal sequences could be remnants of rocks deposited within the arc-trench gap of an active continental margin, more or less coevally with volcanic and sedimentary rocks farther inland on the continent: the Hardanger and Telemark sequences in the South Norway segment; and the Gräsmark and Dala supracrustals in Southwest Sweden.

Plate movements associated with the proposed rotation of the Baltic Shield could have involved the shearing off and lateral, differential displacement of marginal blocks along transform, strike-slip, or transpressive fault systems, to produce a mosaic of terranes appearing to represent a number of individual, disconnected orogenic fragments; but which originally formed coherent parts of one cordilleran-type orogenic belt. The various fault-separated, tectonic mega-units and blocks of Southern Scandinavia may constitute such a mosaic of more or less displaced terranes.

In their lithotectonic relations to the adjacent continental plate, displaced terranes form two distinct types: exotic terranes and disjunct, native terranes. Exotic terranes are foreign to the continent where they occur at present; they were built on oceanic crust or originated as parts of other continents, and were rafted on oceanic lithosphere to the continental margin. Disjunct, native terranes are displaced parts of the same continent in which they presently occur (Ben--Avraham & Nur 1983, Johnson 1984, Schermer et al. 1984). A case in point is the Salinian block, outboard of the San Andreas Fault System in California where a southerly derived fragment doubles up the Cretaceous Plutonic Belt of the Cordillera against its Sierra Nevada portion, east of the San Andreas Fault System (Page 1982).

Where terranes become accreted along a subduction zone with major consumption of oceanic lithosphere, slivers of ophiolite or trench-derived sediment prisms and mélanges commonly occur as well (Nur & Ben--Avraham 1982, Ben-Avraham & Nur 1983, Silver & Smith 1983, Schermer et al. 1984). The notable lack of such lithotectonic slivers in South Scandinavia indicates that none of the major shear zones there represents plate collision sutures (Gorbatschev 1980, 1984, Lindh et al. 1981), as has been suggested for some of them (Zeck & Malling 1976, Berthelsen 1980a). Instead, it is tentatively suggested here that they may have sustained early periods of major strike-slip and transpressive movements, related to the proposed, clockwise rotation of the Baltic Shield relative to the Laurentian Shield.

DISCUSSION

No structural evidence for major, strike-slip displacements has, to the author's knowledge, been reported from the South Scandinavian shear zones. They are, however, considered to have had long and complex structural histories (Lindh 1974, Skjernaa & Pedersen 1982, Gorbatschev 1984). Early structural elements may have been completely transposed or obliterated by later episodes of pervasive deformation. The extant structures in the shear zones mostly record compressive dip-slip or oblique-slip movement. Considering a possible rotation of the Baltic Shield amounting to ab. 90°, stress regimes could be expected to show large, transitory variations in direction and magnitude during this process. For example, a development from early strike-slip motion, through transpressive shear with an increasing compressive component (Sanderson & Marchini 1984), to predominant and pervasive compressive shear might be envisaged. A similar, actualistic example of changing

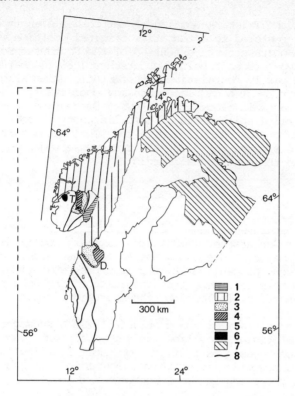

Fig. 3. Tentative restoration of the South Norway Precambrian Region to its approximate position prior to a hypothetical, Sveconorvegian sinistral displacement along the Oslofjord Shear Zone (cf., Fig. 1). The extrapolation of the shear zone northwards from the Oslo Paleorift is arbitrary. Lettering and ornament as in Fig. 1.

stress regimes across a major fault zone in space and time is afforded by the Alpine Fault, New Zealand (Spörli 1980). Hypotheses about stress regime development associated with the Sveconorvegian rotation of the Baltic Shield must remain speculative, as long as the actual poles of rotation for the particular plate movements involved are unknown.

The Sveconorvegian shear zones of Southwest Scandinavia may be of different ages, but their interrelationships are poorly known because critical areas are covered by younger rock complexes. The NW-trending zones east of the Oslo Fjord may be cut by, and older than, the north-trending Oslofjord Shear Zone (Hageskov 1978, 1980b). The northern extension of this zone is assumed to run underneath the Paleozoic Oslo Paleorift and the Vendian Sparagmite Basin north of it, and possibly farther afield underneath the allochthonous, Scandinavian Caledonides (Figs 1, 3).

If, for the sake of argument, a sinistral displacement of the South Norway Precambrian Region of about 500 km is assumed to have taken place along this zone, then the Telemark-Rogaland-Agder cordilleran-type orogenic fragment may be restored to a tentative original position where it would form a north-northwestward extension of the Southwest Scandinavian, cordilleran-type orogenic belt (Gorbatschev 1980, Gorbatschev & Welin 1980), with its partly correlative lithotectonic units (Fig. 3).

Similar, major strike-slip displacements may conceivably have affected the remaining tectonic blocks in Southwest Sweden and Southeast Norway. However, lateral displacements of cordilleran-type, beltiform tectonic elements more or less parallel with the continental margin may not necessarily have disrupted the original, transverse sequence of the main belts relative to the continental margin.

CONCLUSION

Instead of forming a separate orogenic belt outboard of the Southwest Scandinavian orogen of Gorbatschev (1980), the Telemark-Rogaland-Agder block of South Norway may be a disjunct, native terrane; doubling up part of a once continuous orogenic belt. The last pre-separation, rock-forming events within this belt were the deposition of the Seljord Group and the Dala Sandstone, followed by the intrusion into both units of diabase sills.

The Sveconorvegian history of the displaced terrane subsequent to its departure from the original position was very different from that of the remaining, autochthonous part of the Southwest Scandinavian orogen. It involved Sveconorvegian metamorphism, extensive plutonism, and pervasive radiometric resetting in pre-Sveconorvegian rocks.

REFERENCES

Ben-Avraham, Z. & A. Nur 1983:'An introductory overwiev to the concept of displaced terranes.' Can. J. Earth Sci. 20, 994-999
Berthelsen, A. 1980a: 'Towards a palinspastic tectonic analysis of the Baltic Shield.' Int. geol. Congr. Paris C6, 5-21
Berthelsen, A. 1980b: 'Den skandinaviske fortsættelse af Grenville orogenet.' (Abstr.), Geolognytt 13, 8 (Oslo)
Brewer, T.S. & D. Field 1985: 'Tectonic environment and age relationships of the Telemark supracrustals, Southern Norway.' This volume.
Dahlgren, S. 1984: 'Geology of central Telemark area, South Norway.' (Abstr.), NATO advanced Study Inst. Abstracts, Moi (Norway), p. 8
Davies, F.B. & B.F. Windley 1976: 'Significance of major Proterozoic high grade linear belts in continental evolution.' Nature 263, 383-385
Demaiffe, D. & J. Michot 1985: 'Geochronology of the Proterozoic crustal segment of Southern Norway.' This volume.
Dons, J.A. 1960: 'The stratigraphy of supracrustal rocks, granitization and tectonics in the Precambrian Telemark area, South Norway.' Norges geol. Unders. 212 H, 30 pp

Dons, J.A. 1972: 'The Telemark area, a brief presentation.' Sci. Terre 17, 25-29 (Nancy)
Falkum, T. & J.S. Petersen 1980: 'The Sveconorvegian orogenic belt, a case of Late-Proterozoic plate-collision.' Geol. Rdsch. 69, 622-47
Gorbatschev, R. 1980: 'The Precambrian development of Southern Sweden.' Geol. Fören. Sthlm Förh. 102, 129-136
Gorbatschev, R. 1984: 'De sydsvenska "suturerna" och deras roll i sköldens utveckling' (Abstr.), Medd. Sthlms Univ. geol. Inst. 255, 74
Gorbatschev, R. & E. Welin 1980: 'The Rb-Sr age of the Varberg charnockite, Sweden: a reply and discussion of the regional contexts.' Geol. Fören. Sthlm Förh. 102, 43-48
Hageskov, B. 1978: 'On the Precambrian structures of the Sandbukta-Mølen inlier in the Oslo graben, SE Norway.' Norsk geol. Tidsskr. 58, 69-80
Hageskov, B. 1980a: 'The Sveconorwegian structures of the Norwegian part of the Kongsberg-Bamble-Østfold segment.' Geol. Fören. Sthlm Förh. 102, 150-155
Hageskov, B. 1980b: 'Om Sweconorwegiske high strain zoner i området omkring Oslofeltet.' (Abstr.), Geolognytt 13, 25 (Oslo)
Hermans, G.A.E.M., A.C. Tobi, R.P.E. Poorter & C. Maijer 1975: 'The high-grade metamorphic Precambrian of the Sirdal-Ørsdal area, Rogaland/Vest-Agder, S.W. Norway.' Norges geol. Unders. 318, 51-74
Hjelmqvist, S. 1966: 'Beskrivning till berggrundskarta över Kopparbergs Län.' Sver. geol. Unders., ser. Ca 40, 217 pp
Hjelmqvist, S. 1973: 'An old evolution and a young 'model'.' Sver. geol. Unders. C 686, 11 pp
Jackson, G.D. & T.R. Iannelli 1984:'Borden Basin, NW Baffin Island: Mid-Proterozoic rifting and possible ocean opening.' (Abstr.), Geol. Ass. Can. Program w. Abstracts 9, 76
Johnson, S.Y. 1984: 'Evidence for a margin-truncating transcurrent fault (pre-Late Eocene) in Western Washington.' Geology 12, 538-41
Lindh, A. 1974: 'The mylonite zone in SW Sweden.' Geol. Fören. Sthlm Förh. 96, 183-197
Lindh, A. & L. Malmström 1980: 'A possible break in the Mylonite Zone in Värmland, south-western Sweden.' Geol. Fören. Sthlm Förh. 102, 95-103
Lindh, A., Z. Solyom & I. Johansson 1981: 'The question of chemical homogeneity among basic hypabyssals along the Scandinavian protogine zone.' Sver. geol. Unders. C 780, 36 pp
Lundegårdh, P.H. 1977: 'The Gräsmark Formation in Western Central Sweden.' Sver. geol. Unders. C 732, 18 pp
Lundegårdh, P.H. 1980: ' The gneissic granites and allied rocks in Central and Northwestern Värmland, Western Sweden.' Sver. geol. Unders. C 777, 3-23
Lundqvist, T. 1968: 'Precambrian geology of the Los - Hamra region, Central Sweden.' Sver. geol. Unders. Ba 23, 255 pp
Lundqvist, T. 1979: 'The Precambrian of Sweden.' Sver. geol. Unders. C 768, 87 pp
Menuge, J.F. 1985: 'Neodymium isotope evidence for the age and origin of the Proterozoic of Telemark, South Norway.' This volume.

Nur, A. & Z. Ben-Avraham 1982:'Displaced terranes and mountain building' In K.J. Hsü (Ed.): Mountain building processes, 73-84. Academic Press, London, 263 pp
Nyström, J.O. 1980:'Pumpellyit-aktinolit och prehnit-pumpellyitfacies metamorfos i Mellan-Sverige.' (Abstr.), Geolognytt 13, 51 (Oslo)
Nyström, J.O. 1982:'Post-Svecokarelian andinotype tectonic evolution in Central Sweden.' Geol. Rdsch. 71, 141-157
Page, B.M. 1982:'Migration of Salinian composite block, California, and disappearance of fragments.' Am. J. Sci. 282, 1694-1734
Park, G., A. Bailey, A. Crane, D. Cresswell & R. Standley 1979:'Structure and geological history of the Stora Le - Marstrand rocks in western Orust, southwestern Sweden.' Sver. geol. Unders. C 763, 36 pp
Patchett, P.J., G. Bylund & B.G.J. Upton 1978: 'Palaeomagnetism and the Grenville orogeny: new Rb-Sr ages from dolerites in Canada and Greenland.' Earth planet. Sci. Let. 40, 349-364
Piper, J.D.A. 1979: 'A palaeomagnetic survey of the Jotnian dolerites of central-east Sweden.' Geophys. J. R. astr. Soc. 56, 461-471
Piper, J.D.A. 1980:'Analogous Upper Proterozoic apparent polar wander loops.' Nature 283, 845-847
Piper, J.D.A. 1982:'The Precambrian palaeomagnetic record: the case for the Proterozoic Supercontinent.' Earth planet. Sci. Let. 59, 61-89
Poorter, R.P.W. 1976:' Palaeomagnetism of the Svecofennian Loftahammar gabbro and some Jotnian dolerites in the Swedish part of the Baltic Shield.' Physics Earth planet. Interiors 12, 51-64
Priem, H.N.A., N.A.I.M. Boelrijk, E.H. Hebeda, E.A.Th. Verdurmen & R.H. Verschure 1973:'Rb-Sr investigations on Precambrian granites, granitic gneisses and acidic metavolcanics in central Telemark: metamorphic resetting of Rb-Sr whole-rock systems.' Norges geol. Unders. 289, 37-53
Priem, H.N.A., R.H. Verschure, E.A.Th. Verdurmen, E.H. Hebeda & N.A.I.M. Boelrijk 1970:'Isotopic evidence on the age of the Trysil porphyries and granites in eastern Hedmark, Norway.' Norges geol. Unders. 266, 263-276
Samuelsson, L. & K.J. Åhäll 1984: 'Modell av den proterozoiska utvecklingen i Stora Le - Marstrandformationens område i västra Bohuslän.' (Abstr.), Medd. Sthlms Univ. geol. Inst. 255, 191
Sanderson, D.J. & W.R.D. Marchini 1984: 'Transpression.' J. struct. Geol. 6, 449-458
Schermer, E.R., D.G. Howell & D.L. Jones 1984:'The origin of allochthonous terranes: perspectives on the growth and shaping of continents.' Ann. Rev. Earth planet. Sci. 12, 107-131
Sigmond, E.M.O. 1984:'En mega-forkastningssone i Syd-Norge.' (Abstr.), Medd. Sthlms Univ. geol. Inst. 255, 199
Silver, E.A. & R.B. Smith 1983:'Comparison of terrane accretion in modern Southeast Asia and the Mesozoic North American Cordillera.' Geology 11,198-202
Skjernaa, L. & S. Pedersen 1982:'The effects of penetrative Sveconorwegian deformations on Rb-Sr isotope systems in the Rømskog-Aurskog-Høland area, SE Norway.' Precambrian Res. 17, 215-243
Smalley, P.C. & D. Field 1985: Geochemical constraints on the

evolution of the Proterozoic continental crust in southern Norway.' This volume.
Smalley, P.C., D. Field & A. Råheim 1983:'Resetting of Rb-Sr whole-rock isochrons during Sveconorwegian low-grade events in the Gjerstad augen gneiss, Telemark, Southern Norway.' Isotope Geosci. 1, 269-282
Spörli, K.B. 1980:'New Zealand and oblique-slip margins: tectonic development up to and during the Cainozoic.' Int. Ass. Sedimentologists Spec. Publ. 4, 147-170
Stearn, J.E.F. & J.D.A. Piper 1984:'Palaeomagnetism of the Sveconorwegian mobile belt of the Fennoscandian Shield.' Precambrian Res. 23, 201-246
Tobi, A.C. 1965:'Field work in the charnockitic Precambrian of Rogaland (SW Norway).' Geol. Mijnbouw 44, 208-217
Tobi, A.C., G.A.E.M. Hermans, C. Maijer and J.B.H. Jansen 1985: 'Metamorphic zoning in the high-grade Proterozoic of Rogaland/Vest Agder, SW Norway.' This volume.
Torske, T. 1976:'Metal provinces of a cordilleran-type orogen in the Precambrian of South Norway.' Geol. Ass. Can. Spec. Pap. 14, 597-613
Torske, T. 1977:'The South Norway Precambrian Region -- a Proterozoic cordilleran-type orogenic segment.' Norsk geol. Tidsskr. 57, 97-120
Torske, T. 1982:'Structural effects on the Proterozoic Ullensvang Group (West Norway) relatable to forceful emplacement of expanding plutons.' Geol. Rdsch. 71, 104-19
Verschure, R.H. 1985:'Geochronological framework for late-Proterozoic evolution of the Baltic Shield in South Scandinavia.' This volume.
Versteeve, A. 1975:'Isotope geochronology in the high-grade metamorphic Precambrian of southwestern Norway.' Norges geol. Unders. 318, 1-50
Welin, E. & T. Lundqvist 1970:'New Rb-Sr age data for the Sub-Jotnian volcanics (Dala porphyries) in the Los-Hamra region.' Geol. Fören. Sthlm Förh. 92, 35-39
Wilmart, E., J. Hertogen & J.C. Duchesne 1984:'Geochemistry of supracrustals, granite-gneisses and charnockites from the envelope of the Åna-Sira Massif (Rogaland/Vest Agder).' (Abstr.), NATO advanced Study Inst. Abstracts, Moi (Norway), p. 31
Zeck, H.P. & S. Malling 1976:'A major global suture in the Precambrian basement of SW Sweden?.' Tectonophysics 31, T35-T40
Zwart, H. & U.F. Dornsiepen 1978:'The tectonic framework of Central and Western Europe.' Geol. Mijnbouw 57, 624-654

PROTEROZOIC DEVELOPMENT OF BOHUSLÄN, SOUTH-WESTERN SWEDEN

L. Samuelsson and K-I. Åhäll
Geological Survey of Sweden
Kungsgatan 4
S-411 19 Göteborg
Sweden

ABSTRACT. Four groups of intrusives (A-D) have been used as time markers to construct a geological history for the polymetamorphic Proterozoic rocks of Bohuslän, SW Sweden.

The oldest rocks are supracrustals of the Stora Le-Marstrand formation consisting of metagraywackes intercalated with some mafic volcanics. The oldest granitoid plutons (group A) are part of a calc-alkaline suite. The age of this suite and the supracrustals is not known.

The B-group of intrusions is dominated by a second calc-alkaline suite, which has given ages at about 1650 Ma. This is also a minimum age for the A-group of intrusions and the preceeding amphibolite facies metamorphism.

The subsequent period of migmatization and folding is the main gneiss forming event in the region. It occurred before the emplacement of the C-group rocks. The C-group includes basic and granitic rocks, which sometimes display net-veining relations. The granites give ages between 1400 and 1200 Ma.

A 1220 Ma old C-group granite is part of the basement of the Dal group of sediments and basic volcanics. The Dal group is located just to the northeast of the present area and is the only supracrustal sequence deposited in S.W. Sweden in Sveconorwegian (Grenvillian) time.

The Bohus-Iddefjord granite (890 Ma) is the most prominent member of the D-group intrusives, which also include norite with anorthosite and doleritic dykes (WNW-ESE). These rocks are cutting the different deformations and metamorphic alterations, which can be ascribed to the Sveconorwegian orogenic development.

1. INTRODUCTION

The area concerned is the westernmost part of the gneiss region of southwest Sweden (Fig. 1). This region is characterized by c. 1000 Ma K-Ar cooling ages (1). In contrast the eastern part of southern Sweden is dominated by rocks affected by the Svecokarelian orogeny. The Svecokarelian belt has a western border of postorogenic granites and acid volcanics known as the Småland-Värmland granites (2). Apart from

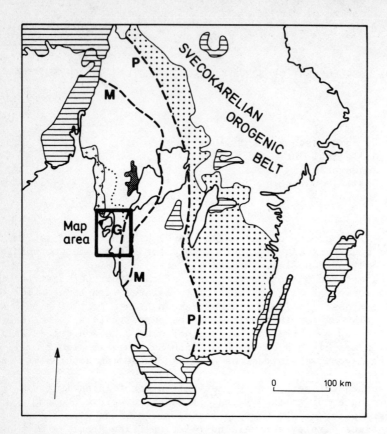

Fig. 1. Sketch map showing the location of the map area. Stipples = Småland-Värmland granites and acid volcanics. Dark grey = Dal group. Crosses = Bohus-Iddefjord granite. Horizontal lines = Phanerozoic rocks. P = Protogine zone ("Sveconorwegian front"), M = Mylonite zone. G = Göta älv shear zone. Bohuslän county is bordered by the dotted line and the Göta älv shear zone.

some minor granites, intrusive activity in the Svecokarelian belt was terminated at about 1600 Ma. The K-Ar blocking temperature of this region was reached at about 1400 Ma (1, 3).

The present boundary between these two regions of different K-Ar closing ages is called the Protogine zone. It is actually a linear belt (some km wide) of ductile and brittle deformation. In places it is also an important lithological boundary. The Protogine zone controlled geological events from the earliest intrusion of the Småland-Värmland granites, at about 1750 Ma (3), to the uplift of the western block in the final stages of the Sveconorwegian orogeny, at about 900 Ma.

Within southwest Sweden, there are several other major zones of ductile deformation, such as the Mylonite Zone and the Göta älv shear zone (Fig. 1 and Fig. 2). Some of these have been active several times

EVOLUTION OF BOHUSLÄN, SOUTH-WESTERN SWEDEN

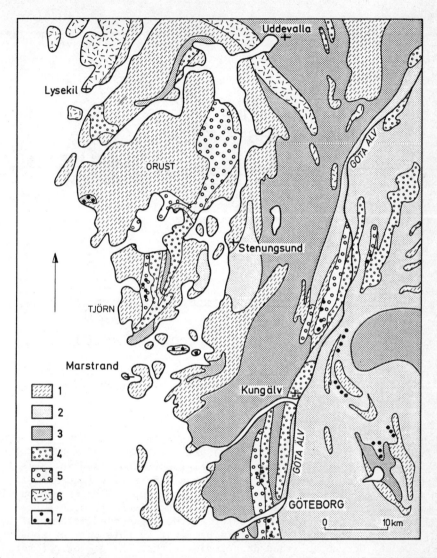

Fig. 2. Lithological map. 1 = Stora Le-Marstrand supracrustal rocks (Åmål supracrustals east of Uddevalla), 2 = tonalite-granite, A-group, 3 = tonalite-granite, B-group, 4 = older augengranite, B-group, 5 = younger augengranite, C-group, 6 = Bohus-Iddefjord granite and pegmatite, D-group, 7 = basic intrusives of different ages. The Göta älv shear zone is outlined by the course of the river.

during the extended (>1650-900 Ma) development of the area. They separate crustal blocks which may have had a different early history but which share a common Sveconorwegian deformation and subsequent uplift.

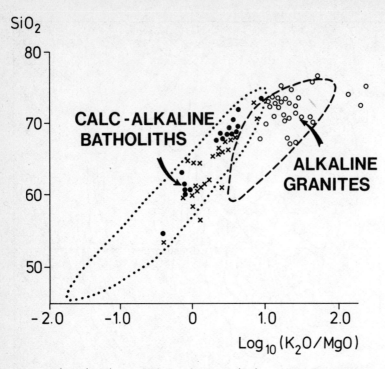

Fig. 3. Log 10(K_2O/MgO) vz. Si_2O diagram (35). Crosses = A-group tonalite-granite. Filled circles = B-group tonalite-granite. Open circles = B-group older augengranite.

Detailed mapping (4,5,6 and unpublished data) in the well exposed Tjörn-Orust and Göteborg areas (Fig. 2) combined with radiometric dating has resulted in the detailed tectonostratigraphy summarised in Table 1. This table of events is based on a polyphase deformation sequence (5) and on the recognition and bracketing of four groups (A-D) of regionally extensive intrusive rocks. These groups of intrusives are each separated by regional metamorphism and deformation. The second metamorphism, termed "Mig 2" is the main gneiss-forming event in this part of southwestern Sweden. Our unpublished data indicate that the geological history in Table 1 is valid for the whole of Bohuslän.

2. DESCRIPTION OF THE MAIN UNITS

The oldest rocks are the supracrustals of the Stora Le-Marstrand formation (7) consisting of metagraywackes intercalated with minor volcanics, mostly mafic, including locally preserved pillow lavas (6). Some early metagabbros, mafic dykes and ultramafic intrusives are found within the supracrustals. No older basement has been recognised.

The oldest granitoid plutons (group A) are part of a calc-alkaline

EVOLUTION OF BOHUSLÄN, SOUTH-WESTERN SWEDEN

Table 1. Sequence of events in Bohuslän, south-western Sweden.

Group of intrusives	Rocks/Metamorphism	Age Ma	Deformation
A	Stora Le-Marstrand formation: greywackes, basic (-acid) volcanics Gabbro and dolerite Tonalite-granite — Mig 1 1 st veining amph. facies	~1700??	
B	Tonalite-granite (Rönnäng) Gabbro: diorite-ultramafite Older augengranite (Ra-granite) Trondhjemite — Mig 2 2 nd veining amph. facies	~1650	D1 — Regional folding: F1 penetrative foliation: S1
M	Dykes: basic-acid — Mig 2+ Migmatites, recrystallization Dykes: granodiorite-pegmatite Gabbro→dolerite-augengranite Dolerite	~1400	D2 — Regional folding: F2 weak foliation: S2 Movements parallel to S2 give diktyonitic structures in mig.
C	Dykes: granite-pegmatite (Granites of the Hästefjorden group): Koster dolerite (Sveconorwegian) — Mig 3 amph. facies some veining, recrystallisation	~1220) ~1090	D3 — Some foliation: S3 local mylonitization F4/F5 — Zones of chevronfolding (2 sets of crenulation-cleavage)
D	Bohusgranite + pegmatite Norite-anorthosite. Dolerite	890	F6 — Local kinks, shears and faults

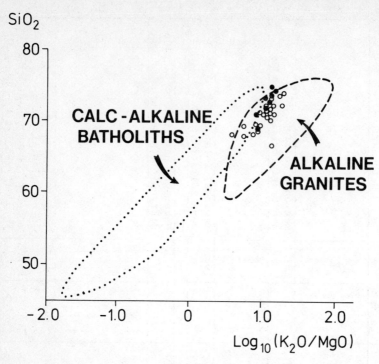

Fig. 4. Log 10(K_2O/MgO) vz. Si_2O diagram (35). Open circles = C-group younger augen granite. Filled circles = D-group Bohus-Iddefjord granite.

suite with a predominantly granodioritic composition (Fig. 3). East of the Göta älv shear zone, gneisses interpreted as A-group granitoids make up most of the bedrock.

The B-group of intrusions is dominated by a second calc-alkaline suite with compositions ranging from tonalite to granite but with tonalites and granodiorites developed most extensively. The B-group also contains intrusions of ultramafic, gabbroic and dioritic compositions, e.g. the Hällevikstrand amphibolite (5, 8).

One of the B-group rocks, the Rönnäng tonalite, has veined and folded xenoliths of Stora Le-Marstrand supracrustals indicating an earlier episode of deformation and metamorphism (4,6,9). In comparison with the A-group calc-alkaline suite, the B-group intrusives contain one less generation of foliation and veining. The calc-alkaline intrusives of the B-group give ages between 1700-1590 Ma (10).

A prominent member of the B-group intrusives is the "older augengranite" which is equivalent to the Ra-granite on the map sheets Göteborg NE and Kungsbacka NE (11, 12). Ra stands for the high γ-radiation, which is displayed by these granites. Their composition (Fig. 3) ranges from granite to alkali-feldspar granite (using the IUGS nomenclature 13). The Ra-granite on Tjörn contains xenoliths of the B-group tonalite and granite. From these localities it is clear that these different B-group intrusives underwent the same periods of deformation and migmatization.

Other probable B-group intrusives are the trondhjemitic sills found in the supracrustals on eastern Tjörn. They have an unusual high Na/K-ratio of about 6, which seems to indicate a source different to that of calc-alkaline suite.

Dykes ranging from basic to granitic compositions intruded after the B- and before the C-group rocks. These so called 'M'-group dykes differ from the younger C-group rocks by being affected by later episodes of the Mig 2 migmatization. At least the basic and intermediary dykes cannot be derived by simple anatexis of crustal material. The M-group dykes thus suggest a break in the metamorphic (deep crustal) evolution in the time between the B- and C-group intrusions.

The C intrusives are a very complex group. They have suffered only Sveconorwegian deformations and metamorphism. The Hästefjorden and Ursand granites are the youngest C-group rocks dated. They were intruded to the north-east of the present area at about 1220 Ma (14, 15). They constitute the basement to the sediments and basic volcanics of the Dal Group (16, 17), which is situated to the west of Lake Vänern (Fig. 1). The Dal Group is the only supracrustal sequence deposited in the area in Sveconorwegian (Grenvillian) times (16).

As the C-group rocks are common and widespread they can be used to discriminate between Sveconorwegian and earlier events. The most prominent members occur in swarms of dykes of different ages and compositions with basic members being the most common. Another important group of rocks is a synchronal suite with gabbroic, doleritic and augen granite members. The Rb-Sr w.r. age of this augen granite is 1416 ± 21 Ma (18).

The Sveconorwegian development involved a sequence of folding and metamorphic events (5). The group D intrusives are late- to postorogenic and include the Bohus-Iddefjord granite, which occupies the coastal area from Lysekil northwards into Norway. Associated pegmatites are sporadically found in a wide area to the east (19, 20, 21).

The Bohus granite is peraluminous and has a restricted range of SiO_2 values (22) and is rich in Th and U (23). The granite is a large, flat-lying, sheet-like body with an eastward subsurface extension of several tens of kilometres (24). Skiöld (25) dated the intrusion of the Bohus granite at 891 ± 34 Ma by the Rb-Sr method.

Norite-anorthosite occurs in two, 500 m wide bodies, with a total east-west length of 5 km, in the southern part of the area. The intrusions are undeformed and brecciate the surrounding gneisses. The anorthosite occurs as a megascopic breccia in the norite. Apparently the differentiation of the original magma into an anorthositic solid fraction and a noritic liquid fraction took place before the emplacement (4, 26).

WNW-striking dikes of dolerite in the southern part of the area also belong to the D group of late Proterozoic postorogenic intrusions. The largest dyke, the Tuve dyke, is 200 m wide in its western part and terminates against the Göta älv shear zone in the east. According to the paleomagnetic characteristics the Tuve dyke seems to have intruded 800-900 Ma ago (27), although Stearn and Piper (28) suggest an age of about 1000 Ma for both the norite and the dolerite.

3. GEOLOGICAL HISTORY

The polymetamorphic nature of the southwest Swedish gneiss region has for a long time hindered efforts to elucidate the geological history. However, in recent years a combination of structural mapping and radiometric dating has resulted in a workable tectonostratigraphic scheme (Table 1).

The Rönnäng tonalite has given the oldest radiometric dates in the area, viz. 1660-1675 Ma (9, 10). From field evidence we regard this rock as belonging to the second group of calc-alkaline intrusions (the B-group), which is separated from the first calc-alkaline suite (the A-group) by a period of folding and metamorphism. Therefore sialic crust existed in the area by 1700 Ma (10).

The oldest recognizable deformation (D1) resulted in isoclinal folds (F1) and a penetrative foliation (S1), which is recognised in the supracrustals and in the A-group rocks. The accompanying metamorphism (Mig 1) was in the amphibolite facies and granitoid leucosomes were formed (5). The radiometric results indicate that this period of folding and metamorphism is older than 1650 Ma (the age of the B-group intrusives).

The second regional folding (F2) is usually the most prominent one in the area. When not reorientated these tight upright folds trend N-S. The folding was synchronous with an amphibolite facies metamorphism, which resulted in abundant veining.

This period of migmatization and folding gave way to a different tectonic regime, when the M-group dykes were intruded. These M-group dykes cut Mig 2 veins and sometimes also F2 folds but are themselves deformed and surrounded by later granitic neosome. The period of migmatization, which postdates the M-group dyke emplacement is called 'Mig 2+'. Many of the M-dykes are multiple intrusions and just a few metres wide.

The M-dyking was apparently followed by a new period of rising temperature sufficient to produce abundant anatectic melts in the Stora Le-Marstrand formation. In areas dominated by intrusive rocks considerable in situ recrystallization took place, which partly or in places completely, obliterated older structures such as foliation, veins and folds. In places discrete granitic bodies were emplaced. A characteristic feature of this interval is the absence of recognisable regional deformation. However, some ductile shearing took place subparallel to the F2 axial planes (S2).

The period, including the Mig 2 and the Mig 2+ events, is responsible for the main metamorphic alterations (veining, foldings and recrystallization) of the bedrock. It took place between 1600 and 1420 Ma. These metamorphic structures were cut by the intrusions of the C-group rocks.

The Sveconorwegian has variable effects across the area. The metamorphic grade reached the amphibolite facies but some mafic C intrusives may be incompletely recrystallised especially in areas of low strain where fluid penetration was less efficient (18). Sveconorwegian shear zones and folds up to a few kilometres in wavelength largely control the outcrop pattern in the area. Shear zones affecting both the C in-

trusives and the older gneisses occur in swarms which are displayed in
the well exposed coastal areas of Tjörn-Orust. While a few larger shear
zones occur, the majority vary in width from a few metres to hundreds
of metres and the spacing between them varies over a somewhat extended
range. Within these zones the C intrusives take up a strong penetrative
foliation and minor amounts of migmatitic leucosome develop in the granites. Little new migmatization is seen in the older gneisses. Outside
the shear zones the Sveconorwegian folds sometimes have an associated
crenulation cleavage and only rarely is a penetrative fabric developed.

The widespread C-dolerites (the "Orust dykes") have yielded a mean
Rb-Sr age of 1087 ± 42 Ma, which was interpreted as dating the metamorphic
peak during the Sveconorwegian (29). This correlates with the Rb/Sr age
of 1028 ± 42 Ma obtained on slates of the Dal Group (25).

In western Orust, Park et al. (5) divided the Sveconorwegian deformation into several phases: a foliation-producing phase (D3) was followed by three foldphases (D4-6) and by the formation of shear belts
and faults. Generally, this sequence seems to be valid for the whole
area, but locally these phases are more or less pronounced or may be
absent.

Among the fold phases, F6 is only developed locally, as minor kink
folds, which in some areas are associated with minor faults. F5 folding
is pronounced in some steep NE trending zones, where the penetrative
planar fabric is sub-parallel to fracture cleavage and small faults. F4
folding is the most common phase throughout the area, developing chevron folds (of 1 m amplitude) with subhorizontal axial planes but generally without a penetrative foliation. However, crenulation cleavage
is common where mica-rich rocks are folded. In some areas there is a
transition from the upright F5 to the recumbent F4 folds but elsewhere
the two are not clearly related.

The heterogeneous imprint of the Sveconorwegian in Bohuslän is well
illustrated by the C-group augengranite, in Tjörn-Orust. In the south,
it is undeformed, while further north towards Orust, it gradually gets
more deformed with increasingly elongated augen combined with a late
overprint of megacrysts. Further N on eastern Orust the augengranite is
wholly gneissic with a foliation (S3) with strongly flattened, elongated
augen, which then have been folded by F4 or F5. Locally, cross-cutting
planar pegmatitic leucosomes are developed along the axial surfaces of
these folds.

The post-tectonic D intrusives, which include the 891 Ma old Bohus
granite (25) and related pegmatites (21), represent the late stages of
the Sveconorwegian evolution.

4. RADIOMETRIC DATING RESULTS: SOME PROBLEMS

Only a few radiometric dates have been published from the present area
but several are in progress. The oldest radiometric date obtained is
from the Rönnäng tonalite. The whole-rock Rb-Sr system of this rock is
disturbed but four selected samples constitute an isochron at 1675 ± 55
Ma with a Sr-initial ratio of 0.7026 ± 0.0016 (9). The U-Pb zircon age
was determined as 1658 ± 34 Ma, but it was interpreted as possibly due

to metamorphic recrystallization of the zircons (10). On structural grounds the Rönnäng tonalite is placed in the B-group. However, the isotopic data from a member of the oldest calc-alkaline group, the A-group granitoid at Stenungsund gives younger isochrons. Work in progress indicates a Rb-Sr and a zircon U-Pb isochron at 1525 and 1540 Ma respectively.

We regard the c. 1530 Ma as a metamorphic age (Mig 2) which might correspond to the metamorphic episode in South Norway at 1536±26 Ma (30). The ages of the oldest granitoids and the supracrustals are thus still unknown.

Several U-Pb studies on granitoids elsewhere in SW Sweden (10,32,33) have given ages close to 1530 Ma, sometimes with much younger (c. 1400 Ma) Rb-Sr ages (e.g. Lane granite, 10, 31, 32). In the present area, a B intrusive on western Orust, the Hällevikstrand amphibolite has given a Rb-Sr isochron of 1432±92 Ma which originally was interpreted as age of emplacement (29). While the inferred (c. 1530 Ma) age for the Mig 2 metamorphism is just within error of the Hällevikstrand isochron it is not yet clear what significance such young ages from B-group intrusives might have. The Hälleviksstrand body is presently being re-investigated by Claesson, Daly and Åhäll.

The timing of the onset of Mig 2 is still not known, but its termination seems to be better understood now that more C-group rocks have been dated. The Askim augen granite is dated at 1360 Ma (34). This is somewhat lower than the Rb-Sr age of 1416±21 Ma, which is obtained from a corresponding C-group augen granite on Tjörn (18). It seems likely that the Mig 2+ metamorphic event ended before or around 1420 Ma.

At Älgön-Brattön in the Marstrand area (Fig. 2) the intrusion of norite with anorthositic xenoliths occurred after the plastic deformation of the country rock. Thus it intruded at about the same time as or later than the Bohus granite. Stearn and Piper (28) concluded that the norite--anorthosite intruded at about 1000 Ma. We consider 800-900 Ma as a more realistic date and this would explain some paleomagnetic correlation difficulties defined by Stearn and Piper. A paleomagnetic age of 800-900 Ma was reported by Abrahamsen (27) for one WNW-ESE dolerite in the Göteborg area. This dyke (at Tuve in the northern part of Göteborg) is also included in the interpretation by Stearn and Piper (28).

5. CONCLUDING REMARKS

The area discussed is large enough to have a significant influence on any evolutionary model proposed for the south-western part of the Baltic shield. Sveconorwegian deformations and metamorphism are recognized all the way eastwards to the Protogine zone. However, older "Middle- to Lateproterozoic" events are more important for the growth of the crust in the area.

The major ductile shear zones of south-western Sweden are also critical to any model. They may separate crustal blocks characterized by the dominance of different metamorphic episodes during the extended Proterozoic development of the area.

6. ACKNOWLEDGEMENTS

This paper is based on the current mapping program of the Geological Survey of Sweden (Samuelsson and coworkers) and additional studies by Åhäll in connection with his doctoral thesis. Radiometric dating projects are in progress in cooperation with E. Welin and S. Claesson at the Laboratory for Isotopgeology in Stockholm and S. Daly at University College, Dublin. Preliminary results from these studies have been used in this paper. We also acknowledge stimulating discussions on structural problems with R.G. Park, Keele University, England and A. Crane, University of Aberdeen, Scotland. I. Starmer, University College, London and S. Daly improved the manuscript and kindly corrected the English.

7. REFERENCES

1 Magnusson, N.H., 1960: 'Age determinations of Swedish Precambrian rocks'. Geologiska Föreningens i Stockholm Förhandlingar 82, 407-432.

2 Magnusson, N.H., 1960: 'The Swedish Precambrian outside the Caledonian Mountain Chain'. In 'Description to accompany the Map of the Pre-Quaternary rocks of Sweden'. Sveriges Geologiska Undersökning Ba 16.

3 Åberg, G., 1978: 'Precambrian geochronology of south-eastern Sweden'. Geologiska Föreningens i Stockholm Förhandlingar 100, 125-154.

4 Bergström, L., 1963: 'Petrology of the Tjörn area in western Sweden'. Sveriges Geologiska Undersökning C 593, 1-134.

5 Park, R.G., Bailey, A., Crane, A., Cresswell, D. & Standley, R., 1979: 'Structure and geological history of the Stora Le-Marstrand rocks in western Orust, southwestern Sweden'. Sveriges Geologiska Undersökning C 763, 1-36.

6 Åhäll, K.-I., 1984: 'Pillow Lava in the Stora Le-Marstrand formation, southwestern Sweden'. Geologiska Föreningens i Stockholm Förhandlingar 106.

7 Lundqvist, Th., 1979: 'The Precambrian of Sweden'. Sveriges Geologiska Undersökning C 768, 1-87.

8 Daly, J.S., Park, R.G. & Cliff, R.A., 1979: 'Rb-Sr ages of intrusive plutonic rocks from the Stora Le-Marstrand belt in Orust, S.W. Sweden'. Precambrian Res. 9, 189-198.

9 Welin, E. & Gorbatschev, R., 1978: 'Rb-Sr isotopic relations of a tonalitic intrusion on Tjörn Island, south-western Sweden'. Geologiska Föreningens i Stockholm Förhandlingar 100, 228-230.

10 Welin, E., Gorbatschev, R. & Kähr, A.-M., 1982: 'Zircon dating of polymetamorphic rocks in south-western Sweden'. Sveriges Geologiska Undersökning C 797.

11 Samuelsson, L., 1982: 'Map of solid rocks Göteborg NE'. Sveriges Geologiska Undersökning Af 130.

12 Samuelsson, L., 1982: 'Description to the map of solid rocks Kungsbacka NE'. Sveriges Geologiska Undersökning Af 124.

13 IUGS Subcommision of the Systematics of Igneous Rocks. 1973: 'Classification and Nomenclature of Plutonic Rocks. Recommendations'. N. Jb. Miner. Mh. 1973, H4.

14 Welin, E. & Gorbatschev, R., 1976: 'The Rb-Sr age of the Hästefjorden granite and its bearing on the Precambrian evolution of south-western Sweden'. Precambrian Res. 3, 187-195.
15 Gorbatschev, R. & Welin, E., 1975: 'The Rb-Sr age of the Ursand granite on the boundary between the Åmål and 'Pregotian' mega-units of south-western Sweden'. Geologiska Föreningens i Stockholm Förhandlingar 97, 379-381.
16 Gorbatschev, R., 1977: 'Correlation of Precambrian supracrustal complexes in south-western Sweden and the sequence of regional deformation events in the Åmål tectonic mega-unit'. Geologiska Föreningens i Stockholm Förhandlingar 99, 336-346.
17 Gorbatschev, R., 1980: 'The Precambrian development of southern Sweden'. Geologiska Föreningens i Stockholm Förhandlingar 102, 129-136.
18 Åhäll, K.I. & Daly, J.S., this volume: 'Late presveconorwegian magmatism in the Östfold-Marstrand belt in Bohuslän, S.W. Sweden'.
19 Sundius, N., 1952: 'Kvarts, fältspat och glimmer samt förekomster därav i Sverige'. Sveriges Geologiska Undersökning C 520.
20 Brotzen, O., 1961: 'On some age relations in the Pre-Cambrian of south-western Sweden'. Geologiska Föreningens i Stockholm Förhandlingar 87, 227-252.
21 Welin, E. & Blomqvist, G., 1964: Age measurements on radioactive minerals from Sweden'. Geologiska Föreningens i Stockholm Förhandlingar 86, 33-50.
22 Asklund, B., 1950: 'Kosteröarna, ett nyckelområde för västra Sveriges prekambriska geologi'. Sveriges Geologiska Undersökning C 517, 1-56.
23 Landström, O., Larsson, S.Å., Lind, G. & Malmqvist, D., 1980: 'Geothermal investigations in the Bohus granite area in southwestern Sweden'. Tectonophysics 64, 131-162.
24 Lind, G., 1982: 'Gravity interpretation of the crust in south-western Sweden'. Geol. Institute, Univ. of Gothenburg Publ. A 41.
25 Skiöld, T., 1976: 'The interpretation of the Rb-Sr and K-Ar ages of Late Precambrian rocks in south-western Sweden'. Geologiska Föreningens i Stockholm Förhandlingar 98, 3-29.
26 Johansson, M., Kjörnsberg, L., Lennblad, I., Ljungstedt, G. & Lund, L.I., 1984: 'Gravimetric measurements on the islands of Älgön and Brattön in the archipelago of western Sweden'. Geologiska Föreningens i Stockholm Förhandlingar 105, 185-190.
27 Abrahamsen, N., 1974: 'The paleomagnetic age of the WNW-striking dykes around Gothenburg, Sweden'. Geologiska Föreningens i Stockholm Förhandlingar 9b, 163-170.
28 Stearn, J.E.F. & Piper, J.D.A., 1984: 'Palaeomagnetism of the Sveconorwegian mobile belt of the Fennoscandian Shield'. Precambrian Res. 23, 201-246.
29 Daly, J.S., Park, R.G. & Cliff, R.A., 1983: 'Rb-Sr isotopic equilibrium during Sveconorwegian (= Grenville) deformation and metamorphism of the Orust dykes, S.W. Sweden'. Lithos 16, 307-318.
30 Field, D. & Råheim, A., 1981: 'Age relationships in the Proterozoic highgrade gneiss regions of southern Norway'. Precambrian Res. 14, 261-275.
31 Welin, E. & Gorbatschev, R., 1978: 'Rb-Sr age of the Lane granite in

south-western Sweden'. Geologiska Föreningens i Stockholm Förhandlingar 100, 101-102.
32 Skiöld, T., 1980: 'Granite intrusions in the Proterozoic supracrustals of the Ellenö area, south-western Sweden'. Geologiska Föreningens i Stockholm Förhandlingar 102, 201-205.
33 Persson, P.O., Wahlgren, C.H. & Hansen, B.T., 1983: 'U-Pb ages of Proterozoic metaplutonics in the gneiss complex of southern Värmland, south-western Sweden'. Geologiska Föreningens i Stockholm Förhandlingar 105, 1-8.
34 Welin, E. & Samuelsson, L., in manuscript: Rb-Sr and U-Pb studies of granitoid plutons in the Göteborg region, southwestern Sweden'.
35 Rogers, J.J.W. & Greenberg, J.K., 1981: 'Trace elements in continental-margin magmatism: Part III. Alkali granites and their relationship to cratonization'. Geological Soc. of America Bull. Part II, 92, 57-93.

LATE PRESVECONORWEGIAN MAGMATISM IN THE ÖSTFOLD-MARSTRAND BELT, BOHUSLÄN, SW SWEDEN.

K.I.Åhäll[1] & J.S.Daly[2]
[1]Geologiska Institutionen, Göteborgs Universitet,
S-412 96 Göteborg, Sweden.
[2]Geology Department, University College, Belfield,
Dublin 4, Ireland.

ABSTRACT. The Östfold-Marstrand belt in Bohuslän is dominated by migmatized supracrustals of the Stora Le-Marstrand formation and several generations of Presveconorwegian intrusives. Members of the youngest group (the C intrusives) are widespread and give ages ranging from about 1400 to 1200 Ma. They include granodiorite - granite sheets, gabbro - dolerite massifs, bimodal augen granite - dolerite bodies, discrete massifs of augen granite, and basic to acid dykes possibly of two generations.
    The C group separate Sveconorwegian from earlier events and is thus an important structural marker and a key element to evaluate the Sveconorwegian metamorphism and deformation.

1.   INTRODUCTION

A widespread group of late Proterozoic intrusives are a key element in the correlation of Sveconorwegian deformation and metamorphism in SW Sweden and adjacent parts of Norway. These intrusives were termed the "C group" by Samuelsson and Åhäll (1), who placed them in a detailed regional chronology for the Bohuslän area of SW Sweden (Fig. 1). Bohuslän is dominated by migmatised supracrustals of the Stora Le-Marstrand formation (2), which continues north into Östfold in Norway and together with several generations of intrusives makes up the Östfold-Marstrand belt (Fig. 1).
    The C intrusives include a wide range of rock types which have been grouped together essentially on structural grounds (Table 1 of ref. 1), i.e. they cut an already gneissose basement and were first deformed and metamorphosed in the Sveconorwegian at about 1100 Ma (3). In southern Bohuslän two early C granites were emplaced at about 1400 Ma (this paper, Welin and Samuelsson pers. comm.). Their regional structural correlatives, the Hästefjorden granites (4), range in age up to ca 1200 Ma. Deformed C group rocks are in turn cut by post-tectonic intrusives including the 900 Ma old Bohus-Iddefjord granites.
    In this paper we describe the C rocks from the well exposed coastal area of Bohuslän. Based on the available literature and using the above geochronological and structural criteria we have attempted to correlate

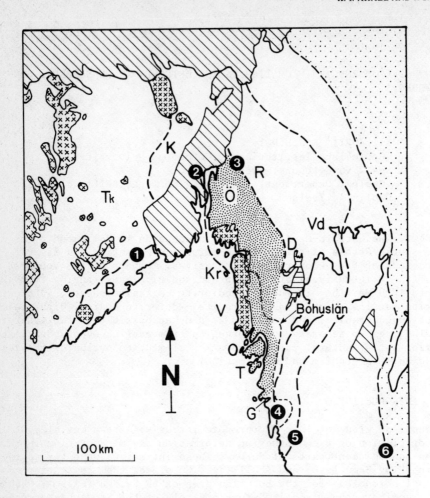

Fig. 1. Sketch map showing the location of Bohuslän and the major Sveconorwegian structures in SW Sweden and SE Norway. Short dashes = boundary of Bohuslän. Long dashes = Sveconorwegian structures: 1 = Kristiansand-Bang shear zone, 2 = Fjordzone, 3 = Dalsland Boundary Thrust, 4 = Göta Älv shear zone, 5 = Mylonite Zone, 6 = Protogine Zone ("Sveconorwegian Front"). Fine stipple = Östfold-Marstrand belt, coarse stipple = Svecokarelian Province, horizontal lines = Dal Group, diagonal lines = Phanerozoic, crosses = 900 Ma granites. G = Göteborg, T = Tjörn, O = Orust, V = Väderöarna, Kr = Koster, Ö = Östfold, R = Römskog-Aurskog-Höland, D = Dalsland, Vd = Värmland, K = Kongsberg, Tk = Telemark, B = Bamble.

the C group throughout Bohuslän, and suggest possible correlations in Dalsland and the adjacent part of Värmland in SW Sweden, and in the W. Östfold, Römskog-Aurskog-Höland, Bamble and Kongsberg areas of Norway. If our correlation is accepted it goes some way to resolving the debate concerning the recognition and regional distribution of Sveconorwegian orogenic events.

2. LITHOLOGIES

The C intrusives are a diverse group which include tonalite-granite dykes, gabbro-dolerite massifs, bimodal augen granite-dolerite sheets with subordinate hybrids, discrete bodies of augen granite and basic to acid dykes of various types. These rocks have been deformed and metamorphosed to varying degrees during the Sveconorwegian. They are summarised in Table 1. The various rock types are found throughout the area but their mutual proportions vary: Koster (Fig. 1) is dominated by the youngest C-dolerite (the Koster Dolerites (5), Väderöarna by granodiorite dykes and by the second mafic dyke suite and Tjörn-Orust by a 30 km long augen granite sheet and abundant granite-pegmatite dykes. Generally, C intrusives make up 10-25% of the bedrock.

Although there are clear genetic links between certain members, the grouping is essentially structural, i.e. those intrusives that post-date the Mig 2+ metamorphic episode (Table 1, of ref. 1) and which are deformed during the Sveconorwegian are placed in the C group.

Table 1. Sequence of C magmatism in Bohuslän, SW Sweden.

Second regional deformation and metamorphism	1600-1420 Ma (1)
tonalite-granite sheets, pegmatites	
gabbro-dolerite-augen granite association	
e.g. Stigfjorden augen granite	1420 Ma
e.g. Askim granite	1360 Ma
mafic dykes (with some silicic members)	
granitic and pegmatitic dykes	
(possibly = Hästefjorden granites)	1220 Ma
Minor shearing	
Koster dykes, dominantly mafic	
(= Kattsund dykes)	
Third regional deformation and metamorphism (Sveconorwegian)	ca 1090 Ma

The entire area is affected to a varying extent by Sveconorwegian deformation (1). In places discrete shear zones are separated by belts of low strain (usually tens to hundreds of metres in width). In such belts the C intrusives may have preserved igneous textures and distinct cross-cutting contacts with the surrounding gneisses. Occasionally, the primary igneous mineralogy is partly preserved in such low strain areas especially in the cores of the larger bodies where fluid penetration was least efficient. Generally however, the C intrusives have a thoroughly metamorphic appearance. Usually a penetrative fabric is developed and even in weakly deformed bodies the mafic rocks have the amphibolite facies assemblage: hbl-plag-(gt)-(bt)-(q). Some of the intermediate to acidic rocks develop minor amounts of granitic leucosome in shears and along the axial planes of Sveconorwegian folds.

2.1. Tonalite-granite sheets

Sheets and dykes of tonalitic to granitic composition are common over the whole area (cf the younger granites of Park et al. (6)). Various members of the suite crosscut each other in random order reflecting a complex emplacement history. Granitic bodies are often composite with coarse pegmatitic borders. The more mafic bodies are all fine-grained and homogeneous with a well defined tonalite-granodiorite composition (A, Table 2). Such dykes are usually not more than 5 m wide and are often the oldest within this group of C rocks.

2.2. Gabbro-dolerite-augen granite association

The oldest mafic C intrusive is also found throughout the area. It occurs in a complex association of basic and acidic rocks including net--veined and hybridised types. This bimodal suite is found over a distance of 200 km from Koster in the north to Kungsbacka, south of Göteborg. The oldest member of the suite is a medium-grained, massive, rather Mg-rich gabbro, which in some bodies grades into a dolerite. More commonly the dolerite forms discrete dykes with chilled margins against the gabbro. On Koster, the gabbro exhibits igneous layering.

On the small islands southwest of Tjörn (Fig. 1) a gabbro-dolerite body is associated with a megacrystic granite (now augen granite). The augen granite has brecciated the gabbro and the more primitive part of the dolerite while the more evolved Fe-rich dolerite dykes have been net veined by the same augen granite. Hybrids formed between the augen granite and the dolerite sometimes occur as dykes and also as enclaves within the augen granite.

Geochemistry (Table 2) and field relations show a gradation from Mg-rich (9% MgO) to more Fe-rich (14% Fe2O3) gabbro and dolerite dykes following a tholeiitic trend. While cooling, some of the latest dolerites were intruded by the porphyritic granite, resulting in extensive netveining. Variably corroded K-feldspar megacrysts and rounded (often bluish) quartz grains both in the hybrids and in the adjacent dolerite indicate that some mechanical mixing took place between the porphyritic granite and at least the Fe-rich mafic magma.

In places augen granite makes up the largest area of C intrusives.

Table 2. Selected geochemical data on C intrusives from Bohuslän, SW Sweden.

(%)	A	B	C	D	E	F	G	H	RANGE
SiO_2	68.2	71.1	70.2	72.8	48.8	51.6	51.7	47.2	44.5 –52.6
TiO_2	0.5	0.4	0.7	0.4	1.2	2.1	2.1	2.2	0.5 – 3.6
Al_2O_3	16.0	14.1	13.7	13.0	18.0	15.3	13.6	14.7	13.4 –17.3
tot Fe (Fe_2O_3)	4.6	4.6	5.3	3.4	11.7	14.2	14.7	14.8	10.4 –18.1
MnO	<0.1	<0.1	0.07	0.04	0.17	0.20	0.21	0.20	0.16– 0.27
MgO	1.6	1.2	0.9	0.4	8.6	6.3	6.9	7.0	4.6 –10.0
CaO	4.0	3.2	2.1	1.4	10.6	9.4	9.8	9.7	7.2 –11.6
Na_2O	3.9	4.3	2.6	2.6	2.7	2.9	2.5	1.4	0.5 – 2.7
K_2O	1.8	1.8	4.9	5.6	0.4	0.6	1.1	1.4	0.2 – 3.2
P_2O_5	0.14	0.10	0.19	0.09	0.15	0.41	0.23	0.4	0.05– 0.95
(PPM)									
Rb	69	69	155	280	<2	<2	40	49	5–146
Sr	360	390	135	70	175	310	150	150	57–307
Y	18	18	60	80	17	11	17	39	15– 64
Zr	240	300	400	350	110	170	160	170	22–327
Nb	26	30	33	28	13	16	17	11	<2– 21

A = most mafic member of the tonalite-granodiorite suite,
B = typical granodiorite of the tonalite-granodiorite suite,
C = 'dark' varieties of the Stigfjorden augen granite,
D = 'light' varieties of the Stigfjorden augen granite,
E = typical Mg-rich gabbro,
F = typical Fe-rich gabbro,
G = typical net-veined dolerite,
H = mean and range of 58 analyses of mafic and ultramafic Orust dykes (25).

The Stigfjorden augen granite may be followed for 30 km, from S. Tjörn where it occurs as discrete thin sheets which coalesce northwards to form a 2 km thick body through Stigfjorden and northeastwards into the Myckleby body of NE Orust (7). Towards the north the associated basic rocks diminish but hybrid xenoliths occur sporadically within the augen granite. This is reflected in a general sympathetic geochemical variation (Table 2) from a 'dark' more Fe-rich variety to a 'light' more acidic type. Locally on Tjörn the augen granite carries large angular xenoliths of the country rock gneisses. These contain older granitoid C dykes which are thus shown to predate the augen granite.

2.3. Later C dykes

Net-veined dolerites and associated granites are cut by two sets of mafic dykes which may be distinguished from one another in only a few areas. These mafic dykes are separated by a suite of granitic and pegmatitic dykes which might be coeval with granites of the Hästefjorden group which form larger massifs to the east. In the Koster area the younger set (the Koster dykes (8)) are clearly later than some local deformation which affects the earlier C rocks. Most of the Koster dykes have the usual Sveconorwegian amphibolite facies mineralogy but they also include some well-preserved olivine dolerites. Generally it is not possible to place any particular dyke in its proper position on Table 1 since all members of the C group are not often present together. The Orust dykes (3) belong in this category.

3. AGE OF THE C MAGMATISM

For most of Bohuslän a good field-based chronology has been established (1). Several projects are now in progress with the aim of calibrating this sequence using Rb-Sr, U-Pb and Sm-Nd techniques.

Work on the C group is still at an early stage. Rb-Sr data on the Stigfjorden augen granite, Tjörn-Orust, indicate an age of emplacement of 1416 +/- 21 Ma (IR=0.713 +/- 1, MSWD=1,6), while the Askim granite in the Göteborg area has given an U-Pb zircon age of 1357 +/- 30 Ma (Welin & Samuelsson, pers. comm., 1983). As these granites both are members of a widespread group of corresponding C augen granite bodies, this clearly indicates significant granitic magmatism of C age before the emplacement of the Hästefjorden granites.

Besides the latter ca 1220 Ma group of granites no data are yet available for the emplacement age of the younger members of the C group. However, a younger limit for the C magmatism in Bohuslän is suggested by the ca 1090 Ma age of the Sveconorwegian metamorphism (3).

4. DISCUSSION

Using the same structural and geochronological criteria as for the Bohuslän area above, late Presveconorwegian intrusions corresponding to the C group may be identified in adjacent parts of the Sveconorwegian

Province. Rocks meeting these criteria have recently been described from the Bamble and Kongsberg sectors (9) e.g. the Vinor gabbros, the coarse Telemark granite and the Vatnås granite; from western Östfold (10) e.g. the Moss-Filtvet augen gneiss and the Royken porphyry; from the Römskog-Aurskog-Höland area (11) e.g. the Römskog red orthogneisses; from Dalsland and adjacent part of Värmland 12,13,14,15,16,17,18, e.g. the Hästefjorden granites (4).

Age data from these areas and Bohuslän suggest that emplacement of the C intrusives at least took place between ca 1420 Ma and 1210 Ma. It is not clear if emplacement was continuous or episodic but one group, the Hästefjorden granites, clusters around ca 1220 Ma (10,14,15,16,17,18, 19). Younger limits for the C magmatism outside Bohuslän are given by the ca 1015 Ma estimate for the Sveconorwegian deformation in Östfold (10) and ca 1030 Ma for the deformation of the Dal Group which itself post-dates the C magmatism in Dalsland (20).

In detail there are differences between the various areas not least in the level of the orogen now exposed, i.e. higher grade Östfold-Marstrand belt in the west now juxtaposed with low grade rocks of the Dal Group to the east. Precise correlations are probably impossible and much more isotopic dating is needed. Nevertheless, some features in common are outlined here.

1. The bimodal magmatism, exemplified from the Tjörn area by the gabbro-augen granite association, is locally found throughout Bohuslän. A possible correlative is the Gjerstad-Morkheia complex of S. Norway (21). Smalley et al. (22) consider its age to be pre 1250 Ma and probably post 1350 Ma and thus somewhat younger than the 1420 Ma and 1360 Ma ages from Bohuslän.

2. The repeated intrusion of mafic C rocks is a prominent feature in Bohuslän where three different sets are separated by other types of intrusives (Table 1). Hageskov & Pedersen (10) also recognised a threefold subdivision in Östfold, viz: layered gabbro and basic dykes, the Moss Globulith (23) and the Kattsund dykes. The latter are equivalent to the Koster dykes (8). Similarly Starmer (9) separates three phases of Vinor gabbro (='hyperites') i.e. his early, main and late phases in the Bamble and Kongsberg areas.

3. Except for some shearing which occurs in the Koster area before the emplacement of the Koster dykes (Hageskov pers. comm. 1983), there is no other evidence for any structural break during the emplacement of the C rocks in Bohuslän. However, in the Bamble and Kongsberg areas of S. Norway significant regional deformation interrupts the C sequence (9). Some of the younger C rocks overlap in time with the widespread ca 1260-1190 Ma old basic intrusives in the N. Atlantic region. These rocks have been attributed to early Sveconorwegian continental rifting and are sometimes taken as a time marker for the onset of Sveconorwegian events. An earlier start to this tectonic regime would be consistent with the protracted history of anorogenic magmatism going back to ca 1400 Ma in the mid-continental region of the U.S.A. (24). When the timing of events in Scandinavia is known with better precision a redefinition of the Sveconorwegian should be possible in which the C group intrusives will play a key role.

5. CONCLUSIONS

The age of Sveconorwegian and earlier events as well as the question of different Sveconorwegian periods of deformation is still too uncertain to allow precise correlations between Bohuslän and adjacent areas. However, some important common features are:
1. The existence of a cratonic gneissose basement by about 1420 Ma. into which the C rocks were intruded.
2. Varied late Presveconorwegian igneous activity with several episodes of basic intrusives, bimodal magmatism in the 1420-1300 Ma interval and the widespread emplacement of red granites of the Hästefjorden group in the period 1300-1200 Ma.
3. The available radiometric data fall into two groups at about 1350 Ma and 1220 Ma. However, more geochronology is needed before this appearent bimodal age pattern is confirmed.
4. As the C intrusives are widespread and often occur as dykes, they are an extremely valuable structural marker in separating the Sveconorwegian from the earlier events.
5. The evidence for a metamorphic event of at least amphibolite facies conditions in the interval 1220-900 Ma is indisputable for Bohuslän, western Dalsland and adjacent parts of Värmland and probably all of Östfold. Higher levels of the orogen at greenschist facies grade are represented by the Dal Group in eastern Dalsland.

6. ACKNOWLEDGEMENTS

We gratefully acknowledge the financial support by the Swedish Natural Science Foundation (NFR) and the University College in Dublin. We are also indebted to L.Samuelsson, Geological Survey of Sweden, Göteborg, R.G.Park, Keele University, and C.Smalley, Inst. for Energy Technology, Oslo, for stimulating discussions and improvements of the manuscript.

7. REFERENCES

(1) Samuelsson, L. and Åhäll, K.I., this volume.
(2) Lundqvist, T. 1979., Sver. Geol. Unders. C 768.
(3) Daly, J.S., Park, R.G. and Cliff, R.A. 1983., Lithos 16, pp.307-318.
(4) Gorbatschev, R. 1975., Geol.Fören.Stockh.Förh. 97 pp. 107-114.
(5) Asklund, B. 1950., Sver.Geol.Unders. C 517.
(6) Park, R.G., Bailey, A., Crane, A., Cresswell, D. and Standley, R. 1979., Sver.Geol.Unders. C 763.
(7) Berthelsen, A. and Murthy, T.N.N. 1970., Sver.Geol.Unders. C 649.
(8) Hageskov, B. Meddelanden från Stockholms Universitets Geol. Inst. Nr 255, pp. 83.
(9) Starmer, I.C. this volume
(10) Hageskov, B. and Pedersen, S. 1981, Bull.Geol.Soc. Denmark 29, pp. 119-128.
(11) Skernaa, L. and Pedersen, S. 1982, Precambrian Res. 17, pp.215-243.
(12) Gorbatschev, R. 1977, Geol.Fören.Stockh.Förh. 97, pp.336-346.

(13) Gorbatschev, R. 1980., Geol.Fören.Stockh.Förh. 102, pp.129-136.
(14) Gorbatschev, R. and Welin, E. 1975., Geol.Fören.Stockh.Förh. 97, pp. 379-381.
(15) Welin, E. and Gorbatschev, R. 1976., Precambrian Res. 3, pp.187-195.
(16) Zeck, H.P. and Wallin, B. 1980., Contrib.Mineral.Petrol. 74, pp.45-53.
(17) Welin, E., Lundegårdh, P.H. and Kähr, A.-M. 1982., Geol.Fören. Stockh.Förh. 103, pp.514-518.
(18) Persson, P.-O., Wahlgren, C.-H. and Hansen, B.T. 1983., Geol.Fören. Stockh.Förh. 105, pp.1-8.
(19) Jacobsen, S.B. and Heier, K.S. 1978., Lithos 11, pp.257-276.
(20) Skiöld, T. 1976., Geol.Fören.Stockh.Förh. 98, pp.3-29.
(21) Milne, K.P. and Starmer, I.C. 1982.,Con.Min.Petr. 79, pp.381-393.
(22) Smalley, P.C., Field, D. and Råheim,A. 1983., Isotope Geoscience 1, pp.269-282.
(23) Berthelsen, A. 1970, Norges Geol.Unders. 266, pp.70-85.
(24) Emslie, R.F. 1978, Precambrian Res. 7, pp.61-98.
(25) Daly, J.S. 1978, Unpublished Ph.D. thesis. University of Keele, England.

THE WEST UUSIMAA COMPLEX, FINLAND: AN EARLY PROTEROZOIC THERMAL DOME

L. Westra and J. Schreurs
Institute of Earth Sciences, Free University
De Boelelaan 1085
1081 HV Amsterdam
The Netherlands

ABSTRACT. The West Uusimaa Complex in SW Finland represents a simple case of prograde, transgressive granulite-facies metamorphism.
The granulites occur in supracrustal and intrusive rocks of the Svecokarelian of the Baltic Shield. The tectonics are characterized by three succesive phases of deformation, of which the second (D2) forms the dominant fold geometry of the area. The hypersthene isograd that defines the boundary of the complex, cuts through the regional trend of F2 axes and the lithology. The present surface shows one crustal level of regional metamorphism.
The application of various geothermometers gives consistent results: the temperature averages about 650 °C in the amphibolite-facies rocks and increases to over 800 °C in the West Uusimaa Complex. The T increase of 100 to 150 °C takes place in a narrow zone of only 4 km wide.
Fluid-inclusion data and solid phase geobarometry indicate pressures of 3 to 5 kbar for the amphibolite- as well as the granulite domain. The amphibolite- to granulite-facies transition is virtually isobaric and the granulite-facies complex is interpreted as a thermal dome.
The West Uusimaa Granulite Complex could have originated by local 'hot spots' in a regional amphibolite-facies regime. This would account for the extremely low pressure of the granulites in comparison to other granulite-facies areas (Newton, this meeting).
It is suggested that the West Uusimaa Complex is only one of a series of thermal domes in a E-W oriented linear zone, possibly as a precursor of the Rapakivi granites in southern Finland.

INTRODUCTION

The early Proterozoic Svecokarelian orogeny in Finland consists of two main linear metamorphic belts. The NNW

trending Karelian belt in the east, parallel to the boundary of the Archean basement in eastern Finland, and the E-W trending Svecofennian belt in the south.
According to Simonen and Vorma (1978) and Simonen (1980) metamorphism took place under low-pressure amphibolite--facies conditions mainly. Granulite-facies metamorphism would occur in very restricted areas, such as the West Uusimaa Complex (Parras, 1958) in the Svecofennian belt, and the Sulkava area described by Korsman (1977) at the transition of the Karelian and Svecofennian belts. Gaál (1982) showed that in the Svecokarelian belts a medium-grade zone, greenschist to low amphibolite facies, and a high--grade zone, high amphibolite- to granulite facies, can be distinguished.
Regional studies show that granulite-facies areas, similar to the West Uusimaa Complex, are much more common in the high-grade zone than is generally known (Figure -1-). As such they form an integrated part of the Sveocokarelian orogen and should be one of the controlling parameters in reconstructing the thermotectonic evolution of this orogeny.

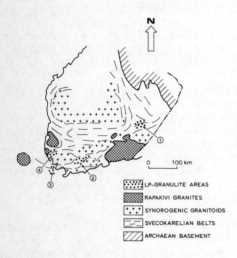

Figure –1–: Sketch map of the Svecokarelian metamorphic belt in southern Finland with some granulite-facies domains:
-1- The Haukivesi area (Gaál and Rauhamäki, 1971) and the Rantasalmi- Sulkava area (Korsman, 1977).
-2- The West Uusimaa Complex (Parras, 1958).
-3- The Gullkrona region (Edelman, 1960; Schellekens, 1980).
-4- The Uusikaupunki area (Hietanen, 1947; Kays, 1976).

Since 1979 a team of the Free University, Amsterdam, studied the West Uusimaa Complex in more detail. It is the aim of this paper to present a condensed general overview of the results of this work. It will be shown that the complex probably represents a low pressure thermal dome, due to an increased heat flow and the influx of CO_2-rich fluids in a late stage of the thermotectonic evolution.

STRUCTURAL GEOLOGY

The West Uusimaa Complex is a part of the high-grade Kemiö - Orijärvi - Lohja - Järvenpää belt in SW Finland (Figure -2-). The belt consists of a series of volcano - sedimentary rocks, intruded by a pre-tectonic gabbro - tonalite - granodiorite suite, that is in our opinion syngenetic with the volcanic rocks. The supra- and infracrustal rocks are subsequently intruded by syntectonic granitoids of anatectic origin. This belt of mainly supracrustal rocks is wedged between huge masses of granitoids, possibly of a similar origin as the smaller bodies in the belt.

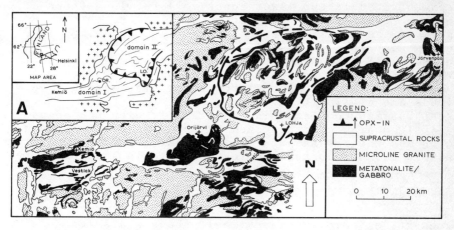

Figure -2-: Simplified geological map of the Svecofennian belt in SW Finland. Modified after Härme (1960).
 -2a-: Tectonic map with the major fold axes and foliations.

Structurally, this belt can be divided into two essentially different domains. From Kemiö to Orijärvi - Lohja, the belt has east-west striking, subvertical foliations with large isoclinal folds that show subvertical axial planes and low-angle fold axes. From Orijärvi - Lohja northeastwards, the supracrustal belt has a diamond shape , characterized by strongly varying foliations, more open large scale folds and lensoid or "eye"-shaped structures.
Three succesive stages of ductile deformation are clearly reconstructable, more or less corresponding to other areas in the Svecokarelides (cf: Gaál, 1982). The first phase (D1) is responsible for the formation of the dominating foliation. The second phase (D2) probably controls the dominating folds. The large scale eye structures are interpreted as gigantic boudins ("mega-boudins"), superposed on F2 folds and formed late during the D2 episode. The third deformation (D3) is characterised by open, disharmonic

folds with steep NW and NE trending axes. F3 folds are best developed in the pinch areas of the mega-boudins (Schreurs and Westra, in prep.).
Figure -2- shows the trends of east-west striking B2 axes in the Kemiö - Orijärvi domain, swinging to NE in the domain north of Orijärvi - Lohja and back to EW near Järvenpää. F3 folds can be recognized in this illustration by the refolding of F2 axes in large scale folds. FF2 axes undulate subhorizontally, indicating that no major tilt of the crust after folding (D2) has taken place. However, narrow shear zones, postdating D2, with mylonites, ultramylonites and pseudotachylites are recognized in numerous localities. They are concentrated in two major zones, (1) the long (tens of kilometers) Kisko shear zone, striking NE close to the boundary of the domain north of Orijärvi - Lohja , (2) the Orijärvi shear zone that strikes NW a few km NE of Orijärvi. Small NE and NW striking shear zones are found in the whole Orijärvi - Lohja - Järvenpää domain. Direct indications of displacement direction or displacement magnitude are difficult to obtain.
The West Uusimaa Complex coincides largely with the second structural domain, characterised by the mega-boudins, open F2 folds and the NE and NW striking shear zones. This suggests a relation between deformation style and granulite--facies metamorphism.

METAMORPHISM

The high-grade Kemiö - Orijärvi belt shows a distinct metamorphic zoning on Kemiö island (Dietvorst, 1981). Dietvorst distinguishes three different zones in pelitic gneisses. With increasing metamorphic grade succesively: (1) the muscovite + quartz zone, (2) the biotite + sillimanite + alkalifeldspar zone and (3) the cordierite + alkalifeldspar zone. Temperature and pressure estimates range from 560 °C, 3 kbar for the upper boundary of the muscovite + quartz zone to 670 °C, 4 kbar for the cordierite + alkalifeldspar zone at partial water pressures of $0.4\ P_{total}$. The zone boundaries, however, are interpreted as tectonic contacts rather then isograds.
Dietvorst' cordierite + alkalifeldspar zone continues eastwards as far as the transition zone of amphibolite- to granulite facies of the West Uusimaa Complex. Here the assemblage garnet + cordierite appears in pelitic gneisses. The actual boundary of the West Uusimaa Complex is defined as the Hypersthene - IN isograd in metavolcanic and meta-igneous rocks with an intermediate composition. Hypersthene (+/- diopside) are formed at the expense of hornblende and biotite. Almost coinciding with the hypersthene isograd are (1) the first occurrence of wollastonite in marbles near Lohja, (2) the breakdown of

anthophyllite to hypersthene + quartz in Mg-rich rocks (3) the already mentioned appearence of the cordierite + garnet assemblage in pelitic gneisses and (4) meionite (Parras, 1958) replacing calcite and plagioclase in calc-silicate rocks.

These mineral reactions do not represent tectonic contacts as on Kemiö island. The isograd crosscuts the lithology, the main foliation and the axes of the dominating (F2) folds (Figure -2-). Near Lohja, the Complex boundary is affected by a F3 fold. The transition zone is probably displaced along a right lateral shear zone in the north.

Anatectic granitoids within the supracrustal belt are concentrated in the West Uusimaa Complex and its immediate surrounding amphibolite-facies domain. They carry numerous rafts of volcano- sedimentary gneisses along the general strike directions. The granites can be rich in garnet and biotite but they do never contain hypersthene. Xenoliths of appropriate compositions show the same mineral reactions and changes in mineral chemistry as the surrounding supra- and infracrustal rocks.

All observed metamorphic reactions in the transition zone of the Complex are prograde. The granulite assemblages grow at the expense of the older amphibolite-facies assemblages. Porphyroblasts of hypersthene overgrow the dominating amphibolite-facies foliation (S1). Prograde reactions in various rock types across the Complex boundary have been studied in detail. (Schipper, 1982; Schreurs, 1985-a).

The structural and petrographic data so far allow the following conclusions:

(1) The West Uusimaa Complex represents a unique case of simple prograde granulite-facies metamorphism with a fully developed transition zone in rocks of various compositions.
(2) There is no evidence of major tilting of the crustal block that contains the Complex, neither has block faulting of any consequence been established. The narrowness of the transition also argues against tilting to explain the facies transition, whereas the presence of a real transition argues against block faulting. This suggests a homogeneous uplift of the crust after D3. The present erosion surface represents one crustal level of peak metamorphism. The granulites do not show a deeper level of the crust.
(3) Metamorphism in the amphibolite-facies domain started pre to syn D1 and possibly continued during D2. Granulite-
-facies metamorphism in the West Uusimaa Complex started post D1 and continued during D2. Both phases thus represent one continuous episode of regional metamorphism.
(4) Coinciding isograds indicate local steepening of isothermal surfaces or metamorphic telescoping.
(5) Shear zone activity postdates all metamorphic episodes and major deformations (D1-D2 and D3). "Young" displacements along the shear zones in West Uusimaa have obliterated most

evidence of older activity in these zones. However, remobilisation of older shear zones is a possibility which is presently investigated.

GEOTHERMOMETRY AND GEOBAROMETRY

An extensive analytical program of mineral chemistry and fluid inclusions has been carried out by the junior author (Schreurs). In this paper the main results of this work are given and the major implications are discussed.

Quantitative geothermometry and geobarometry has been performed with various independent methods: mineral reactions and exchange geothermometers. The most reliable results are obtained by comparing the temperature limits set by the different methods for small metamorphic domains. Garnet biotite thermometry can only be applied after thoroughly understanding the zoning in garnet and, to a lesser extent, in biotite (Schreurs 1985-b). The Ferry and Spear (1978) calibration gives temperatures which give the best correlation with the other methods: 710-820 °C for the West Uusimaa Complex and 550 to 650 °C for the adjacent Orijärvi - Lohja amphibolite-facies domain. The grossular or spessartine content of garnet do not have an apparent influence on the (temperature sensitive) Fe-Mg disttribution between garnet and biotite. High Ti contents of biotite do not seem to have a strong influence on this distribution. In the transition zone, calculated temperatures show a rapid increase of 100 to 150 °C (Figures -3- and -4-).
Two pyroxene thermometry (Lindsley, 1983) applied on samples of the West Uusimaa Complex gives temperatures between 720 - 800 °C (Figures -3- and -4-). Application of this thermometer is only possible if retrogressive Ca-resetting of coexisting ortho- and clinopyroxene is taken into account.
Spear (1980) developed an empirical thermometer based on the distribution of Na between plagioclase and calcic amphibole. Temperatures calculated according to this method give 630 to 660 °C for the amphibolite-facies domain and 730 to 780 °C for the West Uusimaa Complex (Figures -3- and -4-).
Experiments indicate that the breakdown of amphiboles to ortho and clinopyroxene takes place above 700 °C, even at very low water pressures at pressures above 2 kbar (see Gilbert et al.,1982).
The assemblage garnet + cordierite (+/-biotite) + sillimanite indicates metamorphic conditions of 700 to 800 °C at pressures between 3 and 5 kbar and a low water pressure (Newton and Haselton, 1981, Lonker, 1981; Martignole and Sisi, 1981).
Garnet hypersthene barometry (Newton and Perkins, 1982) gives pressures between 3.8 and 5.8 kbar for samples of the

West Uusimaa Complex.

Fluid inclusions provide another possibility to estimate metamorphic conditions (Schreurs, 1984). Two groups of inclusions are identified in the whole region.
(1) "Early" inclusions that are predominantly carbonic (CO_2) in the granulite domain and aqueous in the amphibolite--facies domain. The densities of the CO_2 inclusions from high to low grade vary between 0.85 - 0.95 gr/cc.
(2) "Late" inclusions; they are mainly carbonic with densities less than 0.80 gr/cc.

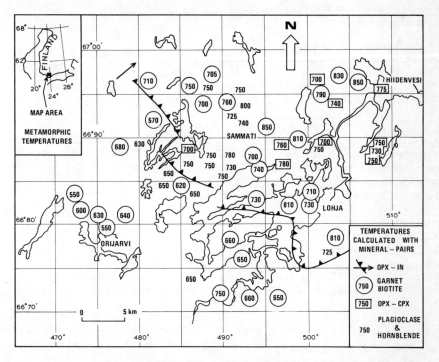

Figure –3–: Regional distribution of metamorphic temperatures based on garnet – biotite (Ferry and Spear, 1978), ortho– and clinopyroxene (Lindsley, 1983) and plagioclase – hornblende (Spear, 1980).

The early inclusions can be related to peak metamorphism. The density of the CO_2 inclusions show a general tendency to increase with decreasing metamorphic grade. The corresponding pressures of metamorphism are calculated using independently established metamorphic temperatures. The pressures corresponding to the observed densities at the estimated temperatures do not show a significant variation between the amphibolite facies and the granulite-facies domains (Figure –5–). Peak metamorphic pressure estimates

range between 3.5 and 4.5 kbar, which is in agreement with
pressure estimates based on mineral assemblages and mineral
barometry (3- 5 kbar).

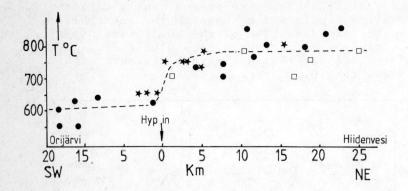

Figure -4-: Metamorphic temperatures along a profile from Orijärvi (SW)
to lake Hiidenvesi (NE) based on:
- garnet - biotite (Ferry and Spear, 1978), black dots
- ortho- and clinopyroxene (Lindsley, 1983), open squares
- plagioclase - hornblende (Spear, 1980), stars.

In summary; mineral chemistry and fluid inclusions support
the following conclusions:
(1) The West Uusimaa Complex has been formed by a virtually
isobaric temperature increase of approximately 100 to 150
°C, from 550/650 °C to 700/825 °C. This temperature increase
takes place in a zone of only a few km wide (Figures -3- and
-4-).
(2) This and the presence of predominantly CO_2-rich fluid
inclusions supports the idea that the Complex developed as a
thermal dome by an increased heat flow, which is tentatively
related to the influx of hot CO_2-rich fluids.

DISCUSSION

Structural evidence suggests that the granulite-facies event
in the West Uusimaa Complex lasted till after the main
deformation (D2). The amphibolite- to granulite-facies
transition can not be explained by tilting and thus
revealing a deeper section of the crust. The transition zone
is too narrow. Furthermore no evidence could be established
for tilting or block faulting. Block faulting contradicts
with the presence of a real transition zone. Therefore it
seems reasonable to assume that the studied area got into
its present position by more or less homogeneous uplift. The

present surface presents one crustal level of peak
metamorphism. This is supported by the geobarometrical data
that show no significant pressure variation between the
amphibolite-facies and the granulite-facies domains.

Mineral chemistry and fluid inclusions all convincingly show
that the granulite-facies event must represent a thermal
event enhanced or probably even caused by the influx of hot
CO_2 fluids.

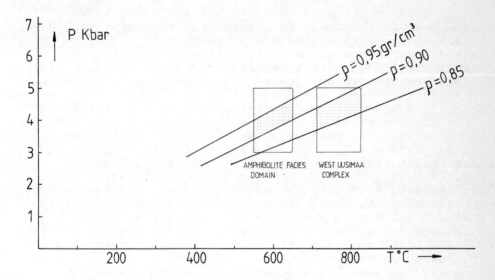

Figure -5-: P - T grid with CO_2 isochores (lines of equal density) based
on densities of CO_2 bearing fluid inclusions of the West Uusimaa Complex
and the adjoining amphibolite-facies domain. P - T boxes of the two
domains are also indicated.

Geothermometry and geobarometry indicate a considerable
temperature increase over a short distance (100 to 150 °C
over 2 to 4 km) at low pressures of 3 to 5 kbar. This
accounts for the telescoping of mineral isograds at the
boundary of the Complex.
The West Uusimaa Complex can be regarded as a low pressure
thermal dome.

The Complex measures at the present surface a width of
approximately 50 km from SE to NW. The depth of erosion,
according to geobarometrical data, corresponds to pressures
of 3 to 5 kbar or 12 to 15 km. Taking 650 °C as an average
for the amphibolite-facies domain implicates an average
geothermal gradient for the upper crustal layer in the
amphibolite-facies domain of 45 to 55 °C/km. For the

granulite domain, with an average temperature of aproximately 750 °C, the temperature gradient would be in the range 50 to 65 °C/km. The vertical deflection of the isothermal surfaces in the thermal dome is in the order of magnitude of kilometers.

The West Uusimaa Complex is an example of the heterogeneous distribution of temperature and fluid activities in parts of the continental crust. Is the thermal dome tectonically controlled, e.g. by shear zones that provide pathways for hot fluids from the deep crust (Drury and Holt, 1980) ? Or is the heat flow and CO_2 influx related to mantle hot spots or mafic intrusions in the deeper crust ?
The gravimetric map of southern Finland (Simonen, 1980) shows a distinct positive anomaly associated with the West Uusimaa Complex. This suggests the presence of a large mafic pluton(s) at a deeper crustal level.
An interesting aspect of the granulites in southern Finland is the close spatial association with Rapakivi granites (Vorma, 1976). The time range for Svecokarelian metamorphism is about 1.85 - 1.7 Ga, in which the granulite-facies metamorphism represents the youngest episode. Rapakivi granites have been dated at 1.7 to 1.5 Ga (Simonen, 1980). This suggests that both the granulite domes and the Rapakivi granites could have been formed during one thermal event, by local hot spots that give high level, low pressure granulites and deeper level Rapakivi melts. The emplacement of these granites could well have taken place some tens of million years later.
Vorma (1976) discussed a possible relation between the Svecofennian granulites and the post-orogenic Rapakivi granites. The granites would be formed from "dry" anatectic melts from deeper crustal levels, with the granulites as refractory residues. This idea is not supported by our data. A preliminary study of Schipper (1982) shows that the rocks of the West Uusimaa Complex have a similar chemistry as their amphibolite-facies equivalents. They are not depleted as would be expected for anatectic residuals.
According to Emslie (1978), the heat source for the formation of Proterozoic Rapakivi granites in general could be large anorthosite plutons. The Finnish Rapakivi granites are accompanied by only a small minority of anorthosites. The present erosion level, however, represents a high level (12 to 15 km) of the original Svecofennian protolith. Anorthosites and possibly other mafic plutons, could well be present at deeper crustal levels.

ACKNOWLEDGEMENTS

The junior author acknowledges a grant of the Netherlands Organization of Pure Research (ZWO).

REFERENCES

Dietvorst, E J L, 1981. Pelitic gneisses from Kemiö, southwest Finland, a study of retrograde zoning in garnet and spinel. Ph.D. Thesis, Free University Amsterdam.
Drury, S A and Holt, R W, 1980. 'The tectonic framework of the South India craton: a reconaissance involving LANDSAT imagery'. Tectonophysics, 65, T1-T15.
Edelman, N, 1960. 'The Gullkrona region, SW Finland'. Bull Comm Geol Finlande, 187, 187 pp.
Emslie, R F, 1978. 'Anorthosite massifs, Rapakivi granites and late Proterozoic rifting in N.America'. Precam Res, 7/1, 61-98.
Ferry, J M and Spear, F S, 1978. 'Experimental calibration of the partitioning of Fe and Mg between biotite and garnet'. Contrib Mineral Petrol, 66, 113-117.
Gaál, G and Rauhamäki, E, 1971. 'Petrological and structural analysis of the Haukivesi area between Varkaus and Savonlinna, Finland'. Geol Surv Finland Bull, 43, 265-337.
Gaál, G, 1982. 'Proterozoic evolution and late Svecokarelian plate deformation of the Central Baltic Shield'. Geol Rundschau, 71, 158-170.
Gilbert, M C, Helz, R T, Popp, R K and Spear, F S, 1982. 'Experimental studies of amphibole stability'. In: Veblen, D R and Ribbe, P H (eds), Amphiboles, petrology and experimental phase relations. Reviews in Mineralogy, 9B, Min Soc Am.
Hietanen, A, 1947. 'Archean geology of the Turku district in southwestern Finland'. Bull Geol Soc Am, 58, 1019-1084.
Härme, M, 1960. The geological map of Finland, Turku, sheet B1.
Kays, M A, 1976. 'Comparative geochemistry of migmatized, interlayered quartzofeldspathic and pelitic gneisses: a contribution from rocks of southern Finland and northeastern Saskatchewan'. Precam Research, 3, 433-462.
Korsman, K, 1977. 'Progressive metamorphism of the metapelites in the Rantasalmi - Sulkava area, southeast Finland'. Geol Surv Finland Bull, 290.
Lindsley, D H, 1983. 'Pyroxene thermometry'. Am Mineral, 68, 477-493.
Lonker, S W, 1981. 'The P-T-X relations of the cordierite - garnet - sillimanite - quartz equilibrium'. Am J Sci, 281, 1056-1090.
Martignole, J and Sisi, J C, 1981. 'Cordierite-Garnet-H2O Equilibrium: A Geological Thermometer, Barometer and Water Fugacity Indicator'. Contrib Mineral Petrol, 77, 38-46.
Newton, R C and Haselton, H T, 1981: 'Thermodynamics of the garnet - plagioclase - Al_2SiO_5 - quartz barometer', in Newton, R C, Navrotsky, A and Wood, B J (eds).

Thermodynamics of minerals and melts. Advances in physical geochemistry, 1, 131-148. Springer Verlag.

Newton, R C and Perkins, D III, 1982: 'Thermodynamic calibration of geobarometers based on the assemblages garnet - plagioclase - orthopyroxene (clinopyroxene) - quartz'. Am Mineral, 67, 203-222.

Parras, K, 1958. 'On the charnockites in the light of a highly metamorphic rock complex in SW Finland'. Bull Comm Geol Finlande, 181.

Schellekens, J H, 1980. 'Application of the garnet - cordierite geothermometer and geobarometer to gneisses of Attu, SW Finland; an indication of P and T conditions of the lower granulite facies'. N Jb Miner Monatsh, 1, 11-19.

Schipper, N J, 1982. 'Prograde granulite facies metamorphism in the West Uusimaa Complex, SW Finland'. Internal report Free University, Amsterdam, Inst. Earth Sc, 68 pp.

Schreurs, J, 1984. 'The amphibolite-granulite facies transition in West Uusimaa, S.W. Finland. A fluid inclusion study'. J. Metam. Geol., 2, 327-341.

Schreurs, J, 1985-a. 'The West Uusimaa low pressure thermal dome, SW Finland'. Ph.D. Thesis, Free University Amsterdam, 179p.

Schreurs, J, 1985-b. 'Prograde metamorphism of metapelites, garnet - biotite thermometry and prograde changes of biotite chemistry in high-grade rocks of West Uusimaa, SW Finland. Lithos, in press.

Simonen, A and Vorma, A, 1978. 'The Precambrian of Finland'. Metam. map of Europe. 1:2500000. Explanatory text. Subcomm. for the Cartography of the Metam. belts of the World. Leiden-Unesco, Paris, 20-27.

Simonen, A, 1980. 'The Precambrian in Finland'. Geol Surv Finland Bull, 304, 58 pp.

Spear, F, 1980. 'NaSi=CaAl exchange equilibria between plagioclase and amphibole: an empirical model'. Contrib Mineral Petrol, 72, 33-41.

Vorma, A, 1976. 'On the petrochemistry of Rapakivi granites with special reference to the Laitila massif, southwestern Finland'. Geol Surv Finland Bull, 285, 98 pp.

GEOCHRONOLOGICAL FRAMEWORK FOR THE LATE-PROTEROZOIC EVOLUTION OF THE BALTIC SHIELD IN SOUTH SCANDINAVIA

R.H. Verschure
Z.W.O. Laboratorium voor Isotopen-Geologie
De Boelelaan 1085
1081 HV Amsterdam
The Netherlands

ABSTRACT. Two major periods with orogenic and magmatic activities are responsible for the formation of the Late Proterozoic crust underlying large parts of Norway and Sweden, the Gothian between about 1.70 Ga and 1.20 Ga ago and the Sveconorwegian between about 1.20 Ga and 0.85 Ga ago. The main crustal accretion was achieved during the early, orogenic part of the Gothian period between about 1.70 Ga and 1.55 Ga ago. From about 1.60 Ga to 1.20 Ga ago the Gothian period had an anorogenic, ensialic character. During the Gothian period two major structural elements were formed, (1) in the east, the north trending Trans-Scandinavian Småland-Värmland Granitic Belt, mainly comprising a suite of 1.70 Ga to 1.60 Ga old granites and acid volcanics, and (2) in the west, the Southwestern Gneiss Region made of gneisses, granites, migmatites and supracrustal intercalations with ages between about 1.70 Ga to 1.20 Ga. Sveconorwegian metamorphism, which appears to increase in grade towards the west, has severely affected the Gothian rocks of both structural elements. The polyorogenic development of southern Scandinavia makes an unequivocal unravelling of the geology extremely difficult, if not impossible. It is argued that there have been high-grade metamorphic events during both the Gothian and the Sveconorwegian orogenic periods. Moreover, it is postulated that some published Rb-Sr whole-rock isochrons, especially those of coarse-grained metamorphic rocks, have no geological meaning. These meaningless isochrons often reflect the effects of isotopic disequilibration due to a partial metamorphic recrystallization of primary mineralogies, in combination with non-representative sampling.

1. GENERAL OUTLINE OF THE BALTIC SHIELD

The Precambrian Baltic Shield is only partially exposed. To the south and east the Shield gradually disappears under a thick autochthonous Phanerozoic sedimentary cover, to the northwest it was incorporated in the Phanerozoic Caledonian mobile belt and to the southwest it is covered by the sea. From northeast to southwest (Fig. 1) three main orogenic areas can be distinguished: the Archaean area characterized by

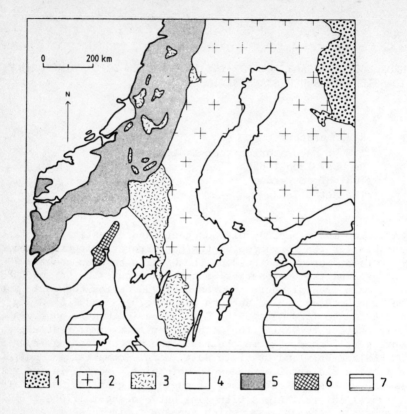

Fig. 1. Orogenic and epeirogenic areas of the Baltic Shield (after A. Lindh, 1982)

1. Archaean orogenic area (> 2.50 Ga)
2. Svecokarelian orogenic area (2.50 - 1.75 Ga)
3. Gothian-Sveconorwegian orogenic area (1.70 - 0.85 Ga) (Trans-Scandinavian Småland-Värmland Granitic Belt)
4. Gothian-Sveconorwegian orogenic area (1.70 - 0.85 Ga) (Southwestern Gneiss Region)
5. Caledonian orogenic area (0.60 - 0.30 Ga)
6. Oslo-Graben epeirogenic area (0.25 - 0.20 Ga)
7. Autochthonous basement cover (< 0.60 Ga)

rocks older than about 2.50 Ga, the Svecokarelian area with rocks yielding ages between 2.50 Ga and 1.75 Ga, and the Gothian-Sveconorwegian area with a complex history involving two orogenic periods. For the oldest of these periods, from about 1.70 Ga to 1.20 Ga ago, the designation "Gothian orogenic period" has been proposed (e.g. Gorbatschev, 1980). This concept of a Gothian orogenic belt fringing the southwestern margin of the Svecokarelian orogenic area implies a return to the old concept of Backlund (1936), Wahl (1936) and Magnusson (1936), and sets aside Magnusson's (1962) assumption of a much older Pre-Svecokarelian continental nucleus, the so-called "Pregothian nucleus", in southwestern Sweden. The second orogenic period, from about 1.20 Ga to 0.85 Ga, is designated as the Sveconorwegian, Dalslandian or Grenville orogeny. The Gothian-Sveconorwegian orogenic belt in southern Scandinavia has been correlated by workers in both Europe and North America with the Pre-Grenville-Grenville orogenic belt in North America (e.g., Wynne-Edwards and Hasan, 1970, Bridgwater and Windley, 1973, Ueno et al., 1975, Sturt et al, 1975, Morris and Roy, 1977, Patchett and Bylund, 1977, Patchett et al., 1978, Van Breemen et al., 1978, Max, 1979, Berthelsen, 1980, Falkum and Petersen, 1980, Poorter, 1981, Gower and Owen, 1984). Gorbatschev (1980) postulated that the Sveconorwegian orogeny is spatially, but also causally a direct predecessor of the Phanerozoic Caledonian orogeny with a similar orogenic style and the same direction of thrusting.

Towards the end of the Precambrian the Baltic Shield became peneplanated and overlain by Eocambrian and Phanerozoic sediments. During the subsequent Caledonian orogeny, from about 0.60 Ga to 0.40 Ga ago, these sedimentary rocks, together with parts of their Precambrian substratum (Reymer et al., 1980), were thrust in eastern direction upon the Shield and its autochthonous sedimentary cover. Finally, from about 0.25 Ga to 0.20 Ga ago, the southwestern part of the Baltic Shield was subjected to a phase of epeirogenesis during which the Oslo Graben with its alkaline igneous rocks was formed. Erosion and glacial abrasion on the gradually emerging Baltic Shield led to the present geological configuration.

2. THE GOTHIAN-SVECONORWEGIAN POLYOROGENIC AREA

The Gothian-Sveconorwegian polyorogenic area of southern Sweden and Norway is one of those typical Precambrian areas (Smithson et al., 1971) characterized by vast stretches of gneisses, augengneisses, migmatites and granites intercalated with amphibolites, quartzites and marbles, and subordinate amounts of recognizable low- to medium-grade sedimentary and volcanic supracrustal rocks. Two important structural elements can be distinguished:
(1) the north trending Trans-Scandinavian Småland-Värmland Granitic Belt of more than 1000 km in length, partially covered by the overthrust nappes of the Caledonian mountain chain (e.g., Gorbatschev, 1980) and
(2) the Southwestern Gneiss Region west of this belt.
Both structural elements have been initiated 1.7 Ga to 1.6 Ga ago during the Gothian orogeny, after the termination of the Svecokarelian orogeny. They fringe the southwestern margin of the Svecokarelian area.

2.1. The Trans-Scandinavian Småland-Värmland Granitic Belt

The Småland-Värmland Belt in Sweden and Norway consists mainly of acid plutonic and volcanic rocks and strikes oblique to the older Svecokarelian structures (Gorbatschev, 1980, Nyström, 1982). The eastern part of the Småland-Värmland Belt shows Svecokarelian enclaves and the volcanic rocks unconformably overlie Svecokarelian rocks (e.g. Lundqvist, 1968), while the western part is delimited by a major tectonic zone. The ages of the intrusions and volcanics vary from about 1.70 Ga to 1.63 Ga (e.g. Welin et al., 1966, 1977, Priem et al., 1970, Oen, 1982). On Norwegian territory, in the Trysil area, the part of the Småland-Värmland Belt not covered by Caledonian nappes is only some 70 km long, but its presence underneath the Caledonian mountains is betrayed by several tectonic windows. The influence of the Sveconorwegian and the Caledonian metamorphism in the Trysil area and the adjoining Dala area in Sweden is manifested by Sveconorwegian and Caledonian mineral ages (Priem et al., 1968, 1970). The effects of the Sveconorwegian orogeny in the Småland-Värmland Belt are, however, much less intense than in the Southwestern Gneiss Region. Sveconorwegian influences do not, as is generally assumed (e.g. Welin and Blomqvist, 1966, Gorbatschev, 1980), "stop" at the so-called Sveconorwegian or Grenville Front, a major tectonic zone between the Southwestern Gneiss Region and the Småland-Värmland Granitic Belt, but actually pass through it. Sveconorwegian mineral age-resetting is not only wide-spread in the Småland-Värmland Belt, but has also been observed in the adjoining part of the Svecokarelian area east of the belt (Verschure, 1981).

2.2. The Southwestern Gneiss Region in Sweden

2.2.1. The Gothian Orogenic Period

During the last decades geochronologists have studied the Swedish part of the Southwestern Gneiss Region more systematically than the Norwegian part (e.g. Gorbatschev and Welin, 1975, 1980, Welin and Gorbatschev, 1976, 1978, Skiold, 1976, Åberg, 1978, Welin et al., 1982). In the Swedish part several high-grade metamorphic supracrustal formations are distinguished, for example the Åmål and the Stora Le - Marstrand formations that have been deposited before about 1.68 Ga ago on an unknown substratum (Gorbatschev, 1979). These formations were metamorphosed and later invaded by granites of the Åmål-I type about 1.68 Ga ago, and again about 1.60 Ga ago by granites of the Åmål-II type (Welin and Gorbatschev, 1980, Welin et al., 1982). The emplacement of the 1.60 Ga granites was accompanied by high-grade metamorphism and migmatitization. Folding connected with the metamorphism follows predominantly northerly trends. The climax of tectonic, metamorphic and magmatic activity was reached about 1.60 Ga ago (Gorbatschev, 1980). After the termination of this comprehensive orogenic phase in southern Scandinavia, the whole region underwent an essentially anorogenic, ensialic evolution until the onset of the Sveconorwegian orogeny (Gorbatschev, 1980).

2.2.2. The Late-Gothian Anorogenic Interlude

During the interlude between the main phase of the Gothian and the Sveconorwegian orogeny several phases of basic and acid igneous magmatism have been identified in both structural elements, along with compressive as well as tensional tectonic activity. Gorbatschev (1980) interprets these features as "peripheral or extra-orogenic manifestations of distant orogenic events". About 1.55 Ga ago the Småland-Värmland Belt and its adjoining areas became invaded by hyperite dolerites of the Breven-Hällefors type (Patchett, 1978, Welin et al., 1980). About 1.54 Ga ago the Lane-type granites intruded the Southwestern Gneiss Region (Welin et al., 1982) and about 1.45 Ga ago the northern part of the Swedish Southwestern Gneiss Region underwent amphibolite-grade metamorphism connected with granite emplacement, migmatitization, folding and mylonitization. At the same time the more southern part, the Varberg Region south of Göteborg, underwent granulite-grade metamorphism associated with the intrusion of charnockites (Welin and Gorbatschev, 1978b, Park et al., 1979). In the southern Småland-Värmland Belt the emplacement of the Götemar granite took place about 1.33 Ga ago (Åberg, 1978). The emplacement of the Hästefjord granites in the Swedish Southwestern Gneiss Region about 1.22 Ga ago (Welin and Gorbatschev, 1976) was accompanied by metamorphism and extensive alkali-basaltic magmatism (Patchett et al., 1978, Gorbatschev, 1980). The latter magmatism is not confined to Scandinavia, but is found on both sides of the North Atlantic. According to Patchett (1978) this major magmatic phase is indicative of widespread fracturing of continents in the North Atlantic Region. This alkaline magmatism that may be designated as the "North Atlantic Region-Type" could be interpreted to herald the onset of a new period of orogenesis, the Sveconorwegian or Grenville orogeny. Poorter (1981) inferred from palaeomagnetic data for the period between about 1.20 and 1.00 Ga a drift of the Baltic Shield relative to the North American Shield of about 30 degrees lattitude and longitude combined with a clockwise rotation of about 90^0. According to Poorter this considerable tectonic reorganisation is obviously related to the Sveconorwegian-Grenvillian orogeny.

2.2.3. The Sveconorwegian Orogenic Period

The Sveconorwegian period in Scandinavia comprises a semicontinuous, polyphase tectonic, igneous and metamorphic evolution between about 1.20 Ga and 0.85 Ga ago. In Sweden, the onset of the Sveconorwegian orogeny is characterised by the deposition of the Dal Group and the Jotnian Supracrustals, mainly low-grade sedimentary-volcanic sequences. The Dal Group Supracrustals must have been deposited on the eroded Hästefjord granites between about 1.20 Ga and 1.10 Ga ago (Skiold, 1976). In the same time-span (Welin, 1966) the Jotnian Supracrustals of the Swedish Dala region accumulated on the eroded surface of the Småland-Värmland Granitic Belt (Mulder, 1971, Patchett, 1978). Both in Sweden and in Norway comprehensive thrusting and faulting is testified by the development of important new shear zones, the reactivation of older Gothian shear zones, and the formation of extensive nappe translations (e.g., Zeck and Malling, 1976, Berthelsen, 1980). Large-scale metamorphism, migmatitization and anatexis took place about 1.20 Ga to 0.90 Ga ago, mainly in the southwestern part of Sweden and southern Norway. Geochronologically this is witnessed by the opening of the K-Ar mineral systems, the partial or complete opening of the Rb-Sr systems of rocks and minerals, and the episodic loss of radiogenic lead from zircons (Welin et al., 1982). These Sveconorwegian activities were concluded by the intrusion of large masses of granite and pegmatite of the "Bohus-type", about 0.90 Ga ago. Post-orogenic fracturing, uplift and cooling of the Crust are reflected by the intrusion of the "Sveconorwegian Tectonic Zone-Type" dolerites between about 1.10 Ga and 0.90 Ga ago, parallel to the Sveconorwegian tectonic zones in central and southern Sweden (Patchett, 1978), and by mineral ages between about 0.90 Ga and 0.85 Ga. This marks the close of the Sveconorwegian orogeny.

3. THE MAJOR TECTONIC ZONES IN SOUTH SCANDINAVIA

The occurrence of major, north trending, linear tectonic zones is a characteristic feature of the Southwestern Gneiss Region. From east to west the following tectonic zones are distinguished (Fig. 2): the Sveconorwegian Front (also designated as Gothian Front, Grenville Front, Protogine Zone or Småland Suture), the Mylonite Zone, the Dalsland Boundary Thrust, the Kristiansand-Bang Shear Zone (also known as Great Breccia or Kristiansand-Porsgrun Shear Zone), and the Mandal-Ustaoset Line (e.g., Bugge, 1928, Magnusson, 1962, Lindh, 1974, Berthelsen, 1980, Hageskov, 1980, Sigmond, 1984). It is believed that these tectonic zones have been active repeatedly, from Gothian time till after the formation of the Permian Oslo Graben. The role of these major structures is not yet fully documented, but they serve to divide the Gothian-Sveconorwegian area into convenient geographical "sectors" or crustal blocks (Fig. 2). Differential vertical and tilting movements are responsible for raising different crustal levels to the present-day surface. For example granulite-facies rocks from the deeper Crust in the Rogaland-Vest Agder (e.g., Michot, 1960, Michot and Michot, 1969, Hermans et al., 1975), and the Kongsberg and Bamble Sectors (e.g., Bugge, 1936, Touret, 1968, Morton et al., 1970, Starmer, 1972, Jacobsen and Heier, 1978), and greenschist-facies rocks from the higher Crust in the Telemark Sector, especially in its median part (e.g., Barth and Dons, 1960, Dons, 1960, 1972). In other sectors mainly amphibolite-grade rocks from intermediate crustal levels are exposed.

4. THE SOUTHWESTERN GNEISS REGION OF NORWAY

4.1. The Gothian and Sveconorwegian Orogenic Periods

The unravelling of the Gothian-Sveconorwegian geological evolution in Norway is particularly cumbersome because of the influence of the Caledonian orogeny and the Permian Oslo Graben epeirogeny, which both affected the Swedish area to a much lesser degree. Moreover, Zeck and Wallin (1980) have noted that throughout the Gothian-Sveconorwegian polyorogenic area of southern Scandinavia the Sveconorwegian metamorphic activity appears to increase to the west. According to these authors, the internal high-grade part of the Sveconorwegian orogenic belt should

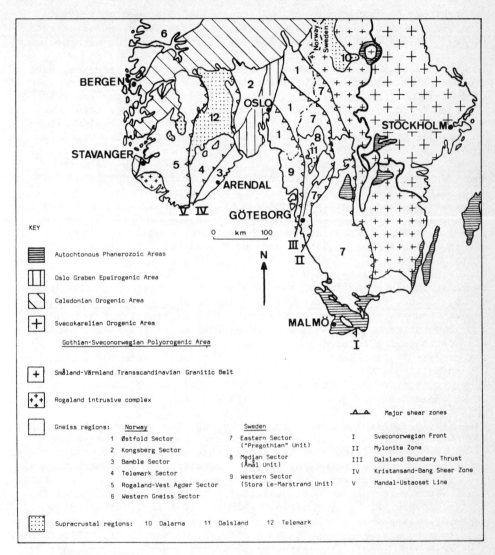

Fig. 2. Geological framework southern Scandinavia

have been situated in south Norway. Like in Sweden, the Norwegian Southwestern Gneiss Region is mainly composed of a monotonous sequence of augengneisses, migmatites and granitic gneisses interlayered with thin amphibolitic lenses and layers, quartzites and marbles. Low- to medium-grade metasedimentary and metavolcanic supracrustal formations are also present, in the Telemark and in the Rogaland-Vest Agder Sector. In the Telemark Sector these metasediments and metavolcanics form the so-called "Telemark-Suite" that gives the impression to be "swimming in a vast sea of granites and granitic gneisses" (Dons, 1960). Such granites and granitic gneisses occur all over southern Norway. They are often rather loosely designated as "Telemark Gneisses" or "Telemark Granite-Gneisses".

The type section of the Telemark Suite lies in central Telemark, east of the Mandal-Ustaoset Line. The Suite is divided into three major superimposed lithostratigraphic groups (Dons, 1960, 1972): from top to bottom, the Bandak Group, the Seljord Group and the Rjukan Group. Each Group has a thickness exceeding 2000 m. Both on a regional and a local scale there are angular unconformities between the groups, presumably reflecting folding and erosion. The first and the last group are very similar, rhyolitic and basaltic lavas and tuffs interbedded with quartz-rich sediments and polymictic conglomerates. The Seljord Group is composed of pure white and locally red quartzites, some bands of shale and locally conglomerates. Sedimentary structures are remarkably well preserved (Singh, 1968, 1969). Numerous thick gabbroic sills follow the bedding planes over long distances. Where the Seljord Group is lacking, the distinction between the Bandak and the Rjukan Group is difficult.

The sequence in the type-area east of the Mandal-Ustaoset Line offers the most complete section of the Telemark Suite. West of the Mandal-Ustaoset Line, in the Rogaland-Vest Agder Sector, only rocks of the Bandak Group have been found. Sigmond (1978) distinguishes here several supracrustal areas. In these western extensions of the Bandak Group the migmatitic gneiss forms a true basement, but in the type area east of the Mandal-Ustaoset Line the boundary relations between the Telemark Supracrustal Suite and the Telemark Gneiss are unclear. At many places they are obscured by intrusions and faults; elsewhere the contacts seem to be concordant and gradational (Avila Martins, 1969).

Ploquin (1972, 1980) was able to trace geochemically the Rjukan Group acid volcanics from the type area southwards into the Telemark Gneiss area. It has been speculated by Ploquin that the Telemark Supracrustals, notwithstanding the increasing grade of metamorphism and anatexis, can also be traced southwards into the high-grade and very high-grade rocks of the Bamble Sector and southwestwards into the Rogaland-Vest Agder Sector. Anyhow, it has become increasingly clear that supracrustal sedimentary and volcanic rocks form an important, if not the most important crustal component in Norway (e.g., Morton et al., 1970, Touret, 1969, Ploquin, 1980). It remains to be proved that long-range correlations between widely separated occurrences of supracrustal rocks, showing different grades of metamorphism, are feasible.

On the geological sketch map of southern Norway (Fig. 3) the principal Rb-Sr whole-rock isochron results from various sources (Table 1) are indicated. Although the number of age determinations is still rather small, no evidence of a Svecokarelian Crust is apparent in southern Norway, a similar situation as in Sweden. The importance of the Gothian orogenic period is testified by a growing number of dispersed Gothian age determinations (e.g., O'Nions and Baadsgaard, 1971, O'Nions and Heier, 1972, Versteeve, 1975, Jacobsen and Heier, 1978, Wielens et al.,

Fig. 3 Geological Sketchmap southern Norway (after A. Ploquin et al. 1972) with principal whole-rock Rb-Sr isochron ages.

1: Oslo Graben (Palaeozoic) - 2: Caledonides (Palaeozoic) - 3: Supracrustal Regions - 4: Granites - 5: Gneiss Regions (gneiss, augengneiss, migmatite); H: Hyperites - 6: Charnockitic Regions; F: "Farsundites; A: Anorthosites; M: Mangerites; N: Norites; a: Arendalites - 7: Muscovite-out isograd (2) and Orthopyroxene-in isograd (1) - 8: Main shear-zones - 9: Mylonite-zones - 10: Carbonatites and explosion breccias.

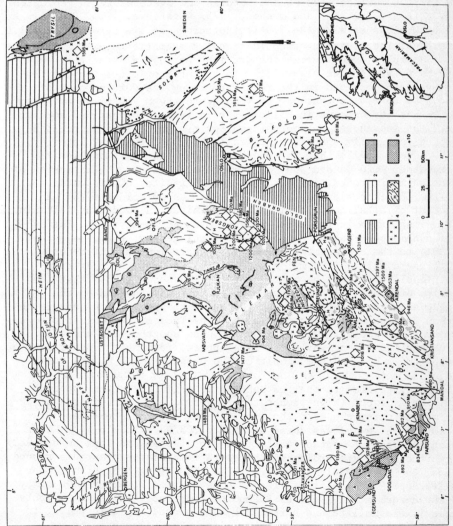

1981, Priem and Verschure, 1982). The available data indicate that the crustal evolution in south Norway started in Gothian times about 1.6 Ga ago, possibly later than in Sweden. The highest (Early Gothian) ages, about 1.60 Ga, were observed in the eastern part of southern Norway, while the lowest (Late Gothian) ages, about 1.45 Ga are found in the western part. It also remains a matter of speculation whether this signifies a real trend, indicating a later crustal accretion in the west than in the east, or whether it reflects Sveconorwegian and/or Caledonian reworking.

The southern part of the Rogaland-Vest Agder sector is relatively well-dated. Here, a preliminary geochronological framework could be established for the decipherment of the Gothian and Sveconorwegian geological evolution (e.g., Pasteels and Michot, 1975, Versteeve, 1975, Pasteels et al, 1979, Wielens et al., 1981). All isotopic ages lie between about 1.5 Ga and 0.8 Ga. The oldest ages could signal a magmatic as well as a high-grade metamorphic event during the Gothian period, an event that was labeled M_0 by Wielens et al. (1981). No textural or mineralogical features have been observed in the rocks providing the Gothian ages of about 1.5 Ga to 1.4 Ga that may be related to the M_0 event, apart from the fact that it concerns in all cases relatively fine-grained rocks. However, the subsequent metamorphic phases M_1, M_2, M_3 and M_4 could be unravelled texturally and mineralogically. It has been postulated by, e.g., Versteeve (1975) that the phase of high-grade metamorphism M_1 occurred about 1.20 Ga ago and was immediately followed by intrusion of alkaline magmatic rocks, the pyroxene-syenite suites of Gloppurdi and Botnavatn. The intrusion of these alkaline rocks coincides with the emplacement of alkaline Hästefjord-type granites in Sweden and alkaline doleritic magmatism in the whole of the North Atlantic realm. The subsequent metamorphic phases M_2 and M_3 were related by, e.g., Maijer et al. (1981) to the emplacement of the two successive very high-temperature magma pulses (Rietmeijer, 1979) that formed the Bjerkreim-Sokndal lopolith. It has been speculated by Duchesne (1984) that the generation of the lopolithic magma in its turn was induced by the prolonged emplacement of the huge Egersund-Ogna anorthosite body in the time-span from about 1.50 Ga to 0.95 Ga ago. The earlier magmatic pulse, making up the lower part of the lopolith, was anorthositic and (leuco) noritic in composition, while the later pulse, forming the upper part of the lopolith, was (quartz)monzonitic. The phases M_2 and M_3 are interpreted by Maijer et al. (1981) as "regional contact metamorphism" around the lopolith, related to the two successive magmatic pulses, respectively. The M_2 event (high-temperature/intermediate pressure granulite-facies metamorphism close to the lopolith and amphibolite-faces metamorphism farther away) was dated at about 1.05 Ga by whole-rock Rb-Sr isochron analyses of contact-anatectic veins (Priem and Verschure, 1982) and by zircon U-Pb discordias in the contact-metamorphic aureole around the lopolith (e.g., Wielens et al., 1981). The M_3 metamorphism (lower granulite/upper amphibolite facies) has been dated at about 0.90 Ga by Rb-Sr whole-rock studies and Rb-Sr and/or K-Ar analyses of metamorphic minerals like osumilite and hornblende. The second magmatic pulse, to which M_3 is related, was dated by whole-rock Rb-Sr isochron (e.g.,

Wielens et al., 1981) and zircon U-Pb analyses of the upper part of the lopolith (e.g., Pasteels et al., 1979); they also yield ages of about 0.90 Ga.

Although the M_2 and M_3 phases of metamorphism in the Rogaland-Vest Agder Sector coincide with the Sveconorwegian period of metamorphism in the whole of southern Sweden and Norway, they seem to be unique for this part of southern Scandinavia because of their intimate relation with the emplacement of the Egersund-Ogna and Bjerkreim-Sokndal magmatic complexes.

On the whole the role of the very-high-grade granulite-facies metamorphism is much more prominent in Norway than in Sweden. Apart from the Rogaland-Vest Agder Sector, granulites are found at several places, for example in the Kongsberg and Bamble Sectors of southeastern Norway e.g., Bugge, 1943, Touret, 1969, 1971, Starmer, 1972). This can be attributed either to regional metamorphic effects in relation with high-temperature mafic intrusions (Touret, 1969), or to the denudation of deep crustal rocks as a consequence of differential uplift or tilting of particular crustal segments.

The Sveconorwegian evolution in Norway ended, just as in Sweden, by the intrusion of impressive volumes of granitic and pegmatitic magma, mainly in the form of late- and postorogenic granite plutons and pegmatitic bodies, about 0.90 Ga ago. These Bohus-type granites and pegmatites are often regarded as resulting from anatectic melting of the Telemark Gneiss (e.g., Dons, 1960, 1972, Barth and Reitan, 1963, Sylvester, 1964, Smithson, 1965, Menuge, 1982). Mineral ages ranging from 0.90 Ga to 0.80 Ga indicate, like in Sweden, subsequent uplift and cooling of the Gothian-Sveconorwegian crustal segment.

In southern Norway upper-intercept ages defined by U-Pb zircon discordia plots are predominantly in the range of 1.10 Ga to 0.96 Ga (Swainbank, 1965, O'Nions and Baadsgaard, 1971, Pasteels and Michot, 1975, Wielens et al., 1981), whereas in southern Sweden higher upper-intercept ages are found, ranging between 1.68 Ga to 1.54 Ga (Åberg, 1978, Welin et al., 1982). This difference might be ascribed to the increase in western direction of the metamorphic grade during the Sveconorwegian period, leading to a complete resetting of U-Pb zircon systems. Many lower intercepts, however, both in southern Norway and in southern Sweden, display ages between about 0.45 Ga and 0.30 Ga. Wielens et al. (1981) tentatively interpreted the latter ages in terms of episodic loss of radiogenic lead due to a moderate increase in temperature during the Caledonian orogeny, up into the range of the greenschist- and pumpellyite-facies M_4 (Verschure et al., 1980). Verschure (1981) stressed that this low-grade metamorphic event is not limited to southern Norway and Sweden, but can also be found elsewhere in Scandinavia. Sauter et al. (1983) explained the retrogradation in terms of burial metamorphism and the depression of the crust after the emplacement of the Caledonian mountain chain. Andriessen and Bos (1984) and Andriessen et al. (1985) relate apatite fission-track ages of about 0.25 Ga to 0.12 Ga in southwestern

Table 1. Principal whole-rock Rb-Sr isochron ages and initial $^{87}Sr/^{86}Sr$ ratios of the Late Proterozoic Crust in southern Norway

Rock	Sector	Age and initial ratio	Reference
Grey orthogneiss	Østfold Sector	1614 ± 78 Ma 0.7028 ± 0.0006	Skjernaa & Pedersen, 1982
Porphyry and granite	Småland-Värmland Belt	1595 ± 63 Ma 0.7089 ± 0.0052	Priem et al., 1970
Enderbitic granulite	Kongsberg Sector	1.58 ± 0.05 Ga 0.70236 ± 0.00006	Jacobsen & Heier, 1978
Granitic gneiss	Kongsberg Sector	1.56 ± 0.04 Ga 0.7014 ± 0.0018	Jacobsen & Heier, 1978
Bamble charnockitic gneiss	Bamble Sector	1535 ± 123 Ma 0.70349 ± 0.00928	Field & Råheim, 1979
Northern Levang dome gneiss	Bamble Sector	1531 ± 38 Ma 0.6998 ± 0.0020	O'Nions & Baadsgaard, 1971
Dioritic gneiss	Kongsberg Sector	1.52 ± 0.05 Ga 0.70249 ± 0.00006	Jacobsen & Heier, 1978
Bamble metasediment	Bamble Sector	1509 ± 96 Ma 0.70529 ± 0.00334	Field & Råheim, 1981
Ormakam-Moldhesten granite from Caledonian nappe	Rogaland-Vest Agder Sector	1501 ± 125 Ma 0.7015	Andresen & Heier, 1975
Røldal granodioritic gneiss in Caledonian nappe window	Rogaland-Vest Agder Sector	1488 ± 30 Ma 0.703 ± 0.006	Berg, 1977
Langeidvatn augengneiss	Rogaland-Vest Agder Sector	1477 ± 35 Ma	Sigmond, 1978
Drangsdal charnockitic migmatite	Rogaland-Vest Agder Sector	1453 ± 60 Ma 0.7138 ± 0.0018	Versteeve, 1975
Soyland granitic gneiss	Rogaland-Vest Agder Sector	1420 ± 35 Ma 0.7057	Pasteels & Michot, 1975
Tvedestrand granite	Bamble Sector	1397 ± 57 Ma 0.72527 ± 0.00924	Field & Råheim, 1981
Uvdal granitic gneiss	Telemark Sector	1376 ± 38 Ma 0.7052 ± 0.0031	Priem et al., 1972
Vatnaas granite	Kongsberg Sector	1.37 ± 0.02 Ga 0.70620 ± 0.00048	Jacobsen & Heier, 1978
Red orthogneiss	Østfold Sector	1277 ± 20 Ma 0.7070 ± 0.0008	Skjernaa & Pedersen, 1982
Gjerstad augengeiss	Telemark Sector	∼ 1250 Ma	Smalley et al., 1983
Helgevann granite	Kongsberg Sector	1.20 ± 0.02 Ga 0.708 ± 0.002	Jacobsen & Heier, 1978
Vinor gabbro	Kongsberg Sector	1.20 ± 0.05 Ga 0.70240 ± 0.00012	Jacobsen & Heier, 1978
Gloppurdi Pyroxene-syenite	Rogaland-Vest Agder Sector	1180 ± 70 Ma 0.7103 ± 0.0032	Versteeve, 1975
Kviteseidvatn Telemark Gneiss	Telemark Sector	1105 ± 23 Ma 0.7086 ± 0.0002	Priem et al., 1973
Meheia granite	Kongsberg Sector	1.07 ± 0.01 Ga 0.715 ± 0.002	Jacobsen & Heier, 1978
Bamble granite	Bamble Sector	1053 ± 20 Ma 0.71572 ± 0.00082	Field & Råheim, 1979
Høvringsvatn granite-monzonite	Rogaland-Vest Agder Sector	1016 ± 42 Ma 0.748 ± 0.003	Pedersen, 1973
Rollag granite	Telemark Sector	1003 ± 30 Ma 0.7220	Killeen & Heier, 1975
Holum granite	Rogaland-Vest Agder Sector	980 ± 34 Ma 0.7045 ± 0.003	Wilson et al., 1977
Flå granite	Telemark Sector	994 ± 28 Ma 0.7093	Killeen & Heier, 1975
Grimstad granite	Bamble Sector	946 ± 66 Ma 0.7034	Killeen & Heier, 1975
Bjerkreim-Sokndal quartz monzonite	Rogaland-Vest Agder Sector	928 ± 50 Ma 0.7075 ± 0.0028	Wielens et al., 1981
Seterdalen granite	Østfold Sector	925 ± 56 Ma 0.7105 ± 0.0026	Skjernaa & Pedersen, 1982
Lyngdal hornblende granite	Rogaland-Vest Agder Sector	912 ± 38 Ma 0.7054 ± 0.0038	Pedersen & Falkum, 1975
Kleivan granite	Rogaland-Vest Agder Sector	910 ± 7 Ma 0.7053 ± 0.0002	Petersen, 1977
Herefoss granite	Bamble-Telemark Sector	909 ± 26 Ma 0.7051	Killeen & Heier, 1975
Bessefjell granite	Telemark Sector	904 ± 16 Ma 0.7066	Killeen & Heier, 1975
Hidra charnockitic dikes and pegmatites	Rogaland-Vest Agder Sector	892 ± 25 Ma 0.7085 ± 0.0007	Pasteels et al., 1979
Bohus granite	Østfold Sector	890 ± 35 Ma 0.711 ± 0.003	Skiold, 1976
Vrådal granite	Telemark Sector	888 ± 46 Ma 0.7064 ± 0.0055	Priem et al., 1973
Iddefjord-Bohus granite	Østfold Sector	881 ± 35 Ma 0.9098	Killeen & Heier, 1975
Farsund charnockite	Rogaland-Vest Agder Sector	834 ± 41 Ma 0.7128 ± 0.0009	Pedersen & Falkum, 1975

GEOCHRONOLOGY OF THE BALTIC SHIELD IN SOUTH SCANDINAVIAN

Table 2. Tentative comparative time-scale of the Gothian-Sveconorwegian polyorogenic area

	Ga	SWEDEN	NORWAY
	< ~0.85 Ga	Denudation	Denudation
↑ SVECONORWEGIAN OROGENIC PERIOD	~0.85 Ga	Final uplift, fracturing and cooling Crust	Final uplift, fracturing and cooling Crust
	~0.90 Ga	Intrusion postorogenic granites and pegmatites of Bohus-type	Intrusion postorogenic granites and pegmatites of Bohus-type
	~1.00 Ga	Start uplift and fracturing Crust, dolerite intrusion (Sveconorwegian Tectonic Zone-Type)	Fracturing Crust, dolerite intrusion (Sveconorwegian Tectonic Zone-Type) (Rogaland-Vest Agder Sector, Østfold Sector)
	~1.10 - 0.90 Ga	Regional metamorphism, (shear)faulting, folding W-directed thrusting and nappe formation	Regional metamorphism, (shear)faulting, folding Emplacement Bjerkreim-Sokndal lopolith (Rogaland-Vest Agder Sector) Resetting "Telemark Gneiss"
	~1.20 - 1.10 Ga	Deposition Dal and Jotnian sediments and volcanics	Inferred deposition Seljord and Bandak Group of Telemark Supracrustals (Telemark and Rogaland-Vest Agder Sectors)
	~1.22 Ga	Denudation	Denudation
↓	~1.22 Ga	Development of tensional stress regime, rifting Intrusion alkaline dolerites (North Atlantic Region-Type) Intrusion alkaline granites (Hästefjord-type)	~1.18 Ga Intrusion of Gloppurdi pyroxene syenite (Rogaland-Vest Agder Sector) ~1.20 Ga Hyperite intrusion (Bamble Sector) Vinor gabbro intrusion (Kongsberg Sector) ~1.20 Ga Helgevann granite intrusion (Kongsberg Sector)
← GOTHIAN ANOROGENIC ENSIALIC INTERLUDE →	~1.35 Ga	Intrusion anorogenic Götemar granite in Småland-Värmland Belt	~1.28 Ga Intrusion protolith Red orthogneiss (Østfold Sector)
↑ GOTHIAN OROGENIC PERIOD	~1.50 - 1.30 Ga	Metamorphism, migmatitisation, (shear)faulting, magmatism, folding North Southwestern Gneiss Region: granite intrusion amphibolite-grade metamorphism South Southwestern Gneiss Region: charnockite intrusion (1.45 Ga Varberg charnockite) granulite-grade metamorphism	Metamorphism, migmatitisation, (shear)faulting, magmatism, folding ~1.40 Ga Tvedestrand granite intrusion (Bamble Sector) ~1.42 Ga Søyland granitic gneiss protolith intrusion (Rogaland-Vest Agder Sector) ~1.45 Ga Forming Drangsdalen charnockitic migmatite (Rogaland-Vest Agder Sector) ~1.48 Ga Forming Langeidvatn augengneiss (Rogaland-Vest Agder Sector) ~1.49 Ga Forming Røldal granodioritic gneiss (Rogaland-Vest Agder Sector)
	~1.54 Ga	Granite intrusion (Lane-type)	~1.52 Ga Forming dioritic gneiss (Kongsberg Sector) ~1.53 Ga Forming northern Levang gneiss dome (Bamble Sector) ~1.54 Ga Forming charnockitic gneiss (Bamble Sector)
	~1.55 Ga	Intrusion hyperite dolerites (Breven-Hällefors-type)	Basic intrusions (Rogaland-Vest Agder Sector)
	~1.60 Ga	Folding, metamorphism, migmatitisation, magmatism Intrusion granites (Åmål II-type)	~1.56 Ga Forming granitic gneiss (Kongsberg Sector) ~1.58 Ga Forming enderbitic granulite (Kongsberg Sector)
	~1.68 Ga	Granite intrusion (Åmål I-type) in Southwestern Gneiss Region	
↓	~1.70 - 1.63 Ga	Granite intrusion (e.g. Dala Granites) and deposition acid volcanics in Småland-Värmland Belt	~1.63 Ga Inferred deposition Rjukan Group acid volcanics of Telemark Supracrustals in Southwestern Gneiss Region (Telemark Sector) Deposition Trysil acid volcanics and intrusion of Trysil granites in Småland-Värmland Belt (Trysil Area)
	> ~1.68 Ga	Deposition of Åmål-Stora Le-Marstrand supracrustals in Southwestern Gneiss Region	

Norway to the uplift of the crust as the result of denudational unloading after the termination of crustal downwarping. The geological relationships are very similar throughout the whole Precambrian basement of Norway and Sweden. It is thus not necessary to invoke fundamental geological differences between southern Sweden and Norway, nor to postulate the engagement of far-travelled continental masses in large-scale plate collisions. Geotectonic modelling to interpret the Precambrian geologic evolution of southern Scandinavia in terms of Phanerozoic plate tectonics, either invoking Himalayan continent-continent, or Andean ocean-continent collisions, has been endeavoured by many authors (e.g., Zeck and Malling, 1976, Torske, 1977, Krogh, 1977, Berthelsen, 1980, Falkum and Petersen, 1980), but such hypotheses are not (yet) supported by reliable geological and geochronological evidence.

In Table 2 a correlation is presented between Swedish and Norwegian time-scales for the Gothian-Sveconorwegian geological evolution in south Scandinavia, based upon time-scales published by e.g. Berg (1977) and Welin et al. (1982).

6. GEOCHRONOLOGICAL CONTROVERSIES IN SOUTHERN NORWAY

There are at present two principal controversies regarding the geochronology of southern Norway. They concern (1) the age relationship between the Telemark Supracrustal Suite and the Telemark Gneisses, and (2) the timing of the metamorphic events.

Regarding the first controversy there are two main views: (1) the Telemark Gneiss represents (poly-)reworked and mobilised products of Telemark Supracrustal material (e.g., Barth and Reitan, 1963, Mitchell, 1967, Cramez, 1969, Avilla Martins, 1969, Venugopal, 1970, Stout, 1972, Priem et al., 1973), and (2) the Telemark Gneiss represents the basement upon which the Telemark Supracrustals were deposited, after which both the gneisses and the Supracrustals underwent the same reworking and remobilization, although the latter locally to a much lesser degree (e.g., Saether, 1957, Dons, 1960, 1972, Menuge, 1982).

In a Nicolaysen diagram the Rb-Sr data-points of the Rjukan and Bandak acid volcanics scatter between two lines corresponding to ages of 1.63 Ga and 1.10 Ga, respectively (Priem et al., 1973). These data were interpreted by Priem et al. (1973) to signal (1) a deposition of the Telemark Supracrustals contemporaneously with the about 1.63 Ga old Trysil and Dala porphyries and granites of the Trans-Scandinavian Småland-Värmland Granitic Belt (e.g., Welin et al., 1966, Welin and Lundqvist, 1970, Priem et al., 1970), and (2) a high-grade Sveconorwegian metamorphism about 1.10 Ga ago, causing a resetting to varying degrees of the Rb-Sr whole-rock systems in the Supracrustals and a complete resetting of the Telemark Gneiss protoliths.

Other authors (Berg, 1977, Jacobsen and Heier, 1978, Ploquin, 1980) favour an intermediate hypothesis. According to them only the Rjukan

Group can be regarded as the protolith from which locally a gneissose precursor of the Sveconorwegian Telemark Gneiss was formed during a Gothian, high-grade metamorphic event about 1.50 Ga ago. Gothian ages of about 1.50 Ga for "Telemark-type" gneisses have indeed been reported from several places in southwestern Norway, for example in the Rogaland-Vest Agder Sector (e.g., Berg, 1977, Sigmond, 1978). Such 1.50 Ga old gneisses may have escaped Sveconorwegian reworking. Following this hypothesis the 1.63 Ga old supracrustal protoliths elsewhere were reworked during a Gothian metamorphic event around 1.50 Ga ago, but may then have been reworked for a second time during a Sveconorwegian metamorphic event about 1.10 Ga ago. Another possibility is that locally the 1.63 Ga old supracrustal protoliths escaped Gothian reworking, and underwent only a Sveconorwegian reworking. The local occurrence of supracrustal formations that still show recognizable sedimentary and volcanic textures (Singh, 1968, 1969) may then be explained by the fact that they have not been affected by high-grade Gothian reworking, but only by low- to medium-grade Sveconorwegian metamorphism. T_{Ur}^{Sr} model ages of the Rjukan acid volcanics calculated by Jacobsen and Heier (1978) from the Rb-Sr data published by Priem et al. (1973) fall in the range of 1.00 Ga to 1.60 Ga, supporting the assumption of a Gothian age for the Telemark Supracrustal Suite.

In 1982 Menuge published the results of a Sm-Nd investigation of Telemark Supracrustals. A Sm-Nd Nicolaysen plot of acid and basic Rjukan volcanics produce an isochron age of about 1.20 Ga, if a number of datapoints are omitted. However, an isochron of about 1.60 Ga can be calculated from Menuge's data for the basic lavas of the Bandak Supracrustal Group, which is situated higher in the Supracrustal stratigraphy. It remains therefore still a matter of speculation whether the Supracrustals, or part of them should be regarded as Gothian or Sveconorwegian. There may be an age-difference of some 0.50 Ga between the Rjukan Group (1.63 Ga) and the Bandak Group (1.10 Ga), as was presumed by several authors (e.g., Berg, 1977, Jacobsen and Heier, 1978, Ploquin, 1980) but all Supracrustals of the type area may also be of roughly the same age, either about 1.63 Ga as was concluded by Priem et al. (1973), or about 1.20 Ga as was postulated by Kleppe and Råheim (1979) and Menuge (1982). If a hiatus in the order of 0.5 Ga between the Rjukan and Bandak Group is inferred, the Rjukan Group volcanics could be correlated with the rocks of the Trans-Scandinavian Småland-Värmland Belt. The Seljord and Bandak Groups could then be correlated with the Dal sediments in the Southwestern Gneiss Region, and possibly also with the Jotnian supracrustals overlying the rocks of the Småland-Värmland Belt in central Sweden.

As to the second geochronological controversy, the unravelling of the polymetamorphic events, there are two opposing hypotheses: (1) the Sveconorwegian metamorphism in the Southwestern Gneiss Region included a high-grade to very-high-grade phase that overprinted the older Gothian metamorphism, and (2) the pre-Sveconorwegian, Gothian metamorphism was the most important and intense metamorphic phase, whereas the Sveconorwegian metamorphic imprint was rather weak and insignificant.

The first hypothesis was generally accepted until 1979. Then the proponents of the second hypothesis, Field and Råheim (1979, 1980, 1981) and Smalley et al. (1983) suggested on the base of geochronological work that the high-grade metamorphism and gneissification in the whole of southern Norway occurred about 1.50 Ga ago. During the Sveconorwegian episode, only a large-scale, low-grade hydrous resetting should have taken place about 1.10 Ga ago. Nevertheless the low-grade resetting has induced disturbances of the Rb-Sr whole-rock systems. These disturbances locally gave rise to often remarkably good, but meaningless secondary isochrons, even for suites of samples from one single outcrop (Field and Råheim, 1979, 1980, 1981).

Although the geochronological evidence presented by Field and Råheim seems compelling at first sight, there remain serious objections. For example, it can be shown that the Sveconorwegian deformation in the Gjerstad augengneiss body in the southernmost Telemark Sector near the Kristiansand-Bang Shear Zone is certainly younger than the intrusion presumed by Smalley et al. (1983) to have taken place about 1.25 Ga ago. However, Touret (1984) showed that the high-grade metamorphic minerals in unstrained coronas of the augengneiss postdate the deformation, which is incompatible with a low-grade, hydrous alteration during the Sveconorwegian. The large-scale migmatitization, anatexis and gneissification in the Telemark Sector about 1.10 Ga ago (Priem et al., 1973) and the generation of huge masses of late and posttectonic granites and pegmatites of the Bohus-type at about 0.90 Ga ago all over southern Norway (e.g., Priem et al., 1973, Killeen and Heier, 1975) can also not be reconciled with a weak, low-grade Sveconorwegian metamorphic event.

One may postulate that the Rb-Sr whole-rock systematics of the older Gothian phase of high-grade metamorphism remained undisturbed locally in south Scandinavia during the very-high-grade or high-grade Sveconorwegian metamorphism, although the mineral phases and in several cases also the whole-rock systems have been reset. This could be true especially in the case of the metamorphism under essentially anhydrous conditions of high-grade, "dry" rocks. Very-high-grade metamorphism is characterized by solid-solid reactions with material transport over only minute distances. Possibly, Field and Råheim (1979, 1980, 1981) measured in the whole-rock Rb-Sr systems of the Bamble Sector a Gothian, 1.5 Ga old phase of the high-grade and very-high-grade metamorphism that is petrographically displayed by the rocks and that is geochronologically shown in mineral systems (e.g., O'Nions et al., 1969, O'Nions and Baadsgaard, 1971). A Sveconorwegian disturbance of the Rb-Sr whole-rock systems has taken place only in samples from zones where aquous fluids were introduced in high-grade, "dry" rocks, as evidenced by nearby granitic and pegmatitic intrusions (Field and Råheim, 1979, 1980, 1981).

7. TRUE VERSUS FALSE Rb-Sr ISOCHRONS

The persistence of older whole-rock Rb-Sr ages during a later phase of metamorphism is not uncommon. Moorbath stated already in 1975 that "good

primary Rb-Sr whole-rock isochrons are commonly observed in gneiss
terrains which underwent much later intense metamorphism". Krogh and
Davis (1973) demonstrated unequivocally on the basis of detailed Rb-Sr
studies on 1.85 Ga old gneisses of the Grenville Province in Ontario
that isotopic and chemical equilibrium was not reached over distances
of even a few centimeters during subsequent Grenville, high-grade,
amphibolite facies metamorphism, some 0.80 Ga later. Wielens et al.
(1981) showed that in the Rogaland Sector rocks, although yielding a
Late Gothian whole-rock Rb-Sr age of 1553 Ma, nothing of the texture and
mineralogy of this metamorphic phase (M_0) has been recognized.

Nevertheless there is a growing body of evidence for resetting of whole-
rock systems, mainly under low-grade, hydrous conditions and apparently
recording the time, or approaching the time, of the metamorphic event
(e.g., Van Schmus and Bickford, 1976, Råheim and Compston, 1977, Black
et al., 1979, Page, 1978, Priem et al., 1978, Field and Råheim, 1979,
1980, Brattli et al., 1983). The process of isotopic resetting or reworking of whole-rock Rb-Sr systems is still not well understood. Whole-
rock systems apparently can become isotopically equilibrated over large
distances during metamorphism and produce reasonably good alignments of
data-points in a Nicolaysen diagram. These alignments may give meaningless ages, but they also may record the time of metamorphism. It is
difficult to understand the mechanism that leads to isotopic disturbances of the whole-rock systems in such a "programmed" way that they
result in often fairly good alignments of data-points and not in a wild
scatter of data-points. A possible solution for this paradox is proposed
by Roddick and Compston (1976) who envisage two situations: (1) preservation of a pre-metamorphic age if regional variations in $^{87}Sr/^{86}Sr$
exist, and (2) recording of the metamorphic age if the $^{87}Sr/^{86}Sr$ variations exist only on a small scale, less than the supposed scale of
strontium isotopic homogenization. The homogenization is thus in fact
only local, but gives the impression of having taken place on a regional
scale. Such an explanation could hold for the apparently Sveconorwegian
whole-rock Rb-Sr resetting of the Gothian Telemark Gneiss protoliths.

An explanation for false, meaningless isochrons may also be found in the
sampling procedure customary in geology, i.e. the collection of suites
of samples from small domains. According to Kleeman (1967) statistical
rules have to be observed rigorously when collecting samples for geochemical study, especially in the case of trace element geochemistry.
The amounts of material that are sampled have to be in accordance with
the particle size of the material. For instance, to obtain a reliable
representative sample with regard to the main elements Si, Al, Ca, K and
Na from a granite with a maximum particle size of 1.0 mm, a sample of
2.5 kg will suffice. For a granite with a maximum particle size of 1.5 cm,
however, this amounts already to about 10.000 kg . For trace elements
residing in accessory minerals even larger quantities have to be sampled.
This evidently puts severe constraints on whole-rock geochemical studies
of medium- and coarse-grained rocks. Poor sampling practice may therefore also play a role in geochronology, particularly in the case of
coarse-grained metamorphosed rocks where primary minerals are partially

replaced by new mineral phases and thereby have gained or lost unequilibrated quantities of mother and/or daughter isotopes. For example, Brooks (1968) has shown on the basis of a detailed study of the weakly metamorphosed Heemskirk Granite in western Tasmania that during the metamorphism the partially recrystallized primary plagioclase and K-feldspar were not equilibrated isotopically with regard to Rb and Sr. It was demonstrated that the isotope relocation in the relict crystals occurs in the alteration products of the plagioclase and the K-feldspar (for example, saussurite and sericite, respectively) and not in the crystals themselves. Relative to their calculated unaltered present-day compositions the most altered plagioclase had gained 15.8 times the radiogenic Sr, 8.8 times the ^{87}Rb and 1.7 times the ^{86}Sr, whereas the sericitized K-feldspar had lost 19, 14 and 34% of these isotopes, respectively. The persistence in relict crystals of the original premetamorphic Rb-Sr and K-Ar systems was also demonstrated by Verschure et al. (1980), who showed that relict Sveconorwegian brown biotites from the basement near the Caledonian overthrust existed side by side with newly formed Caledonian green biotites. Partial replacement of brown biotite by green biotite or the analysis of brown biotite concentrates contaminated with green biotite resulted in partially reset ages.

If a coarse-grained metamorphic granite, for example an augengneiss with large saussuritized plagioclase and/or sericitized K-feldspar crystals, is sampled in the conventional manner (samples of a few kilograms and a selection based on, for example, differences in colour, i.e. the relative amounts of light and dark minerals) this may result in an apparent isochron relationship without geochronological meaning. The "isochron" can be too old with regard to the emplacement age of the original granite if saussuritized plagioclase was the main criterion for sample selection, or too young if the main criterion was sericitised K-feldspar. However, such a "positive" or "negative natural spiking", performed during the selection of the samples in the field, is also "programmed" and can therefore give rise to false isochrons. This probably is the case when more than one isochron can be calculated for coarse-grained samples from a small sampling area (e.g., Field and Råheim, 1979, 1980, 1981, Smalley et al., 1983).

7. CONCLUDING REMARKS

(1) The scarce geochronological data indicate that the south Scandinavian Crust of Sweden and Norway was essentially formed during the early Gothian orogenic period from about 1.70 Ga to 1.55 Ga. This is testified by a growing number of Early Gothian whole-rock Rb-Sr isochrons. Crustal material older than about 1.70 Ga probably will not be found in Norway, as is the case in Sweden.

(2) Thoroughly reworked supracrustal rocks form the backbone of the early Gothian as well as Late Gothian-Early Sveconorwegian Crust. It might be speculated that the relatively low-grade Telemark Supracrustal Groups can be correlated with several highly reworked supra-

crustal formations over large distances in southern Scandinavia. It is still uncertain, however, whether the Telemark Supracrustal Groups were all deposited in Gothian or in Sveconorwegian time, or that they were deposited partly in Gothian, partly in Sveconorwegian time.

(3) The metamorphism and tectonic activity in the Småland-Värmland Belt and in the Southwestern Gneiss Region increases in western direction during the Sveconorwegian orogenic period. The geological and geochronological record has therefore been "disturbed" to a higher degree in Norway than in Sweden. The influence of the Caledonian and Oslo Graben Epeirogeny added also to this process of "obscuring" the older geological record.

(4) It remains as yet uncertain whether there really is a trend in Norway of ages grading from about 1.60 Ga in the northeast to about 1.40 Ga in the southwest. The younger ages around about 1.40 Ga might either represent Late Gothian crustal additions, or reflect partial resetting during Sveconorwegian or Caledonian metamorphic reworking.

(5) During the Sveconorwegian orogenic period crustal formation was mainly restricted to the emplacement of mafic and huge quantities of granitic igneous material.

(6) The high-grade and very-high-grade metamorphism that is intimately related with migmatitization, anatexis and acidic magmatism, may have been induced by heat introduced by mafic magmas such as, for example, hyperite magma.

(7) Meaningless isochrons can be expected in Rb-Sr dating of coarse-grained rocks in polyorogenic areas, when sampling is not performed according to strict statistical rules with regard to grain-size and weight of the samples. The extremely large samples than in such a case have to be taken virtually preclude reliable Rb-Sr whole-rock dating. Hence, the only possibility to date primary whole-rock ages in a polyorogenic area is to investigate fine-grained rock samples that do not show any trace of hydrous alteration.

(8) U-Pb zircon geochronology in poly- and plurimetamorphic terrains will generally be characterized by multi-stage leadloss precluding straightforward geochronological conclusions from discordia patterns.

(9) When high-grade rocks undergo anhydrous, high-grade polymetamorphism, they will essentially keep their primary whole-rock Rb-Sr age record.

(10) The geochronological and geological knowledge in southern Norway and Sweden is still too scarce to warrant sweeping geotectonic modelling in terms of Phanerozoic plate-tectonics, other than as mere speculation.

ACKNOWLEDGEMENTS

The author thanks Prof. Dr. H.N.A. Priem, Prof. Dr. A.C. Tobi and Prof. Dr. J.L.R. Touret for their kind perusal of the text. A special word of appreciation is due to Mrs. M.J.L.H. Petit-Puts for her zest typing the camera-ready manuscript. Financial support received from the Stichting Dr. Schürmannfonds and the Scientific Affairs Division of NATO is also gratefully acknowledged. This work forms part of the research program of the "Stichting voor Isotopen-Geologisch Onderzoek" supported by the Netherlands Organization for the Advancement of Pure Research (Z.W.O.).

REFERENCES

Åberg, G. (1978) 'Precambrian geochronology of south-eastern Sweden'. *Geol. Fören. Stockholm Förh.*, 100, 125-254.

Andresen, A. and Heier, K.S. (1975) 'A Rb-Sr whole-rock isochron date on an igneous rock-body from the Stavanger area, south Norway'. *Geol. Rundschau*, 64, 260-265.

Andriessen, P.A.M. and Bos, A.J. (1984) 'Fission track dating in Norway; An interim report'. Abstract lecture on Eighth European Colloquium on Geochronology, Cosmochronology and Isotope Geology, Braunlage. (also in *Terra Cognita*, 4, p. 191).

Andriessen, P.A.M. and Bos, A.J. (1985) 'Post-Caledonian thermal evolution and crustal uplift in the Eidfjord Area, western Norway'. Submitted to *Isotope Geoscience*.

Avilla Martins, J. (1969) 'The Precambrian rocks of the Telemark Area in the south central Norway VII. The Vrådal Area'. *Norges geol. Unders.*, 258, 267-301.

Backlund, H. (1936) 'Till frågan om granitgrupper, bergkedjeveckningar ock cykelindelning inom Fennoskandia'. *Geol. Fören. Stockholm Förh.*, 58, 346-356.

Barth, T.F.W. and Dons, J.A. (1960) 'Precambrian of southern Norway'. In: *Geology of Norway*. O. Holtedahl (ed). *Norges geol. Unders.*, 208, 6-67.

Barth, T.F.W. and Reitan, P.H. (1963) 'Precambrian of Norway'. In: *The geologic systems: The Precambrian*, Vol. 1, 27-80. K. Rankama (ed). John Wiley and Sons Inc., London.

Berg, O. (1977) 'En geokronologisk analyse av Prekambrisk basement i distrikted Røldal-Haukelisaeter-Valdalen ved Rb-Sr whole-rock metoden, dets plass i den Sørvestnorske Prekambriske Provinsen'. Unpublished Thesis Mineralogisk-Geologisk Museum Oslo, 125 pp.

Berthelsen, A. (1980) 'Towards a palinspastic tectonic analysis of the Baltic Shield'. Mém. BRGM. 6th Colloq. Int. Geol. Congr. Paris, 108, 5-21.

Black, L.P., Bell, T.H., Rubenach, M.J. and Withnall, I.W. (1979) 'Geochronology of discrete structural-metamorphic events in a multiply deformed Precambrian terrain'. Tectonophysics, 54, 103-137.

Brattli, B., Tørudbakken, B.O. and Ramberg, I.B. (1983) 'Resetting of a Rb-Sr total rock system in Rödingsfjället Nappe Complex, Nordland, North Norway'. Norsk Geol. Tidsskr., 62, 219-224.

Breemen, O. van, Halliday, A.N., Johnson, M.R.W. and Bowes, D.R. (1978) 'Crustal additions in late Precambrian times'. In: Crustal evolution in Britain and adjacent regions, 81-106. D.R. Bowes and B.E. Leake (eds), Geol. J. Spec. Issue, 10.

Bridgwater, D. and Windley, B.F. (1973) 'Anorthosites, post-orogenic granites, acid volcanic rocks and crustal development in the North Atlantic Shield during the mid-Proterozoic'. In: Symposium on granite, gneisses and related rocks, 307-317. L.A. Lister (ed), Geol. Soc. S. Africa Spec. Pub., 3.

Bugge, A. (1928) 'En forkastning i det syd-norske grunnfjell'. Norges geol. Unders., 130, 1-124.

Bugge, A. (1936) 'Kongsberg-Bambleformasjonen'. Norges geol. Unders., 146, 117 pp.

Bugge, J.A.W. (1943) 'Geological and petrographical investigations in the Kongsberg-Bamble formation'. Norges geol. Unders., 160, 150 pp.

Brooks, C. (1968) 'Relationship between feldspar alteration and the precise post-crystallization movement of rubidium and strontium isotopes in a granite'. J. Geoph. Res., 73, 4751-4757.

Cramez, C. (1969) 'Evolution structurale de la région Nisser-Vråvatn (Norvège méridionale)'. Norges geol. Unders., 266, 5-35.

Dons, J.A. (1960) 'Telemark supracrustals and associated rocks, I'. In: Geology of Norway, 49-58. O. Holtedahl (ed), Norges geol. Unders., 208.

Dons, J.A. (1972) 'The Telemark area: a brief presentation'. Science de la Terre, 17, 25-29.

Duchesne, J.A. (1984) 'Massif anorthosites: Another partisan review'. In: Feldspars and feldspathoids, 411-433. W.L. Brown (ed), D. Reidel Publishing Company.

Falkum, T. and Petersen, J.S. (1980) 'The Sveconorwegian orogenic belt, a case of Late-Proterozoic plate-collision'. *Geol. Rundschau*, 69, 622-647.

Field, D. and Råheim, A. (1979) 'Rb-Sr total rock isotope studies on Precambrian charnockitic gneisses from south Norway: Evidence for isochron resetting during a low-grade metamorphic-deformational event'. *Earth. Planet. Sci. Lett.*, 45, 32-44.

Field, D. and Råheim, A. (1980) 'Secondary geologically meaningless Rb-Sr isochrons, low $^{87}Sr/^{86}Sr$ initial ratios and crustal residence time of high-grade gneisses'. *Lithos*, 13, 295-304.

Field, D. and Råheim, A. (1981) 'Age relationships in the Proterozoic high-grade gneiss regions of southern Norway'. *Precambrian Res.*, 14, 261-275.

Field, D. and Råheim, A. (1982) 'Reply to "Discussion and Comment" by Weiss and Demaiffe'. *Precambrian Res.*, 22, 157-161.

Gorbatschev, R. (1979) 'The basement of the Åmål supracrustals, south-western Sweden'. *Geol. Fören. Stockholm Förh.*, 101, 71-73.

Gorbatschev, R. (1980) 'The Precambrian development of southern Sweden'. *Geol. Fören. Stockholm Förh.*, 102, 129-136.

Gorbatschev, R. and Welin, E. (1975) 'The Rb-Sr age of the Ursand granite on the boundary between the Åmål and "Pregothian" mega-units of southwestern Sweden'. *Geol. Fören. Stockholm Förh.*, 97, 379-381.

Gorbatschev, R. and Welin, E. (1980) 'The Rb-Sr age of the Varberg charnockite, Sweden: a reply and discussion of the regional contexts'. *Geol. Fören. Stockholm Förh.*, 102, 43-48.

Gower, C.F. and Owen, V. (1984) 'Pre-Grenvillian and Grenvillian lithotectonic regions in eastern Labrador - correlations with the Sveconorwegian orogenic belt in Sweden'. *Can. J. Earth Sci.*, 21, 678-693.

Hageskov, B. (1980) 'The Sveconorwegian structures of the Norwegian part of the Kongsberg-Bamble-Østfold segment'. *Geol. Fören. Stockholm Förh.*, 102, 150-155.

Hermans, G.A.E.M., Tobi, A.C., Poorter, R.P.E. and Maijer, C. (1975) 'The high-grade metamorphic Precambrian of the Sirdal-Ørsdal area, Rogaland, Vest Agder, S.W. Norway'. *Norges geol. Unders.*, 318, 51-74.

Jacobsen, S.B. and Heier, K.S. (1978) 'Rb-Sr isotope systematics in metamorphic rocks, Kongsberg sector, south Norway'. *Lithos*, 11, 257-276.

Killeen, P.G. and Heier, K.S. (1975) 'Radioelement distribution and heat production in Precambrian granitic rocks, southern Norway'. *Det Norske Videnskaps-Akademi 1. Mat. Naturv. Klasse. Skrifter Ny Serie*, 35, 32 pp.

Kleeman, A.W. (1967) 'Sampling error in the chemical analysis of rocks'. *J. Geol. Soc. Australia*, 14, 43-47.

Kleppe, A. and Råheim, A. (1979) 'Rb-Sr investigations on Precambrian rocks in the central part of Telemark. Examples on partial and complete resetting of Rb-Sr whole-rock systems'. Abstract lecture on Sixth European Colloquium on Geochronology, Cosmochronology and Isotope Geology, Lillehammer, 1979.

Krogh, E.J. (1977) 'Evidence of Precambrian continent collision in western Norway'. *Nature*, 267, 17-19.

Krogh, T.E. and Davis, G.J. (1973) 'The effect of regional metamorphism on U-Pb systems in zircon and comparison with Rb-Sr systems in the same whole-rock and its constituent minerals'. *Yb. Carnegie Instn. Wash.*, 72, 601-610.

Lindh, A. (1974) 'The mylonite-zone in southwestern Sweden (Värmland). A re-interpretation'. *Geol. Fören. Stockholm Förh.*, 96, 183-197.

Lundegårdh, P.H. (1971) 'Neue Gesichtspunkte zum schwedischen Präkambrium'. *Geol. Rundschau*, 60, 1392-1405.

Lundqvist, Th. (1968) 'Precambrian geology of the Los-Hamra region central Sweden'. *Sver. Geol. Unders.*, Ba 23, 255 pp.

Lundqvist, Th. (1979) 'The Precambrian of Sweden'. *Sver. Geol. Unders.*, C 768, 1-87.

Magnusson, N.H. (1936) 'Om cykelindelningen i det svenska urberget'. *Geol. Fören. Stockholm Förh.*, 58, 102-108.

Magnusson, N.H. (1962) 'The Swedish Precambrian outside the Caledonian mountain chain'. In: Description to accompany the map of the Pre-Quarternary rocks of Sweden. *Sver. Geol. Unders.*, Ba 16, 5-66.

Maijer, C., Jansen, J.B.H., Hebeda, E.H., Verschure, R.H. and Andriessen, P.A.M. (1981) 'Osumilite, a 970 Ma old high-temperature index mineral of the granulite facies metamorphism in Rogaland, S.W. Norway'. *Geol. Mijnbouw*, 60, 267-272.

Max, M.D. (1979) 'Extent and disposition of Grenville tectonism in the Precambrian crust adjacent to the North Atlantic'. *Geology*, 7, 76-78.

Menuge, J.F. (1982) 'Nd isotope studies of crust-mantle evolution: The Proterozoic of south Norway and the Archaean of southern Africa'. Unpublished thesis, University of Cambridge, 257 pp.

Michot, J. and Michot, P. (1969) 'The problem of anorthosites: the South-Rogaland igneous complex, southwestern Norway'. *New York State Mus. Sci. Serv. Mem.*, 18, 399-423.

Michot, P. (1960) 'La géologie de la catazone: le problème des anorthosites, la palingenèse basique et la tectonique catazonale dans le Rogaland méridional (Norvège méridional)'. *Norges geol. Unders.*, 212, 1-54.

Michot, P. (1969) 'Geological environment of the anorthosites of south Rogaland, Norway'. *New York State Mus. Sci. Serv. Mem.*, 18, 411-423.

Mitchel, R.H. (1967) 'The Precambrian rocks of the Telemark area in south central Norway V. The Nissedal supracrustal series'. *Norsk Geol. Tidsskr.*, 47, 295-332.

Moorbath, S. (1975) 'Geological interpretation of whole-rock isochron dates from high grade gneiss terrains'. *Nature*, 255, 391.

Morris, W.A. and Roy, J.L. (1977) 'Discovery of the Hadrynian polar track and further study of the Grenville problem'. *Nature*, 266, 689-692.

Morton, R.D., Batey, R. and O'Nions, R.K. (1970) 'Geological investigations in the Bamble Sector of the Fennoscandian Shield south Norway'. *Norges geol. Unders.*, 263, 1-72.

Mulder, F.G. (1971) 'Paleomagnetic research in some parts of central and southern Sweden'. *Sver. Geol. Unders.*, C 653, 1-56.

Nyström, J.O. (1982) 'Post-Svecokarelian andinotype evolution in central Sweden'. *Geol. Rundschau*, 71, 141-157.

Oen, I.S. (1982) 'Isotopic age determinations in Bergslagen, Sweden: II The Filipstad-type granite of Rockesholm, Grythyttan Area'. *Geol. Mijnbouw*, 61, 305-307.

O'Nions, R.K., Morton, R.D. and Baadsgaard, H. (1969) 'Potassium-argon-ages from the Bamble sector of the Fennoscandian shield in south Norway'. *Norsk Geol. Tidsskr.*, 49, 171-190.

O'Nions, R.K. and Baadsgaard, H. (1971) 'A radiometric study of polymetamorphism in the Bamble region, Norway'. *Contrib. Mineral Petrol.*, 34, 1-21.

O'Nions, R.K. and Heier, K.S. (1972) 'A reconnaissance Rb-Sr geochronological study of the Kongsberg area south Norway'. *Norsk Geol. Tidsskr.*, 52, 143-150.

Page, R.W. (1978) 'Response of U-Pb zircon and Rb-Sr total rock and mineral systems to low-grade regional metamorphism in Proterozoic igneous rocks, Mount Isa, Australia'. *J. Geol. Soc. Austr.*, 25, 141-164.

Park, G., Bailey, A., Crane. A., Cresswell, D. and Standley, R. (1979) 'Structure and geological history of the Stora Le-Marstrand rocks in western Orust, south-western Sweden'. *Sver. Geol. Unders.*, **C 763**, 1-36.

Pasteels, P. and Michot, J. (1975) 'Geochronologic investigation of the metamorphic terrain of southwestern Norway'. *Norsk Geol. Tidsskr.*, **55**, 111-134.

Pasteels, P., Demaiffe, D. and Michot, J. (1979) 'U-Pb and Rb-Sr geochronology of the eastern part of the south Rogaland igneous complex, southern Norway'. *Lithos,* **12**, 199-208.

Patchett, P.J. and Bylund, G. (1977) 'Age of Grenville Belt magnetisation: Rb-Sr and palaeomagnetic evidence from Swedish dolerites'. *Earth Planet. Sci. Lett.,* **35**, 92-104.

Patchett, P.J. (1978) 'Rb/Sr ages of Precambrian dolerites and syenites in southern and central Sweden'. *Sver. Geol. Unders.*, **C 747**, 1-63.

Patchett, P.J., Bylund, G. and Upton, B.G.J. (1978) 'Palaeomagnetism and the Grenville orogeny: New Rb-Sr ages from dolerites in Canada and Greenland'. *Earth Planet. Sci. Lett.,* **40**, 349-364.

Pedersen, S. (1973) 'Age determinations from the Iveland-Evje Area, Aust Agder'. *Norges geol. Unders.,* **300**, 33-39.

Pedersen, S. and Falkum, T. (1976) 'Rb/Sr age determinations of granitic rocks from the Precambrian of Agder, south Norway'. Abstract lecture on Fourth European Colloquium on Geochronology, Cosmochronology and Isotope Geology, Amsterdam.

Pedersen, S., Berthelsen, A., Falkum T., Graversen, O., Hageskov, B., Maaløe, S., Petersen, J.S., Skjernaa, L. and Wilson, J.R. (1978) 'Rb/Sr dating of the plutonic and tectonic evolution of the Sveconorwegian Province, southern Norway'. In: R.E. Zartman (ed) *Short Pap. 4th Int. Conf. on Geochronology, Cosmochronology and Isotope Geology. U.S. Geol. Surv. Open-File Rep.*, **78**, **701**, 329-331.

Petersen, J.S. (1977) 'The migmatite complex near Lyngdal, southern Norway and related granulite metamorphism'. *Norsk Geol. Tidsskr.*, **57**, 65-83.

Ploquin, A. (1972) 'Le granite acide d'Åmli Norvège méridionale: transformation des laves acides du Tuddal (Formation inférieure des séries de roches supracrustales du Telemark'. *Sciences de la Terre,* **17**, 83-95.

Ploquin, A. (1980) 'Étude géochimique et pétrographique du complexe de gneiss, migmatites et granites du Telemark-Aust Agder (Précambrien du Norvège du Sud. Sa place dans l'ensemble épizonal à catazonal profond du Haute Telemark au Bamble'. *Sciences de la Terre*. Memoire **38**.

Poorter, R.P.E. (1981) 'Precambrian palaeomagnetism of Europe and the position of the Balto-Russian Plate relative to Laurentia. In: *Precambrian Plate Tectonics*, 599-622. A. Kröner (ed) Elsevier Scientific Publishing Company, Amsterdam.

Priem, H.N.A., Verschure, R.H., Verdurmen, E.A.Th., Hebeda, E.H. and Boelrijk, N.A.I.M. (1970) 'Isotopic evidence on the age of the Trysil porphyries and granites in eastern Hedmark, Norway'. *Norges geol. Unders.*, 266, 263-276.

Skiold, T. (1976) 'The interpretation of Rb-Sr and K-Ar ages of late Precambrian rocks in southwestern Sweden'. *Geol. Fören. Stockholm Förh.*, 98, 3-29.

Skjernaa, L. and Pedersen, S. (1982) 'The effects of penetrative

Priem, H.N.A., Boelrijk, N.A.I.M., Hebeda, E.H., Verdurmen, E.A.Th. and Verschure, R.H. (1972) 'Rb-Sr investigations in the Uvdal graniticgneissic complex: Evidence of an "Elsonian" event in the Precambrian of southern Norway'. *Progress Report Z.W.O. Laboratorium voor IsotopenGeologie, September 1, 1970 - August 31, 1972*, 111-115.

Priem, H.N.A., Boelrijk, N.A.I.M., Hebeda, E.H., Verdurmen, E.A.Th. and Verschure, R.H. (1973) 'Rb-Sr investigations on Precambrian granites, granitic gneisses and acidic metavolcanics in central Telemark. Metamorphic resetting of Rb-Sr whole-rock systems'. *Norges geol. Unders.*, 289, 37-53.

Priem, H.N.A., Boelrijk, N.A.I.M., Hebeda, E.H., Schermerhorn, L.J.G., Verdurmen, E.A.Th. and Verschure, R.H. (1978) 'Sr isotope homogenization through whole-rock systems under low-greenschist facies metamorphism in Carboniferous pyroclastics at Aljustrel (southern Portugal)'. *Chem. Geol.*, 21, 307-314.

Priem, H.N.A. and Verschure, R.H. (1982) 'Review of the isotope geochronology of the high-grade metamorphic Precambrian of S.W. Norway'. *Geol. Rundschau*, 71, 81-84.

Råheim, A. and Compston, W. (1977) 'Correlations between metamorphic events and Rb-Sr ages in metasediments and eclogite from Western Tasmania'. *Lithos*, 10, 271-289.

Reymer, A.P.S., Boelrijk, N.A.I.M., Hebeda, E.H., Priem, H.N.A., Verdurmen, E.A.Th. and Verschure, R.H. (1980) 'A note on Rb-Sr wholerock ages in the Seve Nappe of the central Scandinavian Caledonides'. *Norsk Geol. Tidsskr.*, 60, 139-147.

Rietmeyer, F.J.M. (1979) 'Pyroxenes from iron-rich igneous rocks in Rogaland, S.W. Norway'. *Geologica Ultratrajectina*, 21, 341 pp.

Roddick, J.C. and Compston, W. (1976) 'Strontium isotopic equilibration: a solution to a paradox'. *Earth Planet. Sci. Lett.*, 34, 238-246.

Saether, E. (1957) 'The alkaline rock province of the Fen area in southern Norway'. *K. Norsk Vid. Selsk. Skr.*, 1, 1-150.

Sauter, P.C.C., Hermans, G.A.E.M., Jansen, J.B.H., Maijer, C., Spits, P. and Wegelin, A. (1983) 'Polyphase Caledonian metamorphism in the Precambrian basement of Rogaland/Vest Agder, S.W. Norway'. *Norges geol. Unders.*, 38, 7-22.

Schmus, W.R. van, and Bickford, M.E. (1976) 'Rotation of Rb-Sr isochrons during low-grade events'. Abstract lecture on Fourth European Colloquium on Geochronology, Cosmochronology and Isotope Geology, Amsterdam.

Sigmond, E.M.O. (1978) 'Beskrivelse till det berggrunnsgeologiske kartbladet Sauda 1: 250.000'. *Norges geol. Unders.*, 341, 94 pp.

Sigmond, E.M.O. (1984) 'En Megaforkastningssone i Syd Norge'. Abstract lecture Sixteenth Geological Wintermeeting Norden, Stockholm 9-13 January, 1984.

Singh, I.B. (1968) 'Lenticular and lenticular-like bedding in the Precambrian Telemark Suite, southern Norway'. *Norsk Geol. Tidsskr.*, 48, 165-170.

Singh, I.B. (1969) 'Precambrian rocks of the Telemark Area in south central Norway VI. Primary sedimentary structures in Precambrian quartzites of Telemark, southern Norway'. *Norsk Geol. Tidsskr.*, 49, 1-31.

Skiold, T. (1976) 'The interpretation of Rb-Sr and K-Ar ages of late Precambrian rocks in southwestern Sweden'. *Geol. Fören. Stockholm Förh.*, 98, 3-29.

Skjernaa, L. and Pedersen, S. (1982) 'The effects of penetrative Sveconorwegian deformation on Rb-Sr isotope systems in the Romskog-Aurskog-Höland area, S.E. Norway'. *Prec. Research*, 17, 215-243.

Smalley, P.C., Field, D. and Råheim, A. (1983) 'Resetting of Rb-Sr whole-rock isochrons during Sveconorwegian low-grade events in the Gjerstad augengneiss, Telemark, southern Norway'. *Isotope Geoscience*, 1, 269-282.

Smithson, S.B. (1965) 'The nature of the "granitic" layer of the crust in the southern Norwegian Precambrian'. *Norsk Geol. Tidsskr.*, 45, 113-133.

Smithson, S.B., Murphy, D.J. and Houston, R.S. (1971) 'Development of an augengneiss terrain'. *Contrib. Mineral Petrol.*, 33, 184-190.

Starmer, I.C. (1972) 'Polyphase metamorphism in the granulite facies terrain of the Risør area, south Norway'. *Norsk Geol. Tidsskr.*, 52, 43-71.

Stout, J.H. (1972) 'Stratigraphic studies of high-grade metamorphic rocks east of Fyresdal'. *Norsk Geol. Tidsskr.* **52**, 33-41.

Sturt, B.A., Skarpenes, O., Ohanian, A.T. and Pringle, R. (1975) 'Reconnaissance Rb/Sr isochron study in the Bergen Arc System and regional implications'. *Nature*, **253**, 595-599.

Swainbank, I. (1965) 'Zircon geochronology of the Norwegian basement'. Annual Progr. Rep. 10, Contract AT (30-1)-1669, Lamont Geological Observatory, App. C., 22 pp.

Sylvester, A.G. (1964) 'The Precambrian rocks of the Telemark area in south central Norway III. Geology of the Vrådal granite'. *Norsk Geol. Tidsskr.*, **44**, 445-482.

Torske, T. (1977) 'The south Norway Precambrian region, a Proterozoic cordilleran-type orogenic segment'. *Norsk Geol. Tidsskr.*, **57**, 97-120.

Touret, J. (1969) 'Le socle Précambrien de la Norvège méridionale'. Thesis University Nancy, *C.N.R.S. A.O.*, **2902**, 316 pp.

Touret, J. (1971a) 'Le facies granulitique en Norvège méridionale I. Les associations minéralogiques'. *Lithos*, **4**, 239-249.

Touret, J. (1971b) 'Le facies granulitique en Norvège méridionale II. Les inclusions fluides'. *Lithos*, **4**, 423-436.

Touret, J. (1984) 'Excursion Log. The Telemark Supracrustals and the Telemark Gneisses'. In: Excursion Guide of the South Norway Geological Excursion, 9-42. C. Maijer (ed). NATO Advanced Study Institute, Moi, 1984.

Ueno, H., Irving, E. and McNutt, R.H. (1975) 'Paleomagnetism of the Whitestone anorthosite and diorite, the Grenville polar track and relative motions of the Laurentian and Baltic shields'. *Can. J. Earth Sci.*, **12**, 209-226.

Venugopal, D.V. (1970) 'Geology and structure of the area west of Fyresvatn, Telemark, southern Norway'. *Norges geol. Unders.*, **277**, 53-59.

Verschure, R.H., Andriessen, P.A.M., Boelrijk, N.A.I.M., Hebeda, E.H., Maijer, C., Priem, H.N.A. and Verdurmen, E.A.Th. (1980) 'On the thermal stability of Rb-Sr and K-Ar biotite systems: Evidence from coexisting Sveconorwegian (Ca 870 Ma) and Caledonian (Ca 400 Ma) biotites in S.W. Norway'. *Contrib. Mineral. Petrol.*, **74**, 245-252.

Verschure, R.H. (1981) 'The extent of Sveconorwegian and Caledonian metamorphic imprints in Norway and Sweden'. Abstract lecture on Seventh European Colloquium on Geochronology, Cosmochronology and Isotope Geology, Jerusalem 1981. (also in *Terra Cognita*, **2**, 1982, 64-65).

Versteeve, A. (1975) 'Isotope geochronology in the high-grade metamorphic Precambrian of southwestern Norway'. *Norges geol. Unders.*, 318, 1-50.

Wahl, W. (1936) 'Om granitgrupperna och bergskedjeveckningarna i Sverige och Finland'. *Geol. Fören. Stockholm Förh.*, 58, 90-101.

Weiss, D. and Demaiffe, D. (1983) 'Age relationships in the Proterozoic high-grade gneiss regions of southern Norway: Discussion and comment'. *Precambrian Res.*, 22, 149-155.

Welin, E. (1966) 'The absolute time scale and the classification of Precambrian rocks in Sweden'. *Geol. Fören. Stockholm Förh.*, 88, 29-33.

Welin, E. and Blomqvist, G. (1966) 'Further age measurements on radioactive minerals from Sweden'. *Geol. Fören. Stockholm Förh.*, 88, 3-18.

Welin, E., Blomqvist, G. and Parwel, A. (1966) 'Rb/Sr whole-rock age data on some Swedish Precambrian rocks'. *Geol. Fören. Stockholm Förh.*, 88, 19-28.

Welin, E. and Lundqvist, T. (1975) 'K-Ar ages of Jotnian dolerites in Västernorland county, central Sweden'. *Geol. Fören. Stockholm Förh.*, 97, 83-88.

Welin, E. and Gorbatschev, R. (1976) 'The Rb-Sr age of the Hästefjorden granite and its bearing on the Precambrian evolution of southwestern Sweden'. *Precambrian Res.*, 3, 187-195.

Welin, E., Gorbatschev, R. and Lundegårdh, P.H. (1977) 'Rb-Sr dating of rocks in the Värmland granite group in Sweden'. *Geol. Fören. Stockholm Förh.*, 99, 363-367.

Welin, E. and Gorbatschev, R. (1978a) 'Rb-Sr age of the Lane granites in southwestern Sweden'. *Geol. Fören. Stockholm Förh.*, 100, 101-102.

Welin, E. and Gorbatschev, R. (1978b) 'The Rb-Sr age of the Varberg charnockite, Sweden'. *Geol. Fören. Stockholm Förh.*, 100, 225-227.

Welin, E. and Gorbatschev, R. (1980) 'An Rb-Sr age of the Åmål granite at Åmål, Sweden'. *Geol. Fören. Stockholm Förh.*, 100, 401-403.

Welin, E. and Kähr, A.M. (1980) 'The Rb-Sr and U-Pb ages of a Proterozoic gneissic granite in central Värmland, western Sweden'. *Sver. Geol. Unders.*, C 777, 24-28.

Welin, E., Lundegårdh, P.H. and Kähr, A.M. (1980) 'The radiometric age of a Proterozoic hyperite diabase in Värmland, western Sweden'. *Geol. Fören. Stockholm Förh.*, 102, 49-52.

Welin, E., Gorbatschev, R. and Kähr, A.M. (1982) 'Zircon dating of polymetamorphic rocks in southwestern Sweden'. *Sver. Geol. Unders.*, C 795, 34 pp.

Wielens, J.B.W., Andriessen, P.A.M., Boelrijk, N.A.I.M., Hebeda, E.H., Verdurmen, E.A.Th. and Verschure, R.H. (1981) 'Isotope geochronology in the high-grade metamorphic Precambrian of southwestern Norway: New data and reinterpretations'. *Norges geol. Unders.*, 359, 1-30.

Wilson, J.R., Pedersen, S., Berthelsen, C.R. and Jacobsen, B.M. (1977) 'New light on the Precambrian Holum granite, south Norway'. *Norsk Geol. Tidsskr.*, 57, 347-360.

Wynne-Edwards, H.R. and Hasan, Z.U. (1970) 'Intersecting orogenic belts across the North Atlantic'. *Am. J. Sci.*, 268, 289-308.

Zeck, H.P. and Malling, S. (1976) 'A major global suture in the Precambrian basement of S.W. Sweden'. *Tectonophysics*, 31, 35-40.

Zeck, H.P. and Wallin, B. (1980) 'A 1220 ± 60 M.Y. Rb-Sr isochron age representing a Taylor-convection caused recrystallization event in a granitic rock suite'. *Contrib. Mineral. Petrol.*, 74, 45-53.

ISOTOPE GEOCHRONOLOGY OF THE PROTEROZOIC CRUSTAL SEGMENT OF SOUTHERN
NORWAY : A REVIEW.

D. Demaiffe and J. Michot
Laboratoires Associés
Géologie-Pétrologie-Géochronologie
Université Libre de Bruxelles
Avenue F. Roosevelt, 50 - 1050 Brussels - Belgium

ABSTRACT. In the Sveconorwegian Province of S.Norway, most geochronological data are in the range 1200-900 Ma. Several deformation events associated with metamorphic episodes have been identified. The paroxysm of the metamorphism has been dated at 1050-1000 Ma. Evidence of a pre-Sveconorwegian basement (1500-1800 Ma?) is well established in the eastern part of the mobile belt while in the western part (Rogaland - Agder) only few ages older than 1200 Ma have been reported. Intrusions of magmatic rocks took place at different periods during the tectono-metamorphic history. Late-to post-tectonic (980-900 Ma) magmatic activity is represented by the intrusions of granitic plutons in the whole Sveconorwegian Province and by the large anorthositic bodies in the Rogaland district.

1. INTRODUCTION

Recent structural, geochemical and geochronological researches on the Baltic shield have introduced new views on the development of that part of the North Atlantic crustal segment during the Precambrian. Similar field and laboratory work has been undertaken in the opposite side of the ocean.
This paper will focus on the late Proterozoic of the Baltic shield, the Sveconorwegian Province, now considered as the extension of the Canadian Grenville Province. After a short review of the Grenville Province, an attempt is made to synthesize the structural and geochronological data, both on the metamorphic and magmatic rocks of the Sveconorwegian Province.

2. THE GRENVILLE PROVINCE

In a recent geochronological and structural synthesis, Baer (1981) subdivided the late Proterozoic 1400-700 Ma interval into four periods with maximum activities at 1300 Ma, 1150 Ma and 950 Ma. The pre-Grenvillian history, before 1350 Ma is not very well known, nevertheless it can be shown that Grenvillian events affected a preexisting (Proterozoic and Archean) continental crust.

The first, pretectonic, stage has been interpreted as the early fracturing of the North Atlantic craton, it is associated with dyke swarms, rift and graben tectonics and extrusion of alkaline lavas. It lasted until about 1100 Ma ago.

The next, syntectonic, stage has been related to the "peak of high-grade metamorphism in the Grenville province and to a deformation fabric at high angle to the Grenville front" (Baer, op.cit., p.374). This major metamorphic and tectonic event has been dated at 1150 Ma (Silver and Lumbers, 1965). This age has been obtained both outside and inside the Grenville Province, on different magmatic bodies (Wanless et al, 1972, 1973). It probably reflects a thermal event and possibly dates the metamorphism (or reactivation) of anorthosite plutons ("Morin event" of Baer, 1981).

The third, late tectonic, stage is considered as the Grenvillian event sensu stricto ; structurally, it is characterized by "thrusting and block-wise uplifts in the Grenville front zone" (Baer, op.cit., p.376) and geochronologically by the closure of all K-Ar systems (900 ± 100 Ma) indicating a "general uplift accompanied by a thickening of the crust".

The emplacement age of the anorthositic plutons is quite controversial. In the Labrador Province, they seem to predate the Grenvillian events : ages between 1500 and 1400 Ma have been reported (Krogh and Davis, 1969; Emslie, 1978 ; Ashwal and Wooden, 1983). The continuity of the anorthosite belt extending from Labrador through Grenville into the Central United States points for Emslie to an unique anorthosite event at 1500-1400 Ma. However, U/Pb zircon data (Silver, 1969) and Rb-Sr whole rock isochrons (Barton and Doig, 1977) on the Adirondacks and Morin anorthosites give ages in the range 1150-1100 Ma. These two groups of ages could be explained either by the existence of two distinct and independent anorthositic episodes or by the reworking at 1150-1100 Ma of a 1500 Ma old anorthositic intrusion.

3. THE SVECONORWEGIAN PROVINCE

The remnant of the Grenville belt in the Baltic Shield is called the Sveconorwegian Province which is delineated by its eastern front against the Svecofennian platform in SW Sweden (Berthelsen, 1980) and is cut off obliquely to the NW by the southern border of the Caledonides. Focusing on the "significance of some low-angle thrust zones", Berthelsen (1980, p.6) regards them as "the most important key to the understanding of the architecture" in that part of the Baltic Shield ; he also first made a distinction between the Canadian Grenville front bordering to the North and West the Grenville mobile belt and the Sveconorwegian front, extending from Skåne to lake Vättern (Magnusson et al, 1963), which limits to the East the Scandinavian part of that mobile belt.

In Southern Norway, the Sveconorwegian Province has been divided by the Oslo rift zone (Ramberg, 1976).

3.1. The Sveconorwegian east of the Oslo rift

This easternmost part of the Sveconorwegian province is outlined by the Great mylonite zone separating the Svecofennian platform (1750-1700 Ma) from the Sveconorwegian province (Fig.1a). In that area, detailed structural analyses have led Berthelsen (1980) to recognize three distinct crustal segments separated by two major west-dipping thrust zones.

The eastern segment (in Sweden) is mainly composed of Pregothian gneisses and some supracrustals (Stora Le-Marstrand formation), including the 1660 Ma old Värmland granite (Lundegarth, 1977). To the west, it is bounded by a thrust running along the Central Värmland mylonite zone (Magnusson, 1937 : Lindh, 1974 ; Berthelsen, 1978). Samuelsson and Åhäll (this volume) have established a preliminary sequence of events in this area.

The median segment includes the Pregothian and the Gothian sequence (Åmål supracrustals in Sweden) dated at 1700-1200 Ma (Gorbatschev, 1979; Welin and Gorbatschev, 1976a, b, c). These older rocks are unconformably overlain by a 1075-1025 Ma sedimentation stage (Skiold, 1976) considered as a Sveconorwegian flysch (Kappebo formation and Dal Group). Skjernaa and Pedersen (1982) have recently presented a thorough geochronological framework for the Norwegian part of this segment. The oldest reported ages fall in the range 1850-1710 Ma ; they are only "model ages" whose meaning is however unclear since they have been calculated on only 2 augen gneisses assuming a Sr isotopic initial ratio of 0.703. Grey and red orthogneisses have been dated respectively at 1631 ± 53 and 1277 ± 20 Ma. The major deformational-metamorphic episode, causing migmatization and folding of the augen gneisses has been dated at 1000-970 Ma.

The western segment, the Østfold slab, is separated from the median one by the Dalsland mylonitic to blastomylonitic boundary thrust. The basement complex of this segment consists of highly deformed and migmatized 1700-1320 Ma old tonalitic to granitic rocks (Welin and Gorbatschev, 1978). In the western part of this segment, younger granitic orthogneisses intruded the basement at 1320 to 1200 Ma and were migmatized at 1015Ma (Hageskov and Pedersen, 1981).

Post-tectonic magmatic activity has been reported for both the median and the Østfold slab segments. Granitic and granodioritic intrusions constitute the Blomskog belt (924 Ma ; Pedersen et al., 1978) east of the Dalsland boundary thrust and the Bohus-Iddefjord-Flå batholith which cuts accross the Oslofjord shear zone. Syndeformational amphibolite facies metamorphism took place at 1015 Ma in the Fjord zone (Hageskov and Pedersen, 1981)

3.2 The Sveconorwegian West of the Oslo rift

The largest part of the Sveconorwegian orogenic belt outcrops to the West of the Oslo rift zone (Fig.1b). A major NE-SW structural lineament - the Friction Breccia (= the Porsgrunn - Kristiansand fault zone) which is a km wide zone of blastomylonites, separates the Bamble-Kongsberg segment from the Rogaland-Agder-Telemark subprovince.

3.2.1. Bamble-Kongsberg segment. The paroxysm of the regional metamor-

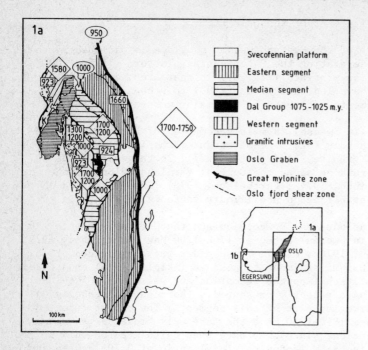
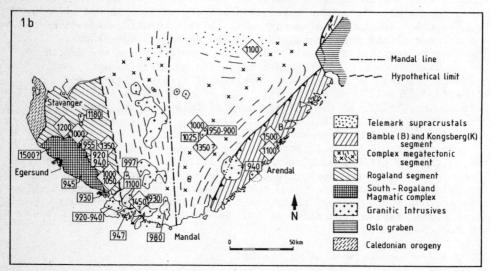

Fig. 1 Sketch map of the Sveconorwegian Province in Southern Norway and Sweden. Geochronological data (in Ma) for the magmatic (□) and metamorphic (◇) rocks.
1A: East of the Oslo Rift (modified from Berthelsen, 1980)
1B: West of the Oslo Rift (modified from Falkum and Petersen, 1980)

phism in that area has been fixed at 1100 Ma by the numerous geochronological (K/Ar, Rb/Sr, U/Pb) data obtained by O'Nions et al (1969) and O'Nions and Baadsgaard (1971). This interpretation has however been questioned more recently by Field and Råheim (1979 a & b, 1981) and Field et al (this volume). From Rb-Sr whole rock isochron data on Bamble charnockitic orthogneisses and metasediments, they tentatively proposed that the main high-grade metamorphic episode took place at about 1500 Ma, before the Sveconorwegian cycle and that these rocks were only slightly affected by a low-grade Sveconorwegian event at 1100-1000 Ma ; this event is related to the intrusion of undeformed granite sheets at 1060 Ma. Large post-tectonic granitic bodies (Herefoss and Grimstad) intruded at 940 Ma (Brueckner, 1972).

In the Kongsberg sector, Jacobsen and Heier (1978) also obtained a 1580 Ma old isochron on enderbitic granulites.

3.2.2. Rogaland-Agder-Telemark subprovince. This large area constitutes the western part of the Sveconorwegian belt. Detailed mapping and geophysical investigations have shown the presence of an extended N-S fracture or thrust zone - the Mandal line (Sigmond, 1983) - of which the precise structural (geotectonic) significance is not yet established. The western limb of this line is underlined by a very important granitic batholith belt.

3.2.2.a. Geochronology of the metamorphic domain.
In the Telemark sector, as Priem et al (1973) reported, supracrustals were deposited around 1570 Ma ; their recrystallization and local anatectic melting took place during the high-grade Dalslandian (= Sveconorwegian) metamorphism at about 1100 Ma (see also Verschure, this volume). However, Smalley et al (1983) argue from their study of the Gjerstad augen gneisses that the Telemark gneisses were formed before 1250 Ma but that the Sveconorwegian event was only of low-grade, causing a resetting of Rb-Sr whole rock isochron.
In the Evje area (Aust-Agder), Pedersen (1981) reported an estimated age of 1350 Ma (errorchron) for the oldest orthogneiss (the Sletthei augen gneiss) while the penultimate major regional deformation is situated between 1120 and 1025 Ma. The last deformational episode is younger than 1025 Ma and is probably contemporaneous with the last regional metamorphism dated at 1000-990 Ma.
In the Vest-Agder district, near Flekkefjord, Falkum (1966 and this volume) distinguished five regional deformation phases ; the first three phases (D1-D3) are characterized by large scale isoclinal folds and granulite facies metamorphism. The D3 phase, dated at 1100 Ma (Pedersen et al, 1978) has been refolded at 990-1000 Ma by the fourth tectonic phase (D4 : N-S trending axes and W-vergent folds) which was associated with an amphibolite facies metamorphism (Falkum and Pedersen, 1979). The D5 phase induced open and gentle folds without any recrystallization.
In the Rogaland district, P. Michot (1956, 1960, 1969) described large and complex refolded nappe structures in the metamorphic terrane to the North of the large anorthositic masses. He distinguished three phases of folding in that deep catazonal complex. The first phase is a recumbent folding system with NS trending axes and W-vergent folds (Falkum's D2?) ;

the synkinematic anatectic granites of that phase have been dated at
1200-1250 Ma by both U-Pb on zircons and Rb-Sr whole rock isochrons
(J. Michot and Pasteels, 1972 ; Pasteels and J. Michot, 1975). The
second phase consists of recumbent folding along E-W axes and N-vergent
folds (Falkum's D4 ?). It is considered as the climax of the Sveconorwegian regional metamorphism and is well dated at 1000-1050 Ma by concordant monazites and upper intercepts of zircon concordia chords from
augen gneisses and granitic gneisses (Pasteels and Michot, 1975 ;
Versteeve, 1975 ; Wielens et al, 1981). The last tectonic phase gives
E-W trending, vertical axial planes, changing progressively to a N-S
direction towards the SE (Falkum's D5 ?). This last event seems to have
taken place 950-970 Ma ago (osumilite cooling age ; Maijer et al, 1981),
an age similar to those obtained on the late-tectonic intrusives (see
next section).
Working in the nearby Sirdal-Ørsdal area, to the NE of the Rogaland
anorthositic province, Tobi (1965) and Hermans et al (1975) concluded to
the existence of four phases of folding (D1 to D4) in the complex and
intricate metamorphic (migmatitic) terrane, lacking the typical nappe
structure described by P. Michot to the North. The oldest phase is an
isoclinal folding presumably related to a first tectonic cycle. Younger
structural features with N-S or NW-SE trending axes and E-dipping axial
planes were considered, at least partly, as contemporaneous with the
intrusion of the basal unit of the Bjerkreim-Sokndal lopolith (see
later).
In the same time, three high-grade metamorphic phases (M1 to M3) have
been recognized mainly in the metapelites (Maijer et al, 1981 ; see also
Jansen et al and Tobi et al, this volume). At present, the relation
between the folding phases and the metamorphic stages is not clarified.
Nevertheless, geochronological data (Versteeve, 1975 ; Wielens et al,
1981 ; Priem and Verschure, 1982) have allowed a tentative interpretation of the tectono-metamorphic successions. An age close to 1200 Ma has
been assigned to the M1 metamorphic stage (upper amphibolite facies ?).
The M2 and M3 high-grade (granulite) metamorphic stages are thought to
be due to the thermal effect related to the intrusion of the Bjerkreim-
Sokndal lopolith : the M2 phase is induced by the intrusion of the
oldest anorthositic phase of the lopolith at 1050 Ma (see discussion in
the later section) while the M3 phase is related to the emplacement of
the late quartz monzonite (= quartz mangerite) of that lopolith. This M3
phase could possibly be linked with the third phase of P. Michot.
In   Rogaland, older "age indications" (up to 1500 Ma) have been
obtained (Pasteels and Michot, 1975 ; Versteeve, 1975 ; Wielens et al,
1981 ; Weis and Demaiffe, 1983a) and will be discussed in a later
section.
Finally, the last M4 phase of incipient retrograde metamorphism (pumpellyite-prehnite facies ; greenschist facies) dated at 450-400 Ma (K/Ar
and Rb/Sr or green biotites ; Verschure et al, 1980) has been related to
the nearby Caledonian orogenic belt.

3.2.2.b Geochronology of the magmatic intrusives
Intrusions of various types of magmatic rocks took place during the
different episodes of the tectono-metamorphic evolution in the Rogaland-
Agder province.

In northern Rogaland, synkinematic granites (= birkremites) were generated during the first major deformation phase 1200-1250 Ma ago (Pasteels and Michot, 1975) while the pyroxene syenitic complexes of Gloppurdi and Botnavatnet intruded just after this phase (1180 Ma : Versteeve, 1975).
In Agder, porphyric granites, subsequently deformed to augen gneisses (Feda) were intruded at 1110 Ma (Pedersen et al, 1978). Other acidic intrusives were emplaced late in the deformation history, just before (Homme granite : 997 Ma ; Falkum and Pedersen, 1979) or after (Holum granite : 980 Ma ; Wilson et al, 1977) the D4 phase of Falkum.
In the Evje area, the Fennefoss granite (later transformed to augen gneiss) intruded at 1025Ma while post-kinematic magmatic activity is represented by the Høvringsvatn monzonite-granite complex (950-900 Ma; Pedersen, 1981).
In South Rogaland, a very large and composite magmatic complex (>2000 km²) mainly composed of anorthosites-norites and associated jotunites (= monzonorites), mangerites and charnockites intruded the gneissic series. Several different anorthositic bodies have been identified and described (P.Michot, 1960 ; J.Michot and P.Michot, 1969 ; De Waard et al, 1974). The field relations and the general sequence of intrusions have been analyzed in the earlier papers and in the Guide Book of the present meeting (Duchesne and Michot, 1984). Only a brief summary of these previous data together with new data and/or interpretations are given below (Figs. 2 and 3).
The magmatic complex is composed of the following units :
- the S-Haaland anorthosito-leuconoritic basement (Ha) and the N.Haaland para-anatectic unit
- the Egersund-Ogna anorthositic pluton (Eg-Og)
- the Helleren (He) and Aana-Sira (AS) anorthosito-leuconoritic plutons
- the Bjerkreim-Sokndal anorthosito-norito-quartz-mangeritic lopolith (Bk-Sk) and the related Eia-Rekefjord (E-R) jotunitic intrusion and its equivalent bordering the AS massif to the East (apophysis)
- the Hidra intrusion (Hi) and the Garsaknatt Outlier (G)
- the Farsund charnockite (F), the Lyngdal granodiorite (L) and the Kleivan granite (K).

The large massif-type anorthositic bodies (Eg-Og, He, AS) constitute the largest part of this province. They consist essentially of coarse-grained anorthositic-noritic rocks ; they are characterized by the presence of Al-rich orthopyroxene megacrysts and giant plagioclases. These complex massifs may result
(i) from the reworking of an older anorthositic basement of unknown age, now represented by the S-Haaland basement complex (J.Michot, 1961). The N-Haaland anorthosito-leuconoritic complex has been considered in that respect as a largely reworked part of this basement and the Helleren body as the resulting product of its basic palingenesis generating anatectic leuconorite and leaving behind a residual anorthosite.
(ii) from the intrusion of a diapirically rising crystal mush affected by a synemplacement deformation (Maquil and Duchesne, 1984 ; Duchesne et al, this volume). In this mechanism, the leuconoritic

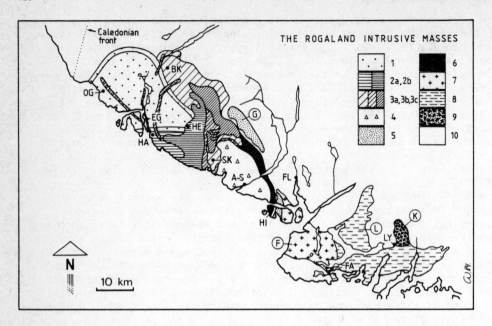

Fig. 2   Geological sketch map of the South Rogaland Igneous Complex (from Michot and Michot, 1969, modified).
1. Egersund Ogna (Eg-Og) anorthosite ;
2. Haaland (2a) and Helleren (2b) anorthosite ;
3. Bjerkreim-Sokndal lopolith (Bk-Sk) : anorthositic basal unit (3a),
Eia-Rekefjord (E.R.) monzonoritic (= jotunitic) intrusion and related dykes (3b),
mangerites and quartz mangerites (3c) ;
4. Aana-Sira (A-S) anorthosite ;
5. Hidra (Hi) and Garsaknatt (G) leuconorites ;
6. Apophysis ;
7. Farsund (F) charnockite ;
8. Lyngdal (L) granodiorite ;
9. Kleivan (K) granite ;
10. Surrounding granulite facies gneisses

border zone of the Eg-Og body was consolidated, foliated and then
enclosed (as blocky inclusions) in the still buoyant central anorthositic part.
The beginning of this complex story, whatever the exact mechanism, is
not known : it could possibly be correlated with the 1500±300 Ma old
event obtained by the Pb-Pb method applied to large orthopyroxene and
plagioclase megacrysts from Eg-Og (Weis and Duchesne, 1983). According
to P. Michot, the end-stage of the Eg-Og emplacement appears from field
relations to be synkinematic, it is related to the first (N-S) folding
phase.

The Haaland anorthosito-leuconoritic basement (J.Michot, 1957), the
oldest unit of this magmatic complex actually appears as a folded unit
consisting of a layered sequence of anorthositic and leuconoritic
granofels with variable thickness, leuconoritic and noritic gneisses
containing small lenses of granulated orthopyroxene and gneissic leuconorite. In general, the contacts between the anorthositic and leuconoritic or noritic rocks are fairly sharp ; the transition between the different types of leuconorites is otherwise very gradual. On a large
scale, the structure shows wide undulations. On a small scale, it is
characterized by the orientation of the ferromagnesian minerals and of
the small anorthositic lenses parallel to the general foliation which in
some places is isoclinally folded, the axial planes being nearly horizontal.

The Bjerkreim-Sokndal (Bk-Sk) layered lopolith has been abundantly
described previously (P.Michot, 1960, 1965 ; J. and P.Michot, 1969 ;
Duchesne, 1970, 1972, 1984). This lopolith was intruded at the very
beginning of the second tectonic phase during which the nappe structure
deformation took its E-W trending axes orientation. It was continuously
deformed, into a syncline, contemporaneously with the updoming of the
Eg-Og mass (P. Michot, 1960). The differentiation process started with a
rythmically layered sequence of anorthosite-norite. No direct isotopic
ages are available for this first differentiation stage. Indirect estimation has been obtained (Pasteels et al, 1979) : a monazite sampled in
a granitic gneiss (Ollestad) at 5 meters of the contact with the layered
norites (rythm III of P.Michot) yielded concordant U-Pb ages at 955 Ma.
This age probably corresponds to the resetting of the monazite U-Pb
system due to the thermal effect related to norite intrusion[*]. Wielens
et al (1981) and Priem and Verschure (1982) reported ages of about 1050
Ma (upper intercepts of discordia chords) for zircons separated from
high-grade metamorphic rocks (M2 metamorphic phase). They related this
age to the intrusion of the older phase of Bk-Sk although, in our
opinion, it could as well be related to final emplacement of the much
larger Eg-Og body.

The previously described Helleren and Aana-Sira (Krause and Pedall,
1980) plutons are quite similar. The former cuts across the layered
anorthosite-leuconorite sequence of Bk-Sk, and was emplaced, as the

---

[*] Maijer (pers. comm.) suggested that this age could correspond to the
resetting of the monazite during the M3 phase and not to the intrusion
age of Bk-Sk.

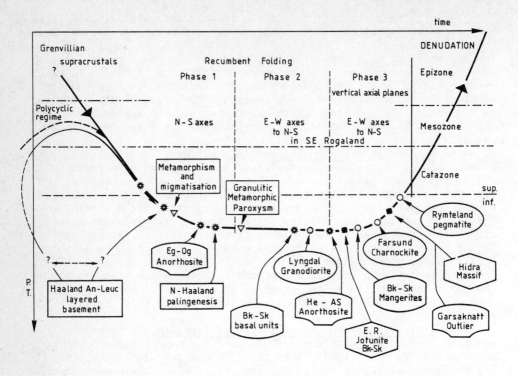

Fig. 3 Tectogram showing the relative timing of the magmatic intrusives together with the tectono-metamorphic evolution.

latter, before the intrusion of the jotunitic phase, and just before the end of the last deformation period.
A system of jotunitic (monzonoritic) dykes (Duchesne et al, 1984) intruded all the large anorthositic bodies ; they are probably related to the Eia-Rekefjord intrusion and to the jotunites occurring in the Apophysis bordering the AS body to the East (Demaiffe, 1972) . This jotunitic phase has been dated at 940-950 Ma (Pasteels et al, 1979). The last acidic differentiation phase of the Bk-Sk lopolith - the quartz mangerite - was dated at 930-950 Ma by both U-Pb on zircons (Pasteels et al, 1970, 1979) and Rb-Sr whole rock isochron on samples collected at close proximity (Wielens et al, 1981).
The Hidra leuconoritic body (Demaiffe et al, 1973 ; Demaiffe, 1977 ; Demaiffe and Hertogen, 1981 ; Weis and Demaiffe, 1983b) is a separate unit, directly intruded in the gneissic envelope (like the Garsaknatt Outlier). It consists essentially of medium grained leuconorite grading to a coarse grained orthocumulate anorthosite at its centre and to a fine grained, locally porphyritic, chilled facies of monzonoritic composition. A stockwork of thin (up to 1 m) charnockitic dykes and granitic pegmatite lenses crosscut the massif. The Hidra body is clearly a late-to post-tectonic intrusion : it lacks the typical proto-clastic structure of most anorthositic massifs and moreover, small leuconoritic dykes injected the gneisses and the nearby Farsund charnockite. U-Pb zircon data on the monzonoritic facies and on the charnockite and Rb-Sr whole rock isochron on the acidic dykes give ages close to 930 Ma (Pasteels et al, 1979).

The Southeastern tip of the magmatic province consists of a very large acidic intrusion, long considered as a single unit, the so-called farsundite (Barth, 1960 ; Middlemost, 1968). Detailed studies have shown the composite nature of this body : three intrusions separated at least locally by gneiss septa (Beelen, 1971 ; Falkum et al, 1972) have been described : the Farsund charnockite to the West, the Lyngdal granodiorite to the East and the differentiated Kleivan granite to the North. These bodies are typically late-to post-tectonic intrusions.
The Lyngdal granodiorite has been dated at 947 Ma (U-Pb discordia chord; Pasteels et al, 1979) which agrees with the Rb-Sr whole rock isochron (932±38 Ma ; Pedersen and Falkum, 1975). The Kleivan granite yields a similar age : 930±7 Ma (Petersen and Pedersen, 1978). The Farsund charnockite gives a rather ill-defined (clustering of data points) Rb-Sr whole rock isochron age of 852±41 Ma (Pedersen and Falkum, 1975), while, in the concordia diagram, the least discordant (not aligned) zircon fractions give apparent ages between 920 and 940 Ma (Pasteels et al, 1979). (This age indication is younger than that of the Lyngdal granodiorite which agrees with the field observations).
The large Rymteland pegmatitic body occurring within the Lyngdal granodiorite marks the end of the magmatic activity in that area: uraninite from this pegmatite gives concordant U-Pb ages at 914±6 Ma (Pasteels et al, 1979).

A Caledonian overprint is visible on both the magmatic and metamorphic rocks : it is of very low (pumpellyite prehnite) to low (greenschist) metamorphic grade and has been dated at 400 Ma (Rb-Sr and K-Ar on green biotites, U-Pb lower intercept ages ; Verschure et al, 1980).

3. DISCUSSION

The previous review of the geochronological data for Southern Norway has shown that many reported ages corresponding to tectono-metamorphic events and magmatic activity fall, as revealed by the large number of values for both the U-Pb and Rb-Sr systems, in the range 1200-900 Ma, the <u>Sveconorwegian period</u> (Table I).

Table I : Tentative geochronological column for both the magmatic and metamorphic rocks of the Rogaland-Agder area during the Sveconorwegian period.

Ages (Ma)	Magmatism	Metamorphism
		biotite (cooling age)
900	acidic plutons around Farsund Hidra leuconorite Bk-Sk mangerite,	
950	Bk-Sk layered norites	third metamorphic phase (M3) *(granulite LT, amphibolite?)*
		osumilite (cooling age)
	Holum granite	
1000	Eg-Og anorthosite	second metamorphic phase (M2)
	(final emplacement ?)	*(granulite HT)*
1050		synemplacement deformation
	porphyritic granite, later deformed to augengneisses (Feda)	
1100		
	Gloppurdi & Botnavatnet syenitic complexes	
1200		first metamorphic phase (M1) *(granulite-amphibolite?)*
1500 ?	Eg-Og anorthosite : first stage Haaland anorthositic basement	

In the whole area, an important tectono-metamorphic event corresponding to the paroxysm of the Sveconorwegian metamorphism has been dated at 1000-1050 Ma. Earlier events, close to 1100 Ma and 1200 Ma have also been recognized, particularly in the western part of the belt (Rogaland, Agder). The end of the Sveconorwegian period is characterized by the late-to post-tectonic (950-900 Ma) intrusions of important volumes of magmatic rocks, mainly of granitic type in the eastern zone and in Agder and of anorthositic-charnockitic type in the western zone (Rogaland).

However, as it was already noted by Berthelsen (1980), the western and eastern subprovinces of the Sveconorwegian mobile belt are quite different. In particular, in the eastern part (East of the Oslo rift), many ages are higher than 1200 Ma which points to the probable existence of a pre-Sveconorwegian (1500-1800 Ma) basement. The same situation is observed for the Bamble district. In the western part (Agder-Rogaland) pre- Sveconorwegian ages have also been reported but they are much more scarce :
- a few zircons from metasedimentary rocks have $^{207}Pb/^{206}Pb$ ages in the range 1500-1400 Ma (Pasteels and Michot, 1975) ;
- the garnetiferous migmatites data points scatter widely in a Rb-Sr isochron diagram between two reference lines at 1500 and 1000 Ma (Wielens et al, 1981). Six charnockitic migmatites taken within a stretch of some 200 m in Drangsdalen give an "age estimate" of 1450 Ma (Versteeve, 1975) ;
- a Pb-Pb whole rock isotopic study on Rogaland granulite facies gneisses yields a linear array which, if interpreted as an isochron, gives an "age indication" of 1359±120 Ma (Weis and Demaiffe, 1983a). The significance of this "age" is that of an U-depletion event related to a high-grade metamorphism.

These higher ages may result :
(i) from a pre-Sveconorwegian high-grade metamorphic episode implying then the existence of a pre-Sveconorwegian basement
(ii) from incomplete resetting and/or isotopic rehomogenization of pre-Sveconorwegian rocks (mainly sediments, but possibly also volcanics (Versteeve, 1975)) during the Sveconorwegian metamorphism.

Whatever the correct explanation, pre-Sveconorwegian supracrustal material seems to have been present (nearly) everywhere in Southern Norway.

The nature of the Sveconorwegian event itself has been questioned : was it principally an episode of thermal regeneration (Welin and Gorbatshev, 1976c) or was it a true tectono-metamorphic event with penetrative deformation and metamorphic recrystallization. The detailed structural and geochronological data obtained both on the eastern zone (Berthelsen, 1980 ; Hageskov and Pedersen, 1981 ; Skjernaa and Pedersen, 1982) and on the western zone (Michot and Michot, 1969 ; Falkum, 1966 ; Falkum and Pedersen, 1979 ; Pedersen et al, 1979 ; Hermans et al, 1975) show the development of penetrative Sveconorwegian structures on both regional (large) and local (small) scales.

The metamorphic grade of the Sveconorwegian period has also been subject to much debate. From their data on the Arendal charnockitic orthogneisses, Field and Råheim (1981, 1983) and Field et al (this volume) concluded that the granulite facies metamorphism was pre-Sveconorwegian (1500 Ma) in the Bamble district and that the superimposed Sveconorwegian (1060 Ma) event was only of very low-grade associated with the intrusions of granitic sheets, without significant deformation. Comparing their data to those obtained in SW Sweden (Lundqvist, 1980 ; Welin et al, 1980, 1982) Field and Råheim suggest (p.161) that "the main high-grade gneiss forming event may have been pre-Grenvillian (= pre-Sveconorwegian) in all sectors (of southern Norway) and that there may have been no regional high-grade reworking during the period 1.2-0.9 Ga ago". This interpretation for the Bamble has however been questioned recently : Hagelia (1985) has dated a Sveconorwegian high-grade metamorphic event in the charnockitic Ubergsmoen augengneiss (1077±64 Ma) and Starmer (this volume) presented evidence for a tectono-metamorphic event which reached amphibolite facies grade during the Sveconorwegian time (1100-1000 Ma) in the Kongsberg and Bamble sectors.

In the Rogaland-Agder district, it is sufficiently well established (see discussion in Weis and Demaiffe, 1983a) that the main Sveconorwegian metamorphism is of high-grade type (granulite facies in South Rogaland grading to amphibolite facies towards the North and the East). In South-eastern Norway, Skjernaa and Pedersen (1982) have shown that this area was repeatedly deformed and metamorphosed under high-and medium-grade conditions, accompanied by migmatization after 1270 Ma (possibly after 1160 Ma), that is during the Sveconorwegian period. In Central Telemark, Priem et al (1973) also noted that the area recrystallized under Dalslandian (= Sveconorwegian) high-grade (almandine amphibolite facies) metamorphism around 1050 Ma.

Consequently, except very locally in the Bamble area, it appears that the Sveconorwegian metamorphism was of high-grade type everywhere (Table II). In that respect, it is interesting to note that Field and Råheim themselves stated as a possible explanation that "the 1536±26 Ma age given by the charnockitic orthogneisses is not in fact recording the time of granulite facies (re)crystallization but a discrete, earlier magmatic event" (p.268). In that hypothesis, the age of the high-grade event in the Bamble is not specified but can conceivably be Sveconorwegian also.

The Precambrian crustal segment of Southern Norway is thus possibly a polymetamorphic (polycyclic?) terrane in which an old (1500-1800 Ma?) basement was deformed and recrystallized several times under high-grade metamorphic conditions during the Sveconorwegian period (1200-900 Ma).

Table II : Metamorphic grade during the Sveconorwegian

Rogaland :	- high T granulite to amphibolite facies (Michot and Michot, 1969 ; Tobi et al, this vol.)
Agder :	- amphibolite facies (Falkum and coworkers)
Kongsberg-Bamble :	- amphibolite to granulite facies (Touret, Hagelia, Starmer) - low grade (Field and Råheim)
Østfold :	- amphibolite facies with migmatization (Berthelsen, Pedersen)

The metamorphic grade increased towards the Southwest where high-T granulite facies mineral assemblages (Maijer et al, 1981 ; Tobi et al, this volume) are found close to the large Rogaland anorthositic complex. The final emplacement of the large magmatic bodies on the eastern border occurred syn-or late-tectonically during the Sveconorwegian. However, the possibility of a Sveconorwegian (= Grenvillian) reworking of an older (1500 Ma?) anorthositic basement cannot be demonstrated with the available geochronological data but can neither be ruled out (see also Duchesne et al, 1984).
In the Agder-Rogaland province, the Sr-isotopic composition of the Sveconorwegian magmatic rocks (granitic intrusions or rocks of the anorthosite-charnockite suite) is usually low ($^{87}Sr/^{86}Sr$ = 0.7029 to 0.7075 ; Michot and Pasteels, 1968 ; Demaiffe et al, 1974 ; Pedersen and Falkum, 1975 ; Demaiffe, 1977 ; Duchesne and Demaiffe, 1978 and this vol.). These low values suggest an origin in the upper mantle or in the LIL depleted lower crust. Pb-isotopic data (Weis and Demaiffe, 1983b ; Weis and Duchesne, 1983; Weis, this volume) confirm this interpretation. In the Pb/Pb diagram, the Hidra monzonoritic border facies (the most radiogenic rock so far analyzed in   Rogaland) together with the Eg-Og giant plagioclases and orthopyroxene megacrysts (equilibrated at high P and T, Maquil and Duchesne, 1984) define a trend which may be representative of the mantle. Moreover, contamination processes by lower crustal material, represented by the less radiogenic granulite facies gneisses, occurred during the magmatic differentiation and is particularly important in the late stage charnockitic liquid.
These plutonic rocks are actually in a high-grade domain which has recrystallized at a pressure of 4-5 Kb (Swanenberg, 1980 ; Jansen et al, this volume) which corresponds to a depth of 15-20 km (previous estimates give slightly higher values, in the range 6 to 7 Kb ; Henry, 1974). The geophysical data (Sellevoll, 1973) show that the Moho discontinuity under SW Norway is actually situated at a depth of 28-30 km. Consequently, the Sveconorwegian continental crust was about 50 to 60 km thick. Moreover, from gravimetric data, Smithson and Ramberg (1979) have estimated at 4 to 5 km the thickness of the anorthosite sheet, so that there is still more than 20 km of crustal rocks, probably in the high pressure granulite facies, beneath the anorthosites.
Since Dewey and Burke (1973), the crustal thickening of the Grenville Province and its Sveconorwegian continuation is classically considered as a deeply eroded Precambrian analogue of the Tibetan platform which results from continental collision and basement reactivation.
In detail however, there is not yet a consensus on the geological evolution of the Grenville Province as a whole; several models have been proposed : millipede model of ductile plate tectonics (Fig.4a) (Wynne-Edwards, 1976), cordilleran-type orogenic segment (Torske, 1977), plate collision with subduction (Fig.4b) (Berthelsen, 1980 ; Falkum and Petersen, 1980), intracontinental deformation by "plate-jams" without subduction (Baer, 1981).
The crustal structure actually observed in South Rogaland could also be explained by the superposition of two high-grade continental blocks (J.Michot, 1984) (fig.4c) : the segment with the anorthosite complex could have been overthrust   on an "older" block which has thus been

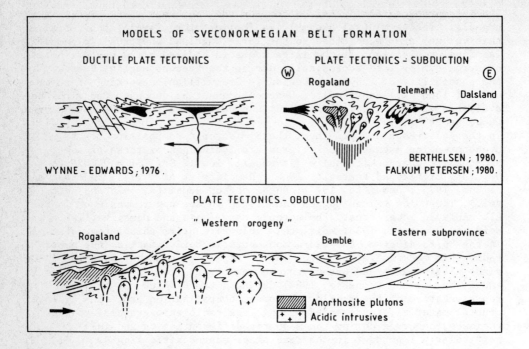

Fig. 4   Schematic hypothetical models for the crustal thickening in the Sveconorwegian (= Grenvillian) Province.
A - Ductile Plate Tectonics (Wynne-Edwards, 1976)
B - Plate collision with subduction (Berthelsen, 1980 ; Falkum and Petersen, 1980)
C - Plate collision with obduction

rejuvenated. The limit between these two units is not identified at the surface : it could possibly be located at the augen gneiss formation, i.e. Feda Augengneiss, which appears as a deep-seated blastomylonitic rock. Dewey and Burke (1973) indeed proposed that the Precambrian reactivated terranes have been deeply eroded to a "cryptic suture-charnockite-anorthosite level" (p.691)
In that hypothesis, the metamorphic complex east of the Rogaland segment represents the old reactivated basement characterized by many granitic intrusions. In the western part of the underthrust block, the rejuvenation process might explain the crosscutting relations of the late granitic intrusives of crustal affinities into an upper block which has evolved under deeper conditions of upper mantle or lower crustal signature.
The thrusting may be correlated with the "Great Mylonite zone" considered as the last episode of the Sveconorwegian event in the eastern zone ; this event has been dated at 950 ± 50 Ma by the intrusion of dyke swarm related to "the accumulation of thrust-sheets coming from the West onto the Craton" (Baer, 1981, p.378). More structural data are needed to substantiate this model.

ACKNOWLEDGEMENTS

Dr Touret, the NATO ASI Director, gave considerable editorial assistance and made helpful suggestions to improve the manuscript. Drs Duchesne and Verschure critically read a first draft of this paper.
This work was supported by a FRFC grant from the Belgian "Fonds national de la Recherche Scientifique".

REFERENCES

ASHWAL L. and WOODEN J.L. (1983) - 'Sr and Nd isotope geochemistry, geologic history and origin of the Adirondack anorthosite'. Geochim. Cosmochim. Acta, 47, 1875-1885.
BAER A.J. (1981) - 'A Grenvillian model of Proterozoic plate tectonics'. In : KRONER (Ed) : Precambrian Plate tectonics. Develop. in Precamb. Geol., n° 4, Elsevier, 353-385.
BARTH T.F.W. (1960) - 'The Precambrian of Southern Norway'. In : OLTEDAHL O. (Ed) : Geology of Norway. Nor. Geol. Unders., 208.
BARTON J. and DOIG R. (1977) - 'Sr-isotopic studies of the origin of the Morin anorthosite complex, Quebec, Canada'. Contr. Min. Petrol., 61, 219-230.
BEELEN R. (1971) - 'Contribution à l'étude pétrologique de la farsundite'. Mém. Lic. Sciences, ULB.
BERTHELSEN A. (1978) - 'Himalayan and Sveconorwegian tectonics, a comparison'. In : SAKLANI P.S. (Ed) : Tectonic geology of the Himalayas, 287-294.
BERTHELSEN A. (1980) - 'Towards a palinspastic tectonic analysis of the Baltic Shield'. Mém. BRGM, 6th Coll. Int. Geol. Cong., Paris 1980, 108, 5-21

BRUECKNER H.K. (1972) - 'Interpretation of Rb-Sr ages from the Precambrian and Paleozoic rocks of southern Norway'. Amer. Journ. Sci., 272, 334-358.
DEMAIFFE D. (1972) - 'Etude pétrologique de l'apophyse SE du massif de Bjerkrem-Sogndal (Norvège méridionale)'. Ann. Soc. Géol. Belg., 95, 255-269.
DEMAIFFE D. (1977) - 'Les massifs satellites anorthosito-leuconoritiques d'Hidra et de Garsaknatt : leur signification pétrogénétique'. Ann. Soc. Géol. Belg., 100, 167-174.
DEMAIFFE D., DUCHESNE J.C., MICHOT J. and PASTEELS P. (1973) - 'Le massif anorthosito-leuconoritique d'Hidra et son faciès de bordure'. C.R. Acad. Sci. Paris, 277, 17-20.
DEMAIFFE D., MICHOT J. and PASTEELS P. (1974) - 'Time relationship and Sr isotopic evolution in the magma of the anorthosite-charnockite suite of South Norway'. Intern. Coll. Geochron. Cosmochron. and Isot. Geol., Paris, abst.
DEMAIFFE D. and HERTOGEN J. (1981) - 'REE geochemistry and Sr isotopic composition of a massif-type anorthositic-charnockitic body : the Hidra massif (Rogaland, SW Norway)'. Geochim. Cosmochim. Acta, 45, 1545-1561.
DE WAARD D., DUCHESNE J.C. and MICHOT J. (1974) - 'Anorthosites and their environment'. In : BELLIERE J. and DUCHESNE J.C. (Eds) : Géologie des domaines cristallins. Centen. Soc. Géol. Belg., Liège, 323-346.
DEWEY J.F. and BURKE K.C. (1973) - 'Tibetan, Variscan and Precambrian basement reactivation : products of continental collision'. J. Geol., 81, 683-692.
DUCHESNE J.C. (1970) - 'Microtextures of Fe-Ti oxide minerals in the South Rogaland anorthositic complex (Norway)'. Ann. Soc. Géol. Belg., 93, 527-544.
DUCHESNE J.C. (1972) - 'Iron-titanium oxide minerals in the Bjerkrem-Sogndal massif'. J. Petrol., 13, 57-81.
DUCHESNE J.C. (1984) - 'The Bjerkreim-Sokndal lopolith'. In : C. MAIJER (Ed) : Excursion guide of the South Norway geological excursion, 94-101.
DUCHESNE J.C. and DEMAIFFE D. (1978) - 'Trace elements and anorthosite genesis'. Earth Planet. Sci. Lett., 38, 249-272.
DUCHESNE J.C., MAQUIL R. and DEMAIFFE D. (1985) - The Rogaland anorthosites : facts and speculations (this volume).
DUCHESNE J.C. and MICHOT J. (1984) - 'The Rogaland intrusive masses : an introduction'. In : C. MAIJER (Ed) : Excursion guide of the South Norway geological excursion, 77-83.
DUCHESNE J.C., ROELANDTS I., DEMAIFFE D. and WEIS D. (1984) - 'Petrology, trace element geochemistry, Sr and Pb isotopes in monzonorites related to Rogaland anorthosites (SW Norway) : evidence of variable crustal contribution'. Contr. Mineral. Petrol. (in press).
EMSLIE R.F. (1978) - 'Anorthosite massifs, Rapakivi granites and late Proterozoic rifting of North America'. Precamb. Res., 7, 61-98.
FALKUM T. (1966) - 'Structural and petrological investigations in the Precambrian metamorphic and igneous charnockite and migmatite complex in the Flekkefjord area, S. Norway'. Nor. Geol. Unders., 242, 19-25.
FALKUM T. (1985) - 'Geotectonic evolution of Southern Scandinavia in light of a late-Proterozoic plate collision' (this volume)
FALKUM T. and PEDERSEN S. (1979) - 'Rb-Sr age determination on the

intrusive Precambrian Homme granite and consequences for dating the last regional folding and metamorphism in the Flekkefjord region, SW Norway'. Nor. Geol. Tidsskr., 59, 59-65.
FALKUM T. and PETERSEN J.S. (1980) - 'The Sveconorwegian orogenic belt: a case of late-Proterozoic plate collision'. Geol. Rundsch., 69, 622-647.
FALKUM T., WILSON J.R., ANNIS M.P. FREGERSLEV S. and ZIMMERMANN H.D. (1972) - 'The intrusive granites of the Farsund area, South Norway'. Nor. Geol. Tidsskr., 52, 463-465.
FIELD D. and RÅHEIM A. (1979a) - 'Rb-Sr total rock isotope studies on Precambrian charnockitic gneisses from south Norway : evidence for isochron resetting during a low-grade metamorphicdeformational event'. Earth Planet. Sci. Lett., 45, 32-44.
FIELD D. and RÅHEIM A. (1979b) - 'A geologically meaningless Rb-Sr total rock isochron'. Nature (London), 282, 497-499.
FIELD D. and RÅHEIM A. (1981) - 'Age relationships in the Proterozoic high-grade gneiss regions of Southern Norway'. Precamb. Res., 14, 261-275.
FIELD D. and RÅHEIM A. (1983) - 'Age relationships in the Proterozoic high-grade gneiss regions of Southern Norway : reply to "Discussion and comment" by WEIS and DEMAIFFE. Precamb. Res., 22, 156-160.
FIELD D., SMALLEY P.C., LAMB R.C. and RÅHEIM A. (1985) - 'Geochemical evolution of the 1.6-1.5 Ga old amphibolite-granulite facies terrain, Bamble sector, Norway : dispelling the myth of regional Grenvillian orogenic reworking'. (this volume)
GORBATSCHEV R. (1979) - 'The basement of the Åmål supracrustals, SW Sweden'. Geol. Fören. Stockholm Förh, 101, 71-73.
HAGELIA P. (1985) - 'Rb-Sr dates of the Ubergsmoen augengneiss and its country rock, Bamble, South Norway : Evidence for a Sveconorwegian (Grenvillian) high-grade event superimposing ca. 1500 Ma old crust'. (this volume)
HAGESKOV B. and PEDERSEN S. (1981) - 'Rb/Sr whole rock age determinations from the western part of the Ostfold basement complex, SE Norway'. Bull. Geol. Soc. Denmark, 29, 119-128.
HENRY J. (1974) - 'Garnet-cordierite gneisses near the Egersund-Ogna anorthositic intrusion, SW Norway'. Lithos, 7, 207-216.
HERMANS G.A., TOBI A.C., POORTER R.P. and MAIJER C. (1975) - 'The high-grade metamorphic Precambrian of the Sirdal-Ørsdal area, Rogaland/Vest Agder, SW Norway'. Nor. Geol. Unders., 318, 51-74.
JACOBSEN S.B. and HEIER K.S. (1978) - 'Rb-Sr isotope systematics in metamorphic rocks, Kongsberg sector, S.Norway'. Lithos, 11, 257-276.
JANSEN J.B., SCHEELINGS M. and BOS A. (1985) - 'Geothermometry and geobarometry in Rogaland and preliminary results of the Bamble area, S. Norway' (this volume).
KRAUSE H. and PEDALL G. (1980) - 'Fe-Ti mineralizations in the Ana-Sira anorthosite, S.Norway'. In : JAAKO SIIVOLA (Ed) : "Metallogeny of the Baltic Shield", Geol. Surv. Finland, 307, 56-83.
KROGH T. and DAVIS G.L. (1969) - 'Geochronology of the Grenville Province'. Carnegie Inst. Washington, Yearb., 67, 224-230.
LINDH A. (1974) - 'The Mylonite zone in SW Sweden (Värmland). A re-interpretation'. Geol. Fören. Stockholm Förh., 96, 183-187.

LUNDEGARTH P.H. (1977) - 'The Gräsmark formation of western Central Sweden'. Sver. Geol. Unders., Ser C, 732, 1-17.
LUNDQVIST T. (1979) - 'The Precambrian of Sweden'. Sver. Geol. Unders., Ser C, 768, 1-87.
MAGNUSSON N.H. (1937) - 'Den centralvärmlandske mylonitzonen och dess fortsätning i Norge'. Geol. Fören. Stockholm Förn., 59, 205.
MAGNUSSON N.H., LUNDQVIST G. and REGNELL G. (1963) - 'Sveriges geologi'. 4th ed., Scand. Univ. Books, Svenska Bokförlaget, 698 p.
MAIJER C., ANDRIESSEN P.A., HEBEDA E.H., JANSEN J.B. and VERSCHURE R.H. (1981) - 'Osumilite, an approximately 970 Ma old high-temperature index mineral of the granulite facies metamorphism in Rogaland, SW Norway'. Geol. Mijnb., 60, 267-272.
MAQUIL R. and DUCHESNE J.C. (1984) - 'Géothermométrie par les pyroxènes et mise en place du massif anorthositique d'EgersundOgna (Rogaland, Norvège méridionale)'. Ann. Soc. Géol. Belg., 107, 27-49.
MICHOT J. (1957) - 'Un nouveau type d'association anorthositenorite dans la catazone norvégienne (Egersund)'. Ann. Soc. Géol. Belg., 80, 449-461.
MICHOT J. (1961) - 'Le massif complexe anorthosito-leuconoritique de Haland-Helleren et la palingenèse basique'. Acad. Roy. Belg., Classe des Sciences, Mém., 2e série, XV,1, 95 p.
MICHOT J. (1984) - 'The Rogaland segment: an obducted precambrian mantle based crust' First EGT Workshop : The northen segment, Copenhague, October 83, European Science Foundation, 1984.
MICHOT J. and PASTEELS P. (1968) - 'Etude géochronologique du domaine métamorphique du SW de la Norvège'. Ann. Soc. Géol. Belg., 91, 93-110.
MICHOT J. and MICHOT P. (1969) - 'The problem of anorthosites: the South-Rogaland igneous complex, SW Norway'. In ISACHSEN Y.W. (Ed): "Origin of anorthosites and related rocks", N.Y. St. Mus. Sci. Serv., Mem 18, 399-410.
MICHOT J. and PASTEELS P. (1972) - 'Aperçu pétrologique et géochronologique sur le complexe éruptif du Rogaland méridional et sa couverture métamorphique'. Sci. Terre, XVII, 195-214.
MICHOT P. (1956) - 'La géologie des zones profondes de l'écorce terrestre'. Ann. Soc. Géol. Belg., 80, 19-73.
MICHOT P. (1960) - 'La géologie de la catazone: le problème des anorthosites, la palingenèse basique et la tectonique catazonale dans le Rogaland méridional (Norvège méridional)'. Norges Geol. Unders., 212a, 1-54.
MICHOT P. (1965) - 'Le magma plagioclasique'. Geol. Rundsch., 55, 956-976.
MICHOT P. (1969) - 'Geological environments of the anorthosites of South South-Rogaland, Norway. In ISACHSEN Y.W. (Ed) : "Origin of anorthosites and related rocks", N.Y. St. Mus. Sci. Serv., Mem 18, 411-423.
MIDDLEMOST E. (1968) - 'The granitic rocks of Farsund, South Norway'. Nor. Geol. Tidsskr., 48, 81-99.
O'NIONS R.K., MORTON F.B. and BAADSGAARD H. (1969) - 'Potassium-argon ages from the Bamble sector of the Fennoscandian shield in South Norway'. Nor. Geol. Tidsskr., 99, 171-190.
O'NIONS R.K. and BAADSGAARD H. (1971) - 'A radiometric study of polymetamorphism in the Bamble region, Norway'. Contr. Miner. Petrol., 34, 1-21.
PASTEELS P., MICHOT J. and LAVREAU J. (1970) - 'Le complexe éruptif du Rogaland méridional. Signification pétrogénétique de la farsundite et de

la mangérite quartzique des unités orientales ; arguments géochronologiques et isotopiques'. Ann. Soc. Géol. Belg., 93, 453-476.
PASTEELS P. and MICHOT J. (1975) - 'Geochronologic investigation of the metamorphic terrain of SW Norway'. Nor. Geol. Tidsskr., 55, 111-134.
PASTEELS P., DEMAIFFE D. and MICHOT J. (1979) - 'U-Pb and Rb-Sr geochronology of the eastern part of the South Rogaland igneous complex, S. Norway'. Lithos, 12, 199-208.
PEDERSEN S. (1981) - 'Rb/Sr age determinations on late Proterozoic granitoids from the Evje area, S. Norway'. Bull. Geol. Soc. Denmark, 29, 129-143.
PEDERSEN S. and FALKUM T. (1975) - 'Rb-Sr isochrons for the granitic plutons around Farsund, S. Norway'. Chem. Geol., 15, 97-101.
PEDERSEN S., BERTHELSEN A., FALKUM T., GRAVERSEN O., HAGESKOV B., MAALOE S., PETERSEN J.S., SKJERNAA L. and WILSON J.R. (1978) - 'Rb-Sr dating of the plutonic and tectonic evolution of the Sveconorwegian Province, S. Norway'. U.S. Geol. Surv. Open File Rept., 78-701, 329-331.
PETERSEN J.S. and PEDERSEN S. (1978) - 'Strontium-isotope study of the Kleivan granite, Farsund area, S. Norway'. Nor. Geol. Tidsskr., 58, 97-107.
PRIEM H.N.A. and VERSCHURE R.H. (1982) - 'Review of the isotope geochronology of the high-grade metamorphic Precambrian of SN Norway'. Geol. Rundsch., 71, 81-84.
PRIEM H.N.A., BOELRIJK N.A., HEBEDA E.H., VERDURMEN E.A. and VERSCHURE R.H. (1973) - 'Rb-Sr investigations on Precambrian granites, granitic gneisses and acidic metavolcanics in central Telemark : metamorphic resetting of Rb-Sr whole rock systems'. Nor. Geol. Unders., 289, 37-53.
RAMBERG I.B. (1976) - 'Gravity interpretation of the Oslo Graben and associated igneous rocks'. Nor. Geol. Unders., 325, 1-194.
SAMUELSSON L. and ÅHALL K.I. (1985) - 'Preliminary sequence of events in Stora Le-Marstrand formation, Bohuslän - SW Sweden'. (this volume).
SELLEVOLL M.A. (1973) - 'Mohorovicic discontinuity beneath Fennoscandia and adjacent parts of the Norwegian and the North sea'. Tectonophys., 20, 359-366.
SILVER L.T. (1969) - 'A geochronologic investigation of the anorthosite complex, Adirondacks Mountains, New York'. In : ISACHSEN Y.W. (Ed) : "Origin of anorthosites and related rocks", N.Y. St. Mus. Sci. Serv., Mem 18, 399-410.
SILVER L.T. and LUMBERS S.B. (1965) - 'Geochronologic studies in the Bancroft-Madoc area of the Grenville Province, Ontario, Canada'. (Abstr) Geol. Soc. Amer., Ann. Mtg, p. 153.
SIGMOND E.M.O. (1983) - Communic. to the EUGENO-S. Workshop, Copenhague.
SKIÖLD T. (1976) - 'The interpretation of the Rb-Sr and K-Ar ages of late Precambrian rocks in SW Sweden'. Geol. Fören. Stockholm Förh., 98, 3-29.
SKJERNAA L. and PEDERSEN S. (1982) - 'The effects of penetrative Sveconorwegian deformations on Rb-Sr isotope systems in the Römskog-Aurskog-Höland area, SE Norway'. Precamb. Res., 17, 215-243.
SMALLEY P.C., FIELD D. and RÅHEIM A. (1983) - 'Resetting of Rb-Sr whole rock isochrons during Sveconorwegian low-grade events in the Gjerstad augengneiss, Telemark, S. Norway'. Isot. Geosci., 1, 269-282.
SMITHSON S.B. and RAMBERG I.B. (1979) - 'Gravity interpretation of the

Egersund anorthosite complex, Norway : its petrological and geothermal significance'. Geol. Soc. Amer. Bull., 90, 199-204.
STARMER I.C. (1985) - 'The evolution of the South Norwegian Proterozoic as revealed by the major and mega-tectonics of the Kongsberg and Bamble sectors'. (this volume)
SWANENBERG H.E. (1980) - 'Fluid inclusions in high-grade metamorphic rocks from S.W. Norway'. Geol. Ultraiectina, 25, 145 pp.
TOBI A.C. (1965) - 'Field work in the charnockitic Precambrian of Rogaland (SW Norway)'. Geol. Mijnb. 44, 207-217.
TOBI A.C., HERMANS G.A. and MAIJER C. (1985) - 'The high-grade metamorphism in the Proterozoic of Rogaland-Vest Agder, SW Norway'. (this volume).
TORSKE T. (1977) - 'The South Norway Precambrian region - a Proterozoic cordilleran type orogenic segment'. Nor. Geol. Tidsskr., 57, 97-120.
VERSCHURE R.H. (1985) - 'Geochronological framework for Late-Proterozoic evolution of the Baltic Shield in South Scandinavia'. (this volume).
VERSCHURE R.H., ANDRIESSEN P.A., BOELRIJK N.A., HEBEDA E.H., MAIJER C., PRIEM H.N. and VERDURMEN E.A. (1980) - 'On thermal stability of Rb-Sr and K-Ar biotite systems : evidence from coexisting Sveconorwegian (ca 870 Ma) and Caledonian (ca 400 Ma) biotites in SW Norway'. Contr. Mineral. Petrol., 74, 245-252.
WANLESS R.K., STEVENS R.D., LACHANCE G.R. and DELABIO R.N. (1972) - 'Age determinations and geologic studies, K-Ar isotopic ages report 10'. Geol. Surv. Can., Pap., 71-2.
WANLESS R.K., STEVENS R.D., LACHANCE G.R. and DELABIO R.N. (1973) - 'Age determinations and geologic studies, K-Ar isotopic ages report 11'. Geol. Surv. Can., Pap., 73-2.
WEIS D. and DEMAIFFE D. (1983a) - 'Age relationships in the Proterozoic high-grade gneiss regions of Southern Norway : discussion and comment'. Precamb. Res., 22, 149-155.
WEIS D. and DEMAIFFE D. (1983b) - 'Pb isotope geochemistry of a massif-type anorthositic body : the Hidra massif (Rogaland, S.W. Norway)'. Geochim. Cosmochim. Acta, 47, 1405-1413.
WEIS D. and DUCHESNE J.C. (1983) - 'Pb isotopic compositions in the Egersund-Ogna anorthosite (S.W. Norway) : an indication of a Gothian magmatic event'. (Abstr) Terra Cognita, 3, 140.
WELIN E. and GORBATSCHEV R. (1976a) 'Rb-Sr age of granitoid gneisses in the "Pre-Gothian" area of SW Sweden'. Geol. Fören. Stockholm Förh., 98, 378-381.
WELIN E. and GORBATSCHEV R. (1976b) - 'A Rb-Sr geochronological study of the older granitoids in the Åmål tectonic mega-unit, SW Sweden'. Geol. Fören. Stockholm Förh., 98, 374-377.
WELIN E. and GORBATSCHEV R. (1976c) - 'The Rb-Sr age of the Hästefjorden Granite and its bearing on the Precambrian evolution of southwestern Norway'. Precamb. Res., 3, 187-195.
WELIN E. and GORBATSCHEV R. (1978) - 'Rb-Sr age of Lane granites in SW Sweden'. Geol. Fören. Stockholm Förh, 100, 101-102.
WELIN E., GORBATSCHEV R. and KAHR A.M. (1982) - 'Zircon dating of polymetamorphic rocks in SW Sweden'. Sv. Geol. Unders., 797, 1-34.
WIELENS J.B.W., ANDRIESSEN P.A., BOELRIJK N.A., HEBEDA E.H., PRIEM H.N., VERDURMEN E.A. and VERSCHURE R.H. (1981) - 'Isotope geochronology in the

high-grade metamorphic Precambrian of SW Norway : new data and reinterpretation'. Nor. Geol. Unders., 359, 1-30.
WILSON J.R., PEDERSEN S., BERTHELSEN C.R. and JAKOBSEN B.M. (1977) - 'New light on the Precambrian Holum granite, S. Norway'. Nor. Geol. Tidsskr., 57, 347-360.
WYNNE-EDWARDS H.R. (1976) - 'Proterozoic ensialic orogenesis : the millipede model of ductile plate tectonics'. Amer. J. Sci., 276, 927-953.

NEODYMIUM ISOTOPE EVIDENCE FOR THE AGE AND ORIGIN OF THE PROTEROZOIC OF TELEMARK, SOUTH NORWAY.

Julian F. Menuge
Department of Geology
University College
Belfield
Dublin 4
Ireland

ABSTRACT. Nd isotope data indicate derivation of 0.9 Ga old post - tectonic granites from acid gneisses. Both granites and gneisses consist of material accreted ~1.5 Ga ago from a depleted mantle source ($\epsilon Nd(T)$ ~5.5). A Sm-Nd whole - rock isochron age of 1190 +/- 37 Ma from the Rjukan metavolcanics is believed to date their extrusion; the initial ratio ($\epsilon Nd(T)$ = 2.1 +/- 0.7) suggests depleted - mantle - derived magma contaminated by 1.5 Ga old crust. Basic Bandak volcanics also comprise depleted - mantle - derived magma ($\epsilon Nd(T)$ ~7) variably contaminated by older acid rocks. The acid Bandak volcanics were derived essentially from 1.5 Ga old crust. Combined with published results, the data suggest gradual mantle depletion and complementary crustal growth for at least 3.8 Ga.

1. INTRODUCTION

The mantle source regions of Archaean and Phanerozoic crust are now fairly well characterized in terms of their Nd isotope ratios. Most Archaean rocks (reviewed by Carlson et al. (1)) had mantle sources of near - chondritic $143Nd/144Nd$ ratio ($\epsilon Nd(T)$ = +1 to +3). In contrast, many studies have shown that most Phanerozoic mantle sources of oceanic and continental crust had $\epsilon Nd(T)$ = +7 to +13. This has led to the widely held view that the Sm/Nd ratio of these mantle regions has increased from an approximately chondritic value since the early Archaean, consistent with depletion of lithophile elements in general by extraction of partial melts into a growing continental crust, e.g. (3,4).

Several studies indicate depleted mantle with ϵNd up to +8, 0.6 - 0.8 Ga ago (5,6,7,8), but the period 0.8 - 2.5 Ga ago is poorly known. Values up to ϵNd = +5, 1.3 Ga ago (53) and +4, 1.8 Ga ago (10) are amongst the highest reported. Modern MORBs provide uncontaminated samples of the source regions of Phanerozoic continental crust, and Archaean rocks usually reveal negligible contamination, presumably due to the absence of significantly older crust, but except for a few late Proterozoic ophiolites, Proterozoic mantle - derived rocks have mostly suffered substantial crustal contamination which is difficult to quantify. A relatively large number of analyses will be required to demonstrate convincingly the Nd isotope evolution of the Proterozoic mantle.

Most intracrustal processes, including magmagenesis, involve relatively minor fractionation of Sm/Nd ratios compared to the change resulting from mantle partial melting preceding crust formation. This allows newly accreted crust to be distinguished isotopically from rocks derived by reworking of substantially older crust. Most Archaean rocks analysed, both acid and basic, were derived either directly from the mantle or from a young (<100 Ma old) crustal precursor (1), whereas Phanerozoic igneous activity involved a major, often dominant, crustal component, usually of Proterozoic age (11,12,13). The relative importance of accretion and reworking during the Proterozoic is not well known, although a few studies, e.g. (10), suggest that both processes were of major significance.

Apart from its regional implications, the Nd isotope study presented here provides further information both on Proterozoic crust - mantle evolution and on the relative volumes of mantle - derived and crust - derived magmas generated in the Proterozoic.

2. PREVIOUS WORK

Geological descriptions of Telemark (14-19) (Fig. 1) and previous geochemical and isotopic studies (Table i) are summarized in this section. Three major structural units have been recognised in the area.

(a) A basement complex of mainly acid gneisses.
(b) The Telemark Supracrustals, a metamorphosed and deformed sequence of acid and basic volcanics and clastic sediments, with a maximum thickness of at least 6km (16).
(c) Post - tectonic granites accompanied by minor intermediate and basic intrusions.

2.1. Acid gneisses

In western Telemark, the gneisses have been divided into two groups. The Flothyl Complex consists of partly - migmatized granitic to granodioritic orthogneisses, and the Botsvatn Complex comprises mainly biotite gneisses which may be derived from andesitic and dacitic supracrustals. In this area, the gneisses form a basement to the Telemark Supracrustals (18). Whole - rock Rb-Sr isochron ages around 1500 Ma (Table i) from western Telemark gneisses are believed to date high - grade regional metamorphism (20,21). The low initial $^{87}Sr/^{86}Sr$ ratios suggest a pre - 1500 Ma crustal residence of <100 Ma.

The central Telemark gneisses are also predominantly granitic to granodioritic (17). However, gradual transitions between gneisses and Telemark Supracrustals indicate that some of the gneisses were derived from the Telemark Supracrustals by migmatization and anatexis (22-26). In both western and central Telemark, the gneisses were metamorphosed to

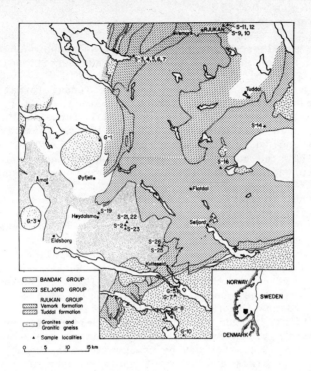

Fig. 1. Simplified geological map of central Telemark, modified after Dons (14), showing sample localities.

Rock type	Age(Ma) +/- 2 S.E.	Published interpretation
Acid gneisses	1520 +/- 30	Metamorphism (20)
	1477 +/- 35	Metamorphism (21)
	1107 +/- 24	Metamorphism/anatexis (30)
	1106 +/- 26	(31)
Post - tectonic granites	916 +/- 40	Emplacement (20)
	901 +/- 23	Emplacement (21)
	888 +/- 48 *	Emplacement (30)
	895 +/- 38	Emplacement (31)

Table i. Published Rb-Sr geochronology of Telemark. 87Rb decay constant = .0142 per Ga (49). * = mineral isochron; others are whole - rock isochrons.

amphibolite facies, followed by widespread retrogression to greenschist assemblages; in western Telemark at least two phases of deformation preceded deposition of the Telemark Supracrustals (18,22,25).

2.2. The Telemark Supracrustals

The Telemark Supracrustals comprise three lithostratigraphic units, the Rjukan, Seljord and Bandak groups, separated by unconformities (14,15), each up to at least 2 km thick (16).

The Rjukan group has been subdivided into the Tuddal formation and the apparently conformably overlying Vemork formation, each typically several hundred metres thick (14,15). The Tuddal formation consists almost entirely of acid metavolcanics, and the Vemork formation comprises basic lavas and tuffs with minor metasediments. Major element analyses indicate that the basic lavas are dominantly tholeiitic, with basaltic andesite to andesite compositions suggesting sialic contamination (27).

The Seljord group consists of a basal conglomerate overlain by mature quartzites with sedimentary structures characteristic of shallow marine deposition (28). Minor schists and calcareous sandstones occur mainly in the upper part of the group.

Acid volcanics, basic lavas, quartzites and other sediments make up the Bandak group. In western Telemark, where the Rjukan and Seljord groups are absent, it reaches a thickness of 5 km (18). The metabasalts have tholeiitic to alkaline compositions, but their high K contents suggest metasomatic alteration (27,29). Amphibolitized basic sills and dykes up to 600m thick intrude the Rjukan and Seljord supracrustals (14). As their mineralogy and major element chemistry are similar to some Bandak metabasalts (27), they may be comagmatic with them.

The Telemark Supracrustals and basement gneisses were deformed and metamorphosed to greenschist or amphibolite facies during the Sveconorwegian orogeny, when migmatization and anatexis of some gneisses and supracrustals also occurred (14,22,25). Whole - rock Rb-Sr isochron ages of 1107 +/- 24 Ma (30) and 1106 +/- 26 Ma (31) probably date these events (Table i), providing a minimum age for the Telemark Supracrustals.

2.3. Post-tectonic granites

Porphyritic granites up to 10 km in diameter were intruded into the gneisses and Telemark Supracrustals around 900 Ma ago (Table i). The quartz - monzonitic composition of the Vrådal granite (32) may be typical. Pre - emplacement crustal histories of >100 Ma are likely for the granites, which have initial 87Sr/86Sr ratios >.705.

3. ANALYTICAL METHODS

2 - 5 kg samples were collected from the localities in central Telemark shown in Fig. 1. Sample preparation and separation of Sm and Nd closely followed Hooker et al. (33). Two chemical separations were carried out for each sample, one unspiked to measure the 143Nd/144Nd ratio, the other spiked with 149Sm and 146Nd to determine the concentrations of Sm and Nd. The Nd blank was <100 pg, and no blank corrections are required.

The normalization procedures used are summarized in Table ii. Repeated analysis of an internal Nd standard yielded between - run variation in 143Nd/144Nd ratio similar to the within - run statistics of sample analyses. Isochron calculations follow York (57).

4. RESULTS AND INTERPRETATION

4.1. Acid gneisses and post - tectonic granites

Three acid gneisses and three post - tectonic granites from six separate localities were analysed. In Table ii their Nd isotope ratios, presented as ϵNd values, are shown for 1500 Ma ago, the Rb-Sr whole - rock isochron age obtained from gneisses in western Telemark (Table i). Despite a present day variation in ratio of about 4ϵ, two gneisses and two granites have indistinguishable ϵNd(1500) values with a mean of 5.5. The third granite sample has a slightly higher ϵNd(1500) value of 6.5 +/- 0.4, and the third gneiss sample has a much higher value of 11.2 +/- 0.4.

The common Nd isotope ratio of four granites and gneisses 1500 Ma ago strongly suggests that the granites were derived from acid gneisses (900 Ma ago, Table i) with negligible (<5%) change of Sm/Nd ratio, despite resetting of Sr isotopes. This may indicate a high degree of partial melting in the source regions of the granitic magmas.

The acid gneiss Rb-Sr ages around 1500 Ma may date a metamorphic episode (Table i), but Sm-Nd whole - rock systems are usually little affected by metamorphism, e.g. (34-36). Therefore an igneous event 1500 Ma ago is likely, and the high ϵNd(1500) values of 5.3 +/- 0.4 to 5.7 +/- 0.4 are consistent with this event being accretion of the gneiss precursors from a mantle source previously depleted by extraction of crust - forming melts.

The granite sample with ϵNd(1500) = 6.5 +/- 0.4 (G-3) was probably also derived from acid gneisses, but with accompanying reduction of Sm/Nd ratio; a slightly higher initial ratio is another possible explanation. The value of ϵNd(1500) = 11.2 +/- 0.4 for gneiss sample G-5 is implausibly high, and except in the unlikely event of a large decrease in its Sm/Nd ratio, an age of 1500 Ma or more can be discounted. Assuming an ultimate source in a depleted mantle region, accretion 1.1 - 1.2 Ga ago, when the Telemark Supracrustals were formed (see sections 4.2 and 4.3), is consistent with the data, e.g. G-5 has ϵNd (1150) = 6.6 +/- 0.4. This gneiss, from the Kviteseid area (Fig. 1) might therefore have been derived by anatexis of Telemark Supracrustals. In any case there are two types of acid gneisses of different crustal residence age.

4.2. Rjukan group metavolcanics

Five Vemork formation metabasalts from one locality and four Tuddal formation acid metavolcanics from two localities were analysed. Three acid samples define an isochron with T = 1190 +/- 38 Ma, ϵNd(T) = 2.2 +/- 0.7 and MSWD = 0.11 (Fig. 2a). The reason for the large difference in Sm/Nd ratio between S-9 and S-10, which were collected <100m apart, is not clear from petrographic examination, and may reflect variation in

Sample	Sm(ppm)1	Nd(ppm)1	143Nd/144Nd	147Sm/144Nd*	ϵNd(T)~
Acid gneisses(pl,qz,mi,bi,sph,op+/-hbl+/-ap+/-zr)					T=1500Ma
G-5	15.758	100.08	.512209+/-18	.09518	11.2+/-.4
G-7	7.6139	37.278	.512207+/-20	.12347	5.7+/-.4
G-10	5.0459	31.984	.511928+/-16	.09536	5.6+/-.4
Post-tectonic granites(mi,pl,qz,bi,chl,musc,ap,op,sph)					T=1500Ma
G-1	20.139	106.53	.512096+/-16	.11428	5.3+/-.4
G-3	6.7381	51.138	.511815+/-18	.07965	6.5+/-.4
G-8	15.625	111.66	.511811+/-16	.08458	5.4+/-.3
Rjukan basic volcanics (pl,ep,chl,qz,op,cal+/-bi+/-hbl)					T=1190Ma
S-3	6.7830	28.252	.512351+/-22	.14514	2.2+/-.5
S-4	5.9397	25.175	.512320+/-10	.14263	2.0+/-.2
S-5	5.7139	24.250	.512319+/-20	.14244	2.0+/-.4
S-6	3.7593	15.713	.512304+/-11	.14463	1.4+/-.3
	(3.8933	16.358	.512309+/-12	.14387)	
S-7	10.710	49.136	.512232+/-15	.13177	2.0+/-.3
Rjukan acid volcanics (mi,pl,qz,musc,bi,op+/-hbl+/-ap)					T=1190Ma
S-9	5.2050	17.441	.512622+/-13	.18043	2.2+/-.3
S-10	5.0159	27.472	.512074+/-12	.11036	2.1+/-.3
S-11	8.3129	33.523	.512440+/-14	.14991	3.3+/-.3
	(8.2979	33.520	.512464+/-20	.14965)	
S-12	9.3940	44.435	.512213+/-12	.12780	2.2+/-.3
Seljord metaquartzite (qz,musc,chl,op,pl,mi)					T=1190Ma
S-14	10.214	47.563	.512211+/-18	.12982	1.9+/-.4
Bandak basic volcanics (pl,ep,hbl,op+/-qz+/-bi+/-cal)					T=1150Ma
S-2	4.9904	17.805	.512679+/-14	.16945	4.8+/-.3
S-23	5.0436	18.312	.512672+/-20	.16652	5.1+/-.4
S-25	18.522	79.386	.512499+/-12	.14105	5.5+/-.3
	(18.925	80.917)			
S-26	7.7566	29.960	.512533+/-20	.15652	3.9+/-.4
Bandak acid volcanics (qz,musc,pl,mi,cal,op+/-chl+/-ep)					T=1150Ma
S-19	7.5200	39.183	.512107+/-18	.11602	1.5+/-.4
S-21	4.4211	21.078	.512140+/-18	.12679	0.6+/-.4
S-22	7.3123	36.835	.512146+/-14	.12000	1.7+/-.3
Metabasite dyke (hbl,pl,qz,bi,op)					T=1150Ma
S-16	3.2761	10.102	.512985+/-16	.19606	6.9+/-.4

Table ii. Sm-Nd isotope data. 1, +/- 2%. *, +/- .2%. ~, T varies as shown for each group. All errors quoted as 2 standard errors of the mean. 143Nd/144Nd ratios normalized to 146Nd/144Nd = .7219, BCR-1 143Nd/144Nd = .512664 +/- 7. Sm and Nd concentrations calibrated against CIT standard n(Sm/Nd)b (50). Replicate analyses in parentheses. ϵ notation (51) based on (143Nd/144Nd)CHUR = .512638, (147Sm/144Nd)CHUR = .1966 (2).

type and abundance of minor REE - rich phases. Fig. 2b shows the almost identical isochron obtained from the acid samples of Fig. 2a and four basic samples. S-6 (basic) and S-11 (acid) lie off the isochron, and have been omitted from regression calculations. All the volcanics have suffered retrogressive metamorphism, but no difference in style or extent of alteration has been detected which might explain the deviation of these two samples from the isochron.

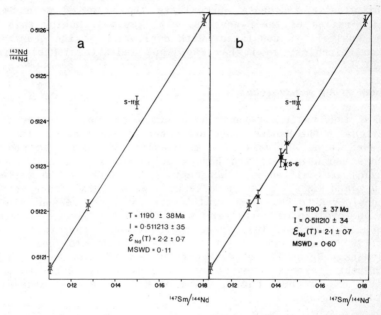

Fig. 2. a, Sm-Nd isochron diagram of Rjukan group acid metavolcanics. S-11 omitted from regression calculations. b, isochron diagram for acid (O) and basic (●) Rjukan metavolcanics. S-11 and S-6 omitted from regression calculations.

The age of 1190 Ma is interpreted as the extrusion age of the volcanics, requiring that differences in age and initial ratio of acid and basic volcanics are smaller than the isochron uncertainties. This age interpretation rules out derivation from Telemark Supracrustals of those acid gneisses >=1500 Ma old, but is consistent with such a derivation for sample G-5, as discussed previously. For the age to correlate instead with a metamorphic event, redistribution of Nd isotopes over >100m is required, in contrast to results from other metavolcanic sequences (34,35,38). Furthermore no other evidence of an 1190 Ma metamorphism in Telemark is known.

The initial ratio ($\epsilon_{Nd}(1190) = 2.1 +/- 0.7$) indicates a slightly depleted magma source. About half the Rjukan volcanics are metabasites, requiring a major mantle - derived component. The acid gneisses and post - tectonic granites reveal depleted mantle having $\epsilon_{Nd}(1500)$ ~5.5, which would have evolved to $\epsilon Nd(1190) > 5.5$ assuming a greater than chondritic Sm/Nd ratio. Therefore this mantle could not be the sole source of the Rjukan volcanics. A mantle source with $\epsilon Nd(1190) = 2.1$ could be invoked, but the large volume of acid volcanics favours a mixed crust - mantle source. Basic magma derived from mantle with $\epsilon Nd > 5.5$ and acid magma derived from 1500 Ma old crust ($\epsilon Nd(1190) \sim 1 - 3$) could yield the Rjukan volcanics, with most of the Nd derived from the crustal component.

One Seljord group metaquartzite was analysed. Sm/Nd and $^{143}Nd/^{144}Nd$ ratios of clastic sediments approximate the weighted mean Sm/Nd and $^{143}Nd/^{144}Nd$ ratios of their sources, e.g. (39,40). Making this assumption, the analysis is consistent with derivation of the quartzite from the Rjukan volcanics, as it has $\epsilon Nd(1190)$ indistinguishable from them (Table ii).

4.3. Bandak group metavolcanics

Analyses of four basic and three acid metavolcanics collected from various horizons in the Bandak group are shown on an isochron diagram (Fig. 3). Included is an analysis of a metabasite dyke which is probably of similar age and origin to the Bandak metabasalts. If S-21 (acid) and S-25 (basic) are omitted an isochron with T = 1667 +/- 36 Ma, $\epsilon Nd(T)$ = 6.8 +/- 0.7 and MSWD = 1.61 is defined. However this linear correlation is not considered to have age significance for three reasons. In western Telemark, metavolcanics correlated with the Bandak group lie unconformably on acid gneisses (18). These gneisses would therefore have to be at least 1700 Ma old and unrelated to the 1500 Ma old gneisses of central Telemark. Secondly the Rjukan group age of 1190 Ma would have to reflect a post - extrusive event, which is unlikely as argued previously. Finally the isochron $\epsilon Nd(T)$ value of 6.8 +/- 0.7 is unusually high for 1670 Ma old rocks.

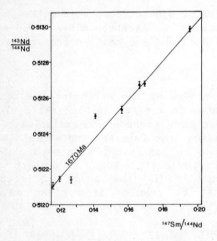

Fig. 3. Sm-Nd isochron diagram of acid (O) and basic (●) Bandak group metavolcanics and metabasite dyke (■), showing 1670 Ma reference line.

The preferred explanation of the linear array is by variation of initial 143Nd/144Nd ratio. If the age of the Bandak volcanics lies between the 1190 +/- 37 Ma age of the Rjukan group and the 1107 +/- 24 Ma metamorphic age (30), an age of 1150 Ma can be assumed for the Bandak group and initial ratios calculated (Table ii). The acid and basic samples have ϵNd(1150) = 0.6 - 1.7 and 3.9 - 6.9 respectively. The linear correlation of initial ratio with 1/Nd (Fig. 4) indicates that two - component bulk mixing can explain the variation in initial ratio of four of the five basic samples. If component 1 is assumed to be a depleted - mantle - derived magma, with 143Nd/144Nd ratio and Nd concentration equal to those of S-16, then a plausible range of Nd concentration of component 2 allows calculation of bulk mixing proportions of the two components. For Nd concentrations of 30 - 150 ppm in component 2, the corresponding 143Nd/144Nd ratios are .51135 - .51129, and the mean proportion of component 2 in the four linearly correlated samples is 6.5 - 45 % (Table iii). Depleted - mantle - derived magma is therefore the dominant component of these samples. One basic sample (S-25) lies far from the linear array due to its unusually high Nd concentration of 80 ppm. It may have crystallized from a magma of the same origin as the other basic volcanics, but following REE enrichment by fractionation.

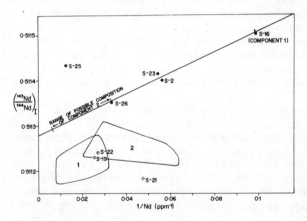

Fig. 4. Bulk mixing diagram of Bandak group acid (O) and basic (●) metavolcanics and metabasite dyke (■). Field 1 = 1500 Ma old crust, field 2 = Rjukan group. 143Nd/144Nd ratios calculated for 1150 Ma ago.

The initial ratio of component 2 of .51129 - .51135 scarcely overlaps with the range exhibited by pre - existing crust 1150 Ma ago (Fig. 4), so the origin of this component remains unknown. However the three acid Bandak volcanics, which do not lie on the basic volcanics linear trend, all have initial ratios similar to that evolved by 1500 Ma old gneisses 1150 Ma ago, and may therefore represent magmas derived entirely from them.

Sample number	Percentage of component 2	
	Model 1	Model 2
S-16	0	0
S-2	39	6
S-23	41	6
S-26	100	14
Mean	45	6.5

Table iii. Two - component bulk mixing calculations for basic Bandak volcanics. Component 1 has $143Nd/144Nd$ = .511505, Nd = 10.1 ppm. In model 1, component 2 has $143Nd/144Nd$ = .51135, Nd = 30 ppm. In model 2, component 2 has $143Nd/144Nd$ = .51129, Nd = 150 ppm. $143Nd/144Nd$ ratios calculated for 1150 Ma ago.

5. DISCUSSION

5.1. Regional correlations

The major episode of crustal accretion in Telemark 1.5 - 1.6 Ga ago recorded by acid gneisses and post - tectonic granites, and the lack of evidence for older continental crust, permits a similar interpretation for other parts of south Norway. In the Kongsberg and Bamble areas, Rb-Sr whole - rock isochron ages up to 1.5 - 1.6 Ga have been interpreted as dating high - grade metamorphism, with low initial $87Sr/86Sr$ ratios suggesting a pre - metamorphic crustal history of <100 Ma (41,42).

The intermittent volcanic activity which produced the basic lavas of the Rjukan and Bandak groups 1.1 - 1.2 Ga ago may be related to many other basic intrusions. In particular, numerous basic dyke swarms of mid to late Proterozoic age occur in Scandinavia, Greenland, Canada and the northern U.S.A.. Few have well - constrained ages, but some were intruded 1.1 - 1.2 Ga ago (43,44). The Telemark volcanics may therefore form part of a major addition of depleted - mantle - derived material to the pre - Sveconorwegian (pre - Grenvillian) crust during a period of crustal extension.

5.2. Evolution of Proterozoic crust and mantle

A compilation of initial Nd isotope ratios of igneous and metamorphic Proterozoic rocks is presented in Fig. 5. Assuming the existence of a discrete depleted mantle source region, its evolution may be demonstrated by the highest ϵNd values obtained at any time, because all other significant magma sources are expected to have had lower ϵNd values. However the curve thus obtained may deviate from the true evolution path for several reasons, including crustal contamination of magmas, contamination of magmas by material derived from undepleted or enriched mantle, regional heterogeneity of depleted mantle, small scale heterogeneity of depleted mantle, and analytical and geological uncertainties in age and initial ratio.

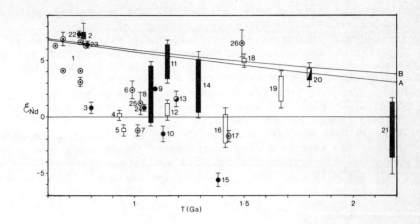

Fig. 5. Initial Nd isotope ratios of basic (solid symbols) and acid (open symbols) Proterozoic rocks, where possible relative to 143Nd/144Nd = .51264 for BCR-1. Curves A and B (10,45) are estimates of depleted mantle evolution. Data from (6-10,52-56, this paper and unpubl. data).

Despite these uncertainties, useful estimates of Proterozoic depleted mantle evolution can be obtained. Two published estimates shown in Fig. 5 (10,45) fit the existing data well except for their slight underestimates of late Proterozoic ϵNd. The gradual increase of ϵNd with time precludes earlier theories of crust - mantle evolution requiring roughly constant volumes of crust and mantle throughout earth history, e.g. (46). Theories involving crustal growth with time, whether continuous (10) or episodic (47), are consistent with the data. The apparently smooth increase of ϵNd rules out major discontinuities in the rate of net crustal accretion during the Proterozoic, supporting continuous rather than episodic evolution. The data are insufficient to determine whether the Proterozoic rate of crustal accretion differed substantially from that of the Archaean and Phanerozoic.

6. CONCLUSIONS

In central Telemark the main episode of crustal accretion 1.5 - 1.6 Ga ago is represented by acid gneisses. The mantle source of this new crust was already substantially depleted (ϵNd(1500) ~5.5) due to prior extraction of partial melts to form crust elsewhere. The continental reconstruction of Piper (48) places south Norway at the margin of the Proterozoic supercontinent 1.5 Ga ago, consistent with Andean - type crustal accretion. Other gneisses were derived by migmatization and anatexis of part of the Telemark Supracrustals sequence, as has been inferred from field and geochemical evidence (22-26,37). The acid gneisses therefore cannot be regarded as a single geological unit, and further Nd isotopic analyses are needed to characterize the various types, especially in western Telemark and the apparent transition zone between gneisses and the Rjukan group in central Telemark.

The Telemark Supracrustals volcanics were extruded about 1.2 - 1.1 Ga ago in an anorogenic environment. They consist of mixtures of depleted - mantle - derived basic magma (ϵNd ~7) with older crustal material probably derived from 1.5 Ga old gneisses.

Post - tectonic granites intruded around 0.9 Ga ago included no newly accreted material, and were derived entirely from 1.5 Ga old gneisses with little change in Sm/Nd ratio.

There is no compelling evidence for any accretion from chondritic or enriched mantle sources during evolution of the Telemark crust. The data further constrain Proterozoic depleted mantle evolution and are consistent with globally continuous growth of the Earth's continental crust.

ACKNOWLEDGEMENTS

I thank T.Brewer for assistance in the field and for useful discussion of the results. J.S.Daly and the two reviewers provided constructive criticism which led to substantial improvement of the manuscript. This research was carried out while the author was a research student at Cambridge University, and was supported by N.E.R.C and the Royal Society. I am particularly grateful to R.K.O'Nions for supervision and encouragement.

REFERENCES

1. R.W.Carlson, D.R.Hunter and F.Barker 1983, Nature 305, 701-704.
2. S.B.Jacobsen and G.J.Wasserburg 1980, Earth Planet. Sci. Lett. 50, 139-155.
3. P.Richard, N.Shimizu and C.J.Allègre 1976, Earth Planet. Sci. Lett. 31, 269-278.
4. R.K.O'Nions, P.J.Hamilton and N.M.Evensen 1977, Earth Planet. Sci. Lett. 34, 13-22.
5. F.Y.Bokhari and J.D.Kramers 1981, Earth Planet. Sci. Lett. 54, 409-422.
6. C.J.Hawkesworth, J.D.Kramers and R.Mc.G.Miller 1981, Nature 289, 278-282.
7. H.J.Duyverman, N.B.W.Harris and C.J.Hawkesworth 1982, Earth Planet. Sci. Lett. 59, 315-326.
8. S.Claesson, J.S.Pallister and M.Tatsumoto 1984, Contrib. Mineral. Petrol. 85, 244-252.
9. A.R.Basu and H.S.Pettingill 1983, Geology 11, 514-518.
10. D.J.DePaolo 1981, Nature 291, 193-196.
11. P.J.Hamilton, R.K.O'Nions and R.J.Pankhurst 1980, Nature 287, 279-284.
12. C.J.Allègre and D.Ben Othman 1980, Nature 286, 335-342.
13. M.T.McCulloch and B.W.Chappell 1982, Earth Planet. Sci. Lett. 58, 51-64.
14. J.A.Dons 1960, Norges Geol. Unders. 208, 49-58.

15   J.A.Dons 1960, Norges Geol. Unders. 212, 1-30.
16   J.A.Dons 1972, Sciences de la Terre 17, 25-29.
17   T.F.W.Barth and P.H.Reitan 1963, in : The Precambrian Vol.1, K.Rankama (ed.), Wiley Interscience, 27-80.
18   E.M.O.Sigmond 1978, Norges Geol. Unders. 341, 1-94.
19   C.Oftedahl 1980, Norges Geol. Unders. 356, 3-114.
20   Ø.Berg 1977, Unpubl. hovedoppgave, Universitetet i Oslo.
21   van der Wel, reported in ref. 18.
22   R.H.Mitchell 1967, Norsk Geol. Tidss. 47, 295-332.
23   J.Avila Martins 1969, Norges Geol. Unders. 258, 267-301.
24   C.Cramez 1970, Norges Geol. Unders. 266, 5-35.
25   D.V.Venugopal 1970, Norges Geol. Unders. 268, 57pp..
26   J.H.Stout 1972, Norsk Geol. Tidss. 52, 23-41.
27   B.Moine and A.Ploquin 1972, Sciences de la Terre 17, 47-80.
28   I.B.Singh 1969, Norsk Geol. Tidss. 49, 1-31.
29   T.Prestvik and F.M.Vokes 1983, Norges Geol. Unders. 378, 49-63.
30   H.N.A.Priem, N.A.I.M.Boelrijk, E.H.Hebeda, E.A.Th.Verdurmen and R.H.Verschure 1973, Norges Geol. Unders. 289, 37-53.
31   A.Kleppe and A.Råheim 1979, Europ. Colloq. on Geochron. 6, Lillehammer, 53.
32   A.G.Sylvester 1964, Norsk Geol. Tidss. 44, 445-482.
33   P.J.Hooker, P.J.Hamilton and R.K.O'Nions 1981, Earth Planet. Sci. Lett. 56, 180-188.
34   P.J.Hamilton, N.M.Evensen, R.K.O'Nions, H.S.Smith and A.J.Erlank 1978, Nature 279, 298-300.
35   P.J.Hamilton, R.K.O'Nions, N.M.Evensen, D.Bridgwater and J.H.Allaart 1978, Nature 272, 41-43.
36   P.J.Hamilton, N.M.Evensen, R.K.O'Nions and J.Tarney 1979, Nature 277, 25-28.
37   A.Ploquin 1980, Sciences de la Terre Mem. 38, 389 pp..
38   P.J.Hamilton, R.K.O'Nions and N.M.Evensen 1977, Earth Planet. Sci. Lett. 36, 263-268.
39   A.G.Herrman 1970, in : Handbook of Geochemistry, vol. 2, part 5, K.H.Wedepohl (ed.), Springer, 57-71.
40   S.M.McLennan, B.J.Fryer and G.M.Young 1979, Geochim. Cosmochim. Acta 43, 375-388.
41   S.B.Jacobsen and K.S.Heier 1978, Lithos 11, 257-276.
42   D.Field and A.Råheim 1979, Earth Planet. Sci. Lett. 45, 32-44.
43   P.J.Patchett 1978, Sver. Geol. Unders. C747, 3-63.
44   P.J.Patchett, G.Bylund and B.G.J.Upton 1978, Earth Planet. Sci. Lett. 40, 349-364.
45   D.Ben Othman, M.Polvé and C.J.Allègre 1984, Nature 307, 510-515.
46   R.L.Armstrong 1968, Rev. Geophys. Space Phys. 6, 175-199.
47   S.Moorbath 1978, Phil. Trans R. Soc. Lond. A288, 401-413.
48   J.D.A.Piper 1982, Earth Planet. Sci. Lett. 59, 61-89.
49   R.H.Steiger and E.Jäger 1977, Earth Planet. Sci. Lett. 36, 359-362.
50   G.J.Wasserburg, S.B.Jacobsen, D.J.DePaolo, M.T.McCulloch and T.Wen 1981, Geochim. Cosmochim. Acta 45, 2311-2324.
51   D.J.DePaolo and G.J.Wasserburg 1976, Geophys. Res. Lett. 3, 249-252.

52 D.J.DePaolo and G.J.Wasserburg 1976, Geophys. Res. Lett. 3, 743-746.
53 L.D.Ashwal and J.L.Wooden 1983, Geochim. Cosmochim. Acta 47, 1875-1885.
54 A.Zindler, S.R.Hart, and C.Brooks 1981, Earth Planet. Sci. Lett. 54, 217-235.
55 R.A.Cliff, C.M.Gray and H.Huhma 1983, Contrib. Mineral. Petrol. 82, 91-98.
56 H.S.Pettingill, A.K.Sinha and M.Tatsumoto 1984, Contrib. Mineral. Petrol. 85, 279-291.
57 D.York 1969, Earth Planet. Sci. Lett. 5, 320-324.

THE ROGALAND ANORTHOSITES: FACTS AND SPECULATIONS

J.C.DUCHESNE [1], R.MAQUIL [1], D.DEMAIFFE [2]
[1] L.A. Géologie, Pétrologie, Géochimie
Université de Liège.
[2] L.A. Géologie, Pétrologie, Géochronologie
Université Libre de Bruxelles.

ABSTRACT. The massif-type anorthosites of the province were emplaced as crystal mushes of plagioclase, lubricated by noritic liquids crystallizing along a P-T gradient, and containing aggregates of plagioclase and/or Al-rich orthopyroxene megacrysts, formed in a magma chamber at the base of a thickened crust. Synemplacement deformations were produced in the envelope and within the plutons, where they started in the magmatic stage and ended in the solid stage. The anorthositic suite can be accounted for by three magma types, the compositions of which are basaltic, jotunitic and charnockitic. Each magma generates part of the suite, with some overlapping. The basaltic magma is mantle-derived in an undefined geodynamic environment. The alkali to alkali-calcic jotunitic magma is generated as distinct batches with variable crustal signatures, due to contamination by deep-crustal material or direct partial melting of this material. The charnockitic magma can be produced by fractionation of the jotunites but can also result from direct partial melting in granulite facies conditions. For both magma types partial melting is triggered by the hot anorthositic diapirs en route to their final level of emplacement.

1. INTRODUCTION

The Egersund province in Rogaland (S.W. Norway) is a well-exposed Proterozoic anorthositic complex, with features typical of many Grenvillian massifs. It has been studied for some 40 years by Paul Michot, who built up a well-known model of geological evolution (Michot P., 1960a; Michot J. and Michot P., 1969), which has given rise to the concepts of *anorthositic basement*, *marginal intrusion*, and *foundation orogen* (P.Michot, 1960b, 1965a). It is on this basis that petrological and geochemical studies were later developed, mainly at the Universities of Liège and Brussels in Belgium. It is the purpose of the present paper to review the current state of knowledge on the various units of the province and to focus on those aspects which might enlighten the perennial anorthosite problem.

The province is made up of three large massif-type anorthosites (Egersund-Ogna, Håland-Helleren, and Åna-Sira massifs), two smaller leuconoritic bodies (Hidra and Garsaknatt), a layered lopolith (Bjerkreim-Sokndal), and three large acidic plutons (the Farsund charnockite, the Lyngdal granodiorite and the Kleivan granite). An important part of the intrusives - almost three times the area now visible - extends under the North Sea (Sellevol and Aalstad, 1971).

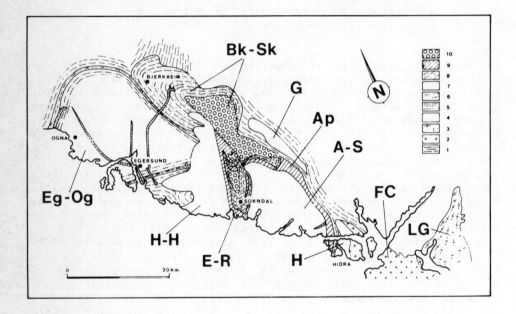

Fig.1 : General geological map of the Rogaland igneous complex and southeastern satellites (after Michot and Michot, 1969). Eg-Og = Egersund-Ogna body; H-H = Håland-Helleren massif; A-S = Åna-Sira body; Bk-Sk = Bjerkreim-Sokndal lopolith; E-R = Eia-Rekefjord intrusion; Ap = Apophysis; FC = Farsund charnockite; LG = Lyngdal hornblende-granodiorite; G = Garsaknatt body; H = Hidra body.
Legend : 10 = mangerite and quartz mangerite; 9 = jotunite; 8 = layered anorthosite, leuconorite and norite of the Bjerkreim-Sokndal lopolith; 7 = leuconorite and anorthosite; 6 = norite, locally migmatitic; 5 = foliated anorthosite and leuconorite; 4 = anorthosite; 3 = Lyngdal hornblende granodiorite; 2 = Farsund charnockite; 1 = surrounding gneisses.

2. THE VARIOUS INTRUSIVE BODIES

2.1. The Egersund-Ogna massif

The Egersund-Ogna massif (Eg-Og fig.1) (Michot P., 1960a; Duchesne and Maquil, 1981) is an anorthositic dome, 20 km in diameter, made up of a granulated, equal-sized (1-3 cm), homogeneous (unzoned) plagioclase (An_{40}-An_{50}). The anorthosite frequently contains ill-defined plagioclase phenocrysts and also meter-sized *sub-ophitic aggregates of megacrysts* (5-50 cm) of plagioclase (An_{55}) (fig.2) and Al-rich orthopyroxene (mega-opx) (7-9 % Al_2O_3; En_{23-27}; 600-950 ppm Cr; plagioclase exsolutions) (fig.3). These aggregates are more abundant in the central part of the massif. A remarkable feature, called the anorthosito-noritic complex, is the occurrence of these aggregates included and grossly oriented in a medium-grained leuconoritic matrix, in which the opx is markedly different in Al and Cr contents (2-3% Al_2O_3; <200 ppm Cr) from the mega-opx (fig.3).

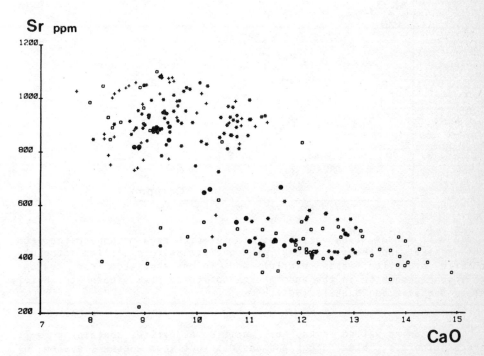

Fig.2 : Sr versus CaO in plagioclases from the Egersund-Ogna massif. Legend : (star) : phenocrysts from the central part of the margin; (cross) : matrix plagioclases from the central part; (open circle) : idem from the margin; (filled circle) : inclusions (from Duchesne and Maquil, 1981).

A marginal zone of the massif (1-3 km thick) (Michot P., 1939; 1957; 1960a) is made up of *foliated* anorthosites and leuconorites, concordant with the foliation in the adjacent metamorphic gneisses of the envelope. Mega-opx are conspicuously stretched, deformed, and granulated along the foliation plane. The overall lithological composition of the margin is leuconoritic. Striking differences with the central part exist in the mineral composition and association (Duchesne, 1968; Duchesne and Maquil, 1981). Meta-troctolites (Fo_{70} + An_{67}) are occasionally present; undeformed mega-opx are 3.5-6% Al_2O_3, with Cr content of 600-1500 ppm (fig.3); plagioclases vary from An_{45} to An_{74} with a Sr content that can be much lower (300-450 ppm) than in the central part (800-1000 ppm) (fig.2); differences also appear in the Ba and K content of the plagioclases. Initial Sr isotopic ratios (fig.4) are somewhat higher in the marginal rocks (up to 0.7045) than in the centre (0.7035) (Michot and Pasteels, 1969; Duchesne and Demaiffe, 1978); the mega-plagioclases show the lowest values in the whole area (0.7029) (Demaiffe, unpublished).

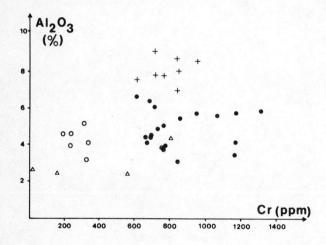

Fig.3 : Cr versus Al_2O_3 variations in Al-rich mega-orthopyroxene from the central anorthosite (cross), the gneissic margin (filled circle), noritic pegmatite (open circle) and in opx separated from medium-grained leuconorite in the anorthosite-norite complex (open triangle) (from Duchesne and Maquil, 1981).

Sharp-walled dykes of noritic pegmatite, containing Fe-Ti oxide minerals, also occur and are distinct from another system of noritic dykes, devoid of oxide minerals and merging into small intrusions and anastomozing into the anorthosito-noritic complex.

A highly significant fact is the occurrence of *foliated blocky inclusions* of meta-leuconorite within the central part of the massif. The plagioclases of these inclusions are geochemically identical to those of the margin. This fact together with the similarity of textures leaves little doubt as to whether the inclusions and the marginal rocks have the same origin.

In short, the Egersund-Ogna massif is a typical mantled dome (P.Michot, 1957).

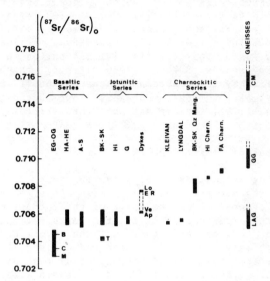

Fig.4 : Initial Sr isotopic compositions in Rogaland rocks.
Abbreviations : EG-OG: Egersund-Ogna (M: plagioclase megacrysts; C: central anorthosite; B: margin); HA-HE: Håland-Helleren; A-S: Åna-Sira; BK-SK: Bjerkreim-Sokndal (T: Tjörn chilled facies); HI: Hidra; G: Garsaknatt; Dykes: Lo: Lomland dyke; E.R.: Eia-Rekefjord intrusion; Ve: Vettaland dyke; Ap: Apophysis; FA charn.: Farsund charnockite; HI charn.: Hidra charnockitic dykes; BK-SK Qz Mang. : Bjerkreim-Sokndal quartz mangerites; LAG: Liland augen gneiss; GG: granulitic gneisses; CM: charnockitic migmatites.

2.2. The Håland-Helleren and Åna-Sira massif

The Håland-Helleren massif, also of the massif-type (H-H fig.1), is a composite intrusion (J.Michot, 1961a,b). The Håland body is made up of banded and foliated leuconorites and anorthosites similar to the Egersund-Ogna margin, notably in that it contains granulated Al-rich mega-opx. It locally grades into an igneous-looking association with multiple evidence of a filter-press mechanism acting on a crystal mush lubricated by noritic liquids. The Helleren body, which diapirically pierces the other one, has many common features with the central part of Egersund-Ogna, namely the Al-rich mega-opx, the blocky inclusions of foliated leuconorite and the anorthosito-noritic complex. It is in these bodies that the *leuconoritic anatexis* and *basic palingenesis* were defined years ago by P. Michot (1955) and J.Michot (1960).

The third large massive anorthosite - the Åna-Sira body (A-S fig.1) - recently mapped by Krause and co-workers (Krause and Pedall, 1980; Krause et al., in press), is also very similar to the central part of Egersund-Ogna. The foliated margin is lacking, the massive anorthosite being directly in steep contact with surrounding rocks. Also present are numerous foliated inclusions, sometimes remarkably folded (see fig.19 in Barth and Dons, 1960), Al-rich mega-opx with plagioclase and spinel exsolutions (Duchesne and Maquil, 1981), and the anorthosito-noritic complex - i.e. the "*Egersund-Ogna trilogy*". Preliminary Sr isotopic data indicate initial Sr ratios in Håland, Helleren and Åna-Sira around 0.7055-0.7060 (fig.4). Moreover the massifs are characterized by the presence of numerous Fe-Ti oxide orebodies (see Krause et al., in press; Hubeaux, 1960; Duchesne, 1973; Roelandts and Duchesne, 1979), the largest of them still in operation (Tellnes).

2.3. The Bjerkreim-Sokndal massif (BK-SK, fig.1)

It is a layered syncline with an axis dipping in such a way that a complete sequence of charnockitic rocks - from anorthosite to quartz mangerite - can be observed from the floor of the intrusion (NE of Bjerkreim) to the roof (central part of the massif). Absence of ultramafic layers in the deepest part of the syncline has been demonstrated by gravity measurements (Smithson and Ramberg, 1979).

The rocks show the typical structure of *cumulates*: planar lamination, small-scale rhythmic (modally graded) banding, intermittent layering, etc. Microscopically, they can be defined as adcumulates, although granulation of the plagioclase usually blurs the cumulate structure. The continuity of some layers of typical lithology or structure (index horizons) has permitted P.Michot (1960a, 1965a) to divide the lower part of the massif into 5 macro-scale rhythmic units and the upper part, into two more massive units, a mangeritic one and, on top of it but separated from it by a discontinuous septum of xenolithic rocks, a quartz mangeritic unit. Each rhythm starts with anorthosite or leuconorite at the base and grades upwards into mafic-richer rocks. Successive influxes of fresh magma into the chamber have been invoked to explain the recurrences (Duchesne, 1972a). When their effects are cancelled, the complete sequence of rocks can be reconstructed. It corresponds to the so-called anorthosite suite: anorthosite, norite, jotunite (= monzonorite), mangerite and quartz mangerite.

Discontinuous layering (fig.5) (Michot P., 1965a) and cryptic layering are conspicuous. They conform to a closed-system fractional crystallization in dry conditions. Variation in major and trace elements have been described by various authors in the different minerals: plagioclase (Duchesne, 1968, 1971, 1978); pyroxenes and olivine (Duchesne, 1972b; Rietmeijer, 1979; Wiebe, 1984; Duchesne, Denoiseux and Hertogen, this volume ; Fe-Ti oxide minerals (Duchesne, 1970b, 1972a); apatite (Roelandts and Duchesne, 1979); zircon (Caruba et al., in preparation).

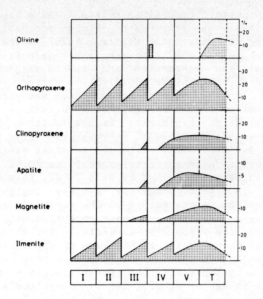

Fig.5 : Evolution of the mafic mineral contents against a schematic representation of the structural height in the Bjerkreim-Sokndal lopolith. The five rhythmic units (I to V) and the transition towards acidic rocks (T) are arbitrary represented by segments of equal length, because they considerably vary in thickness from the axial plane towards the flanks of the syncline (after P.Michot, 1965a; and Duchesne, 1972a; with slight modifications).

Plagioclase varies from An_{50} to An_{30} in the five rhythms. They display a Sr-Ca relationship typically controlled by the variation in the nature of the accompanying cumulate minerals (Duchesne, 1971, 1978). Mesoperthite is the only feldspar in the mangerites. Pyroxenes show a regular increase in Fe/Mg ratio, the most magnesian one being $En_{74}Fs_{25}Wo_2$. Olivine ($Fo_{70}Fa_{30}$) briefly crystallizes in the lower part of the sequence, then reappears in the transition between norites and mangerites, where it quickly varies from Fo_{50} to Fo_8. Quartz mangerites can contain a fayalitic olivine. Fe-Ti oxide minerals (ilmenite and magnetite) smoothly vary in proportion and chemical composition. Subsolidus reaction between the two oxides has partly altered their original compositions. Cr and Ni in the oxides rhythmically decrease in the three lower rhythms. Cumulate apatites display a distinct increase in rare earth elements through the sequence. Zircons from quartz mangerites contain sharply bounded "inherited" U-rich cores.

Strontium isotopic compositions (fig.4) have been determined in the lower anorthosito-noritic part (Michot and Pasteels, 1969; Pasteels et al., 1970; Duchesne and Demaiffe, 1978) where the initial $^{87}Sr/^{86}Sr$ ratio averages 0.7057. On the other hand Pasteels et al. (1979) have shown that the mangerites and quartz mangerites have an initial Sr ratio distinclty higher (0.7105). A better selection of samples, restricted to the quartz mangerites, has yielded better isochrons and lower values of the initial ratio (0.7075 : Wielens et al., 1981; 0.7086 : Weis, this book). Oxygen isotopic ratios were measured by Demaiffe and Javoy (1980). $\delta^{18}O$ values vary from 6.0-6.6 %., with a sample at 4.0 %. . They fall into a normal range for acidic plutonic rocks emplaced in high-grade metamorphic terrains. Pb isotopic compositions have also been determined by Weis (1982, this book).

The massif is almost completely devoid of any chilled margin that would give indication on the nature of the *parental magma* or of the residual liquids at any stage of their evolution. The absence of ultramafic layers on the floor or at depth in the chamber and the overall composition of the minerals at the base of the series rule out a basaltic composition and points to more fractionated, i.e. intermediate compositions. In this respect, the recent discovery of a contact facies at Tjörn (Tobi in Maijer et al.1984) is of particular importance in that it shows the occurrence of a jotunitic rock (fig.7), with typical texture of chilled rocks. Rare earth element distribution as well as Sr isotopic data (0.704) (fig.4) are consistent with what should be expected for parental liquids (Duchesne, Demaiffe and Hertogen, 1985).

2.4. The Eia-Rekefjord intrusion (E-R fig 1), the Egersund dykes and the Apophysis (Ap fig 1).

Jotunites occur either as numerous dykes (particularly in the region of Egersund) or as two intrusions, the Eia-Rekefjord body along the W flank of the Sokndal lobe of the lopolith and the Apophysis at the northern and eastern contact of the Åna-Sira body and the metamorphic envelope (fig.1). This type of magmatism develops later than the consolidation of the massive anorthosites and prior to or simultaneously with the acidic magmatism, as shown by pillow-like inclusions of Wiebe (1980b).

The close space and time relationship of Eia-Rekefjord and the Egersund dykes with the lopolith has led P.Michot (1960a, 1965a) to consider them as offshoots of residual liquid after the formation of the anorthosite-norite sequence of the lopolith. Similarly the Apophysis was considered by J.Michot and P.Michot (1969) and Demaiffe (1972) as the southern extension of the lopolith. Isotopic data (fig.4) failed to confirm these views; the various dykes and the Apophysis (fig.4) have initial $^{87}Sr/^{86}Sr$ ratios higher than the Bjerkreim-Sokndal anorthosites and norites, and also higher than the massif-type anorthosites. Moreover the absence of Eu anomaly in the REE distribution (fig.6) and the high K/Rb ratios (Duchesne et al., 1974; Duchesne et al., 1984) are hardly conceivable in residual liquids after crystallization of plagioclase. In many occurrences, the Fe/Mg ratios are also too low to be comparable with those of liquids in equilibrium with the norites of Bjerkreim-Sokndal (fig.7)

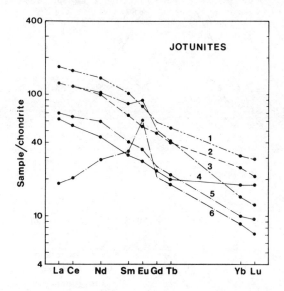

Fig.6 : Selected chondrite-normalized REE distributions of jotunitic rocks.
1. Eia-Rekefjord intrusion (rock JCD66125; Duchesne et al., 1974).
2. Apophysis (fine-grained jotunitic pillow -JCD8112.8.1- Duchesne, Demaiffe and Hertogen, 1985).
3. Lomland dyke (average of 5 analyses; Duchesne et al;, 1984).
4. Hidra border facies (DD200 2/2; Demaiffe and Hertogen, 1981).
5. Tjörn chilled border rock (JCD80.12.3A; Duchesne, Demaiffe and Hertogen, 1985).
6. Eigeröy facies of the Vettaland dyke (JCD7557; Duchesne et al., 1984).
Note : The distribution in the Eigeröy facies is shown here to emphasize its contrast with the distribution in the other jotunitic occurrences. It has been accounted for by partial melting of basic cumulates (Duchesne et al., 1984).

The Eia-Rekefjord intrusion shows some lithological variations from noritic rocks to mangeritic rocks (Wiebe, 1984) (fig.7). The same holds for the Lomland dyke (Duchesne et al. 1984) and the Håland dyke (Duchesne, unpublished) in the Egersund-Ogna massif and for the Tellnes dyke in the Åna-Sira massif (Krause and Pedall, 1981). On the other hand, the Vettaland dyke (Egersund-Ogna massif) shows a monotonous quartz ferronoritic composition along its 20 km length, but strongly variable LIL element contents (Duchesne et al. 1984) (fig.6).

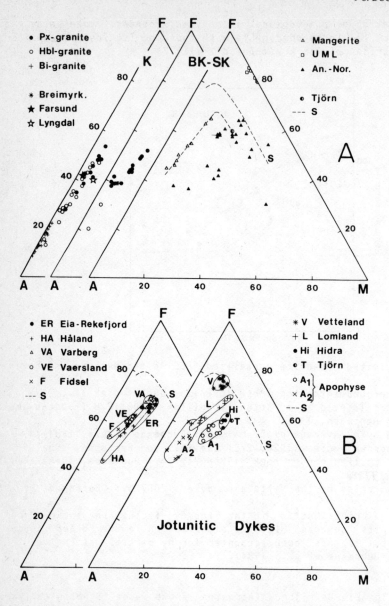

Fig.7 : AFM diagrams of various jotunitic occurrences.
A. Bjerkreim-Sokndal lopolith (BK-SK) and Kleivan granite (K).
Legend : open triangle : mangerite; open square : ultramafic layers (UML) (from Duchesne, Denoiseux and Hertogen, 1985); filled triangle : anorthositic, leuconoritic and noritic cumulates (Duchesne, unpublished); half filled circle : Tjörn chilled jotunite (Duchesne,

Fig.7 (continue):
Demaiffe and Hertogen, 1985); dashed line : liquid line of descent (S for Skaergaard, Wager and Brown, 1968); filled circle : orthopyroxene granite and/or quartz mangerite; open circle : hornblende granite; cross : biotite granite and leucogranite (data from Petersen, 1980a, for Kleivan and from Duchesne, unpublished, for Bjerkreim-Sokndal); asterisk : Breimyrknutan charnockite (Wilmart et al., 1985); open star : Lyngdal hornblende granite; filled star : Farsund charnockite (Wilson and Annis, 1973).
B. Various jotunitic dykes and intrusions.
Sources : Vettaland and Lomland dykes (Duchesne, Demaiffe and Hertogen, 1985); Hidra (Duchesne et al., 1984; Duchesne, Demaiffe and Hertogen, 1985); Tjörn and Apophysis (Duchesne, Demaiffe and Hertogen, 1985) (A1 is the fine-grained jotunitic material; A2 is the porphyroidic mangerite/monzonite). All other data are from Duchesne, unpublished.

These peculiarities have been accounted for by progressive partial melting at a eutectic and suggest a plagioclase cumulate of noritic composition as a source rock.
   The Sr isotopic composition varies from occurrence to occurrence (fig.4) : 0.7061 in the Vettaland dyke and in the fine-grained material of the Apophysis; 0.7077 for the Lomland dyke and the Eia-Rekefjord intrusion (Duchesne et al., 1984). Moreover, the Lomland dyke does not show any variation of the Sr initial ratio in the differentiation.

2.5. The Hidra and Garsaknatt leuconoritic bodies (H and G fig.1)

These two units appear as late- to post-tectonic massifs directly intruded in the metamorphic envelope : although grossly concordant with the regional gneissic structure, they are discordant and transgressive in detail. Small leuconoritic dykes (up to 1 m thick) cut across the gneisses and the nearby Farsund charnockite.
   The *Hidra body* (Demaiffe et al., 1973; Demaiffe, 1977) consists essentially of an undeformed, medium grained (1-3 cm) leuconorite (plagioclase An_{42}-An_{47}) containing large (up to 1m) plagioclase megacrysts, while the mega-opx are typically lacking. Near Itland, lath-shaped plagioclases (5-10 cm) are embedded in a finer-grained (<1 cm) leuconoritic mesostasis. Towards the centre of the massif, the leuconorite locally grades into a coarse-grained (3-10 cm) anorthosite with a typical orthocumulate texture : the large plagioclase crystals (An_{47}-An_{54}) have a zoned antiperthitic rim merging into quartz-K-feldspar granophyric intergrowths. Leucotroctolite (An_{52} + Fo_{65}) occurs locally near Urstad. Rare pyroxenite inclusions displaying igneous lamination have been observed in the leuconorite. The contact with the adjacent gneisses is outlined by a fine-grained (300-500 um), locally porphyritic (blue plagioclase phenocrysts: An_{47}) border facies of jotunitic composition. A stockwork of charnockitic dykes and large granitic pegmatite lenses (famous for their REE minerals; Adamson, 1942) cut across the massif.

Detailed geochemical and isotopic studies (Duchesne et al., 1974; Demaiffe and Hertogen, 1981; Weis and Demaiffe, 1983b) have shown that the sequence of rocks result from a fractional crystallization process of the jotunitic parental magma (no Eu anomaly). The average $^{87}Sr/^{86}Sr$ initial ratio is 0.7055 (10 values). The charnockitic dykes might correspond to the residual liquid of the differentiation (high REE content, large negative Eu anomaly), contaminated by crustal material $(^{87}Sr/^{86}Sr)_o = 0.7086$. Commingling relationship between jotunitic liquid and coarser-grained mangeritic to monzonitic material has been described near the contact with adjacent rocks (Duchesne, Demaiffe and Hertogen, 1985).

The *Garsaknatt Outlier* (Demaiffe, 1977) is also made up of an undeformed medium-grained leuconorite similar to that of the Hidra body. In some places it has a roughly banded appearance and contains abundant lenticular masses (5-10 cm) of orthopyroxene and Fe-Ti oxides. A fine-grained jotunitic border facies has only been observed locally. Numerous blocky inclusions of foliated or massive coarsed-grained anorthosite or leuconorite with Al-rich mega-opx are found in the leuconorite. Small (30 cm thick) charnockitic dykes and some meter-sized pegmatitic pods cut across the massif. Garsaknatt has many features in common with Hidra (same petrographic structure of the leuconorite, charnockitic dykes, jotunitic border facies, with similar REE distribution, Sr initial ratio of 0.7055). On the other hand, the presence of anorthositic blocky inclusions is characteristic of the previously described massif-type anorthosites. However, the exact relationship of the inclusions (xenolith or autolith ?) with the enclosing leuconorite is still to be assessed.

2.6. The Acidic Plutons

The South-eastern extremity of the Rogaland magmatic province is occupied by a very large acidic pluton formerly considered as a single intrusion, called the Farsundite (Barth and Dons, 1960; Middlemost, 1968). Detailed mapping (Pasteels et al., 1970; Falkum et al., 1972; Falkum and Petersen, 1974) has revealed the composite nature of this pluton. Three units have been defined : the Farsund charnockite, the Lyngdal granodiorite and the Kleivan granite.

The large *Farsund charnockite* (FC fig. 1) is a homogeneous, medium-grained undifferentiated massif. Its western contact has been deformed (blastomylonitic recrystallization) at the contact with the Hidra leuconoritic body. The geochemical features of the charnockite (Pasteels et al., 1970; Demaiffe et al., 1979; Petersen, 1980b) - high isotopic initial ratios (0.709)(*), flat REE pattern, sometimes with small Eu depletion - are compatible with an origin by anatexis of crustal rocks.

(*) An initial ratio of 0.713 is reported by Pedersen and Falkum (1975). It has however been questioned (Demaiffe et al., 1979) since it was deduced from an ill-defined isochron giving a much younger age (834 Ma) than the ziron age of 930 Ma (Pasteel et al., 1979). Recalculated at 930 Ma the Farsund charnockite initial ratio is close to 0.709.

The *Lyngdal granodiorite* (LG fig. 1) is also a rather homogeneous large body; it is however geochemically quite distinct from the Farsund charnockite : its low Sr isotopic initial ratio (0.7054; Pedersen and Falkum, 1975) and its negative Eu anomaly (Demaiffe et al., 1979) do not preclude a genetic relationship with the neighbouring anorthositic massifs. However, in view of the limited number of data, the exact origin is far from being understood and anatexis of deep-crustal material remains a possible origin.

The *Kleivan granite* (East of the map, fig. 1) is a well differentiated intrusion (Petersen, 1980a, 1980b; Kornerup-Madsen, 1980); it starts with a pyroxene granite (mineralogically comparable to the Farsund charnockite) and changes gradually through hornblende granite to biotite granite. The low Sr isotopic initial ratio (0.7052; Petersen and Pedersen, 1978) is compatible with this hypothesis.

3. EMPLACEMENT MECHANISM OF MASSIF-TYPE ANORTHOSITES

The Egersund-Ogna massif provides excellent arguments to decipher the mechanism of emplacement of deformed massif-type anorthosites, particularly common in the Grenville Province.

Maquil and Duchesne (1984) have provided much evidence that the foliation in the inner margin of the the Egersund-Ogna massif is not generated by tectonic reworking of a previously consolidated mass by a subsequent regional deformation but simply results from a *synemplacement deformation*, produced by mechanical friction in the mass itself and also by differential mobility of the central part relative to its margin. Three independent lines of evidence are used by Maquil and Duchesne (1984) : i. progressive transition in the field between magmatic fluidal textures and granoblastic foliated textures; ii. occurrences of foliated autolithic inclusions in the central part of the massif; iii. temperatures of recrystallization in the margin higher than in the metamorphic envelope. These arguments indicate that the deformation is intimately linked to the crystallization of the igneous mass itself and does not result from an external cause.

A second important point is the fact that the Egersund-Ogna massif has *not consolidated at constant pressure*. This is supported by the occurrences of Al-rich mega-opx (Duchesne and Maquil, 1981 ; Duchesne, 1984; Maquil and Duchesne, 1984). Maquil and Duchesne (1984) object to Morse's statement (Morse, 1975) that the high Al contents are due to rapid *in situ* crystallization of plagioclase-rich magmas, as they find such process difficult to reconcile with the extreme size of the crystals, the limited range of major and trace element composition, the absence of zoning, the sharp contrast in composition between the megacrysts and the matrix minerals and the low $^{87}Sr/^{86}Sr$ isotopic ratios of the associated mega-plagioclases relative to the embedding matrix. On the contrary, these authors follow Emslie's views (1975), supplemented by Maquil's work (1979), on a strong pressure control on the Al-content of the opx, even if quantitative experimental data are yet to be obtained. This consideration inevitably leads to accept that the magma has gone through a crystallization episode at greater depths (probably 50-60 km, corresponding to 12-15 kb) than the final depth of emplacement (see below).

The Egersund-Ogna massif moreover provides evidence of a crystallization in a *pressure gradient*. Mega-opx, preserved from the deformation, in the marginal rocks, are poorer in Al than in the central part (3.5-6% Al_2O_3 versus 7-9%) with broadly the same Fe/Mg ratio. Moreover the An content of some plagioclases from marginal rocks can reach values as high as An_{74} (Duchesne, 1968; Duchesne and Demaiffe, 1978; Duchesne and Maquil, 1981) and olivine ($Fo_{70}Fa_{30}$) is not rare as a principal mafic mineral. According to the experimental works of Presnall et al. 1978 on the stability of the Fo + An association and of Green (1969) on the liquidus plagioclase composition at high pressure, these compositions suggest lower pressure conditions than in the central part. Final emplacement of the massif took place in still lower pressure conditions. The norites which enclose the agglomerates of plagioclase and opx megacrysts are made up of Al-poor opx (1-2% Al_2O_3) and the last drops of liquids which crystallize in Fe-Ti-rich noritic pegmatitic dykes also give an Al-poor opx (Maquil and Duchesne, 1984). Absence of corona structure resulting from retrograde reaction between olivine and plagioclase (Griffin and Heier, 1973) allows an upper limit of 7 kb to be fixed for the pressure during consolidation and further evolution (Duchesne and Michot, 1974). Deformation and recrystallization in the envelope due to the emplacement also suggest medium pressure conditions: the assemblage garnet - cordierite/osumilite - sillimanite in metapelite has been observed at short distances from the contact (Maquil, unpublished; see also Jansen, this book). Moreover pressure conditions determined in a metapelite adjacent to the Ana-Sira massif with the plagioclase - garnet - sillimanite - quartz assemblage of Newton and Haselton (1980) indicate 4-5 kb (Wilmart, 1985).

It must be emphazised that although the marginal and central rocks can show some differences in mineral composition, the hypothesis that they crystallize from two different magmas must be ruled out. Indeed the transition between centre and margin is everywhere progressive and morerover, plagioclases of composition (Ca, Sr, K, Ba, etc..) similar to those in the central part are frequently found in the margin.

To account for the crystallization in a pressure gradient and also for the deformation of the rocks and minerals, we postulate the *uprise of the mass* during crystallization and a high proportion of crystals relative to liquid, i.e. a mushy stage. Common sense suggests that mantled domes of such large diameter (20 km) compared to the thickness of the average continental crust (30-35 km), have not a root of the same diameter below the level of exposure. Geophysical models based on gravimetric data (Smithson and Ramberg, 1979) point to thickness of about 4 km for the anorthosite. It must therefore be speculated that a final act of telescopic flattening of the uprising system must have operated to bring the root of the diapir nearer to the roof of the pluton and in a central position within it. This process has given the massif its actual areal extension and can as well have generated part of the deformation inside and around the massif.

This interpretation of the the Egersund-Ogna massif body thus confirms Martignole and Schrijver's (1970a, 1970b) model for the emplacement of the Morin massif (Quebec) and for other Grenville-type anorthosites. It emphasizes the fact that the emplacement mechanism is *not necessarily linked to a regional deformation*.

It is pertinent to relate the present model for the emplacement to *geophysical data* on the crust below the province. The Moho discontinuity is presently at a depth of 28 km and the heat flow through the anorthositic units is very low (0.4 to 0.5 HFU), suggesting a crust under the 4-km thick anorthosites made up of U-, Th-, and K-depleted material (Smithson and Ramberg, 1979). Since the depth of emplacement of the various massifs is estimated about 15-20 km, as suggested by the 4-5 kb pressure recorded by geobarometers, an equivalent thickness must have been eroded during isostatic reequilibration to bring the anorthosites to the level of peneplanation. This puts the total thickness of the crust when the anorthositic plutonism occurred at a value around 40-50 km, which might imply a thickening process - yet to be defined - prior to or during the plutonism. It is interesting to note that this value is not far from the depth suggested for the early crystallization of the aggregates of megacrysts of Al-rich opx and plagioclases (50-60 km). It is thus conceivable that the deep-seated magma chamber where the parental magma started crystallizing was situated at the limit between the thickened crust and the upper mantle.

The present model for the emplacement can also account for the variation of initial Sr isotopic ratios in the Egersund-Ogna massif (fig.4) : the low value 0.7029 for the mega-plagioclase points to a mantle origin for the parental magma and to very little or no contamination in the deep-seated magma chamber, the higher values in the common anorthosite and especially in the margin suggest progressive contamination of the mass during the ascent of the diapir.

It must also be pointed out that the material which is penetrated and likely to be melted by the uprising anorthositic crystal mush has inevitably the characteristics of lower crustal material. This is consistent with what will be deduced below from the jotunitic magmatism related to massif-type anorthosites.

4. THE AGE DEBATE: REJUVENATION OR LONG-LASTING CONSOLIDATION

Egersund-Ogna thus appears to have crystallized and emplaced in a single continuous process in which notably the inner margin was generated and deformed prior to final consolidation and arrest in the uprise movement of the central part.

It is tempting to explain the other anorthositic massifs by a grossly similar process with some variants to account for the differences with Egersund-Ogna. The existence of the so-called Egersund-Ogna trilogy - Al-rich mega-opx, foliated inclusions, anorthosito-noritic complex - in Ana-Sira and Helleren massifs inevitably leads to accept them as uprising diapirs, which have broken through and disrupted an early formed gneissic margin, the fragments of which are found as inclusions (Duchesne and Maquil, 1981). This interpretation is however not accepted by Jean Michot (Duchesne and Michot, 1984), particularly in the Håland-Helleren composite body. According to him the gneissic inclusions are fragments of the foliated part of the Håland body, which itself represents an older anorthositic basement, reactivated by an *in situ* leuconoritic anatexis (P.Michot, 1955, J.Michot, 1960a,b), around 1200-1000 Ma. Duchesne and Maquil on the other hand simply consider the foliated part of the Håland massif as an inner margin, similar to that of Egersund-Ogna but with a more complex structure - possibly due to a more apical position in the original dome structure. It must be noted that the transitional structure between deformed and magmatic rocks can equally well be interpreted by both processes, inasmuch as a filter-press mechanism is acting to separate the solid phases from the melt. As noted by Jean Michot (1972,p.26) it is not possible in the field or under the microscope to distinguish between a crystallization and a melting process.

The main argument of Maquil and Duchesne is the occurrence in the Håland massif of deformed Al-rich mega-opx, embedded in a noritic matrix made-up of Al-poor opx besides plagioclase, thus implying a crystallization in a pressure gradient in a quite comparable way as the Egersund-Ogna inner margin. This interpetation has moreover the advantage of circumventing the difficulty of finding the necessary heat to trigger off the *in situ* remelting postulated in the leuconoritic anatexis.

The present model nevertheless stumbles against another constraint. An imprecise Pb-Pb isochron pointing to 1500 $\pm$ 300 Ma has been obtained by Weis and Duchesne (1983) on megacrysts of plagioclases and opx from the central part of the Egersund-Ogna body. If this older age were to be confirmed by other methods it would require either an extremely long-lasting period of crystallization (200 to 500 Ma !) or a reactivation by partial melting without perturbing the Pb-Pb isotopic ratios in some part of the original system. Both hypotheses need to be carefully studied and confronted with geophysical constraints. In these aspects Rogaland is not different from other Grenville anorthosites - e.g. the Marcy massif - where a convincing explanation for "old ages" recently discovered is still awaited (Ashwal et al. 1982, 1983).

In the reactivation hypothesis it is also not inconceivable (Duchesne, 1984) either that a small fraction of partial melting would be sufficient to start a process of diapiric uprise, or that during the ascent of the diapir, small amounts of partial melt would be produced. In experiments in the ternary system albite - anorthite - diopside (Lindsley and Emslie, 1968), a drop in pressure at constant temperature indeed shifts the cotectic line towards the mafic pole of the system and lowers the liquidus temperature, thus provoking the remelting of part of the already solidified phases. The phenomenon could have taken place during the uprise through the lower thickened crust, that is at higher pressures than those prevailing at the final depth of emplacement.

5. THE ANORTHOSITIC SUITE AND THE PARENTAL MAGMA PROBLEM.

In Paul Michot's views (1956, 1960b) the anorthositic suite in Rogaland can be described by the Egersund-Ogna and Bjerkreim-Sokndal massifs, which intruded successively. The anorthosite from the central part of Egersund-Ogna was followed by more mafic-rich rocks, giving rise to the leuconoritic border of the massif, then by the Bjerkreim-Sokndal succession, starting with anorthosito-noritic rocks in the lower part and grading into acidic rocks on top. It is on this basis that Paul Michot forged the concept of a "plagioclasic magma" essentially defined as a basic magma in which plagioclase was the first liquidus mineral (Michot, 1956, 1965a). Paul Michot could not further characterize the magma but postulated that it was formed by assimilation of pelitic material by tholeiitic basalt. Later, he focused on the Bjerkreim-Sokndal lopolith (Michot, 1965a), where the composition of the plagioclasic magma was redefined and estimated as quartz monzonoritic in composition, to account for the relatively high proportion of acidic rocks on top of the sequence. Bjerkreim-Sokndal was a milestone in the evolution of ideas about Rogaland. Indeed the sequence of rocks clearly linked noritic rocks to quartz mangerites in the same series and thus brought a solution to the longstanding debate on the relationship between anorthosite and acidic products.

More recent work has however not confirmed this model. Indeed, as already stated, the Sr isotopic ratios proved to be higher in the acidic rocks than in the lower part of the massif, and zircons in quartz mangerites contain complex U-rich cores. These two independent features point either to *bulk contamination* by exotic material or hybridization of residual liquids by anatectic melts produced by contact anatexis, the discontinuous zone enriched in xenolithic material grossly corresponding to the original roof of the magma chamber. Trace element data are also compatible with such a model (Demaiffe et al. 1979; Duchesne, Denoiseux and Hertogen, 1985). In particular, mangerites have the characteristic positive Eu anomaly of feldspar cumulates and possibly represent a flotation cumulate at the top of the series near the roof.

Much evidence indicates nowadays that the Bjerkreim-Sokndal series actually results from the differentiation of a *jotunitic magma* similar to the border facies of Hidra, the Eia-Rekefjord intrusion, the Apophysis and the dyke system (fig. 6 and 7). The jotunitic series produced by fractional crystallization of jotunitic liquids comprises some anorthosite - now at the basis of the rhythms in Bjerkreim-Sokndal or locally in the central part of Hidra - and some quartz mangerite - above the Bjerkreim-Sokndal mangerites, diluted in the quartz mangerites or as a network of dykes in Hidra -, but these rocks are less abundant than the leuconoritic, noritic and mangeritic members. With such a composition it is difficult to account for the large massive anorthosites as well as for the acidic bodies SE of the complex.

This was already surmised by Duchesne and Demaiffe (1978) who concluded from trace element evidence that a unique magma is unable to produce the whole anorthositic suite. The trace element variations in plagioclase of Egersund-Ogna indeed cover several orders of magnitude, which implies extreme fractionation and would lead to unrealistically small amounts of residual liquids compared to what is observed for intermediate and acidic rocks. Moreover the Cr content of the Al-rich mega-opx (up to 1500 ppm) (Duchesne and Maquil, 1981) is incompatible with the low content of this element in jotunitic liquids (<10 ppm).

In Rogaland no evidence of chilled rocks related to the massif-type anorthosites has been found. This is not surprising in view of the crystal-mush state of the intrusive masses. However, the striking similarity in trace element and major element composition of aluminous mega-opx in Rogaland, and more generally in the Grenville province, with those in Labrador (Emslie, 1980), where a variety of basaltic chilled liquids or marginal rocks have been described (Emslie, 1978; Morse, 1982), points to a *grossly basaltic composition* for the parental magma of the massif-type bodies.

As for the acidic rocks Rogaland provides numerous occurrences clustering around a *charnockitic composition*, enriched in both ferromagnesian and feldspathic components and impoverished in SiO_2 relative to granite (fig.7). Several processes appear to lead to such a composition : firstly, fractional crystallization of jotunitic magma with or without contamination by crustal material; the former illustrated by the Bjerkreim-Sokndal quartz mangerites and the Hidra charnockitic dykes, the latter by the differentiation in the Lomland dyke. Secondly, partial melting of metamorphic rocks in granulite facies conditions, probably triggered off by the heat supplied by the anorthosite massifs. This process is particularly well demonstrated by the Breimyrknutan body, situated between the Åna-Sira and Hidra massifs (Duchesne, Demaiffe and Wilmart, 1984), and which presents a REE distribution incompatible with its derivation from Hidra (lack of Eu anomaly) (Wilmart, Hertogen and Duchesne, 1985).

6. PROCESSES OF DIFFERENTIATION

Two major processes of differentiation are thought to have operated in Rogaland anorthosites : filter-press and gravity-controlled mineral separation.

Many mesoscale structures, commonly interpreted as due to *filter-press* action on a crystal mush, are conspicuous in massif-type anorthosites, especially in the Håland massif, where they were described by Michot and Michot (1969) as anorthosite with "pseudo-xenoliths" of leuconorite. Moreover petrographic and geochemical evidence, clearly demonstrated in Egersund-Ogna, indicate that crystallization and emplacement have taken place in a pressure gradient. Combination of these processes does not usually lead to a clearcut distinction between cumulates, crystal-laden liquids and perfect liquids. It might also lead to some crystal-melt disequilibrium. This greatly hampers the use of REE and other trace elements as petrogenetic indicators. In this connection many sharp-walled dykes of Fe-Ti oxide-bearing pegmatitic norite can readily be interpreted as having crystallized in two stages: a crystal mush at depth and final consolidation of the lubricating liquid at lower pressure (Maquil and Duchesne, 1984). In the same way coarse-grained anorthositic dykes can also be considered as injection of crystal mushes and do not necessarily correspond to perfect liquid, as suggested, among others, by Wiebe (1980a). The net result, therefore, of the differentiation mechanism in massif-type anorthosites is thus the production of noritic liquids, more or less laden with plagioclase crystals, and residual after the formation of the anorthosite, which itself contains crystals formed in a succession of decreasing P-T conditions.

Gravity-controlled mineral separation under moderate constant pressure is the second differentiation mechanism. It is well demonstrated in the Bjerkreim-Sokndal massif where cumulates are piled up in the magma chamber, the latter being intermittently replenished by fresh magma. The process leads to the *in situ* formation of anorthosito-noritic cumulates (not liquids! as it is the case in Egersund-Ogna) possibly by an oscillatory nucleation process (Morse, 1979), grading upwards into mangeritic cumulates. These may have been formed by flotation since the density difference between felspars and the liquid must increase with the progressive Fe enrichment of the liquids. Bjerkreim-Sokndal thus displays a series of mineral assemblages in equilibrium with the series of liquids produced by fractional crystallization of jotunitic liquids. The Hidra body provides a series of anorthositic and leuconoritic cumulates together with the parental magma chilled against the ajacent rock and the contaminated residual acidic liquids of the dyke network (Demaiffe and Hertogen, 1981). In the case of jotunitic dykes and small intrusions the evolution from jotunitic to mangeritic and quartz mangeritic liquids can be described (Duchesne et al, 1984) by subtraction of *hidden noritic cumulates*, not directly observed. Bowen trends at virtually constant Fe/Mg ratios characterize the liquid lines of descent (fig.7), and Fe/(Mg + Fe) ratios can vary from occurrence to occurrence : 0.77 in Hidra, Tjörn and the fine-grained jotunites of the Apophysis, 0.85 in many dykes and the Eia-Rekefjord intrusion, and 0.89 in the quartz noritic dyke of Vettaland (Duchesne et al. 1974; Duchesne et al. 1984; Duchesne, Demaiffe and Hertogen, 1985).

The acidic rocks are less differentiated (but have also been less studied!). A remarkable case of differentiation from charnockite to granite has been demonstrated in the Kleivan granite (Petersen, 1980a,b). Subtraction of hidden cumulates of noritic and mangeritic composition can account for the major and trace element behaviour. Here again the trend develops at constant Fe/Mg ratio.

To summarize, it appears that *three types of magma* have to be invoked to account for the whole anorthositic suite, each one generating part of it, with some overlapping. The first one, grossly basaltic, produces the massif-type anorthosites and a small amount of lubricating noritic liquids firstly without Fe-Ti oxide minerals, then with these minerals. The second one, jotunitic, produces monzonoritic to mangeritic liquids, more or less quartzic, and complementary cumulates dominated by leuconorite and norites, with some anorthosites and mangerites. The third one, charnockitic, can evolve into granitic liquids with production of noritic and mangeritic cumulates.

7. ISOTOPIC CONSTRAINTS ON THE NATURE OF THE SOURCE ROCKS

Isotope geochemistry can shed some light on the anorthosite problem in better defining the source region of the magmas. In Rogaland as in many other anorthositic provinces in high grade metamorphic rocks, the isotopic convergences (especially in Strontium and Oxygen) between the LIL depleted lower crust and the upper mantle suggest a variety of interpretations. The Sr isotopic initial ratios have already been given in the general descriptions of each massif; they are also reported on fig.4. Whole-rock Pb-isotopic compositions for the magmatic rocks and the granulite-facies gneisses have been obtained and discussed by Weis (see Weis and Demaiffe, 1983a, b; Weis and Duchesne, 1983; Weis, this volume). Only a few Nd-isotopic data (expressed as $_{Nd}$ values, i.e deviation in parts per 10^4 of the $^{143}Nd/^{144}Nd$ ratio relative to the chondritic uniform reservoir at the given age) have been obtained for the anorthosite suite and the surrounding gneisses (Menuge, 1982; Ben Othman et al., 1984; Weis and Demaiffe, unpubl.). These isotopic data are synthesized in Table 1.

TABLE 1 : Synthesis of the isotopic compositions of the three magmatic series and of the granulite facies gneisses (references in the text).

	Basaltic series	Jotunitic series	Charnockitic series	Gneisses
$(^{87}Sr/^{86}Sr)_o$	0.7029-0.7060	0.704-0.7077	0.7053-0.709	0.705-0.720
$(^{207}Pb/^{204}Pb)_i$	15.482-15.527	15.524-15.578	15.463-15.514	15.491
ϵ_{Nd}	positive (+5.5 to 0)	positive (+5.7 to 2.2)	close to zero (+1.1 to -0.8)	negative (-0.8 to -2 2 values at -14 and -28)

Note: $(^{207}Pb/^{204}Pb)_i$ stands for the measured ratio corrected back for in situ U decay since the formation time of the different massifs.

The previously described basaltic and jotunitic series display wide, largely overlapping ranges of values for both Sr and Nd isotopic compositions, the $^{87}Sr/^{86}Sr$ initial ratios vary from 0.7029 to 0.7060 and from 0.7051 to 0.7077, respectively, for the two series, while the ϵ_{Nd} values are in the range +5.5 to 0 and +5.7 to +2.2. The lowest $^{87}Sr/^{86}Sr$ ratio (0.7029) reported for the Egersund-Ogna plagioclase megacrysts is compatible with a mantle source for the parental magma, although its value is already higher than that for the depleted mantle 1000 Ma ago (ca. 0.702), assuming a linear evolution of the Sr isotopic composition (Hart and Brooks, 1977). The higher values (up to ca. 0.708) are equally compatible with contamination of a mantle derived magma by crustal material or with direct partial melting of crustal rocks, the latter having a calculated Sr isotopic composition at 1000 Ma in the range 0.705 (Liland augen gneiss) to 0.720 (charnockitic migmatite) (Demaiffe, unpubl.)

The positive ε_{Nd} values for the anorthosites and jotunites imply a source region with a time-integrated Sm/Nd ratio higher than in the chondrites, that is a *LREE depleted source* such as the upper mantle or a basic rock derived from it. This is again in agreement with a basaltic parental magma for the massif-type anorthosites on the basis of trace element geochemistry and with the model of partial melting of LREE depleted basic cumulates recently proposed by Duchesne et al. (1984) for the jotunitic magma. The less positive ($\varepsilon_{Nd} < 2.7$) values reported for the leuconorites (pyroxene-rich anorthosite of Menuge) and for the Al-rich mega-opx are difficult to explain in terms of the differentiation model proposed for the massif-type anorthosite (see above). Indeed the mega-opx is interpreted as an early crystallizing phase. It should therefore be the least contaminated phase and thus have ε_{Nd} at least as high as the anorthosite. More data are needed to clarify this point.

The Pb isotopic composition of the anorthosites and related rocks (Weis and Demaiffe, 1983a; Weis and Duchesne, 1983; Weis, this volume) is distinctly more radiogenic than that of the granulite facies gneisses (Weis and Demaiffe, 1983b). The U/Pb ratio of the source region of the parental basaltic and jotunitic magmas should be higher than rocks in the U depleted lower crust (U/Pb < 0.05). Derivation from the *upper mantle* or from a *mantle-derived basic rock* thus accounts for the Pb isotopic constraints.

Charnockitic rocks directly associated with anorthosite (Bjerkreim-Sokndal quartz mangerites and Hidra charnockitic dykes) appear as strongly contaminated rocks : they have higher $^{87}Sr/^{86}Sr$ initial ratios (0.7075 - 0.7086), lower ε_{Nd} value (+ 1.1 for a Hidra charnockite) and less radiogenic Pb isotopic compositions than the anorthosites. Similar isotopic data also characterize the large southeastern acidic plutons (Kleivan charnockite, Lyngdal granodiorite, etc.), although their Pb isotopic compositions are closer to those of the granulite gneisses. These data are compatible with an origin by *partial melting of crustal material*.

8. OLD PROBLEMS, NEW QUESTIONS

In Rogaland, the grossly basaltic nature of the parental magma of massif-type anorthosites and, consequently the mantle source, which are deduced from the depth of crystallization of the Al-rich mega-opx, their Cr-contents and the occurrence of the basic association An_{67} + Fo_{70} are in agreement with Nd, Sr and Pb isotopic data. The deep-seated magma chamber postulated at the base of the thickened crust is a likely place for a first stage of differentiation leading to the formation of a plagioclase cumulate through crystallization of mafic minerals. It should be recalled that a similar step in the differentiation is also invoked in the case of Labrador anorthosites (Morse, 1982) to account for the production of hyperfeldspathic liquids from basaltic parental magmas. It thus appears that Labrador and Grenville anorthosites had a common history, at least at the beginning of their evolution. However, in their final evolution, it is still ignored for which geodynamic reasons Grenville anorthosites rise as crystal mush and Labrador anorthosites result from the injection of liquid in shallow magma chambers (Duchesne, 1984). In the same way, the alkalic or subalkalic character of the parental magma is difficult to reconstitute and it seems premature to use this criterion to better define the geodynamic environment of formation.

Rogaland's most striking characteristic is probably the remarkable development of jotunitic magmatism, which has constant features - it is alkalic to alkali-calcic (sensu Peacock) (Duchesne et al., 1984), shows a monotonous REE distribution (La/Yb = ca. 5) without Eu anomaly (fig.6), as well as very low U and Th contents -, but also displays a large variation in Fe/Mg ratios and in Sr isotopic characteristics.

A derivation of the jotunite by fractional crytallization can in most cases be precluded and conspicuous evidence of a direct origin by partial melting is provided in the case of the Vetteland dyke. It is thus permitted to assume that partial melting extended to the generation of most, if not all, jotunitic magma batches. The hot anorthositic diapirs are the most likely sources of the necessary heat of fusion during their ascent. Anorthosites will thus act like probes of the lower crust and inversion technique applied to jotunites will permit better to assess its geochemical characteristics. A point unravelled by the present study seems to be that, besides the now classical hypothesis on the origin of anorogenic Proterozoic granites by partial melting produced by the anorthositic plutonism (Anderson, 1982), one must also take into account crustal basic rocks as potential source of anatectic melts to yield jotunitic rocks : Nd isotopes suggest that the Hidra jotunitic parental magma is produced from a depleted basic source, and REE distributions point to a basic cumulate as the source rock of the Vettaland dyke.

REFERENCES

ADAMSON O.J. (1942) :'The granite pegmatites of Hitterö, S.W. Norway.'
    Geol. Fören. Forhand, 64, 97-116.
ANDERSON J.L. (1983) : 'Proterozoic anorogenic granite plutonism of
    North America'. Geol. Soc. Amer., Mem.161, 133-154.
ASHWAL L.D. and WOODEN J.L. (1983) : 'Sr and Nd isotope geochronology,
    geologic history, and origin of the Adirondack Anorthosite'.
    Geochim. Cosmochim. Acta 47, 1875-1885.
ASHWAL L.D. (1982) : 'Mineralogy of mafic and Fe-Ti oxide-rich
    differentiates of the Marcy anorthosite massif, Adirondacks,
    New York'. Amer. Miner. 67, 14-27.
BARTH T.F.W. and DONS J.A. (1960) : 'The Precambrian in Southern
    Norway'. In Geology of Norway (Ohltedahl C. ed.) Norges
    Geol.Unders. 208, 6-67.
BEN OTHMAN D., POLVE M. and ALLEGRE C.J. (1984) : 'Nd-Sr composition of
    granulites. Constraints on the evolution of the lower conti-
    nental crust.' Nature 307, 510-515.
CARUBA R., DUCHESNE J.C. and IACCONI P. (1985): 'Zircons in charnocki-

tic rocks from Rogaland (South-West Norway). Petrogenetic implications.' (in preparation).

DEMAIFFE D. (1972) : 'Etude pétrologique de l'apophyse Sud-Est du massif de Bjerkrem-Sogndal (Norvège méridionale)'. *Ann. Soc. Géol. Belg.* 95, 255-269.

DEMAIFFE D. (1977) : 'De l'origine des anorthosites. Pétrologie, géochimie et géochimie isotopique des massifs anorthositiques d'Hidra et de Garsaknatt (Rogaland - Norvège méridionale)'. *Thèse de doctorat, Université Libre de Bruxelles*, 303p.

DEMAIFFE D. and JAVOY M. (1980) :' $^{18}O/^{16}O$ ratios of anorthosites and related rocks from the Rogaland complex (S.W. Norway)'. *Contrib. Mineral. Petrol.* 72, 311-317.

DEMAIFFE D. and HERTOGEN J. (1981) : 'Rare earth geochemistry and strontium isotopic composition of a massif-type anorthositic-charnockitic body : the Hidra massif (Rogaland, S.W.Norway)'. *Geochim. Cosmochim. Acta* 45, 1545-1561.

DEMAIFFE D., DUCHESNE J.C. and HERTOGEN J. (1979) : 'Trace element variations and isotopic composition of charnockitic acidic rocks related to anorthosites (Rogaland - S.W.Norway)'. In *Origin and distribution of the Elements* (Ahrens L.H., ed.) 417-429, Pergamon Press.

DEMAIFFE D., DUCHESNE J.C., MICHOT J. and PASTEELS P. (1973) : 'Le massif anorthosito-leuconoritique d'Hidra et son faciès de bordure'. *C.R. Acad. Sci. Paris* 277, 17-20 (Série D).

DUCHESNE J.C. (1968) : 'Les relations Sr-Ca et Ba-K dans les plagioclases des anorthosites du Rogaland méridional'. *Ann. Soc. Géol. Belg.* 90, 643-656.

DUCHESNE J.C. (1970b) : 'Microtextures of Fe-Ti oxide minerals in the South Rogaland anorthositic complex (Norway)'. *Ann. Soc. Géol. Belg.* 93, 527-544.

DUCHESNE J.C. (1971) : 'Le rapport Sr/Ca dans les plagioclases du massif de Bjerkrem-Sogndal (Norvège méridionale) et son évolution dans la cristallisation fractionnée du magma plagioclasique'. *Chem. Geol.* 8, 123-130.

DUCHESNE J.C. (1972a) : 'Iron-Titanium oxide minerals in the Bjerkrem-Sogndal massif, Southwestern Norway'. *J. Petrol.* 13, 57-81.

DUCHESNE J.C. (1972b) : 'Pyroxènes et olivines dans le massif de Bjerkrem-Sogndal (Norvège méridionale). Contribution à l'étude de la série anorthosite-mangérite'. *24th Intern. Geol. Congress, Montreal, sect.* 2, 320-328.

DUCHESNE J.C. (1973) : 'Les gisements d'oxydes de fer et titane dans les roches anorthositiques du Rogaland (Norvège méridionale)'. In *Les roches plutoniques dans leurs rapports avec les gîtes minéraux*. Coll. E.Raguin. Masson, Paris, 403 p.

DUCHESNE J.C. (1978) : 'Quantitative modeling of Sr, Ca, Rb and K in the Bjerkrem-Sogndal layered lopolith (S.W.Norway)'. *Contr. Mineral. Petrol.* 66, 175-184.

DUCHESNE J.C. (1984) : 'Massif anorthosites : another partisan review'. In *Feldspars and feldspathoids* (Brown W.S., ed.) Reidel, 411-433.

DUCHESNE J.C. and DEMAIFFE D. (1978) : 'Trace elements and anorthosite genesis'. *Earth Planet. Sci. Lett.* 38, 249-272.

DUCHESNE J.C., DEMAIFFE D. and HERTOGEN J. (1985) : 'Further data on monzonoritic rocks from the Egersund anorthositic province (abstract).'NATO Adv. St. Inst. *The deep Proterozoic crust in the North Atlantic provinces.* S.Norway, July 84.

DUCHESNE J.C., DEMAIFFE D. and WILMART E. (1984) : 'Apophysis, Hidra massif and Envelope'. In *The Precambrian of Southern Norway* (Maijer C. et al., eds) *Norges Geol. Unders.* (in preparation).

DUCHESNE J.C., DENOISEUX B. and HERTOGEN J. (1985) : 'The norite-mangerite transition zone in the Bjerkreim-Sokndal lopolith (S.W. Norway) : immiscibility or fractional crystallization'. (abstract) NATO Adv. St. Inst. *The deep Proterozoic crust in the North Atlantic provinces.* S. Norway, July 84.

DUCHESNE J.C. and MICHOT J. (1974) : 'Anorthosites and their environment : the anorthosite problem with respect to the Norvegian plutons'. In *Géologie des domaines cristallins* (Bellière J. and Duchesne J.C., eds) *Soc. Géol. de Belgique*, 333-346.

DUCHESNE J.C. and MICHOT J. (1984) : 'The Rogaland intrusive masses : an introduction'. In *The Precambrian of Southern Norway* (Maijer, C. et al., eds) *Norges Geol. Unders.*(in preparation).

DUCHESNE J.C. and MAQUIL R. (1981) : 'Evidence of syn-intrusive deformation in South-norwegian anorthosites' (abstract). *Terra Cognita*, Special Issue, 94.

DUCHESNE J.C., ROELANDTS I., DEMAIFFE D., HERTOGEN J., GIJBELS R. and DE WINTER J. (1974) : 'Rare earth data on monzonoritic rocks related to anorthosites and their bearing on the parental magma of the anorthosite suite'. *Earth Planet. Sci. Lett.* 24, 325-335.

DUCHESNE J.C., ROELANDTS I., DEMAIFFE D. and WEIS D. (1984) : 'Petrology, trace-element geochemistry, Sr and Pb isotopes, in monzonorites related to Rogaland anorthosites (S.W.Norway) : evidence of variable crustal contributions'.(Submitted to *Contr. Miner. Petrol.*).

EMSLIE R.F. (1975) : 'Pyroxene megacrysts from anorthositic rocks : new clues to the source and evolution of the parent magmas'. *Can. Miner.* 13, 138-145.

EMSLIE R.F. (1978) : 'Anorthosite massifs, Rapakivi granites, and late Proterozoic rifting of North America'. *Precambr. Res.* 7, 61-98.

EMSLIE R.F. (1980) : 'Geology and petrology of the Harp Lake Complex, Central Labrador : an example of Elsonian magmatism'. *Geol. Surv. Canada, Bull.* 293, 136p.

FALKUM T. and PETERSEN J.S. (1974) : 'A three fold division of the "farsundite" plutonic complex at Farsund, Southern Norway'. *Norsk Geol. Tidsskr.* 54, 361-366.

FALKUM T., WILSON J.R., PETERSEN J.S. and ZIMMERMANN H.D. (1979) : 'The intrusive granites of the Farsund area (S.Norway) and their relations with the Precambrian metamorphic envelope'. *Norsk Geol. Tidsskr.* 59, 125-139.

GREEN T.H. (1970) : 'Experimental fractional crystallization of quartz diorite and its application to the problem of anorthosite

origin'. In *Origin of anorthosite and related rocks* (Isachsen Y.W., ed.) N.Y. State Mus. Sci. Serv. Mem. 18, 23-30.
GRIFFIN W.L. and HEIER K.S. (1973) : 'Petrological implications of some corona structures'. *Lithos* 6, 315-35.
HART S.R. and BROOKS C. (1977): 'The geochemistry and evolution of Early Precambrian mantle'. *Contr. Miner. Petrol.* 61, 109-128.
HUBAUX A. (1960) : 'Les gisements de fer titané de la région d'Egersund, Norvège'. *N. Jb. Miner. Festband Ramdohr*, 94, 926-992.
KRAUSE H., GIERTH E., and SCHOTT W. (1985) : 'Ti-Fe deposits in the South Rogaland Igneous complex, especially in the Anorthosite-massif of Åna-Sira'. *Norges Geol. Unders.* (in the press).
KRAUSE H. and PEDALL G. (1980) : 'Fe-Ti mineralizations in the Ana-Sira anorthosite, Southern Norway'. In *Metallogeny of the Baltic shield* (Joako Siivola, Ed.) *Geol. Surv. Finland* 307, 56-83.
LINDSLEY D.H. and EMSLIE R.F. (1968) : 'Effect of pressure on the boundary curve in the system diopside-albite-anorthite'. *Carnegie Inst. Washington*, Yb. 66, 479-480.
MADSEN J.KORNERUP (1977) : 'Composition and microthermometry of fluid inclusions in the Kleivan granite, South Norway'. *Amer. J. Sci.* 277, 673-695.
MAIJER C., DUCHESNE J.C., FALKUM T., SIGMUND E., TOBI A.C. and TOURET J. (eds) (1984) : 'The Precambrian of Southern Norway'. *Norges Geol. Unders.*
MAQUIL R. (1979) : 'Preliminary investigation on giant orthopyroxenes with plagioclase exsolution lamellae from the Egersund-Ogna anorthositic massif (S.W.Norway)'. *Progress in experimental petrology* :4th Report 1975-1978 *N.E.R.C. Publ.* Ser.D, 144-146.
MAQUIL R. and DUCHESNE J.C. (1984) : 'Geothermométrie par les pyroxenes et mise en place du massif anorthositique d'Egersund-Ogna (Rogaland, Norvège méridionale)'. *Ann. Soc. Géol. Belg.* 107, 27-49.
MARTIGNOLE J. and SCHRIJVER K. (1970a) : 'Tectonic setting and evolution of the Morin anorthosite, Grenville Province, Quebec'. *Bull. Geol. Soc. Finland* 42, 165-209.
MARTIGNOLE J. and SCHRIJVER K. (1970b) : 'The level of anorthosites and its tectonic pattern'. *Tectonophysics* 10, 403-409.
MENUGE J. (1982) : 'Nd isotope studies of crust-mantle evolution : the Proterozoic of South Norway and the Archaean of Southern Africa'. *Ph. D. Thesis*; University of Cambridge, 257p.
MICHOT J. (1960) : 'La palingenèse basique'. *Acad. Roy. Belg., Bull. Cl. Sci. série* 5, 46, 257-268.
MICHOT J. (1961a) : 'Le massif complexe anorthosito-leuconoritique de Haaland-Helleren et la palingenèse basique'. *Acad. Roy. Belg., Cl. Sci., Mém. collect.* in-4, 2e série, T.XV, fasc.1, 95 p.
MICHOT J. (1961b) : 'The anorthositic complex of Haaland-Helleren'. *Norsk Geol.Tidsskr.*41, 157-172.
MICHOT J. (1972) : 'Anorthosite et recherche pluridisciplinaire'. *Ann. Soc. Géol. Belg.* 95, 5-43.
MICHOT J. and MICHOT P. (1969) : 'The problem of the anorthosites. The South Rogaland igneous complex (South Western Norway)'. In

Origin of anorthosites and related rocks (Isachsen Y.W., Ed.), New York State Mus. Sci. Serv. Mem. 18, 399-410.
MICHOT J. and PASTEELS P. (1969) : 'La variation du rapport $(^{87}Sr/^{86}Sr)_o$ dans les roches génétiquement associées au magma plagioclasique (Premiers résultats)'. Ann. Soc. Géol. Belg. 92, 255-262.
MICHOT P. (1939) : 'La couronne d'anorthosite hypersthénifère feuilletée et rubanée du massif anorthositique d'Egersund (Norvège)'. Ann. Soc. Géol. Belg. 62, 547-551.
MICHOT P. (1955b) : 'L'anatexie leuconoritique'. Acad. Roy. Belg., Bull. Cl. Sci., série 5, 41, 374-385.
MICHOT P. (1956) : 'La géologie des zones profondes de l'écorce terrestre'. Ann. Soc. Géol. Belg. 80, 19-73.
MICHOT P. (1957) : 'Constitution d'une dôme de gneiss coiffé en milieu catazonal profond'. Acad. Roy. Belg., Bull. Cl. Sci.,série 5, 43, 23-44.
MICHOT P. (1960a) : 'La géologie de la catazone : le problème des anorthosites, la palingenèse basique et la tectonique catazonale dans le Rogaland méridional (Norvège méridionale)'. Norges Geol. Unders. 212g, 1-54.
MICHOT P. (1960b) : 'Le problème des intrusions marginales'. Geol. Rundschau, 50, 94-105.
MICHOT P. (1965a) : 'Le magma plagioclasique'. Geol.Rundschau, 54, 956-976.
MICHOT P. (1965b) : 'Les orogènes fondamentaux (Die Grundorogene)'. Freiberg. Forsh. C190, 49-62.
MIDDLEMOST E.A.K. (1968) : 'The granitic rocks of Farsund, South Norway'. Norsk Geol. Tidsskr. 48, 81-99.
MORSE S.A. (1975) : 'Plagioclase lamellae in hypersthene, Tikkoatokhakh Bay, Labrador'. Earth Planet. Sci. Lett. 26, 331-336.
MORSE S.A. (1979) : 'Kiglapait geochemistry I : systematics, sampling and density'. J. Petrol. 20, 555-590.
MORSE S.A. (1982) : 'A partisan review of Proterozoic anorthosites'. Amer. Mineralogist 67, 1087-1100.
NEWTON R.C. and HASELTON H.T.(1980) : 'Thermodynamics of the Garnet-Plagioclase-Al_2SiO_5-Quartz geobarometer'. In Thermodynamics of minerals and melts (Newton R.C., et al., eds) Adv. Physical Geochemistry 1, 131-149, Springer.
PASTEELS P., MICHOT J. and LAVREAU J. (1970) : 'Le complexe éruptif du Rogaland méridional (Norvège). Signification pétrogénétique de la farsundite et de la mangérite quartzique des unités orientales; arguments géochronologiques et isotopiques'. Ann. Soc. Géol. Belg. 93, 453-476.
PASTEELS P., DEMAIFFE D. and MICHOT J. (1979) : 'U-Pb and Rb-Sr geochronology of the eastern part of the South Rogaland igneous complex, Southern Norway'.Lithos 12, 199-208.
PEDERSEN S. and FALKUM T. (1975) ; 'Rb-Sr isochrons for the granitic plutons around Farsund, Southern Norway'. Chem. Geol. 15, 97-101.
PETERSEN J.S. (1980a) : 'The zoned Kleivan granite - an end member of the anorthosite suite in Southwest Norway'. Lithos 13, 79-95.

PETERSEN J.S. (1980b) : 'Rare-earth element fractionation and petrogenetic modelling in charnockitic rocks, Southwest Norway'. Contrib. Mineral. Petrol. 73, 161-172.
PETERSEN J.S. and PEDERSEN S. (1978) : 'Strontium isotope study of the Kleivan granite, Southern Norway'. Norsk Geol. Tidsskr. 58, 97-102.
PRESNALL D.C., DIXON S.A., DIXON J.R., O'DONNEL T.H., BRENNER N.L., SCHROCK R.L. and DYCUS D.W. (1978) : 'Liquidus phase relations on the join diopside-forsterite-anorthite from 1 atm to 20 Kbar : their bearing on the generation and crystallization of basaltic magmas'. Contr. Mineral. Petrol. 66, 203-220.
RIETMEIJER F.J.M. (1979) : 'Pyroxenes from iron-rich igneous rocks in Rogaland, S.W. Norway'. Geologica Ultraiectina 21, 341p.
ROELANDTS I. and DUCHESNE J.C. (1979) : 'Rare-earth elements in apatite from layered norites and iron-titanium oxide ore-bodies related to anorthosites (Rogaland, S.W. Norway)'. In Origin and distribution of the elements (Ahrens l.H., ed.) Pergamon, 199-212.
SELLEVOLL M.A. and AALSTAD I. (1971) : 'Magnetic measurements and seismic profiling in the Skagerak'. Marine Geophys. Res. 1, 284-302.
SMITHSON S.B. and RAMBERG I.B. (1979) : 'Gravity interpretation of the Egersund anorthosite complex, Norway : its petrological and geothermal significance'. Geol. Soc. Amer. Bull., part I, 90, 199-204.
WAGER L.R. and BROWN G.M. (1968) : Layered igneous rocks. Oliver and Boyd Ltd, Londres, 588 p.
WEIS D. (1982) : 'La géochimie isotopique du Plomb total comme traceur pétrogénétique : méthodologie et exemples d'applications'. Doctorat thesis, Université Libre de Bruxelles, 353p.
WEIS D. (1985) : 'Genetic implications of Pb isotopic geochemistry in the Rogaland anorthositic complex (S.W. Norway)'. (this book).
WEIS D. and DEMAIFFE D. (1983a) : 'Age relationships in the Proterozoic high-grade regions of Southern Norway : discussion and comment'. Precambr. Res. 22, 149-155.
WEIS D. and DEMAIFFE D. (1983b) : 'Pb isotope geochemistry of a massif-type anorthositic-charnockitic body : the Hidra massif (Rogaland, S.W.Norway)'. Geochim. Cosmochim. Acta 47, 1405-1413.
WEIS D. and DUCHESNE J.C. (1983) : 'Pb isotopic compositions in the Egersund-Ogna anorthosite (S.W Norway) : an indication of a Gothian magmatic event ? (Abstract)'. Terra cognita 3, 140.
WIEBE R.A. (1980a) : 'Anorthositic magmas and the origin of Proterozoic anorthosite massifs'. Nature 286, 564-567.
WIEBE R.A. (1980b) : 'Commingling of contrasted magmas in the plutonic environment : examples from the Nain anorthositic complex'. J.Geol. 88, 197-209.
WIEBE R.A. (1984) : 'Fractionation and magma mixing in the Bjerkrem-Sogndal lopolith (S.W. Norway) : evidence for the compositions of residual liquids'. Lithos 17, 171-188.
WIELENS J.B.W., ANDRIESSEN P.A.M., BOELRIJK N.A.I.M., HEBEDA E.H., PRIEM H.N.A., VERDURMEN E.A.T. and VERSCHURE R.H. (1981) :

'Isotope geochronology in the high-grade metamorphic Precambrian of Southwestern Norway : new data and reinterpretations'. *Norges Geol. Unders.* 359, 1-30.

WILMART E. (1985) : 'Géothermométrie et géobarométrie dans l'enveloppe métamorphique du massif d'Åna-Sira (Vest Agder, Norvège méridionale)'. *Ann. Soc. Géol.* (in the press).

WILMART E., HERTOGEN J. and DUCHESNE J.C. (1985) : 'Geochemistry of supracrustals, granite-gneisses and charnockites from the envelope of the Ana-Sira massif (Rogaland/Vest Agder) (abstract)'. NATO Adv. St. Inst. *The deep Proterozoic crust in the North Atlantic provinces* S.Norway, July 84.

WILSON J.R. and ANNIS M.P. (1973) : 'Granit komplekset ved Farsund, Syd Norge'. *Dansk Geol.Foren.* 66-70.

METAMORPHIC ZONING IN THE HIGH-GRADE PROTEROZOIC OF ROGALAND-VEST AGDER, SW NORWAY

Alex C. Tobi, Gé A.E.M. Hermans, Cornelis Maijer and
J.Ben H. Jansen
Instituut voor Aardwetenschappen
Postbus 80.021, NL 3508 TA Utrecht
Netherlands

Abstract

The high-grade low-pressure metamorphism in Rogaland increases towards the southwest to reach a maximum near the border of the Egersund intrusive complex. The first isograd is the "hypersthene line", marking the appearance of hypersthene in rocks of leucogranitic composition. A few kilometres westward hornblende + quartz are replaced by pyroxenes in mafic rocks, and the amphibole colour grades from green to brown in Ti-bearing rocks. Somewhat west of the hypersthene line garnet + sillimanite give way to cordierite + spinel/magnetite. A second isograd heralds the partial decomposition of garnet to orthopyroxene + cordierite + spinel/magnetite. In rocks richer in Fe, garnet + sillimanite react to spinel/magnetite + quartz. In potash-rich compositions garnet + biotite + quartz ± sillimanite react to osumilite + orthopyroxene + spinel/magnetite. Near the intrusive complex a third isograd marks the incoming of metamorphic pigeonite in leucogranitic rocks.

The temperature range of the whole zoning from the hypersthene line to the complex is estimated at 700-1050°C at a pressure of 4-6 kb, although remnants of a higher-pressure older metamorphism may be present.

Somewhat younger retrograde metamorphism is widespread, and includes decomposition of osumilite and the growth of garnet rims.

Review of the geology

The southern tip of Norway mainly consists of intensely folded banded migmatites with compositions varying from leucogranitic to metabasaltic (Fig. 1). Interfolded with these, extensive areas occur of more homogeneous leucogranitic rocks and augengneisses. Also intercalated are bands of metapelites with a thickness usually varying from a few tens to a few hundreds of metres. These have been called garnetiferous migmatites, because garnet is also present in the leucosomes. The Faurefjell metasediments are characterized by quartzites and siliceous metadolomites, partly depleted in carbonate to form diopsidic rocks and quartz-diopside gneisses. Skarns and metabasalts are also represented in this formation. The southern part of the area covered by Fig. 1 is underlain by the

Egersund intrusive complex, which is extensively described in this volume by Duchesne and co-workers. Smaller magmatic bodies include the Sjelset pyroxene granite, roughly comparable with the Farsund charnockite, the Gloppurdi and Botnavatnet olivine-pyroxene monzonites, older and more intensely folded equivalents of the upper part of the Bjerkreim-Sokndal lopolith, and the folded basic intrusions, representing an older cycle of mafic dykes, perhaps coeval with the Gloppurdi Massif (about 1200 Ma).

The migmatites together with the other formations occurring intercalated in them show a metamorphic gradient from amphibolite to granulite facies towards the west and south-west, which is thought to be mainly due to the Egersund intrusive complex. The pattern we now see could well have been created by an interference of two or more successive stages of metamorphism. The age of the main peak of metamorphism could be around 1050 Ma, coeval with the intrusion of the lower part of the Bjerkreim Sokndal lopolith (Hermans et al. 1975, Kars et al. 1980, Wielens et al. 1980). On the whole the metamorphism is characterized by low pressure. Sillimanite is the only aluminium silicate present in the metapelites: kyanite has never been found, andalusite at only two localities in the granulite facies area, probably as a product of retrograde metamorphism. The assemblage amphibole-almandine is not found in the amphibolite-facies area, neither is the assemblage amphibole-clinopyroxene in the granulite-facies area.

The hypersthene line

The "hypersthene line" was mapped and described by Hermans et al. (1975). It marks the in-coming of hypersthene in leucogranitic components of the migmatites. As such, it is roughly equivalent to the isograd B/C described by Field et al. (this volume) in the Bamble Sector. The incoming of hypersthene in metabasites (isograd A/B of Field et al.) in Rogaland occurs at a considerable distance east of the area of Fig. 1; it is not treated in the present paper. The line is rather a gradational zone than a sharp boundary, partly because of the dependence on rock composition, partly because of the complex metamorphic history of the area. In the field, the granulite-facies area is also characterized by the dark green colour of quartz and the feldspars on fresh rock surfaces. In Ca-poor granitic compositions, most likely biotite and quartz are involved in the hypersthene-producing reaction. Possible reactions are:

$$\text{biot} + Q \rightarrow \text{opx} + \text{alm} + \text{Kf} + H_2O \text{ (Winkler 1976)} \quad (1)$$

$$\text{biot 1} + Q \rightarrow \text{opx} + \text{biot 2} + \text{Kf} + H_2O \quad (2)$$

$$\text{biot} + Q + \text{plag} + V \rightarrow \text{opx} + L \text{ (Clemens \& Wall 1981)} \quad (3)$$

$$\text{biot} + Q + \text{plag} \rightarrow \text{opx} + \text{Kf} + L \text{ (ibid.)} \quad (4)$$

For pressures > 4 kb these reactions do not seem to be very pressure-dependent; temperature estimates range from about 700 to 750°C. Obviously, the first reaction requires the production of garnetiferous charnockites. Although these rocks are occasionally found in Rogaland, especially in the neighbourhood of the metapelites, most charnockites

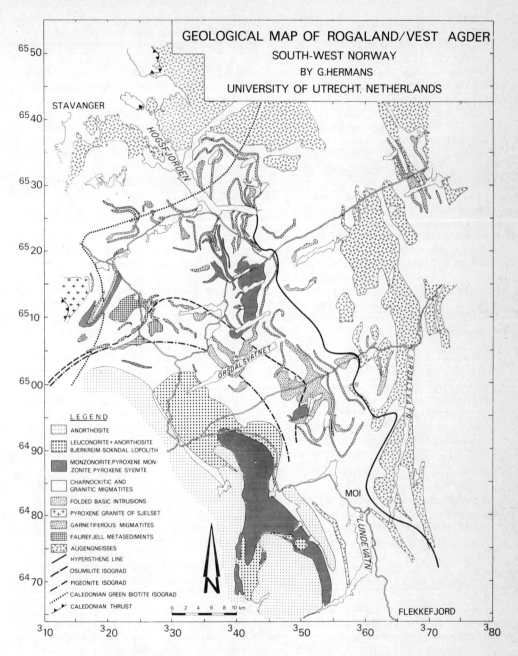

Fig. 1. *Simplified geological map of Rogaland-Vest Agder, with indication of the isograds discussed in this paper. The Gloppurdi monzonite massif is situated north of Örsdalsvatnet. For location of fig. 1 see fig. 2.*

do not contain garnet. Textural indications for pro-grade mineral reactions are scarce, and little analytical work has as yet been done on the environment of the hypersthene line. The second reaction seems appropriate, but so far cannot be proved. In retrograde direction, it seems to be indicated by the occurrence of biotite-quartz symplectites; locally, the orthopyroxene is only partly replaced. The wide-spread serpentinization of orthopyroxene is a later phenomenon.

In reactions (3) and (4) the later history of the melt would be unknown. In most charnockites, there is no textural evidence that orthopyroxene should be older than the other phases. Considering the melt fraction produced by reactions (3) and (4), the extraction of H_2O-bearing granitic melt could have contributed to the origin of charnockites (Newton, this volume). According to Field, Drury & Cooper (1980), the K-poor "arendalites" of Tromøy (Bamble area, S. Norway; see also Field et al., this volume) are enderbitic cumulates from which rapakivi-type granitic magma was extracted. Also, Korstgaard (1979) states that the granitoid rocks in many granulite-facies terrains are often rather enderbites than charnockites. In Rogaland-Vest Agder, (leuco-)granitic rocks continue to occur west of the hypersthene line. However, enderbites west of this line are probably more frequent than tonalites are east of it, so that it remains uncertain to what extent granitic melt could have been removed. CO_2-rich fluid inclusions are abundantly found west of the hypersthene line (Swanenberg 1979), as is the general case in granulite-facies rocks (Touret 1971, 1981, this volume; Wendlandt 1981). This raises the problem whether temperature or introduction of CO_2 is the primary cause of the isograd. The fairly low total pressure (around 5 kb?) probably enables H_2O to disappear as vapour or in a melt. Local deviations from the general pattern of metamorphic zoning could be mainly caused by differences in CO_2/H_2O ratio.

In granitic compositions richer in Ca (not used for determining the hypersthene line), two pyroxenes should form mainly at the expense of amphibole and quartz (de Waard 1965, Binns 1969):

$$\text{amph} + \text{Kf} + Q \rightarrow \text{opx} + \text{cpx} + \text{biot} + \text{plag} + H_2O \qquad (5)$$

$$\text{amph} + Q \rightarrow \text{opx} + \text{cpx} + \text{plag} + H_2O \qquad (6)$$

In prograde charnockites of this composition, brown-green amphibole is in places only found as armoured relics in clinopyroxene, more rarely in orthopyroxene crystals. The colour of the metamorphic amphiboles is also a fairly accurate indicator of metamorphic grade. In the amphibolite-facies part of the migmatites the amphiboles are green or brown-green. West of the hypersthene line, the colour changes within a few kilometres to brown, if the rock contains ilmenite to buffer Ti (Dekker 1978).

In the southern part of the area the hypersthene line seems to be related to the Egersund intrusive complex. West of Tonstad, however, this isograd gradually verges towards the north, to pass east of the Gloppurdi Massif. This part of the hypersthene line was perhaps caused by the intrusion of this massif about 1200 Ma ago. Still less is known about the continuation of the hypersthene line farther to the north. The northernmost locality where fresh charnockite was found is near Prekestolen, at the northern edge of Lysefjorden (just north of the

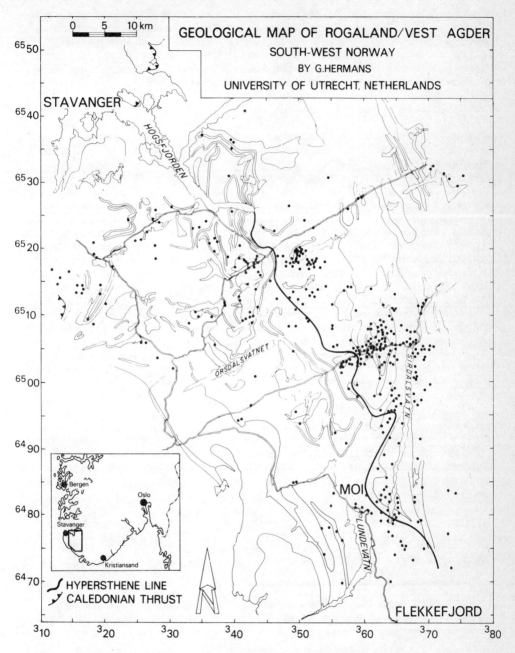

Fig. 2. *Sample locations with stable assemblage amphibole + quartz. The (amphibole + quartz)-out isograd is situated a few kilometres west of the hypersthene line. Westwards the assemblage is mainly conserved in the Sjelset granite and the Gloppurdi massif.*

mapped area). Westward of this locality, the study of high-grade assemblages is increasingly hampered by low retrograde metamorphic overprinting related to the nearby Caledonian Front. In places patches of serpentine could represent hypersthene. More frequently, the original granulite-facies character of the rocks is indicated by the occurrence of brown amphibole, which is less easily retrograded.

The (amphibole + quartz)-out isograd

West of the hypersthene line, the couple amphibole + quartz continues to occur in mafic rocks (amphibolites) over a distance of a few kilometres. Westward, it is replaced by pyroxenes, again following reactions (5) or (6) (see fig. 2). According to Binns (1969) the temperature should be 700-800°C at about 4 kb, depending also on partial water pressure. This isograd is also a transition zone rather than a line, and shows an overlap with the transition zone of the hypersthene line. West of the isograd, the relevant reaction is best demonstrated along contacts between quartz-free amphibolites and leucocharnockites. Here, "dehydration rims" (Schrijver 1973) occur where amphibole is replaced by orthopyroxene ± clinopyroxene. In quartz-free assemblages amphibole continues to occur until the boundary of the intrusive complex is reached. In the Gloppurdi monzonite massif and the Sjelset pyroxene granite, the couple amphibole + quartz remains frequent. As stated above, the granulite facies around the Gloppurdi massif could be coeval with the intrusion. The Sjelset massif intruded after the peak of the granulite-facies metamorphism in the surrounding rocks. Its geochronology is under study. Alternatively, the persistence of the assemblage could be caused by the extreme richness in Fe of both massifs.

The osumilite-in isograd

In the metapelites no significant change is noted near the hypersthene line. In the amphibolite-facies area garnet is younger than fibrolite, because microfolded layers of this mineral often occur enclosed in garnet (fig. 3). Locally, cordierite may be formed together with garnet. The reaction may be:

$$\text{fibrolite} + \text{biot} + Q \rightarrow \text{garn} \pm \text{cord} + \text{Kf} + \text{H}_2\text{O} \qquad (7)$$

Where sillimanite continues to occur outside of the garnet, it has recrystallized to a larger grain size. Some distance west of the hypersthene line, these minerals react to give cordierite and spinel (fig. 4):

$$\text{garn} + \text{sill} \rightarrow \text{cord} + \text{spin/magn.} \qquad (8)$$

This isograd has not yet been mapped. In compositions richer in Fe, sillimanite recrystallizes to spinel/magnetite and quartz in the vicinity of garnet:

$$\text{garn} + \text{sill} \rightarrow \text{spin/magn} + \text{quartz} \qquad (9)$$

The spinel is pseudomorphous after sillimanite and rimmed by quartz (fig. 5). This reaction occurs in the neighbourhood of the osumilite-in isograd.

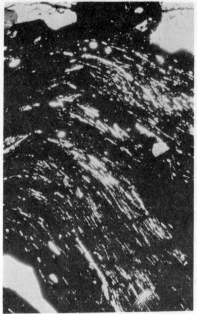

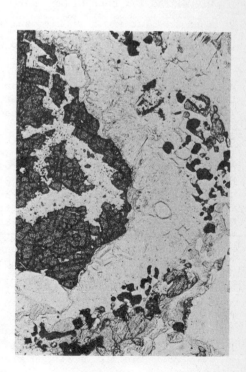

Fig. 3 A, B. Microphotograph of folded fibrolite in garnet. Garnetiferous migmatites, Hunnedalen. Sample Y110. a: one polar. b: crossed polars. Length of photograph: 3 mm.

Fig. 4. Microphotograph of garnet and coarse sillimanite separated by an aggregate of cordierite and spinel. Sample ROG37, one polar. Length of photograph: 4 mm.

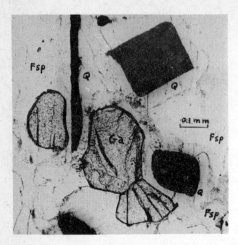 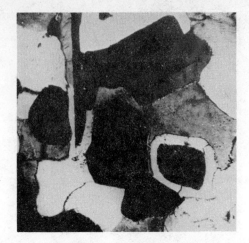

Fig. 5. Microphotograph of the reaction: sillimanite + garnet → spinel/magnetite + quartz.
Sample A94, quartz-spinel granofels. Spinel is pseudomorphous after sillimanite, and surrounded by a mantle of quartz. Spinel and ore are both black on the photograph. a: one polar; b: crossed polars.

An isograd has been drawn where the following reaction occurred:

$$\text{garn} + \text{biot} + Q \pm \text{sill} \rightarrow \text{osum} + \text{spin/magn} + \text{opx} \tag{10}$$

On Fig. 6 localities are indicated where garnet, biotite and quartz occur in contact, and others where osumilite is found. Fig. 7 shows an AFM diagram for this reaction. The osumilite is exceptionally well exposed in a roadcut at Vikesdal near Vikeså. It is a pinkish violet mineral with optical properties similar to cordierite, but with $2V_z = 0 - 30°$. The mineral also differs in the absence of twinning and pleochroic haloes. It was originally described from volcanic rocks; the Rogaland occurrence is the sixth known in the world (Maijer et al. 1981). Its formula is approximately $K(Mg,Fe)_2(Al,Mg)_3(Si_{18}Al_2O_{30})$. The osumilite occurs as fine-grained aggregates with numerous inclusions of spinel/magnetite and orthopyroxene (Fig. 8). It shows characteristic alteration cracks filled with micaceous alteration products. Other inclusions are relict biotite and quartz, the latter with rims of micaceous material. The garnet present in some layers is usually rimmed by plagioclase. At other localities the osumilite contains abundant inclusions of sillimanite (Fig. 9). In this case of excess alumina (cf. Fig. 7) orthopyroxene is not formed. The sillimanite adapts to the higher grade of metamorphism by building Fe^{3+} into the lattice (Evers and Wevers 1984). According to experimental results osumilite is stable at low pressures and at temperatures above 800°C (Hensen 1977). In most localities given in Fig. 6 the osumilite is completely retrograded to a symplectite consisting of cordierite, potash feldspar and quartz (Fig. 10). Another retrogressive reaction is the formation of a second generation of garnet around biotite and spinel inclusions in osumilite.

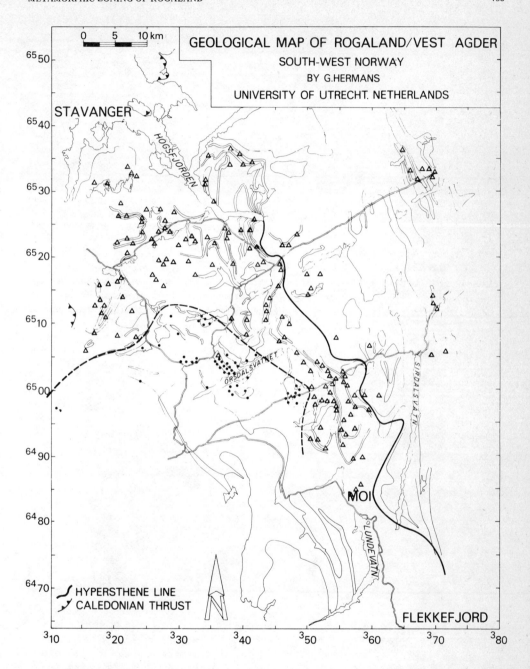

Fig. 6. Reaction isograd garn + biot + Q ± sill → osum + opx + spin/magn (dashed). Triangles: assemblage garn - biot - Q. Dots: assemblage osum - opx - spin/magn.

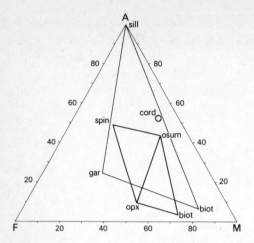

Fig. 7. AFM diagram of the reaction:
garn + biot + Q ± sill → osum + spin/magn + opx
Where biotite is conserved its Fe/Mg ratio increases. In rocks with excess sillimanite orthopyroxene is not formed. Thin line: amphibolite-facies assemblage. Thick line: granulite-facies assemblages.

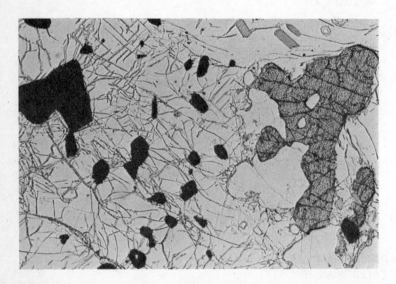

Microphotograph of the reaction:
garn + biot + Q ± sill → osum + spin/magn + opx.
The osumilite shows characteristic alteration cracks filled with micaceous alteration products. It forms an aggregate with inclusions of spinel/magnetite and orthopyroxene (grey), and relict inclusions of biotite and quartz. The quartz is rimmed with serpentine. Sample MA490, Vikesdal. One polar. Length of photograph: 3 mm.

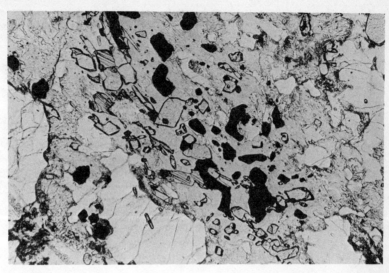

Fig. 9. Microphotograph of osumilite with numerous inclusions of sillimanite and spinel-magnetite. Due to the excess of alumina orthopyroxene is not formed. Sillimanite adapts to the higher grade of metamorphism by building Fe^{3+} into the lattice. Sample GA205, Marsteinsfjell. Length of photograph: 3 mm.

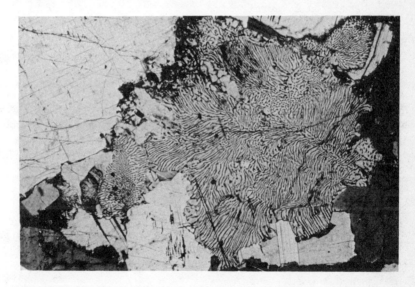

Fig. 10. Microphotograph of a characteristic "myrmekitic" symplectite, here mainly consisting of cordierite and potash feldspar (darker vermicules), and originated by retrogradation of osumilite. Cordierite twin lamellae cross the symplectite. Sample L147. Crossed polars. Lengt of photograph: 3 mm.

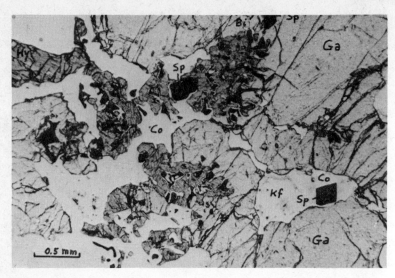

Fig. 11. Microphotograph of the reaction: garnet → hyp + cord + spin/magn. Sample M260, garnet-hypersthene-cordierite granofels, garnetiferous migmatites, Gyadalen. The relative amounts of the reaction products vary in the thin section. Hypersthene and spinel/magnetite are usually separated by cordierite.

It is tempting to ascribe the middle section of the isograd to the intrusion of the lower part of the Bjerkreim-Sokndal lopolith, which dips eastward according to gravimetric data (Smithson & Ramberg 1979). In the west the isograd seems to curve around the Egersund anorthosite. The number of samples is here restricted by the proximity of the Caledonian thrust and the scarcity of metapelites. The few occurrences of osumilite west of the Egersund anorthosite might have been caused by a small adjoining norite intrusion along the margin of the anorthosite (not shown on the map). In the east the isograd bends southward and approaches the lopolith.

The garnet decomposition isograd

Almost coinciding with the osumilite-in isograd, garnet in some rocks of the garnetiferous migmatites starts to decompose into hypersthene, cordierite and spinel (Fig. 11):

garn → hyp + cord + spin/magn. (11)

A rough idea of the reaction may be gleaned from the $SiO_2-(Fe,Mg)O-Al_2O_3$ diagram (Fig. 12). In reality the reaction probably involves more minerals, such as quartz, which is usually absent in the vicinity of the decomposed areas. According to a PT-diagram presented by Hensen and Green (1972) based on experimental results with artificial rocks of comparable

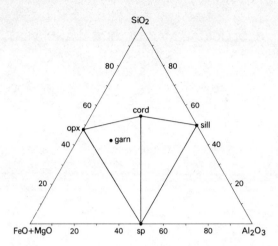

Fig. 12. $SiO_2-(Fe,Mg)O-Al_2O_3$ diagram with indication of minerals occurring in the garnetiferous migmatites.

composition the reaction marks the end of the garnet + cordierite field at the high-temperature side. The temperatures they obtained are probably too high, but the reaction here indicated should have taken place at a temperature of at least 800°C. It is difficult to define the temperature of the peak of metamorphism with geothermometry, since cordierite and spinel/magnetite usually do not show zoning, and may re-equilibrate effectively down to temperatures of about 600°C, the couple spinel/magnetite even below 500°C (Kars et al. 1980, Jansen et al. 1985). The spinel usually is an intense intergrowth of green hercynite and magnetite, but these minerals may also form separate crystals. Complete decomposition of the garnet is rare: the symplectites usually are restricted to roundish patches in the outer margins of the garnets, or at least garnet relics are left in the cores. The sample locations of decomposed garnets of this kind are given in Fig. 13.

Another type of garnet decomposition occurs in mafic bands (opx - plag - amph - biot - ore) in the garnetiferous migmatites. They contain rose garnets with conspicuous white rims consisting of plagioclase with slender radially oriented laths of orthopyroxene and an outer rim of larger orthopyroxene crystals (Fig. 14). It is the only rock type in the Rogaland collection where amphibole and garnet may be found in the same sample, though not in contact, save for a few armoured relicts of amphibole in garnet. We are perhaps dealing here with an older amphibolite facies with somewhat higher pressure than usual, perhaps coeval with the orthopyroxene - sapphirine assemblage found near Ivesdal (Hermans et al. 1976). The garnet decomposition may here be due to the reaction:

$$\text{garn + amph} \rightarrow \text{plag + opx + H}_2\text{O} \tag{12}$$

Some biotite seems also to be involved. This reaction may also occur outside of the osumilite-in isograd.

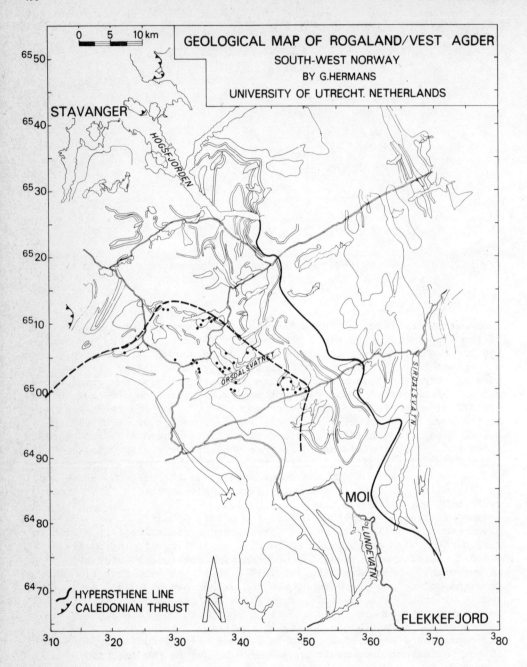

Fig. 13. Sample locations of the assemblage opx - cord - spin (garnet decomposition). The isograd is identical to that of the osumilite-forming reaction (fig. 6).

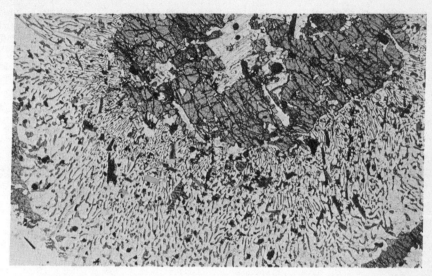

Fig. 14. Microphotograph of garnet replaced by a white rim of basic plagioclase with slender radially oriented orthopyroxene crystals, brown biotite flakes and isometric ore grains. Sample KA103, Gyadalen. One polar. Diameter garnet: 1.5 cm.

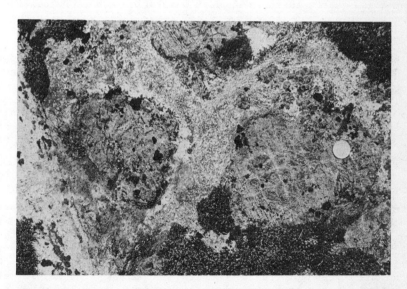

Fig. 15. Field appearance of garnet decomposition in the "Nordåsen garnet norites". Veined patches of red garnet are preserved in the cores. Diameter of coin: 2 cm.

A slightly different decomposition of garnet in mafic rocks is encountered in the "Nordåsen garnet norites" or garnet-hypersthene-plagioclase granofelses (Fig. 15). An initial stage of this decomposition consists of a rim of pure plagioclase, followed by a - sometimes discontinuous - rim of strongly pleochroic hypersthene. Possibly the plagioclase crystallized from a local melt generated by the high-grade metamorphism. Starting from this stage the garnet is then further decomposed to a symplectite consisting of plagioclase and a second-order symplectite of hypersthene and spinel/magnetite (Fig. 16). The reaction could be roughly presented as:

$$\text{garn } (+ \text{ L?}) \rightarrow \text{plag} + [\text{hyp} + \text{spin/magn}] \tag{13}$$

The symplectites replace the garnet as well as the initial plagioclase rim. Retrogressive metamorphism may be evident from the growth of rims of secondary garnet around the components of the symplectites, and around biotite (Fig. 17).

The pigeonite isograd

Extensive studies on the pyroxenes by Rietmeijer (1979, 1980, 1984) revealed, among other things, the occurrence of pigeonite of metamorphic origin in granitic components of the migmatites close to the intrusive complex. The sample locations are indicated in fig. 18. According to the thermometer of Lindsley the peak of metamorphism should have reached temperatures of at least 900°C within this isograd. The pigeonite was later inverted to orthopyroxene.

In a narrow zone close to the intrusive complex intricate K-feldspar-plagioclase-quartz textures suggest exsolution and recrystallization from an original non-stoichiometrical K-Na-Ca-feldspar with the Ca-component built in as "Schwantke - molecule" $Ca_{\frac{1}{2}}AlSi_3O_8$. The location of this isograd is under study.

The isograds discussed in this paper are shown in fig. 1. Apart from the Proterozoic stages of metamorphism, the region has also been affected by retrogressive metamorphism during the Caledonian orogeny (Verschure et al. 1980; Sauter et al. 1983). It is usually found as a quantitatively insignificant recrystallization in pumpellyite-prehnite facies. Near to the Caledonian front the recrystallization is more intense, and the grade increases so that a green-biotite isograd can be drawn (fig. 1).

Concluding remarks

Although several stages of metamorphism have affected the migmatites in Rogaland, it has proved possible to define certain isograds marking boundaries of metamorphic reactions. The first one, the hypersthene line, marking the incoming of hypersthene in leucogranitic rocks, is probably hybrid as far as its age is concerned: around 1050 Ma where it curves around the Egersund intrusive complex, and around 1200 Ma, in part perhaps older, in the northern part of the mapped area. The other isograds are more clearly restricted to the area surrounding the Egersund intrusive complex. They could be caused mainly by the intrusion of the Bjerkreim-Sokndal lopolith.

Fig. 16. Microphotograph of garnet decomposition in the Nordåsen garnet norites. Decomposition of the garnet and its enclosing rim of plagioclase by a symplectite consisting of plagioclase and an intergrowth of hypersthene and spinel/magnetite. The larger crystals of hypersthene and spinel constitute an older outer rim around the garnet. Sample 6.3, Nordåsen. Length of photograph: 13 mm.

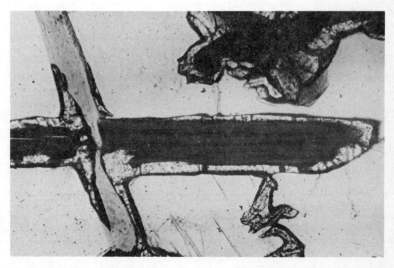

Fig. 17. Microphotograph of rims of garnet (2) around biotite in plagioclase, in the vicinity of the symplectite as depicted in Fig. 16. Sample WA151, Nordåsen. Length of photograph ¼mm.

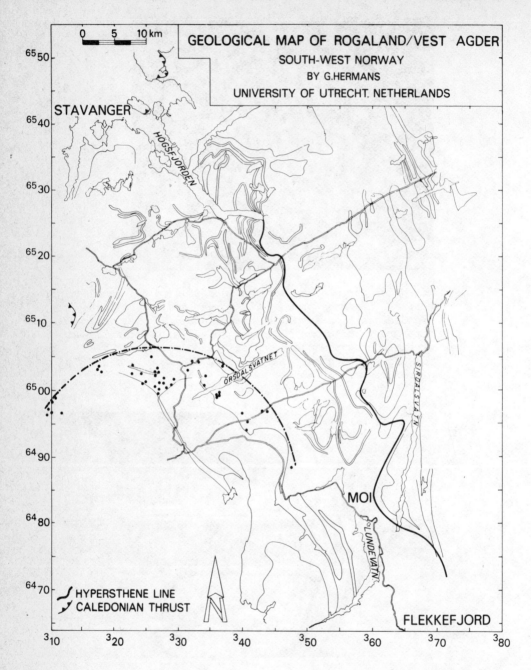

Fig. 18. *Pigeonite-in isograd, with sample locations of (mainly quartzo-feldspathic) rocks containing that mineral.*

Some relict assemblages, e.g. orthopyroxene + sapphirine in Mg-rich quartz-free rocks, and amphibole + garnet in mafic rocks, may represent an earlier metamorphic stage with somewhat higher pressure. Retrograde metamorphism is wide-spread, but seldom blurs the former assemblages completely. Notable examples are the decomposition of osumilite, and the growth of new generations of garnet and other minerals. The Caledonian retrogression is usually quantitatively insignificant, except for the immediate vicinity of the Caledonian front.

The information acquired with isograd study is essential, but not sufficient. It should be complemented with geothermometry and geobarometry (see Jansen et al., this volume). Careful petrologic study is here needed to select stable mineral pairs. Even then it is difficult to reach the highest temperatures because many minerals (pyroxenes, spinel, cordierite, e.g.) re-equilibrate easily during retrogression, often without much microscopical evidence. Unfortunately, the ages of the various stages of Proterozoic metamorphism usually cannot be obtained by isotope geochronology because minerals tend to give cooling ages indicating the final cooling and uplift of the whole terrain. It is only by constantly comparing and evaluating results obtained by all these methods that we may hope to disclose the complex evolution of this region.

Acknowledgement

Substantial support received from the Dr. Schürmann Foundation to carry out the research is gratefully acknowledged.

References

Binns, R.A. (1969) Hydrothermal investigations of the amphibole-granulite facies boundary. Geol. Soc. Austr. Spec. Publ., 2, 341-344.

Clemens, J.D. & Wall, V.J. (1981) Origin and crystallization of some peraluminous (S-type) granitic magmas. Canadian Miner., 19, 111-131.

Dekker, A.G.C. (1978) Amphiboles and their host rocks in the high-grade metamorphic Precambrian of Rogaland/Vest Agder, SW Norway. Thesis Univ. Utrecht, Geol. Ultraiectina, 17, 277 p.

Evers, Th.J.J.M. & Wevers, J.M.A.R. (1984) The composition and related optical axial angle of sillimanites sampled from the high-grade metamorphic Precambrian of Rogaland, SW Norway. N. Jb. Miner. Mh., Jg. 1984, 49-60.

Field, D., Drury, S.A. & Cooper, D.C. (1980) Rare-earth and LIL element fractionation in high-grade charnockitic gneisses, Southern Norway. Lithos, 13, 281-289.

Field, D., Smalley, P.C., Lamb, R.C. & Råheim, A. Geochemical evolution of the 1.6 - 1.5 Ga old amphibolite-granulite facies terrain, Bamble Sector, Norway: dispelling the myth of Grenvillian high-grade reworking. This volume.

Hensen, B.J. (1977) The stability of osumilite in high-grade metamorphic rocks. Contr. Miner. Petrol., 64, 197-204.

Hensen, B.J. & Green, D.H. (1972) Experimental study of the stability of cordierite and garnet in pelitic compositions at high pressure and temperature. II. Compositions without excess aluminosilicate. Contr. Miner. Petrol., 35, 331-354.

Hermans, G.A.E.M., Tobi, A.C., Poorter, R.P.E. & Maijer, C. (1975) The high-grade metamorphic Precambrian of the Sirdal-Örsdal area, Rogaland/Vest-Agder, S.W. Norway, Norges Geol. Unders., 318, 51-74.

Hermans, G.A.E.M., Hakstege, A.L., Jansen, J.B.H. & Poorter, R.P.E. (1976) Sapphirine occurrences near Vikeså in Rogaland, SW Norway. Norsk geol. Tidsskr., 56, 397-412.

Jansen, J.B.H., Scheelings, M. & Bos, A.J. Geothermometry and geobarometry in Rogaland and preliminary results from the Bamble area, S. Norway. This volume.

Kars, H., Jansen, J.B.H, Tobi, A.C. & Poorter, R.P.E. (1980) The metapelitic rocks of the polymetamorphic Precambrian of Rogaland, SW Norway. Part II. Mineral relations between cordierite, hercynite and magnetite within the osumilite-in isograd. Contr. Miner. Petrol., 74, 235-244.

Korstgaard, J.A. (1979) Compositional controls on mineral assemblages in quartzo-felspathic granulite-facies gneisses. N. Jahrb. Miner. Abh., 135, 113-131.

Lindsley, D.H. (1983) Pyroxene thermometry. Amer. Miner., 68, 447-493.

Maijer, C., Andriessen, P.A.M., Hebeda, E.H., Jansen, J.B.H. & Verschure, R.H. (1981) Osumilite, an approximately 970 Ma old high-temperature index mineral of the granulite facies metamorphism in Rogaland, SW Norway. Geol. Mijnbouw, 60, 267-272.

Maijer, C. (ed.) (1984) Excursion Guide of the South Norway Geological Excursion. NATO Advanced Study Institute, Moi, 16-30 July, 1984.

Newton, R.C. Temperature, pressure and metamorphic fluid regimes in the amphibolite facies to granulite facies transition zones. This volume.

Rietmeijer, F.J.M. (1979) Pyroxenes from iron-rich igneous rocks in Rogaland, S.W. Norway. Thesis, Univ. Utrecht. Geologica Ultraiectina, 21, 341 p.

Rietmeijer, F.J.M. (1980) In: Jansen, J.B.H. & Maijer, C.: Mineral relations in metapelites of SW Norway (abstract). Colloquium high-grade metamorphic Precambrian and its intrusive masses of SW Norway, Utrecht, 8-9 May, 1980.

Rietmeijer, F.J.M. (1984) Pyroxene (re)equilibration in the Precambrian terrain of SW Norway between 1030-990 Ma and reinterpretation of events during regional cooling (M 3 stage). Norsk geol. Tidsskr., 64, 7-20.

Sauter, P.C.C., Hermans, G.A.E.M., Jansen, J.B.H., Maijer, C., Spits, P. & Wegelin, A. (1983) Polyphase Caledonian metamorphism in the Precambrian basement of Rogaland/Vest Agder, SW Norway. Norges Geol. Unders., 380, 7-22.

Schrijver, K. (1973) Bi-metasomatic plagioclase-pyroxene reaction zones in granulite facies. N. Jb. Miner. Abh., 119, 1-19.

Smithson, S.B. & Ramberg, I.B. (1979) Gravity interpretation of the Egersund anorthosite complex, Norway: Its petrological and geothermal significance. Bull. Geol. Soc. America, 90, 199-204.

Swanenberg, H.E.C. (1980) Fluid inclusions in high-grade metamorphic rocks from SW Norway. Thesis Univ. Utrecht. Geol. Ultraiectina, 25, 147 p.

Touret, J. (1971) Le facies granulite en Norvège méridionale. I: Les associations minérales. II: Les inclusions fluides. Lithos, 4, 239-249 and 423-436.

Touret, J. (1981) Fluid inclusions in high-grade metamorphic rocks. In: Hollister, L.S. and Crawford, M.J. (eds) Short courses in Fluid Inclusions: Applications to Petrology. Miner. Ass. Canada, short course handbook, vol. 6.

Touret, J. (1985) Fluid regime in southern Norway. This volume.

Verschure, R.H., Andriessen, P.A.M., Boelrijk, N.A.I.M., Hebeda, E.H., Maijer, C., Priem, H.N.A. and Verdurmen, E.A.Th. (1980) On the thermal stability of Rb-Sr and K-Ar biotite systems. Evidence from coexisting Sveconorwegian (ca 870 Ma) and Caledonian (ca 400 Ma) biotites in SW Norway. Contrib. Miner. Petrol., 74, 245-252.

Waard, D. de (1965) A proposed subdivision of the granulite facies. Amer. Journ. Sci., 263, 455-461.

Wendlandt, R.F. (1981) Influence of CO_2 on melting of model granulite facies assemblage: a model for the genesis of anorthosites. Amer. Miner., 66, 1164-1174.

Wielens, J.B.W., Andriessen, P.A.M., Boelrijk, N.A.I.M., Hebeda, E.H., Priem, H.N.A., Verdurmen, E.A.Th. & Verschure, R.H. (1980) Isotope geochronology in the high-grade metamorphic Precambrian of Southwestern Norway: new data and reinterpretations. Norges Geol. Unders., 359, 1-30.

Winkler, H.G.F. (1976) Petrogenesis of metamorphic rocks. Springer Verlag, New York Heidelberg Berlin, 334 p.

GEOTHERMOMETRY AND GEOBAROMETRY IN ROGALAND AND PRELIMINARY RESULTS FROM THE BAMBLE AREA, S NORWAY.

J. Ben H.Jansen, Rob J.P. Blok, Ariejan Bos and Mies Scheelings.
Institute for Earth Science, State University of Utrecht, P.O.Box 80.021, 3508 TA Utrecht, The Netherlands.

Abstract

Mineral textures in Rogaland and Bamble areas give evidence for polymetamorphism, beside the usual low-grade retrogradations in or below the greenschist facies. Due to the fact that geothermobarometry is based on the assumption of chemical equilibrium among minerals, the application has to be treated with an enormous amount of caution in such areas. Especially the presence (or absence) of index minerals must be used as a control upon the derived results. For example, in both areas involved the main Al-silicate is sill[1]. The few occurrences of kya in SE Norway indicate high pressure retrogradation, while the few occurrences of anda in SW Norway presumably point at low pressure retrogradation.
In fact most rocks in Rogaland are reset during the retrograde M3 metamorphism. Part of the chemical rock system, particularly biot, cord, oxide mineral, is reequilibrated in a temperature range of about 550-700 °C at about 3-5 kb total pressure. The conditions are regionally varying. These results are achieved by the use of the garn-cord, garn-biot, garn-opx-plag-qtz, cpx-opx and ilm-ulvø thermo-barometers. In order to establish the conditions of earlier metamorphisms a search for relic equilibrium domains is inevitable, which are not necessarily one restricted part of a thin section with contacting minerals. For the M2 metamorphism the calibrated conditions yielded 700-1050 °C grading to the lower temperatures towards ENE, at pressures of about 4 kb. These estimates are derived by the use of thermobarometry on cpx-opx, ilm-ulvø, fay-qtz-fesil and some garn-opx assemblages. Equilibrium domains of mineral assemblages useful for geothermo-geobarometry upon the M1 metamorphism are not

---

[1] For mineral abbreviations see table I.

unambiguous enough to establish PT-conditions. Based on experimentally derived mineral stability fields some estimates in the temperature range from 5-7 kb can be made. The regional gradients are not well understood yet.
The main metamorphism in Bamble near Tvedestrand took place at temperatures of about 750 $^{\circ}$C and about 7 kb. These results were achieved with garn-cord, garn-biot and garn-opx-plag-qtz thermometry. Only some local chemical resetting is established. The retrogradation was more or less isobaric down to about 600 $^{\circ}$C passing into the kyanite stability field.

Introduction

The title geothermometry and geobarometry in Rogaland etc. was chosen, but, honestly, a better one should have expressed the problems we have had especially in Rogaland to obtain meaningful sets of P and T conditions for each metamorphism. Just putting chemical compositions of contacting or nearby located minerals in the empirically and theoretically derived formulas of the thermometers it is possible to end up with a temperature range from 300 to 1500 $^{\circ}$C with a statistical peak near 750 $^{\circ}$C. Doing the same with the usual barometers it is possible to obtain pressures of 30 kb down to a kind of vacuum of even minus 10 kb. Therefore we like to put your attention first on "isograd mapping", as a very powerful tool in getting a realistic general idea of the metamorphic grade. In Rogaland (Tobi et al., this volume), in Vest-Agder (Petersen, 1980) and in the Bamble area (Touret, 1972, 1973) the transition of the amphibolite facies into the granulite facies is assumed to be marked by the + opx isograd (Fig.1). In Bamble the increasing grade towards the SE is expressed by a 1500 Ma old parallel zonation pattern of the + cord and the + opx isograds (Fig.1b). The - musc isograd is probably caused by a later Sveconorwegian overprint. Parallelism with regard to the older metamorphism is observed here. In Rogaland the metamorphic gradient, as an addition of all effects of several Sveconorwegian metamorphisms is directed towards the SW (Tobi et al., this volume). The zonation is mapped with three isograds (Fig.1a). The lowest one is the + opx isograd at about 750 $^{\circ}$C. Any - musc isograd should be exposed further E if existing. The + opx isograd (for granitic rocks) coincides more or less with the disappearance of amph + qtz. The income of cord is located on the high temperature side of the + opx isograd. It is the reversed situation as in the Bamble area. The second drawn isograd is the + osum at about 800 $^{\circ}$C (Maijer et al.,1981). It coincides with a kind of breakdown-line of alm. Adjacent to the anorthositic masses the third isograd, the + pig was proposed by Rietmeijer (pers.comm.) with an estimated temperature about 900 $^{\circ}$C. "Isograd mapping" should be executed before doing any geothermometry, because the zonation pattern itself may provide information on the focus of heat, on the difference in type of metamorphisms involved and on the metamorphic geogradients.

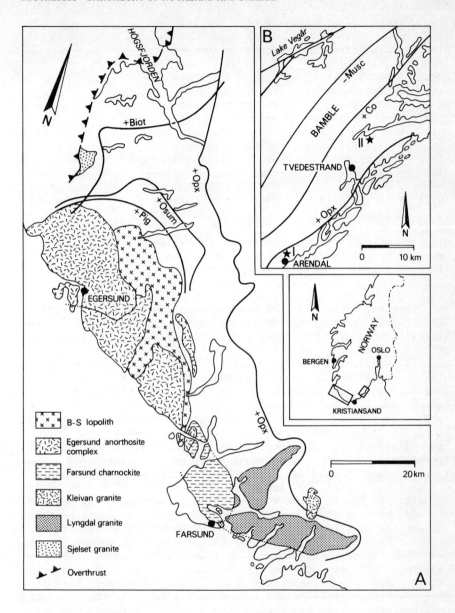

Fig.1. Isograd patterns of Rogaland, Vest-Agder, SW Norway after Petersen (1980) and Tobi et al. (this volume), and of Bamble area, SE Norway after Touret (1972, 1973) and Field et al. (this volume). Sample sites near Arendal I and Tvedestrand II.

Tab.I. List of abbreviations of mineral names.

alm	= almandine	Kfeld	= K-feldspar
amph	= amphibole	korn	= kornerupine
anda	= andalusite	kya	= kyanite
biot	= biotite	liq	= liquid
cc	= calcite	mag	= magnetite
cord	= cordierite	opx	= orthopyroxene
coru	= corundum	osum	= osumilite
cpx	= clinopyroxene	phlog	= phlogopite
ctd	= chloritoid	pig	= pigeonite
diop	= diopside	plag	= plagioclase
fay	= fayalite	qtz	= quartz
fesil	= ferrosillite	sapp	= sapphirine
fo	= forsterite	sill	= sillimanite
garn	= garnet	spin	= spinel
hbl	= hornblende	staur	= staurolite
herc	= hercynite	ulvø	= ulvøspinel
ilm	= ilmenite	vap	= vapour

The isograds directly N of the Egersund anorthosite exhibit a very tight pattern and NE of the leuconoritic phase of the Bjerkreim-Sokndal lopolith a relatively wide pattern. The difference will tell us a lot on the intrusion mechanisms of the magmatic bodies (Duchesne and Maquil, this volume). The extremely wide pattern shown in the northern part of the mapped area (Fig.1a) may point to another, probably older metamorphism. The isograd pattern near the southern part of the lopolith is not yet fully understood, due to lack of metapelite occurrences. Near Flekkefjord and Lyngdal the + opx isograd follows the outlines of the relatively young, postkinematic charnockites and granites (Petersen, 1980). In conclusion it can be stated that most isograds are composites resulted by addition of metamorphic events out of totally different orogenic periods.

Actually in Rogaland the + opx isograd is subhorizontal and NE dipping, whereas the + osum and + pig isograds are supposed to be subvertical or even W-dipping. More accurate isograd mapping must be executed. The Caledonian + biot isograd was caused by an overlying nappe pile of relatively hot thrustsheets upon the basement and its sedimentary Paleozoic cover. The resulting metamorphism is responsible for the reversed vertical geogradients (Andreasson and Lagerblad, 1980). In the northern and western part of the Precambrian basement in the mapped area (Fig.1a) between the overthrust plane and the + biot isograd many rock samples show so much retrogradation that Sveconorwegian minerals are no longer contacting assemblages. Moreover the extreme northern part of the + opx isograd has largely been erased by the Caledonian retrogression.

Index minerals

The simple concept of comparison of the metamorphic grade via index minerals and mineral assemblages with experimental stability fields and with the grade of well-studied terrains in order to estimate P-T conditions, remains a very powerful tool. This comparison must be executed to avoid misinterpretation of the arithmically derived numbers held for P and T conditions (Essene, 1982). The presence (or absence) of index minerals can be used as a control upon the application of geothermometric methods. For example in Rogaland and in the Bamble area the main Al-silicate is sill (Fig.2). Only a few occurrences of kya replacing cord in SE Norway indicate a high pressure retrogradation (Touret , this volume) or it is produced during a separate late metamorphism, probably of Sveconorwegian age. Kya of unknown age is found in the Telemarken supracrustals together with ctd and in the Precambrian basement near Kristiansand (Tobi, pers. comm.). In SW Norway a few occurrences of anda are reported by Huijsmans et al. (1982) and Jansen and Maijer (1980). These findings presumably point to low pressure retrogradation or to a prograde path across the sill-anda reaction in a separate metamorphism. Relict sapp and opx assemblages are described from Bamble area by Touret and de la Roche (1971) and from Rogaland by Hermans et al.(1976). According to Seifert (1974) this assemblage is stable above about 4 kb (Fig.2). Sapp itself is also reported by Scheelings et al. (in prep) near Lyngdal in Vest-Agder. Sill + opx assemblage is scarcely found in Bamble area (Touret, pers.comm.) and does not occur in Rogaland. Sofar we know no staur occurs in the Fe-rich pelitic rocks in both areas. Korn is locally present in the Bamble area (van der Wel, 1973). Seifert (1975) proved that this mineral is a high pressure temperature mineral (Fig.2). The osum in Rogaland is held as indicative for low P and very dry rock systems at temperatures of about 800 $^{\circ}$C (Olesch and Seifert, 1981). So in both terrains the temperature is relatively high. According to Touret and Dietvorst (1983) for Bamble and to Swanenberg (1980) for Rogaland the waterpressure was much smaller than the total pressure. The general impression is that the total pressure in Bamble was probably higher than in the Rogaland area.

Mineral textures

In the last ten years students and staff of the departments of petrology and geochemistry of the State University of Utrecht have analyzed thousands of minerals from Rogaland with microprobes. One of the purposes was of course to model the P T conditions with time. The model is not yet ready, due to the lack of chemical equilibrium as a consequence of the complexity of the polymetamorphic history of each individual rock sample.

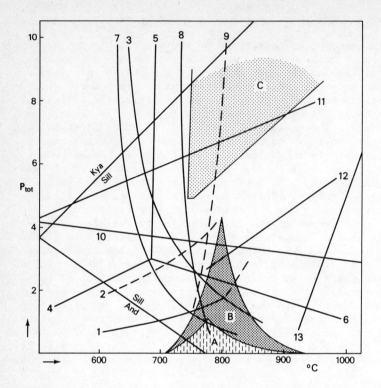

Fig.2. P T diagram with relevant experimental and calculated equilibria involving index minerals and assemblages.
A = stability field of osum (Olesch and Seifert, 1981)
B = idem for X_{H2O} = 0.5 calculated.
C = stability field of korn (Seifert, 1975)
1 = cpx + opx + plag + vap = hbl + qtz (Wells, 1979)
2 = idem for X_{H2O} = 0.5 (Newton, this volume)
3 = basalt solidus $P_{H2O} = P_{tot}$ (ibid)
4 = staur + qtz = cord +sill (Richardson, 1968)
5 = staur + qtz = alm + sill + qtz (ibid)
6 = Fe-cord = herc + qtz + sill (ibid)
7 = H_2O- saturated tonalite solidus (Wyllie, 1977)
8 = phlog + qtz + Kfeld + vap = opx + liq (Wendlandt, 1981)
9 = biot-tonalite solidus (Wyllie, 1977)
10= opx + sapp = cord + spin (Seifert, 1974)
11= Mg-cord = opx + sill + qtz (Newton et al., 1974)
12= garn + cord + sill = spin + qtz (Grew, 1982[a] ;suggested)
13= garn + cord + sill = spin +qtz (Hensen and Green, 1973)
    Al-silicates diagram is from Holdaway (1971)

A selection of mineral textures will clearly show several metamorphisms, beside the low-grade retrogradation in or lower than the greenschist facies of Caledonian age (Sauter et al., 1983). Figure 3 shows the relict M1-assemblage of sapp and opx. Not only M1-garn breaks down to M2 opx-spin-cord symplectites (Tobi et al., this volume), also sapp has been replaced by cord-spin symplectites, around which M1 Al_2O_3-rich opx is often preserved (Hermans et al., 1976). The backward reaction of cord + spin to sapp can be rarely observed as rims of M3-sapp along the interface of cord with spin (Fig.4).

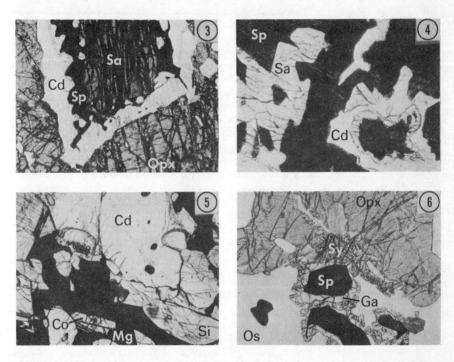

Fig.3. Photomicrograph of relictic M1-sapp (Sa), surrounded by rims of spin (Sp) and cord (Cd) in M1-opx (Opx). (A146, 5 x)

Fig.4. Photomicrograph of newly formed, secondary M3-sapp (Sa) between spin (Sp) and cord (Cd). (J106, 10 x)

Fig.5. Photomicrograph showing two rims of M3-garn (Ga) around spin (Sp) and M2-opx (Opx) in a matrix of Osum (Os). The second rim is symplectitic intergrown with M3-opx (Sy). (MA490, 5 x)

Fig.6. Photomicrograph of M2-sill (Si) with mag (Mg) and coru (Co). The cord (Cd) is newly formed and does not contact coru. Note the grey area in sill where Schiller exsolution is observed. (JM143, 5 x)

In the siliceous dolomites of the Faurefjell formation, complex zoning in diop is studied by Sauter (1983). Relict M1-diop is Al_2O_3-rich and it occurs in a M2-matrix of fo and cc. Late M3-diop contains less Al_2O_3 and it often occurs as rims around fo.
Tobi et al. (this volume) reported the breakdown of M1-garn and the possible recrystallization of garn in relatively medium-grade pelitic rocks during M2. The growth of new generations of M3-garn seems to have locally occurred in two steps (Fig.5).
At conditions below the + osum isograd sill and garn are observed to be synkinematic. Whereas garn often breaks down within the + osum isograd, sill remains stable in pelitic rocks and it shows recrystallization. Chemical resetting to a very high temperature regime parallel to the M2 isograd pattern is observed by Evers and Wevers (1984). The reset sill contains up to 2.3 wt% Fe_2O_3 and the amount seems to increase with temperature. Discrete rims of sill with high Fe_2O_3 content occur around parts with low Fe_2O_3 (Fig.6). Reversed rimming is also analyzed. Presumably three generations of sill are established.
The osum has definitely developed during M2. Relic minerals within osum are for example garn, sill, spin, biot and opx, however some of these minerals may have partly recrystallized in M2. A breakdown reaction of osum into M3 cord-qtz-Kfeld-opx symplectites is frequently observed. So it is extremely difficult to separate relic, high-grade and retrograde assemblages in these rocks. Based on a few osum-bearing handspecimens from Rogaland Grew (1982b) estimated the P T conditions relative to other terrains and he suggested comparable conditions as for the Nain complex, Labrador. Perkins and Newton (1981) calculated for the Nain Complex a pressure of about 3.5 kb in the temperature range of 750-800 °C.
Finally we end up with:
 -1 generation of osum
 -2 generations of fo and sapp
 -3 generations of sill, cpx and opx
 -more than 3 generations of amph, garn, cord, biot and spin.
So we assume at least 3 high-grade metamorphisms. M1 containing every metamorphism before about 1200 Ma, a M2 of about 1050-1100 Ma and a M3 of about 980-950 Ma (Verschure, this volume). Each metamorphism has been high enough in grade to form its own generation of granitic or charnockitic magma with its own type of fractionation sequence ending in aplitic and pegmatitic dikes.

Geothermometry and geobarometry

Due to the fact that geothermometry is based on the assumption of chemical equilibrium between minerals used, the application must be very, very carefully treated. Tobi et al.(this volume) and this paper only show you some textural disequilibria, chemical resetting is hard to show in a photomicrograph. It may occur upon prograde heating or retrograde cooling without any textural change. Even the so called mosaic textures are in Rogaland no indication at all for chemical equilibrium. It remains extremely difficult to prove equilibrium between the minerals even with a set of more than a thousand probe analyses. For geothermometry and geobarometry pelitic rocks are preferred because of the presence of sluggishly reacting Al-rich minerals (Fig.7). But in fact most of the biot, cord, spin and oxide minerals are chemically reequilibrated during the latest metamorphism M3 (Tab.II). The typical M3 rocks with a lot of secondary garn, opx, sill, cord, biot and qtz yield with geothermometry reequilibration temperatures of about 500-700 °C with a

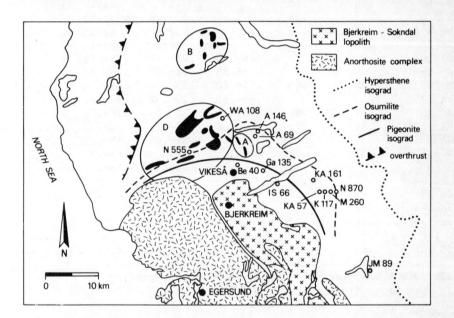

Fig.7. Location map of Rogaland, SW Norway. Outcrops of the Faurefjell formation in areas A, B and D are indicated in black.

decreasing grade towards NE and N at total pressures of about 3-4 kb, locally up to a maximum of 5kb (Fig.11).
In figure 8 the open and closed squares are spin-cord pairs from Rogaland (Henry, 1974 ;Kars et al., 1980). The spin analyses are corrected for mag. All samples are taken from localities near the + osum isograd, so they should show the same K_D equilibrium line if equilibrium were retained. However the analyses of these pairs plot along the stippled trend crossing the theoretical K_D lines. Probably most pairs are reset chemically during M3. Especially

the iron-rich pairs are in a systematic way depending on the X_{Mg} of the mineral. These pairs seem to have a lower closure temperature or they open more easily during a later event(Fig.8).

Tab.II : X_{Mg} for spin, cord, opx, garn, biot and sapp of granofelses from the Faurefjell formation, Rogaland, SW Norway. For locations see figure 7.

mineral sample	spin	cord	opx	garn	biot	sapp
He 14[1)	.19	.69		.28	n.a.	
He 10[1)	.20	.68		.23	n.a.	
N 870	.23	.68		.20-.23	.59-.65	
He 9[1)	.24	.69		.25	n.a.	
He 2[1)	.25	.72		.29	n.a.	
KA 161			.46-.48	.22-.26	.59-.71	
JS 66			.55-.58	.31-.35	.76-.83	
M 260	.30	.80	.57-.60	.29-.34	.69-.73	
GA 135			.59-.60	.29-.34	.79-.82	
Be 40	.31	.81	n.a.	.32-.35	.75-.80	
KA 117	.34	.79	.58-.61	.27-.36	.77	
WA 108	.36	.78-.82	.56-.61	.34-.40	.68-.73	
N 555	.37	.81	.62	.38	.66-.71	
A 69		.82		.36	.68-.83	
JM 136[2)	.40	n.a.	.61			
KA 57	.47	.86	.67-.69		.75-.78	
JM 143[2)	.47	.83			.74	
JM 135[2)	.54	.89	.70		.71	.78
A 146	.55	.89	.78		.80	.80
VH 107[2)	.66	.90	.76-.77		.82-.84	.81

n.a. = not analysed X_{Mg} = Mg/(Fe^{2+}+Mg+Mn)
1) see Henry (1974)
2) location in area A of figure 7

The typical M3 minerals biot, amph and cord are OH-bearing. The disperse distribution of the minerals, practically in all rock types requires an extensive influx of fluid, which probably carried some silica, fluor, potassium and sodium. In the south near Lyngdal the fluid influx may have been related to the late, postkinematic charnockitic and granitic intrusions of Farsund, Lyngdal and Kleivan. Presumably the Sjelset charnockite near Undheim (Fig.1) also belongs to this granitic sequence. Small discordant granitic and charnockitic stocks and pegmatites are considered as postkinematic in Rogaland (Schreurs , pers.comm.). Underneath the lopolith and the anorthosites granulitic crust may be present as it is suggested below the Aana-Sira body by geophysical methods (Smithson and Ramberg, 1979). This granulitic crust may consist of some of these postkinematic charnockitic and granitic bodies (cf. Breimyrknutan charnockite, Maijer, 1984). They may have introduced magmatic fluids which may have even penetrated the anorthositic masses to produce secondary biot and amph (e.g. in the Hidra massif, Tobi, pers.comm.). Radiometric ages, especially those determined by Rb-Sr and K-Ar methods may have been reset by this hardly visible event.

In order to collect P-T estimates for the highest grade metamorphism M2 a search for M3-closed, relict domains of chemical M2 equilibria is necessary. An equilibrium domain is not necessarily one restricted part of the thin section. The information of a relict domain can be retained in separated parts of the various minerals. Generally the contacting rims in rocks from Rogaland are chemically reset to M3 domains. Typically opx and cpx thermometry in combination with ilm-ulvø thermometry is used for this high temperature event. Rietmeijer (1979, 1984) used several pyroxene thermometers for the upper part of the lopolith and his results may be summarized here as a

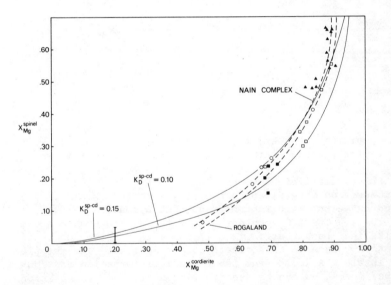

Fig.8. K_D-plot of cord- spin pairs of Rogaland, SW Norway. **solid squares** : data of Henry (1974), **open squares** : data of Kars et al. (1980), **triangles** : new data of Meertens(pers.comm.), **circles** : comparative data on Nain Complex, Labrador of Berg (1977). The hatched lines are best fit through the data points and the full lines show the $K_D^{spin-cord}$ for Mg and Fe of 0.1 and 0.15.

crystallization temperature between 850 and 1050 °C. The lower part of the lopolith crystallized probably at about 1200 °C. Both parts of the lopolith showed pyroxene reequilibration and amphibole cooling temperatures down to 550 °C and that is remarkably close to M3 data. The pressure estimates were about 5-7 kb. A revision after Bohlen et al. (1980) yields a pressure of

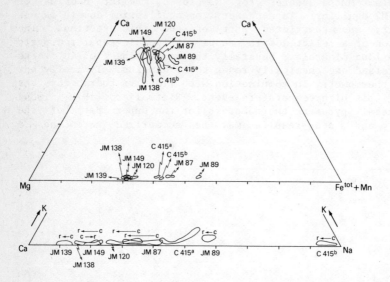

Fig.9. Composition of coexisting cpx and opx (upper part) with plag (lower part). Arrows mark the zoning from core to rim.

about 4-5 kb at relatively low equilibration temperatures.
The thermometry on metavolcanics recently executed (Blok ,1984) is in good agreement with that of Jaques de Dixmude (1978) and Wilmart (1984). The SiO_2 content versus the Zr/TiO_2-ratio shows the subalkaline basalt affinity, if no metasomatism of used elements took place in this impermeable rock type. The sample locations in areas A and B from the Faurefjell formation and JM89 outside that formation can be seen in figure 7. The composition of the associated cpx, opx and plag is plotted (Fig.9). Most plag crystals exhibit an inverse zonation, which may indicate a prograde metamorphic reaction opx + plag to cpx + Ca-rich plag which is regulated by the silica activity (or it could be interpreted as retrograde). To make sure that the opx is metamorphic the composition is plotted on the X_{Fe}^{opx} versus 100x X_{Ca}^{opx} -diagram of Rietmeijer (1983). Using the thermometer of Wood and Banno (1973) on all the possible combinations of opx and cpx in one sample, a temperature estimate for area B gives 816 °C,

for location JM89 near Moi 832 °C, and for area A 852 to 912 °C. The method after Wells (1977) is more or less consistent with these data. In order to improve the temperatures the manganese distribution between cpx and opx is used as a kind of pairing criterium. The distribution line is just a statistical average of all pairs and only the closely plotted pairs were taken into the thermometry. However, no improvement was achieved.

The Ca-distribution thermometer after Kretz (1982) gives cooling temperatures of the pyroxene pairs in the range of 645 down to 375 °C. These temperatures are too low in comparison with the graphical methods after Lindsley (1983) yielding temperatures between 500-700 °C. This means that they are again more indicative for M3 than for M2 temperatures based on the + osum and + pig isograds.

In order to compare our results with those of Jacques de Dixmude (1978) we applied also the Fe-Mg distribution thermometer after Kretz (1963). Jacques de Dixmude (1978) got a temperature of about 1050 °C and a consistent ilm-ulvø temperature of 1000 °C after Buddington and Lindsley (1964) in an area between the + osum and the + pig isograd, where we do not expect temperatures higher than 950 °C. Our results with the method of Kretz (1963) for only area A range from 750 to 1500 °C. Two peaks appear, near 900 °C and near 1050 °C. The high temperatures above 1100 °C are of course meaningless, the peak of 900 °C is debatable, and the significance of the 1050 °C peak is unknown and probably also not

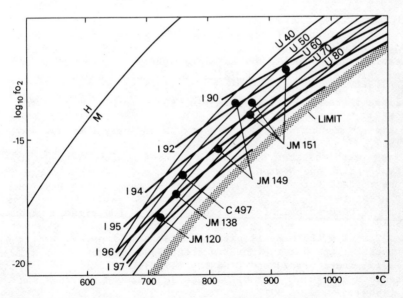

Fig.10. Ilm-ulvø pairs plotted in the temperature-oxygen fugacity field of Spencer and Lindsley (1981).

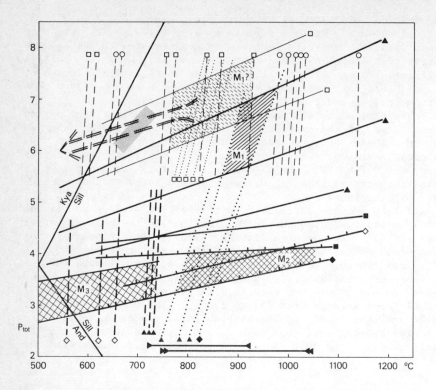

Fig.11. Compilation of PT estimates on mineral assemblages from Rogaland, SW Norway (heavy lines) and of the preliminary results on assemblages from Bamble area, SE Norway (light lines).
triangle WA108, **solid square** JS166, **open diamond** KA161, **solid diamond** GA135, **open square** TV II, **open circle** Ar I.
Broken lines : results of the garn-biot thermometry (Hodges and Spear, 1982).
Dotted lines : results of the garn-opx thermometry (Harley and Green, 1982).
Full lines with hatches : results of the garn-opx-plag-qtz barometry (Perkins and Newton, 1981).
Full lines : idem, but not qtz-bearing, which caused relatively a too high pressure.
Grey box : cord-garn indication for $X_{H2O}=0.5$ (Martignole and Sisi, 1981).
Bracket : ilm-ulvø thermometry after Spencer and Lindsley (1981).
double bracket : cpx-opx thermometry after Wood and Banno (1973), Wells (1977), Kretz (1982) and Lindsley (1983).
Both brackets include data of Jacques de Dixmude (1978) and Wilmart (1984).

realistic. We applied ilm-ulvø thermometry after Spencer and Lindsley (1981) yielding 755 °C for area B and 720-929 °C for area A (Fig.10). These temperatures are particularly achieved in closed, impermeable rocks without retrogradation. The temperatures are comparable with those derived from Wood and Banno (1973) thermometry, although the latter are considered 100 to 150 °C too high by Rietmeijer (1984) and Newton (this volume). The metavolcanic rocks with retrograde M3 development of biot and amph gave all temperatures down to 500 °C with the ilm-ulvø thermometer.

Results of garn-opx-plag-sill barometry within the + osum isograd gave about 4 kb total pressure at 950 °C. The Al_2O_3-content of the opx approaches about 10 wt% in rock locally buffered by cord-spin and garn. Some two feldspar thermometry results after Whitney and Stormer (1977) yielded 750 to 800 °C and one about 1000 °C for osum-bearing rocks. Concluding, it can be stated that M2 reached very high temperatures from about 750 °C near the + opx isograd to about 1050 °C in the vicinity of the lopolith (Fig.11). M1 estimates are very scarce. Some garn-sill-plag-qtz and garn-opx-plag-qtz assemblages sampled east of Stavanger give pressures of about 6-8 kb and garn-biot temperatures of about 550-600 °C. The cord-garn temperatures are in the range of 650-750 °C. The opx-cpx are again in the range of 700-850 °C. The opx in pelitic rocks in this region contains 5-7 wt% Al_2O_3. The Gloppurdi Massif dated to be 1200 Ma old required intrusion pressures of about 7-10 kb according to Rietmeijer (1979). The fesil-fay-qtz barometer after Bohlen et al.(1983) suggests for about 800 °C pressure estimates between 5.5 and 8 kb. Opx + sill and sapp + qtz are never found in Rogaland. Opx + sapp and cord-garn-sill-qtz assemblages are M1 relics in the M2 assemblages yielding pressures in the range of 8-6 kb and temperatures of about 600-700 °C (Fig.11).

The Bamble area near Tvedestrand gives more consistent P-T values than Rogaland, although the number of data is not very large. The preservation of older equilibrium P-T conditions seems to be better than in Rogaland. The calculated temperatures vary from 700 to 900 °C at pressures of about 6-8 kb and are based on garn-biot, garn-cord and garn-opx-plag-qtz geothermo-barometry (Fig.11). The opx in pelitic rocks contains about 4 to 7 wt% Al_2O_3. These results are similar to M1 conditions locally preserved in Rogaland rocks. We assume that this comparison is also valid with respect to the age and to the type of folding.

In spite of the large distance between rhe areas, we suggest that the Bamble area and Rogaland have a common M1. The Bamble remained rather unaffected by later metamorphism and Rogaland was strongly metamorphosed by Sveconorwegian events, causing widespread erasing of M1 assemblages. More data will certainly be needed to confirm the common Pre-Sveconorwegian metamorphism of both areas.

Acknowledgements

We are indebted to profs. Lex Tobi and Jacques Touret for their stimulating advice, their information on mineral occurrences, mineral analyses and their critical review. Gé Hermans and Cees Maijer are thanked for providing us a sample, photomicrographs and guidance with "field" problems. Bob Newton is thanked for reviewing the original draft for the lecture. Chemical analyses were carried out with assistance of students of the Department of Petrology on the microprobe equipment of the Institute of Earth Sciences, which is financially supported by ZWO-WACOM. Financial support for fieldwork, research, and congres attendance was gratefully recieved from the Dr.Schürmann foundation and NATO. Jaco Heenegouwen succeeded in transforming crude drawings into clear figures. Frans Henzen and Hans Schiet gave their help in making photographs and figures.

References

Andreasson, P.G. and Lagerblad (1980) J.Geol.Soc.London, 137, 219-230.
Berg, J.H. (1977) Contrib.Mineral.Petrol., 64, 33-52.
Blok, R.J.P. (1984) M. Sc. thesis Univ. Utrecht, 92p.
Bohlen, S.R., Wall, V.J. and Boettcher, A.L.(1983) In: Advances in physical geochemistry, 3, 141-172.
Buddington, A.F. and Lindsley, D.H.(1964) J.Petrol., 5, 310-335.
Duchesne,J-C. and Maquil(1985) This volume.
Essene, E.J.(1982) Reviews in mineralogy, 10, 153-206. Ed.:J.M.Ferry, Mineral. Soc. of Amer.
Evers, Th.J.J.M. and Wevers, J.M.A.(1984) N. Jb. Miner. Mh., 2, 49-60.
Field, D., Smalley, P.C., Lamb, R.C. and Raaheim, A. (1985) This volume.
Grew, E.S.(1982a) J. Geol. Soc. of India, 23, 469-505.
Grew, E.S.(1982b) Am. Miner., 67, 762-787.
Harley, S.L. and Green, D.H.(1982) Nature, 300, 697-701.
Henry, J.(1974) Lithos, 7, 207-216.
Hensen, B.J. and Green, D.H. (1973) Contr.Mineral.Petrol., 38, 151-166.
Hermans, G.A.E.M., Tobi, A.C., Poorter, R.P.E. and Maijer, C. (1975) Norges Geol. Unders. 318, 51-74.
Hermans, G.A.E.M., Hakstege, A.L., Jansen, J.B.H., Poorter, R.P.E.(1976) Norsk Geol. Tidsskr., 56, 397-412.
Hodges, K.V. and Spear, F.S.(1982) Am. Min., 67, 1118-1134.
Holdaway, M.J.(1971) Amer. J. Sci., 271, 97-131.
Holdaway, M.J. and Sang Man Lee (1977) Contr.Mineral.Petrol. 63, 175-198.
Huijsmans, J.P.P., Barton, M. and Bergen, M.J. van (1982) N.Jb.Miner.Abh., 143, 249-261.
Jacques de Dixmude, S. (1978) Bull. Minéral., 101, 57-65.

Jansen, J.B.H. and Maijer, C.(1980) Colloq. Precambrian, SW Norway, Utrecht (abstract).
Kars, H., Jansen, J.B.H., Tobi, A.C. and Poorter, R.P.E. (1980) Contr. Mineral. Petrol., 74, 235-244.
Kretz, R. (1963) J. Geol., 71, 773-785.
Kretz, R. (1982) Geochim. Cosmochim. Acta, 46, 411-421.
Lindsley, D.H. (1983) Am. Min., 68, 477-493.
Maijer, C.(1984) In : Exc. Guide of the S Norway Geol. Exc. of the NATO Adv. Study Inst., Moi.
Maijer, C., Andriesen, P.A.M., Hebeda, E.H., Jansen, J.B.H. and Verschure, R.H. (1981) Geol. Mijnb., 60, 267-272.
Martignole, J. and Sisi, J.Ch. (1981) Contr. Mineral. Petrol., 77, 38-46.
Newton, R.C. (1985) This volume.
Newton, R.C., Charlu, T.V. and Kleppa, O.J. (1974) Contr. Mineral. Petrol., 44, 295-311.
Olesch, M. and Seifert, F. (1981) Contr. Mineral. Petrol.,76, 362-367.
Perkins III, D. and Newton, R.C. (1981) Nature, 292, 144-146.
Petersen, J.S. (1980) Contr. Mineral. Petrol., 73, 161-172.
Richardson, S.W. (1968) J. Petrol., 9, 467-488.
Rietmeijer, F.J.M. (1979) Ph.D. thesis, Univ. Utrecht, Geol. Ultraiectina, 21, 341p.
Rietmeijer, F.J.M. (1983) Min. Mag., 47, 143-151.
Rietmeijer, F.J.M. (1984) Norsk Geol. Tidsskr., 64, 7-20.
Sauter, P.C.C. (1983) Ph.D. thesis, Univ. Utrecht, Geol. Ultraiectina, 32, 143p.
Sauter, P.C.C., Hermans, G.A.E.M., Jansen, J.B.H., Maijer, C., Spits, P. and Wegelin, A. (1983) Norges Geol. Unders., 380, 7-22.
Scheelings, M., Maijer, C. and Bergen, M. J. van (1985) In prep.
Seifert, F. (1974) J. Geol., 82, 173-204.
Seifert, F. (1975) Amer. J. Sci., 275, 57-87.
Smithson, S.B. and Ramberg, I.B. (1979) Geol. Soc. Am. Bull., 90, 199-204.
Spencer, K.J. and Lindsley, D.H. (1981) Am. Min., 66, 1189-1201.
Swanenberg, H.E.C. (1980) Ph.D. thesis Univ. Utrecht., Geol. Ultraiectina, 25, 147p.
Tobi, A.C., Hermans, G.A.E.M., Maijer, C. and Jansen, J.B.H. (1985) This volume.
Touret, J. (1972) Sciences de la Terre, XVII, 179-193.
Touret, J. (1973) Extrait Coll. Sci. Internat. E.Raquin, 249-260.
Touret, J. (1985) This volume.
Touret, J. and Dietvorst, P. (1983) J. Geol. Soc., 140, 635-649.
Touret, J. and Roche, H. de la (1971) Norsk Geol. Tidsskr., 51, 169-175.
Wel, D. van der (1973) Norsk Geol. Tidsskr., 53, 349-357.
Wells, P.R.A. (1977) Contr. Mineral. Petrol., 62, 129-139.
Wells, P.R.A. (1979) J. Petrol., 20, 187-226.
Wendtlandt, R.F. (1981) Am. Min., 66, 1164-1174.

Whitney, J.A. and Stormer, J.C.Jr. (1977) Contr. Mineral. Petrol., 63, 51-64.
Wilmart, E. (1984) NATO Adv. Study Instit., Moi.(abstract)
Wood, B.J. and Banno, S. (1973) Contr. Mineral. Petrol., 42, 109-124.
Wyllie, P.J. (1977) Tectonophysics, 43, 41-71.

FLUID REGIME IN SOUTHERN NORWAY: THE RECORD OF FLUID INCLUSIONS

Jacques L. R. Touret
Institute of Earth Sciences, Free University
De Boelelaan 1085
1081 HV Amsterdam
The Netherlands

ABSTRACT.

Fluid inclusions have been studied in representative rocks from Southern Norway, notably in the Bamble granulites. On the basis of the earliest fluid inclusions trapped in rock-forming minerals (mainly quartz), five major types of fluid distribution have been recognized: 2-phase aqueous (H_2O dominant, without solid), carbonic (mostly pure CO_2, possible occurrence of N_2 and/or CH_4), mixed 1 (aqueous and carbonic inclusions in comparable amounts, but in separate cavities), mixed 2 (aqueous and carbonic fluids in the same cavity, trapped in the miscible state of the H_2O-CO_2 system), brines (H_2O + solids, NaCl dominant).
Only brines show a relation between a dominant inclusion type and a given protolith; these are especially abundant in 3 well-defined environments: Al-rich metasediments (metapelites), skarns and acid volcanics. The distribution of other types is more related to metamorphic grade: high-density carbonic inclusions are typical for the granulite-facies domain, early 2-phase aqueous inclusions occur almost exclusively in the north-western part of the Bamble and in the Telemark gneiss-granites, mixed (1 and 2) inclusions characterize the complicated transition zone between the amphibolite- and granulite-facies domains north of the orthopyroxene-in isograd.
P-T estimates from fluid inclusions are apparently very different for Bamble (maximum CO_2 density during peak metamorphism) and Rogaland (maximum CO_2 density after the peak of metamorphism). Most of the CO_2 originates from the breakdown of carbonate melts (carbonatites) emplaced as immiscible droplets in deep-seated synmetamorphic intrusives.

Southern Norway is a classical example of amphibolite - granulite facies transition in a Proterozoic terrain.

Although the age of metamorphism remains controversial (see various entries in this volume, notably by R.H. Verschure, D. Demaiffe and J. Michot, D. Field et al.), the pressure and temperature conditions start to be relatively well understood (Jansen et al., this volume). All recent studies emphazise the key importance of fluids, notably H_2O, the fugacity of which decreases suddenly at the granulite - facies boundary. Much information has been derived from the analysis, both experimental and theoretical, of characteristic mineral assemblages; however, since the first discovery of specific, high density CO_2 inclusions in many granulites (Touret, 1971), it has become evident that the direct observation of fluids trapped in rock-forming minerals (notably quartz, plagioclase, pyroxene etc.) can provide a great deal of information. CO_2-rich fluids, sometimes mixed with other species (N_2, CH_4) have later been observed in virtually all granulites in the world (Hollister and Crawford 1981, Roedder 1984), leading to the notion of "carbonic metamorphism", which has provided a new insight into the geology of the lower continental crust (e.g. Newton et al. 1980, Newton, this volume). Various studies dealing with Southern Norway have been issued in a number of publications (see e.g. review in Roedder 1984), but the rapid development of fluid-inclusion techniques, notably the possibility of in-situ, non-destructive analysis by micro--Raman spectroscopy, necessitates a critical reevaluation of earlier data. In this paper, a review of all studies performed in Soutern Norway during the last 15 years will be attempted. It will refine earlier syntheses (notably Touret 1981, 1984) and address three fundamental aspects of fluid investigations in high-grade metamorphic rocks:

i) The distribution of fluid inclusions in amphibolite- and granulite-facies rocks and the relative importance of lithological composition (protolith) and metamorphic grade.

ii) The pressure and temperature estimates derived from fluid inclusions and their comparison with data from solid mineral assemblages.

iii) The origin of the CO_2 fluids.

FLUID-INCLUSION DISTRIBUTION IN AMPHIBOLITE- AND GRANULITE- FACIES ROCKS.

In most metamorphic rocks, quartz, more rarely other minerals (feldspar, garnet, Al-silicates, pyroxene, apatite, etc.) may contain many (up to several millions per cm^3) small (typically 1 to about 30 micrometers) fluid inclusions. The apparently inextricable complexity caused by the great number of inclusions is somewhat compensated by a small number of possible fluids, well described in the specialized literature (Hollister and Crawford 1981, Roedder 1984): H_2O (± dissolved salts, notably NaCl), CO_2, CH_4 and

N_2, either pure or mixed in variable amounts. Solid phases may be associated with the fluids: either crystalline minerals, trapped along with the fluid or generated through oversaturation after the closure of the cavity (daughter mineral), or crystallized remnants of former silicate or carbonate melts ("glass" inclusions). Individual inclusions belong to a certain number of types, classically defined from the nature and relative amounts of phases at room temprature (unless specified otherwise): aqueous, brines, aqueous + CO_2, gaseous (CO_2, N_2, CH_4) etc.

Figure 1 gives some typical examples of fluid inclusions in Bamble granulites. The characteristic high-density CO_2 inclusions are single phase (gaseous) at room temperature and, as was done for years, they can be taken at first sight for holes or empty cavities. But even without the help of specialized equipment (microthermometry), they can be suspected from a well defined negative crystal shape (Fig. 1, N° 5 and 16) and recognized with certainty by exuberant release of gas bubbles during crushing (Fig. 1, N° 1 to 3). Figure 1 shows some features which are not commonly observed in granulites, but which are not exceptional: accidental trapping of silicate minerals (biotite) in CO_2 fluids (N° 5 to 7), occurrence of brines (see also Fig. 3) and possibly remnants of former glass inclusions. This last feature deserves attention, as it has never been described in any detail. But these "former glass inclusions" (Fig. 1 N° 10 and 13), although relatively common, will not be treated in the present paper.

THE 5 MAJOR TYPES OF FLUID-INCLUSION DISTRIBUTION.

Fluid inclusions are so numerous that, in any sample, the representativity of investigated inclusions is always a problem. Fortunately, in most cases one or two types are predominant. This might not be obvious at first sight, but it becomes clearer with practice, especially when only the earliest inclusions are taken into consideration. The criteria are simple and rely on a good dose of common sense (e.g. Touret 1977): "early" inclusions are isolated and their size is variable (often relatively large). They are primary if they can be related to the growth of the host mineral, e.g. in garnet, plagioclase or pyroxene. Late inclusions are conspicuous trail-bound, the trace of which is easily seen under the microscope by changing the focal plane.

Now, if we consider only the earliest inclusions, or more accurately, if we neglect late ones, a total of 7 types of inclusion distribution have been observed. They are all represented in Figure 3, but only 5 types deal explicitely with fluids and will be discussed here in more detail. The 2 remaining ones concern solid inclusions (carbonate + brines

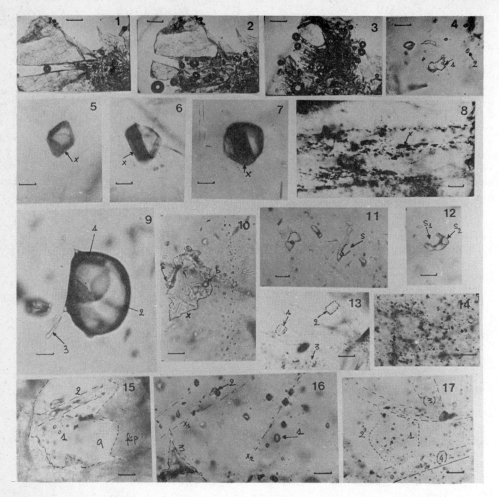

Figure 1: Typical inclusions in granulites from Bamble, Southern Norway.

Photos N° 1, 2, and 3: successive steps in the crushing of a quartz grain in glycerine. Inclusions are visible in the quartz grain as small black dots; they release a number of large gas bubbles which do not dissolve in glycerine (high density non aqueous fluids). N°4: aqueous (with salt cube, 1) and carbonic (2) inclusions in quartz; Na-rich granulite of intermediate composition ("arendalite"), Arendal railway station. N° 5, 6 and 7: Biotite (X) trapped in primary CO_2 inclusions in garnet, thus indicating that biotite was stable during garnet growth. Origin of the samples: 5: metapelite, Tromöy (S.68-152). 6 and 7: Furua granulite complex, Tanzania (Coolen 1980), shown for comparison. N° 8: CH_4 inclusion, often coated with graphite (X), Songe amphibolite (Touret and Dietvorst 1983). **N° 9:** CO_2 inclusions (1: CO_2 gas, 2: CO_2 liquid) with a small quantity of H_2O (3) which would have remained unnoticed in a regular, rounded inclusion. Pegmatite, Naresto near Tvedestrand. **N° 10:** Possible former glass inclusion (X) with remnants of high-quartz, negative crystal shape. Along the boundary of the inclusion, a very small quantity of brine can be seen (b: gas bubble, s: halite). Pegmatite (S 69-28b), Naresto near Tvedestrand. **N° 11 and 12:** NaCl bearing inclusions (brines); s, s_1: halite, s_2: probaly sylvite. Origin of the samples: **11:** Holt "granitel" (magnetite bearing granite). **12:** "charnockitic" gneiss, Tromöy (S 70-45.1), cf. Fig. 8. **N° 13 to 17:** Association of inclusions and elements for a relative chronology (Touret and Dietvorst 1983). **N° 13:** Possible former glass inclusions (1 and 2) and CO_2 inclusions (3) in acid metavolcanics, Arendal. **N° 14:** Cluster of CO_2 inclusions in granulite, Tromöy. **N° 15:** Older, isolated CO_2 inclusion (1) and later, trail-bound H_2O ones (2); q = quartz, fsp = feldspar. **N° 16 and 17:** Idem. 1 to 4 (on the photograph) represent a chronological order (For the principles of observation see e.g. Touret 1977). Length of bar: N° 1 to 3: 0.1 mm, N° 8, 14 to 17: 50 um, others: 5 um.

and carbonates); these will be dealt with later, in the discussion of the origin of the CO_2 fluids.

The 5 types of fluid-inclusion distribution are defined as follows:

1: <u>aqueous</u>: Water (liquid and vapor, 2-phase inclusions) is the only species present in the inclusion. The salinity is variable and daughter minerals are absent. Some halite cubes may occur, but this must remain exceptional.

2: <u>carbonic</u>: High density CO_2 (liquid homogenisation, commonly much lower T_h than room temperature to a minimum of -70 °C, well below the CO_2 triple point) (Swanenberg 1980). Generally pure, they may be mixed with variable quantities of N_2 and/or CH_4; pure methane and/or nitrogen inclusions may also occur, but they are rare and can be neglected as a dominant fluid in a given sample (Swanenberg 1980, Touret and Dietvorst 1983).

3: <u>mixed 1</u>: CO_2 and H_2O are both present, but in seperate inclusions, in roughly comparable amounts and in unclear chronological sequence. If they are contemporaneous, this would mean that they have been trapped in the miscibility gap of the $H_2O - CO_2$ system, at temperatures below approximately 400 °C. This is not impossible, but unlikely. More probably they have not been formed at the same time, but simple observation does not allow to define any chronological order (in most cases, however, dubious indications are in favor of an older origin for CO_2).

4: <u>mixed 2</u>: CO_2 and H_2O are present in the same inclusion (2 concentric bubbles of CO_2 vapor and CO_2 liquid in water, 3 fluid phases at room temperature). The fluid has been

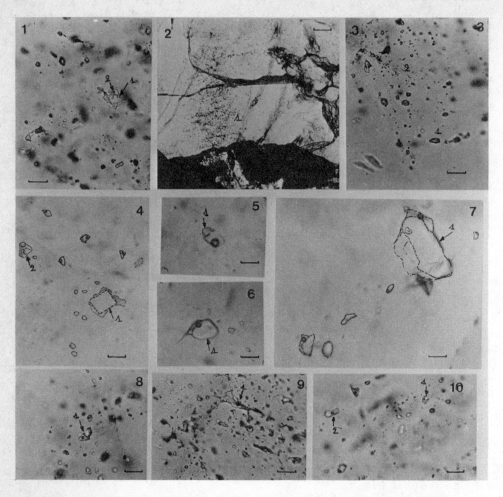

Figure 2: Brines in rocks from the granulite-facies domain (zones C and D, Figure 3).

1 to 7: Skarns, Barbu in Arendal town (sample N° 83.85). **1:** Evidence of brine–CO_2 immiscibility (a: brine, b: gas bubble), dark negative crystal shape: CO_2 inc. **2:** Trails of CO_2 inclusions, (a) in quartz (white) bordering garnet (black) and pyroxene. **3:** CO_2 (a) and brine (b) in the same inclusion cluster, another evidence for brine CO_2 immiscibility. **4, 5, 6** and **7:** examples of brines, illustrating the oversaturation at the time of trapping (variable size of halite, a) and the high density (small gas bubble, b) of the brine. **8 to 10:** 3 examples from brines in metapelites, zones C and D. **8:** Revesand, Tromöy, **9:** Havesöya, Arendal, **10:** cordierite-bearing gneiss, Tvedestrand; a and b: halite-bearing inclusions. Length of the bar: all photographs except N° 2: 5 µm, N°2: 50 µm.

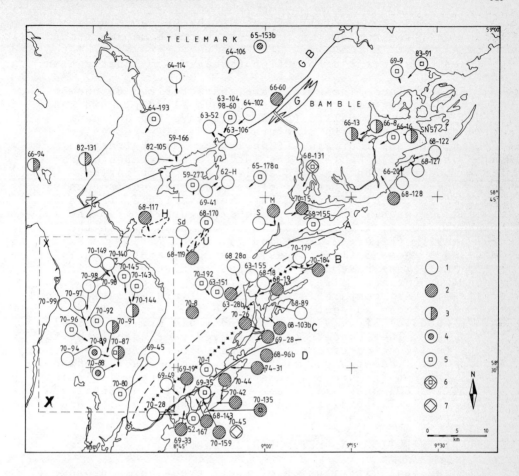

Figure 3: Metamorphic isograds and regional distribution of major inclusion types in Bamble (Southern Norway).

Inclusion type (see text): **1**: <u>aqueous</u>; **2**: <u>carbonic</u>; **3**: <u>mixed 1</u>; **4**: <u>mixed 2</u>; **5**: <u>brines</u>; only these 5 types deal with fluid inclusions. The others (6 and 7) correspond to carbonate melts: **6**: carbonate and brine, Songe hyperite; **7**: carbonate (ferroan dolomite) in Tromöy "charnockitic gneiss", see text and Fig. 8. Metamorphic zones (A to D) after Smalley et al. 1983. **A**: amphibolite facies, no orthopyroxene (Opx). **B**: Opx in basic rocks. **C**: Opx in all lithologies. **D**: LIL depleted "charnockitic gneiss", Tromöy. Each circle corresponds to a sample (hand specimen size) located at the end of the arrow (sample number, coll. J. Touret). **X** (dashed rectangle, lower left): systematic sampling of quartzites along a N-S profile (Touret 1972). **GB**: Great Breccia, **G, H, U**: granulite-facies areas inland (Gjerstad, Hovdefjell, Ubergsmoen), **S, M**: Songe (M: mobilisate) (Touret and Dietvorst 1983).

trapped above the miscibility gap of the system. These inclusions, so abundant in many medium-grade metamorphic terranes, e.g. in the Alps, are here very rare and related to zones of specific deformation (shearing, mylonitization, etc.).

5: brines: These inclusions are characterized by one or several daughter minerals in an aqueous fluid (liquid and vapor) (Figure 2). Halite (NaCl) is always the dominant solid species and, with some practice, easily recognized even in very small crystals (nearly the same refractive index as quartz). NaCl-bearing inclusions have been observed for a long time, but only recently, their systematic occurrence in a small number of well defined environments has been realized. Salinities are very variable, from a minimum of roughly 30 - 40 wt% NaCl (Figure 2, N° 5 and 10), to nearly 100 %, when the salt cube occupies almost the whole cavity (Figure 2, N° 6 and 7).

Besides these 5 fluid types, 2 categories of essentially solid inclusions have been mentioned in Figure 3. Their general appearance (size, shape, grouping in clusters or trails) is very comparable with common fluid inclusions, and many indications suggest that they derive, either from solids deposited from a now vanished fluid or, more likely, from a melt. They concern carbonates, either associated to brines (only one occurrence recorded so far in a hyperite at Akland near Tvedestrand), or to various other solid phases, notably zircon (Figure 8). These solid carbonate inclusions predate the CO_2 ones and, under the name of "carbonatite" melts, they will be discussed more thoroughly at the end of this paper.

FLUID DISTRIBUTION AND METAMORPHIC GRADE.

The regional distribution of the 7 types defined above (5 fluid, 2 solid) is shown in Figure 3. Each point represents a well studied sample which has been retained as representative of a given area. Also indicated are the 4 zones of increasing metamorphic grade defined (among others) by C. Smalley et al. 1983. **A**: Amphibolite facies, **A/B**: orthopyroxene-in isograd, the first occurrence of orthopyroxene in basic rocks, **B**: transition zone ("border migmatites" in the sense of J.A.W. Bugge, 1943), **B/C**: occurrence of orthopyroxene in all lithologies of suitable bulk composition, **C**: granulite facies, **D**: domain of the LIL depleted "charnockitic gneiss" of Tromöy (Field et al. 1980, Smalley et al. 1983). The name "charnockitic gneiss", is rather unfortunate, but so commonly used that it is difficult to abandon it. K-feldspar is conspicuously absent and the rock, of intermediate composition, is certainly more an enderbite than a charnockite. The nature of the C/D boundary is also controversial: D. Field and his co-workers (Field et al. 1980, Smalley et al. 1983) call it an amphibole-out isograd, which delimitates the core of the granulite-facies domain where hornblende is no longer stable. On the contrary, I consider that the scarcity (and not absence) of amphibole in Zone D is mainly due to an unsuitable bulk rock composition and that the limit between C and D is not a metamorphic isograd, but a lithological boundary (also enhanced by a prominent later fault) in an otherwise homogeneous terrain. The most important fact however, convincingly demonstrated by C. Smalley et al., 1983, is that the LIL depleted character of the "charnockitic gneiss" is not caused by metamorphic overprinting, but a direct result of magmatic differentiation. The rocks are synmetamorphic intrusives which have directly crystallized with their present composition under granulite-facies conditions. These results, which brought a major advance in the understanding of the regional geology, do not contradict an earlier hypothesis of B. Moine et al. (1972). They found that the Na-rich, K-poor character of the "charnockitic gneiss" was not restricted to the core of the granulite-facies area, but that it could also be found locally in Zone C (Arendal town) and even, in an attenuated form, at several places of Zone A (Tvedestrand, Songe, domain between Ubergsmoen and Hovdefjell, see Moine et al., 1972, Fig. 3).

A closer look at Figure 3 shows some general simple trends, but, in detail, obvious complexities. The simple trends concern:

i) The abundance of CO_2 inclusions (type 2) in the granulite-facies area (Zone C and D). Carbonic inclusions are also present in the isolated granulite-facies domains in the northern part of the Bamble (G: Gjerstad, U: Ubergsmoen, H: Hovdefjell).

ii) The presence of brines (type 5) in the whole investigated region, both in the granulite- and amphibolite-facies areas. Sporadically occurrences of brines have been noted since the first observations in Southern Norway (Touret 1971), but they have received little attention. Only after some years of experience it became evident that they are so abundant in some specimens that they must be considered as a major fluid type which may coexist with CO_2 in the core of the granulite-facies domain (see below).

Some exceptions to the general rule (amphibolite facies = H_2O, granulite facies = CO_2) may be mentioned:

* Very few cases of aqueous inclusions have been observed in the northern (68-89) and in the southern (70-29) part of Zone C. This might indicate some local uncertainty in the delimitation of the metamorphic isograd or a reintroduction of H_2O during the emplacement of late granites (Bohus type, see Verschure, this volume).

* More significant is the repeated occurrence of carbonic inclusions north of the A/B boundary. This is especially evident if we consider that most type 3 distributions (mixed 1, H_2O and CO_2 in separate inclusions), might very well be type 2 (carbonic) in which the early character of the CO_2 inclusions is not obvious.

Each case should be discussed in detail, but at least two explanations may be proposed. Firstly, the Opx-in isograd might be partly retrograde, it may have fluctuated for a few kilometers after the peak of metamorphism. Secondly, it can be easily demonstrated that most CO_2 inclusions north of the Opx-in isograd are related in some way to partial melting (anatexis) (Touret and Dietvorst 1983, the letters S (Songe amphibolite) and M (mobilisate) in Figure 3 refer to the samples described in detail in this paper). The relation between CO_2 inclusions and granitic "mobilisate" exists at all scales, including granites of batholithic dimensions. I will not attempt here to discuss all aspects of this complicated problem, which has certainly a major influence on the fluid repartition in the continental crust (Touret and Dietvorst 1983); it suffices to point to the remarkable scarcity of mixed $H_2O + CO_2$ inclusions, only present along some zones of specific deformation. At the granulite-facies transition, the change from H_2O- to CO_2- dominant regime is indeed abrupt; it involves very little mixing in a P-T domain where both species are however perfectly miscible.

BRINE-CO_2 IMMISCIBILITY IN GRANULITE FACIES; THE LITHOLOGICAL CONTROL OF BRINES.

Figure 3 shows that, in sharp contrast to the CO_2 and H_2O inclusions, the distribution of brines is <u>not</u> controlled by metamorphic grade. They may occur anywhere in the whole investigated area, also well outside the region depicted in Figure 3 (they are frequent in Telemark quartzites and epimetamorphic supracrustals). Isolated or trail-bound, they are always relatively early or even primary (isolated, irregular inclusions in an unstrained quartz crystal); if carefully searched for, they appear to be fairly abundant and definitely more common in some lithological types than in others. All these features suggest a lithological control and a premetamorphic origin of the brines, although many present characters (shape, densities, etc.) have been acquired during or after the peak of metamorphism. Outside the granulite-facies area, brines are especially abundant in 2 well-defined environments:
  i) Detrital metasediments in the broadest sense. They include quartzites, metapelites and most cordierite-bearing gneisses so typical for the north-eastern part of the Bamble (North of Tvedestrand). A conspicuous example is the famous

cordierite - tourmaline - apatite "pegmatite" at Bjordammen near Kragerö which probably derives from ancient evaporites (Touret 1979) and contains spectacular NaCl-bearing inclusions.

ii) Meta-acidic volcanics, mostly rhyolites and shallow intrusions. Examples of these rocks are well-known in the Telemark supracrustals (e.g. Rjukan group, see Maijer 1984); but they occur also in the higher-grade area, either in the Telemark gneiss-granites (Ploquin's "leptynites" near Amli), or in the Bamble district. An example of an intrusive rock is the magnetite - apatite bearing granite at Holt, locally known as "granitel" and the first iron ore ever mined in Norway, in which possible remnants of glass inclusions (Figure 1) have also been observed.

Similar protoliths occur in the granulite-facies domain, but one deserves to be mentioned: the famous andradite - hedenbergite skarn of Arendal, also related to extensive iron ores. Quartz is relatively rare in these rocks, but not absent and when occurring literally crowded by minute inclusions with large NaCl cubes (Figure 2, N° 1 to 7). Salinities are very high, up to at least 80 wt% (Figure 2, N° 4) and densities also high (small vapor bubble). An interesting feature is the local evidence of immiscibility between brines and CO_2 inclusions (Figure 2, N° 1 and 3: clusters of both inclusion types trapped at the same time). This type of evidence is relatively frequent in many granulite samples. A remarkable feature of the Arendal skarn is less the presence of the CO_2 inclusions, certainly to be expected in a rock which derives from a limestone, than their scarcity: brine inclusions grossly exceed in number the CO_2 ones. This fact confirms an old observation which was already made at the time of the discovery of CO_2 in granulites (Touret 1971): In Southern Norway, the abundance of CO_2 is less related to former sediments than to igneous rocks: basic intrusives, or, in the case of Tromöy (Zone D), to "charnockitic gneiss" of intermediate, enderbitic composition.

PRESSURE AND TEMPERATURE ESTIMATES: PEAK METAMORPHISM AND POSTMETAMORPHIC UPLIFT PATHS.

Many recent papers have been published dealing with the use of fluid inclusions in geothermometry and geobarometry (Hollister and Crawford 1981, Roedder 1984) and only a few words will be given here on the principle, interest and limitation of the method. If we neglect volume variations of the host mineral, fluid inclusions are <u>isochoric</u> (constant density) systems. For a known composition, the determination of the density (or molar volume) defines an isochore which passes through the P-T conditions at the time of trapping. This episode of the "life" of the fluid inclusion is by no means equivalent to peak metamorphic conditions: many inclusions are evidently postmetamorphic and they may also have experienced a complicated history after their formation: partial decrepitation, leakage, "necking-down", many possible causes of perturbation which are abundantly described in the specialized literature (e.g. Roedder 1984). But once this evolution has come to an end, the fluid content is preserved from any external influence and it may stay for millions of years without any contamination. There are indeed many indications that fluid inclusions might be exceptionally well preserved: a cavity containing high--density CO_2, at a pressure of several hundred bars at room temperature, may be distant from another, empty one, by only a few microns. It can be brought to the surface in hot lavas (the internal pressure is then more than 10 kb) without apparent leakage or explosion. On the other hand, in other cases later effects (deformation, recrystallization, etc.) may change drastically the fluid inclusion content. All these points must be carefully evaluated and the representativity of inclusions independently checked (P-T estimates from solid phases) before drawing any conclusion. In short, a correct interpretation of any fluid inclusion data calls for a number of conditions:

i) The fluid system must be sufficiently simple and well known, in order to derive the density from the temperature of phase transitions (notably liquid to vapor). This condition limits in practice possible investigations to 4 pure systems (H_2O, CO_2, N_2 and CH_4) and their mixtures (to some extent).

ii) Each inclusion, a few microns in size, defines its own isochore. The interpretation has to be done at a regional scale; therefore the problem of the representativity of selected isochores, commonly from frequency histograms, is obvious.

iii) Independant P-T data must be available; it is fortunate that the development of the fluid-inclusion techniques came at a time when rapid progress was made in geothermometry - geobarometry based on coexisting minerals.

All these conditions are severely restraining and it would be erroneous to believe that fluid inclusions alone may give reliable P-T estimates in any metamorphic environment. It is clear however that fluid inclusions in granulites present some decisive advantages:

i) Homogenization temperatures are easy to record: phase transitions are well visible, temperature of interest roughly between +31 °C and -60 °C, a domain where measurements are fast and instrumental corrections minimal. Many data can be accumulated in a reasonable time.

ii) For the 4 involved pure systems, P-V-T data are reasonably well known in the whole range of geological interest (Hollister and Crawford 1981).

iii) Most granulites contain sensitive temperature and pressure mineral indicators, notably orthopyroxene- and garnet-bearing assemblages (e.g. Newton 1985).

Many papers have been published recently (Touret 1981, Coolen 1982, Hansen et al. 1984a and b, Roedder 1984, etc.) in which the interested reader can find more details regarding the method. Basic data concerning Southern Norway have been presented in another paper (Touret 1981), but the refinement of their interpretation call for a detailed discussion of 2 fundamental points, namely the purity of the CO_2 fluids and the representativity of homogenization temperature (T_h) histograms.

THE PURITY OF CO_2: **RAMAN** MICROSPECTROMETRIC ANALYSIS.

Homogenization temperature data can be interpreted in terms of density only for pure fluids. Until recently the control of purity could exclusively be done with microthermometry (melting temperature at the triple point of a given pure system, see e.g. Hollister and Crawford 1981). This might be sufficient in most cases, but not always: in CO_2-rich fluids, significant quantities of e.g. N_2 or CH_4 may depress final melting temperatures by only 1 or 2 °C, well within the precision range of the freezing stage (Burrus 1981). The situation has drastically changed with the development of micro-Raman analytical techniques, which make possible accurate, in situ analyses of inclusions in the 1 micron range. Many analyses have been performed on the Southern Norway granulites with the laser-excited, multi-channel detector RAMAN system of the Free University Amsterdam (MICRODIL 28, for technical details and analytical procedure see e.g. Barbillat 1983). In most cases, the CO_2 of the carbonic fluids is very pure; however, a significant number of analyses, roughly 20% of the total, showed variable quantities of CH_4 and/or N_2.

The case of methane is relatively simple: it occurs almost exclusively in graphite-bearing metasediments (mostly metapelites, rarely marbles). It is also detected easily during microthermometric runs, both by the depression of final melting temperature and melting trajectory (temperature difference between initial and final melting) which can exceed 10 °C. The case of nitrogen is far more complicated: N_2-bearing inclusions may be randomly dispersed among pure CO_2 ones (Swanenberg 1980, Touret and Dietvorst 1983) and the melting point depression is nearly unnoticeable (Touret 1982). To ascertain more precisely the effect of N_2 on the homogenization temperature, a detailed study of the sample which had shown the greatest quantity of N_2 has been undertaken (Fig. 4). The investigated sample, situated at Akland near Risör (several other occurrences occur in the vicinity of Tvedestrand) is a cordierite-

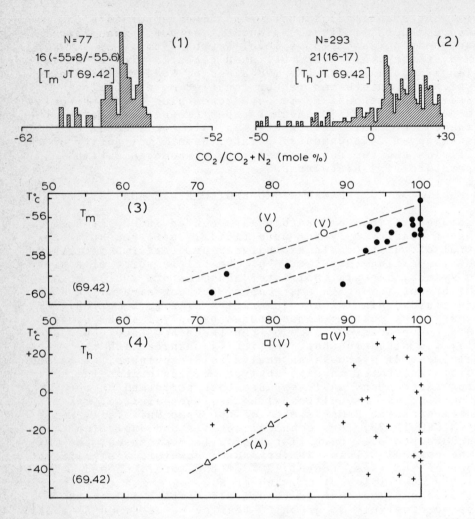

Figure 4: Cordierite-bearing gneiss at Akland near Risör (sample N° 69.42). Relation between melting and homogenization temperatures (T_m and T_h) and composition ($CO_2/(CO_2 + N_2)$) determined by Raman analysis (Microdil 28, Free University Amsterdam, E. Burke, analyst).

1 and 2: T_m and T_h histograms. For each histogram: 1st line (N): total number of measurements, 2nd line: maximum relative frequency in a given temperature interval (example: 1, 16% of all measurements are between -55.6 and -55.8 °C; this procedure calibrates the vertical scale), 3rd line: sample number (same conventions in all histograms, notably in Figure 5).
3 and 4: Relation between T_m (3), T_h (4) and composition. (V): 2 inclusions which homogenize in the vapor phase, all others liquid homogenization. (A): homogenization trend observed in metamorphic dolomites from Northern Tunisia (Guilhaumou et al. 1981).

bearing gneiss containing large (up to several cm) crystals of dark, almost black cordierite, plagioclase (An 30), quartz, phlogopite (pseudomorphs after anthophyllite) and opaque (ilmenite and rutile). Many carbonic inclusions occur in quartz and impurities in the CO_2 had been detected in the earliest stage of fluid investigation in Southern Norway (Touret 1971). They were then mistaken for methane, a hypothesis based partly on the belief that the dark colour of the cordierite was due to dispersed graphite particles; but it turned out that the only gas present besides CO_2 was N_2.

Results of microthermometric measurements and RAMAN analysis, presented in Figure 4, are interesting, partly unexpected and somewhat confusing. Up to 10 mole% N_2 ($CO_2/(CO_2 + N_2)$ molar between 0.9 and 1), there is practically no relation between T_m (final melting), T_h (homogenisation) and composition. Most important, pure CO_2 inclusions (100 in diagrams 3 and 4, Fig. 4) show the complete range of T_m and T_h variation observed in all measurements. The interpretation of T_h variation is unequivocal: different values correspond to density changes, from a maximum of 1.117 g/cm³, the absolute record so far in the Bamble area, to roughly 0.6 g/cm³ (T_h: -40 and +30 °C, respectively). The large (about 5 °C) T_m variation is much more surprising and unexpected: pure CO_2 should melt consistently at -56.6 °C. Note that only one point is near -60 °C. The measurements have been repeated several times, they are reproducable within 0.3 °C. All other points cluster between -55.5 and -56.6 °C, more than the experimental error (with the Chaix--meca freezing stage used in the microthermometric measurements, CO_2 T_m are reproducible within 0.2 °C), but within the range commonly accepted for pure CO_2 (-56.6 ± 1 °C). The cause of the scattering is unknown: possibly metastability during melting or the presence of noble gases (helium ?) undetected by the Raman probe.

A total of 21 inclusions have been analyzed, but only 6 contain more than 10 mole% N_2. Two of them homogenize in the vapor phase (very low density). The others show an important depression of the melting point (2 to 4 °C) and relatively constant T_h between -8 and -20 °C, well above the minimum T_h recorded in pure CO_2 inclusions. This fact indicates that for all inclusions, T_h lowering is due to density variations rather than to mixing. It may suggest that N_2 has itself a low density and, indeed, all pure N_2 inclusions observed so far in the Bamble area have a density close to critical (Touret and Dietvorst 1983). In conclusion, if the melting point remains within 1 °C of that of pure CO_2, possible N_2 contamination will not affect significantly the homogenization temperature. This covers the large majority of carbonic inclusions and practically all those occurring in igneous derivates.

SELECTED T_h HISTOGRAMS IN THE BAMBLE AREA. COMPARISON WITH ROGALAND.

All Bamble samples are more or less comparable to the cordierite-bearing gneiss at Akland: T_h variations are large, composite histograms show a number of frequency maxima (multipeak histograms, Touret 1981) which are very difficult to interpret. In an attempt to clarify the confuse relations indicated by composite histograms (e.g. N° 1 and 2, Fig. 5), 8 samples (N° 3 to 10, Fig. 5) were carefully selected and studied, recording for each sample a limited number of T_h (about 50), but trying systematically to measure the oldest inclusions and to retain one representative of some group in the immediate vicinity. The 50 measurements thus correspond to a total number of at least 150 to 200 inclusions and it has been found that this was sufficient to obtain representative histograms in most samples.

The choice of the 8 samples was dictated by a number of considerations: possible variation within the granulite--facies domain between Tvedestrand and Arendal, comparison of massive rock versus segregation, nature of the protolith (metasedimentary versus igneous) and CO_2 densities outside of the granulite-facies area. T_h histograms for each sample are given in Figure 5; only some questions can be (partly) answered, but the results as a whole are less confused than one could have probably expected.

Firstly, the massive samples from the Arendal region (histograms N° 3, 7, 9 and 10) show many more high density data (T_h below 0 °C) than the composite ones (notably N° 1, which can be taken as the rule for the Bamble area). Note the scarcity of "ultra-high densities" (T_h below -20 °C)

Figure 5: Homogenization temperature (T_h) histograms for CO_2 inclusions in selected samples from the Bamble area (all <u>liquid</u> homogenization; for each histogram, presentation described in Fig. 4). For the comparison of different histograms, note that the horizontal scale varies, but that the emplacement of 0 °C is the same for all histograms.

Location and description of the investigated samples:
1: Average samples of the granulite-facies area (Zones C and D). **2:** Quartzites sampled along a profile north of Arendal, indicated as X, Fig. 1 (dashed rectangle). **3:** Metapelite, massive rock, Revesand, Tromöy. **4:** Garnet-sillimanite gneiss (metapelite), Bakke, Tromöy. **5:** Quartz vein in sample N° 3. **6:** Cordierite-bearing gneiss, cafetaria Tvedestrand. **7:** "Charnockitic gneiss", Tromöy (see Figs. 7 and 8). **8:** Quartzite in Selas paragneiss, Ubergsmoen (Zone A). **9:** Plag-Opx gneiss ("Arendalite"), Arendal railway station. **10:** Idem (9), garnet-bearing type.

FLUID REGIME IN SOUTHERN NORWAY

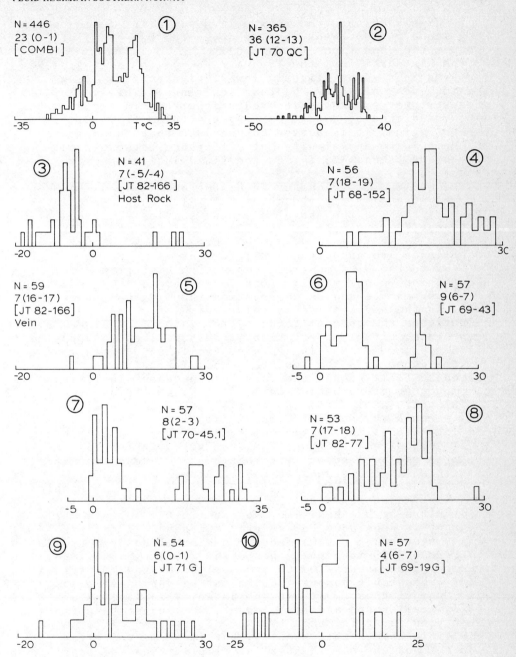

which have been observed in sample 69-42 (Fig. 4) and which correspond to isochores passing above the P-T "box" of peak metamorphic conditions (Figure 6).

The histogram from the Tvedestrand region (histogram N° 6) seems to present slightly lower "high density" peaks (maximum peak at +6/+7 °C) than in Arendal. This would suggest lower pressures in Tvedestrand than in Arendal, but is contradicted by data from Figure 4, also from the Tvedestrand region. It is better to conclude that these results do no indicate any significant difference between Tvedestrand and Arendal as far as the fluid inclusions are concerned.

The maximum T_h frequency in histograms N° 4, 5 and 8 is between 16 and 19 °C, a much higher value (and hence lower density) than in the above mentioned samples. The case of histogram N° 4 (sample JT 68-152) is strange: the rock is very similar to sample JT 82-166 (histogram N° 3) but two other cases are undoubtedly significant:

- Histogram N° 5 must be compared to histogram N° 3: segregation versus massive rock (same sample); CO_2 densities are lower in the quartz segregation than in the massive rock. As the segregation is relatively late, this result illustrates the general tendency for later inclusions to be less dense. Direct observation gives sometimes an opposite indication: some relatively late, trail-bound inclusions may be more dense than the isolated ones. There is also no reason to believe that the rare "ultra-dense" inclusions, as in Figure 4, are systematically older than the dense ones.

The only sample from the amphibolite-facies area (histogram N° 8, sample JT 68-77) contains only low density CO_2 inclusions. This histogram must be compared to histogram N° 2 (composite histogram from quartzites in Zone A) and it confirms the well-known fact that high CO_2-densities are restricted to the granulite-facies area.

Rogaland data (Swanenberg 1980) differ from the Bamble ones in 2 major aspects:

i) "Ultra-high" densities, so rare in Bamble, are very frequent in Rogaland. For several thousands of measurements, Swanenberg (1980) found a maximum frequency at -20 °C, versus 0 °C in Bamble (histogram N° 1, Fig. 5). The lowest T_h, at about -70 °C, are well below the CO_2 triple point (metastable homogenization during supercooling).

ii) Many of these "ultra-high" desnsity inclusions are small, trail-bound and they tend to be younger than the low density ones. Again, this type of observation can be made locally in some Bamble samples, but it seems to be far more frequent and obvious in Rogaland. Implications on possible uplift paths are discussed below.

PEAK METAMORPHIC FLUIDS AND UPLIFT PATHS IN BAMBLE AND IN ROGALAND.

To evaluate the significance of fluid inclusions, representative isochores (Fig. 6) must be compared to independant P-T data obtained from other sources, notably from coexisting solid minerals. Fortunately, granulites contain many sensitive assemblages (garnet-biotite, ortho- and clinopyroxene, quartz - plagioclase - Al-silicate - garnet or orthopyroxene) and they count among the most favourable metamorphic rocks for this type of evaluation (Newton, this volume). Major problems however arise from extensive mineral reequilibration and from the complicated role of water.

P-T data from solid phases are discussed elsewhere in this volume (Jansen et al.) and only the essential points are reported in Figure 6. In Rogaland, 3 distinct episodes of high- grade metamorphism (M_1, M_2, M_3) have been detected. Most striking is the late (about 1050 - 1100 Ma) M_2 phase at high temperature and low pressure (about 3 - 4 kb), followed by M_3 which is still characterized by local granulite-facies assemblages (breakdown of osumilite into cord.- qtz.- Kfspar - Opx. symplectites). The chemical composition of most minerals has been reset during M_3; (M_1, M_2 being only found in isolated remnants and armoured relics) it can therefore be assumed that all fluid inclusions measured by Swanenberg (1980), still the only source of information about Rogaland, are syn- or, for most of them, post M_3. This would lead to a post -metamorphic uplift path labeled R in Figure 6. It corresponds roughly to the model proposed by Swanenberg (1980), with a slight lowering of the pressure in order to exclude the kyanite field (Swanenberg's estimate for the Al-silicate triple point was more than 1 kb higher than the now commonly assumed value, indicated in Figure 6).

Average conditions in Bamble (750 °C, 7 kb, Jansen et al., this volume) correspond to the low-temperature, high-pressure side of M_1. Preliminary data (newton, pers. com. and Touret, unpub.) suggest about 1 kb pressure increase between Tvedestrand and Arendal and also possible higher temperatures in the intrusive rocks of Zone D (Tromöy), but this has to be confirmed by further investigations. It is highly probable that M_1 is common for Bamble and Rogaland, but the age seems to be significantly different: about 1500 Ma in Bamble, 1200 Ma in Rogaland (Verschure, this volume). A younger metamorphic episode which could be the equivalent of M_3 is often inferred in the Bamble, with much controversy regarding its extension and nature: pervasive flooding of aqueous fluids resetting the Rb-Sr ages (Field et al., this volume) or, at least locally, high-grade, granulite-facies assemblages related notably to basic intrusives (e.g. hyperites, see further discussion in Maijer, 1984).

The recent discovery of kyanite in late pegmatitic segregations between Söndeled and Kragerö (C. Maijer, T. Senior, R.H. Verschure, pers. com.) imposes a post-metamorphic P-T trajectory which, in Bamble, passes through the kyanite stability field (B_2, Figure 6). B_2 is strikingly different from the path proposed earlier (B_1, Touret and Dietvorst 1983). We have discussed elsewhere the reasons which may partly reconcile B_1, based on the order of fluid trapping in the molten part of a migmatite and B_2 (Touret and Olsen 1985). The problem remains nevertheless open and we will see that it is not resolved by the present study.

These P-T data must be compared to selected isochores from Figure 5. A total of 6 isochores have been retained, labeled in Figure 6 in terms of CO_2 liquid homogenization temperature. In order of decreasing density, these isochored correspond to the following: (all T_h in °C)

* (-70): maximum density ever recorded in Rogaland (Swanenberg 1980).
* (-40): maximum density ever recorded in Bamble (Sample N° 69-42, Figure 4). Note that this result is exceptional, all other results in the Bamble are above -20 °C (Figure 5).
* T_h between (-20) and (-10): densities bracketting the "high-density peak" in Bamble.
* (+0): Lower limit of the "high-density peak" (approximately).
* (+10): characteristic value of the late, low-density group.

The comparison between this set of isochores and B_1, B_2 and R suggests the following remarks:

i) Isochores bracketting the "high-density peak" in Bamble area pass slightly below Bamble average conditions (B_{Ty}, Figure 6). As amphibole is everywhere a stable mineral during granulite-fiacies conditions, a certain PH_2O must be added to PCO_2 and it has been estimated to about 2 kb (Touret 1974). This would make the high-density CO_2 inclusions precisely compatible with peak metamorphic conditions. Note however that, if B_2 is accepted as post-metamorphic trajectory, high-density CO_2 fluids can be trapped on a very large temperature interval.

ii) B_2 is difficult to reconcile with histograms N° 3 and 5 of Figure 5 (Sample N° JT 82-166, host rock versus quartz segregation). Unless we assume that the segregation has been formed at a temperature below 400 °C, the sharp decrease in density between the rock and the segregation would suggest a B_1 type trajectory, exactly as in the case of the Songe amphibolite (Touret and Dietvorst, 1983). Solid minerals and fluid inclusions give here discordant indications, which can only be resolved by more detailed observations.

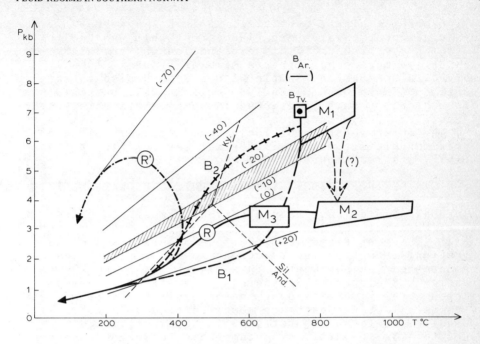

Figure 6: P-T estimates in Bamble and Rogaland. Comparison between solid mineral and fluid inclusion data.

Solid minerals: M_1, M_2, M_3: high-grade metamorphisms in Rogaland (Jansen et al., this volume); (M_1 probably common to Bamble and Rogaland, see text). B_{Tv} (black dot): average conditions in Bamble near Tvedestrand. B_{Ar}: preliminary data near Arendal. And: andalusite, sil: sillimanite, ky: kyanite. Postmetamorphic P-T trajectories: B_1 and B_2: Bamble, R and R': Rogaland (see text). Selected CO_2 isochores (thin solid lines): each isochore is labeled in $T_{h\ (liquid)}$, e.g.(-70) = T_h = -70 °C (d=1.224 g/cm³). CO_2 P-V-T data from Holloway's revised program (Hollister and Crawford 1981).

iii) No trajectory, either in Rogaland or in Bamble, approaches the ultra-high density isochores (-70 or -40). The "Rogaland paradox", first clearly expressed by Swanenberg (1980), is striking: with the late M_2 and M_3 episodes, one would expect fluids in Rogaland to be less dense than in Bamble. On the contrary, they are more dense, both in average and in maximum value. Swanenberg (1980) has shown convincingly that many dense Rogaland inclusions are relatively late. Some process must explain a late reequilibration towards higher densities, as in R', Figure 6. It has been suggested by Swanenberg (1980) that this was due to Caledonian overpressure. This explanation is very attractive, but we must remark that a similar trend is locally observed in the Bamble (Sample N° 69-42) and also in some migmatites (Touret and Olsen 1985). It is well possible that other reasons, possibly detected in some experiments (Pecher and Boullier 1984) cause such a reequilibration. If this is true, the interpretation of inclusion density in terms of P-T metamorphic estimates will be far more difficult than initially assumed (Touret 1981).

THE ORIGIN OF THE FLUIDS: SYNMETAMORPHIC CARBONATE MELTS.

Carbonic fluids have been observed in most granulite--facies areas of the world, but their origin remains controversial. The most direct explanation favours a sedimentary derivation, from prograde reactions in marbles and other carbonate-bearing sediments. This might locally be the case (Glassley 1983), but in Southern Norway direct evidence suggests a relation with deep-seated basic and intermediate intrusives. Isotopic data are complicated and not entirely conclusive. Hoefs and Touret (1975) found very low $\delta^{13}C$ values for CO_2 in early, synmetamorphic inclusions. Retrograded samples give values much closer to a "juvenile" carbon around -7‰. These data were later confirmed by Pineau (1977) and Pineau et al. (1981), who showed that values at -7‰ are always found during heating at 600 °C. They correspond to the destabilisation of minute carbonate microcrystals, systematically present in all granulites.

The $\delta^{13}C$ values confirm the minor importance of CO_2 deriving from sedimentary carbonates; these would give much higher, positive $\delta^{13}C$ values observed e.g. in Naxos (Kreulen 1980). They reveal the widespread occurrence of carbonates with a deep, mantle-derived signature. The first attempts to locate them in the rocks were not very succesful. They were observed in late, open microfractures and thought to be the product of superficial alteration. The publication of a series of papers by D. Field and co-workers (notably Field et al. 1980, Smalley et al. 1983) led to a closer reinvestigation of samples from Zones C and D (Fig. 3). In short, the major results of these papers were a confirmation, from major and trace element geochemistry, that most igneous derived granulites have acquired their characteristic, conspicuously LIL-depleted composition, not during a late metamorphic evolution, but by magmatic differentiation and direct crystallization under granulite-facies conditions. The role played by deep-seated, synkinematic intrusives had been stressed for the late "hyperites", which mark the end of the high-grade metamorphic evolution of the region (Touret 1969). It is also suggested by the complicated uplift paths discussed above: synmetamorphic intrusions can best explain the periods of isobaric cooling which have been proposed for Bamble and Rogaland.

In the studies of the Nottingham group, two major petrographical types have been investigated: metabasites (Smalley et al. 1983) and the high-iron, low-potassium "charnockitic gneisses" of Tromöy (Field et al. 1980), which characterize Zone D (Fig. 3) in the core of the granulite-facies area. This last type was selected for a detailed microscopic investigation for 2 reasons:

i) CO_2 inclusions are locally extraordinarily abundant (tumultuous release of gas bubbles under the crushing stage).

ii) The bulk chemistry of the rock is one of the most typical examples of "depleted granulite" ever described in the literature. A series of thin sections and double polished plates was made of a carefully chosen sample (sample N° 70.45.1, Royrbunten, Tromöy). Two clear objectives were in mind: an attempt to find traces of the magmatic stage and to locate carbonates which could correspond to this.

The results, illustrated in Figures 7 and 8, exceeded the most optimistic expectations. The rock, a typical "arendalite" in the sense of J.A.W. Bugge (1943), contains quartz (less than 10% in volume), plagioclase (An 20-30, at least 50%), orthopyroxene (hypersthene), amphibole (dark green hornblende) and opaques (ilmenite). A well defined metamorphic layering is visible microscopically. Under the microscope, plagioclase occurs as small, equigranular aggregates of crystals with triple junctions at 120° and straight boundaries (annealed granulitic texture). But there are also larger (a few mm) crystals, which contain a remarkable number of solid inclusions: zircons, often with small facets, and perfect euhedral crystals of pyroxene and amphibole (Figure 7, N° 6 to 12). Other samples of a more basic variety (quartz free two-pyroxene granulite) contain garnet, also in perfect euhedral crystals (dodecahedron) in the plagioclase (Figure 7, N° 13 to 15). These spectacular solid inclusions are apparently widespread in the whole granulite--facies area. Most of the crystals are so perfect that one may wonder why they have not been described in greater detail in previous studies. An explination may be that the perfection of the shape is only visible in thick, double polished plates; a normal thin section gives only a less perfect section of the crystal.

The scope of this paper does not permit a detailed description and interpretation of these euhedral solid inclusions. We consider, however, that they represent typical magmatic features for two major reasons: many phenocrysts of present-day superficial lavas contain exactly the same type of euhedral inclusions, which could grow freely in

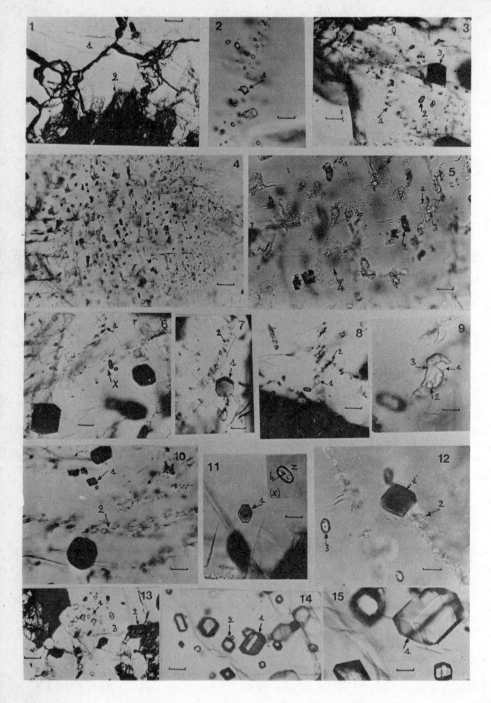

Figure 7: Fluid and solid inclusions in "charnockitic gneisses", Tromöy (Zone D, Figure 3).

1: Quartz segregation in the "charnockitic gneisses" (Numbers on the photograph: 1: quartz, 2: diffuse border with many trails of CO_2 inclusions). **2:** Detail of area 2 in photograph 1 (1: monophase CO_2 inclusion). **3:** Feldspar (plagioclase) in massive gneiss (1: trail of carbonate inclusions, 2: zircons, 3: euhedral pyroxene). **4:** Trail of carbonate inclusions, section parallel to the plane of the trail (1: carbonate). **5:** Detail of photograph 4 (1: unknown isotropic phase, 2: open cavity (squeezed gas bubble ?), 3: euhedral carbonate (iron-rich dolomite), 4: aggregate of calcite micro--crystals, 5: cubic, isotropic phase). **6:** Intersecting carbonate trails perpendicular to the plane of the section (1) and isolated, primary carbonate inclusions close to 3 zircons (X). **7:** Euhedral pyroxene (1) near a trail of carbonate inclusions (2). **8:** Primary carbonate inclusion (1) surrounded by secondary, decrepitated and transposed inclusions (2). **9:** Detail of the primary inclusion (1) shown in photograph 8 (1 and 2: euhedral iron-rich dolomite, 3: aggregate of calcite and unknown isotropic phase) (compare with Figure 8). **10:** Euhedral amphibole (1) and trail of carbonate inclusions (2). **11:** Euhedral pyroxene (1), insert (X): zircon (z) with possible fluid inclusion (b). **12:** Euhedral amphibole (1) and trail of carbonates (2) which stops on the amphibole = pseudosecondary inclusions; 3: zircon. **13:** General aspect of the basic variety of the "charnockitic gneiss", which corresponds to a 2 pyroxene granulite (1: euhedral inclusions in plagioclase (3), 2: ortho- and clinopyroxene). **14** and **15:** Detail of the euhedral inclusions in plagioclase of photograph 13, 1: pyroxene, 2: garnet.
Location of the samples: Photographs 1 to 12: sample N° 70.45.1, Royrbunten, Tromöy. Photographs 13 to 15: sample N° 83.77, Taverna rest., road Fevik, Arendal.
Length of the bar: Photographs N° 1, 3, 13: 40 um. N° 4, 6, 7, 8 and 10: 25 um. Others: 10 um.

the molten magma. In the studied sample, <u>aggregates</u> of several crystals are frequently observed, which suggest immediately movements in a liquid environment. In any case, all these features are in perfect agreement with D. Field et al.'s (1980) conclusions and they support the hypothesis of a direct crystallization of the "charnockitic gneiss" under granulite-facies conditions. It is interesting to note in this respect that, besides zircon, 3 major "magmatic" phases have been observed: orthopyroxene, garnet and amphibole, a fact which suggests that amphibole is stable in the granulite-facies area and that its scarcity (not absence as claimed by C. Smalley et al. 1983) in Zone D is due to chemical reasons and not to specific metamorphic conditions.

CO_2 inclusions are present everywhere in the rock, but they are especially abundant in late, diffuse quartz segregations (Fig. 7, N° 1 and 2); these contain also very small carbonate crystals, often in the immediate vicinity of the

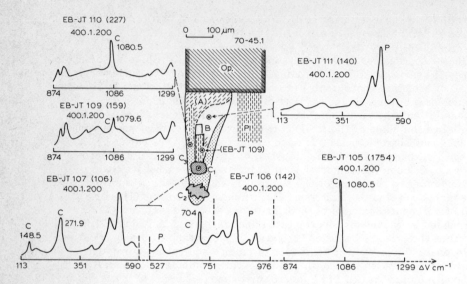

Figure 8: Raman analysis of primary carbonate inclusion in sample N° 70.45.1 (comparable to the inclusion shown in Fig. 7, N° 9).

Op: opaque, Pl: plagioclase (trace of twinning). A, B and C: composite primary inclusion filling a growth defect of the plagioclase. A: aggregate of carbonates and unkown isotropic phase. B: unkown isotropic phase. C_1: idiomorphic carbonate (ferroan dolomite), C_2 and C_3: confuse aggregates of carbonate microcrystals.
Points in a circle: emplacement of the laser beam (1 um in diameter). Raman spectra recorded with MICRODIL 28, Free University, Amsterdam. E. Burke, analyst. Each spectrum is taken with the same analytical conditions: laser power 400 mW, 200 accumulations of 1 sec recording time (indicated by 400.1.200). Above each spectrum is indicated the spectrum number (e.g. upper left: EB-JT 110), followed, in brackets, by the "full scale intensity value" (227), which corresponds to the intensity of the maximum peak and gives an arbitrary, but constant vertical scale for all diagrams. C: characteristic line for carbonates, other line (P, or not labeled): plagioclase.

carbonic inclusions, but mostly in the late, open microcracks. In the massive rock, isolated CO_2 inclusions are present in quartz, but in lesser amount than in the segregation. On the other hand, solid carbonate inclusions are very abundant, especially in plagioclase, decidedly a heaven for solid inclusions. Most of them occur in well-defined, intersecting trails (Fig. 7, N° 3 to 8) and the general appearance, shape, size, etc. are very similar to common fluid inclusions in metamorphic rocks. Particularly impressive is the regular array in the plane of a trail, when its direction is roughly parallel to the plane of the section (Figure 7, N° 4 and 5). Typical primary inclusions have also been observed (Figure 7, N° 6), many of them related to euhedral mineral inclusions, notably opaques. One example is given in Figure 8: the carbonates are situated in a cavity which was protected by magnetite during carbonate growth. The cavity is rather large (about 300 um), but filled with extremely small particles which can only be characterized unambiguously by Raman analysis. Three modes of occurrence have been observed:

i) Small, well crystallized crystals, a few microns in size, slightly yellow under the microscope (one nicol). Some could be analyzed with the Microprobe (Microscan-9, Free University, 20 kV, W. Lustenhouwer, analyst) and found to be a Fe-rich dolomite:

$$Fe_{.238}Mn_{.008}Mg_{.192}Ca_{.516}(CO_3)^{2-}$$

They are apparently the earliest phases in the cavity (C_1, Fig. 8).

ii) Irregular, cryptocrystalline aggregates of calcite (white, lower maximum relief than for dolomite), obviously much disturbed by later alteration and recrsytallization (C_2, Fig. 8).

iii) Very fine-grained association of a carbonate, probably calcite, and an isotropic phase which could not be further characterized. Traces of cubic structure (B, Fig. 8) indicate that this could be halite, but all attempts to search for NaCl with the microprobe have been unsuccessful.

In other carbonate-bearing inclusions, rounded opaque crystals, probably hematite, have also been observed. Zircons (X, Fig. 7, N° 6) and apatite, often also in spectacular euhedral crystals occur also frequently near the carbonates.

The primary nature and general appearance suggest trapping from an immiscible melt which was present at the time of crystallization of the igneous host rock. The notion "carbonatite" comes immediately to the mind, with regard to a number of arguments:

- The magmatic nature of the carbonate and the association carbonate and apatite, eventually also zircon, so common in many carbonatites.
- The composition of the early carbonates (ferroan dolomite, C_2, Fig. 8). Studying the magmatic evolution of the Fen Complex, Andersen (1985) has demonstrated that one major, volatile-rich trend of the carbonatite suite is a ferro - carbonatite (Rauhaugite II of some authors) which contains the same carbonates as C_2; the ferro - carbonatite derives from a deeper Ca-Mg rich carbonatite magma and crystallizes immediately before the separation of a free fluid phase.

— Finally, the isotopic composition of the dispersed carbonates in the granulites with $\delta^{13}C$ near -7‰, close to the carbonatite values.

GRANULITES AND CARBONATITES, THE CO_2 FLOODING PROBLEM.

The preceding example shows that, in Tromöy, gaseous CO_2 was predated by carbonate melts of mantle origin. The modalities of the transition between carbonates and gas (immiscible fluids, breakdown of solid phases, etc.) are unknown and pose many problems, concerning e.g. the sudden decrease in $\delta^{13}C$ values. We can note however that the concept of carbonatite melts as carrier for the CO_2 minimises several difficulties of its mantle derivation. If it is moving as a free gas (CO_2 "wave", Newton et al. 1980), it has been objected (W. Lamb and J. Valley, this volume), that graphite should be deposited and occur in granulites in much larger quantity than actually found. In the present case, the distance of CO_2 migration is much less and if, as suggested by several indications (Kaper, pers. comm.) oxygen fugacity is outside the graphite field at the carbonate / CO_2 boundary, no graphite will be deposited. More work is certainly necessary, but at least the two following questions are obvious: Are these carbonatite melts local or widespread occurrences and how can they be generated in the regional geodynamical context ?

Concerning the first question, several arguments point to a widespread occurrence of carbonatites:

— Several carbonatites are known in Southern Norway: besides Fen, many occurrences have been recorded, both in the Bamble and in Telemark (Ramberg and Barth 1966, Verschure et al. 1983). More than one episode might be involved, e.g. during the eocambrian and late--carboniferous. Some hyperites, dated around 1 Ga, are surrounded by extensive domains with scapolitisation, huge quantities of apatite (e.g. the large Ödegarden deposit) and present at least some affinities with carbonatites. Carbonate generating magmas seem to have been present in Southern Norway during almost one billion years.

— Possibly magmatic carbonate inclusions have been observed in other granulites. They seem to be particularly typical and abundant in high-pressure granulites, e.g. in Northern China (Qianxi granulites, Jahn and Zhang 1984). We have in fact reasons to believe that they are widespread in many granulites.

The second question, namely the origin of the immiscible carbonatite and intermediate silicate melts ("charnockitic gneisses"), remains conjectural. However, a very promising line of research has recently and independently been proposed by J. Otto (1984). This author has obtained experimentally two immiscible liquids, corresponding to 60 wt% quartz-syenite and 40 wt% carbonatite, respectively, by partial melting at 25 kb of a mixture of quartz-eclogite and carbonates. As starting material is taken an equivalent of subducted oceanic crust: basalt + carbonate-bearing sediments. With steeper geotherms, melting would occur at lower pressure and the resulting liquid, more CO_2 rich, corresponds to trondhjemite (about 80%) and carbonatite (20%), an almost ideal parental magma for the "charnockitic gneisses" of Tromöy. Such a model fits well in the regional context: several authors (Torske 1977, Berthelsen 1980) envisage the formation of this segment of the Scandinavian continental crust in a destructive plate-margin environment, above a subducted oceanic crust. Falkum's model of a continental collision and major shear belt (T. Falkum, this volume) is apparently more difficult to reconcile with the present hypothesis. Even not taking into account the time constraints (shear belt deformation might be several hundred years later than the emplacement of the "charnockitic gneisses"), it is not sure, however, that both models are totally incompatible: in a very general way, granulites presently at the surface impose some kind of continental doubling (Vlaar 1985, Newton 1985). In Southern Norway, the Moho is at a normal depth and the initial thickness of the crust might have been of the order of 60 km. Why not a subducted oceanic crust below this huge continental pile ? This might sound and is indeed very complicated and hypothetical, but it could explain why magmas generated during the subduction could not rise to the surface and remained at a lower crustal level.

Acknowledgements

Several drafts of this paper were reviewed or discussed by many colleagues, notably T. Andersen, C. Maijer, R.C. Newton, A.C. Tobi, J. Schreurs and R.H. Verschure. Many microthermometric data were measured by M. Schippers, microprobe and Raman analysis were done by W. Lustenhouwer and E. Burke, respectively; facilities for electron- and laser Raman microprobe analyses were provided by the Free University in Amsterdam and by the WACOM, a working group for analytical chemistry for minerals and rocks; this group is subsidized by the Netherlands Organization for the Advancement of Pure Research (ZWO). The support from NATO and from Dr. Schurman's fund is gratefully acknowledged. Finally, J. and L. Schreurs deserve a special thanks for final typing and editing of the text.

BIBLIOGRAPHY

Andersen, T. (in prep.). 'Whole rock compositional effects of carbonatite reequilibration: Example of Fen, Norway'.

Barbillat, J. 1983. 'Etude d'une seconde génération de microsonde optique a effet Raman mettant a profit les avantages de la détection multicanale'. These Etat, Sci. Phys., Univ. Sci. Techn. Lille, 606, 171 pp.

Berthelsen, A. 1980. 'Towards a palinspastic tectonic analysis of the Baltic shield'. Mém. BRGM, 6th Colloq., Int. Congr. Paris, 108, 5-21.

Bugge, J.A.W. 1943. 'Geological and Petrological investigations in the Kongsberg-Bamble formation'. Norges Geol. Unders., 150, 150 p.

Burrus, R.C. 1981. 'Analysis of phase equilibria in C-O-H-S fluid inclusions'. In: Hollister, L.S. and Crawford, M.L. (eds), Short Course Mineral. Soc. Canada, 6, 39-74.

Coolen, J.J.M.M.M. 1980. 'Chemical petrology of the Furua Granulite Complex, southern Tanzania'. Ph.D. Thesis, Free University Amsterdam, GUA Ser. (Univ. of Amsterdam), Ser 1, 13, 258 p.

Coolen, J.J.M.M.M. 1982. 'Carbonic fluid inclusions in granulites from Tanzania - a comparison of geobarometric methods based on fluid density and mineral chemistry'. Chem. Geol., 37, 59-77.

Demaiffe, D. and Michot, J. 1985. 'Isotope geochronology of the Proterozoic crustal segment of Southern Norway: a review'. This volume.

Falkum, T. 1985. 'Geotectonic evolution of Southern Skandinavia in the light of a late-Proterozoic Plate--collision'. This volume.

Falkum, T. and Petersen, J.S. 1980. 'The Sveconorwegian orogenic belt, a case of late-Proterozoic plate collision'. Geol. Rundschau, 69, 622-647.

Field, D., Drury, A. and Cooper, D.C. 1980. 'Rare earth and LIL element fractionation in high grade charnockitic gneisses, South Norway'. Lithos, 13, 281-289.

Field, D., Smalley, C, Lamb, R.C. and Raheim, A. 1985. 'The 1.6-1.5 Ga-old amphibolite-granulite terrain, Bamble sector, Norway: dispelling the myth of regional Grenvillian orogenic reworking'. This volume.

Glassley, W.E. 1983. 'Deep crustal carbonates as CO_2 fluid sources. Evidence from metasomatic reaction zones'. Contrib. Mineral. Petrol., 84, 15-24.

Guilhaumou, N., Dhamelincourt, P., Touray, J.C. et Touret, J. 1981. 'Etude des inclusions fluides du systeme N_2-CO_2 de dolomites et de quartz de Tunisie septentrionale'. Geochim. Cosmochim. Acta, 40, 657-673.

Hoefs, J. and Touret, J. 1975. 'Fluid inclusion and carbon isotope studies from Bamble granulite, Southern Norway. A preliminary investigation'. Contrib. Mineral. petrol., 52, 165-174.

Hansen, E.C., Newton, R.C. and Janardhan, A.S. 1984a. 'Fluid inclusions in rocks from the amphibolite facies gneiss to charnockite progression in southern Karnataka, India: Direct evidence concerning the fluids of granulite metamorphism'. J. Metam. Geol., in press.

Hansen, E.C., Newton, R.C. and Janardhan, A.S. 1984b. 'Pressures, temperatures and metamorphic fluids across an unbroken amphibolite-facies to granulite-facies transition in southern Karnataka, India'. In: Kröner, A., Goodwin, A.M. and Hanson, G.N. (eds). Archean Geochemistry, Amsterdam, Elsevier, in press.

Hollister, L.S. and Crawford, M.L. (eds) 1981. 'Fluid Inclusions: Applications to Petrology'. Short Course Mineral. Soc. Canada, 6, 304 p.

Jahn, B.M. and Zhang, Z.Q. 1984. 'Archean granulite gneisses from eastern Hebei Province, China: rare earth geochemistry and tectonic implications'. Contrib. Mineral. Petrol., 85, 224-243.

Jansen, J.B.H., Blok, R.P.J., Bos, A. and Scheelings, M. 1985. 'Geothermometry and geobarometry in Rogaland and preliminary results of the Bamble area, Southern Norway'. This volume.

Kreulen, R. 1980. 'CO_2-rich fluids during regional metamorphism on Naxos (Greece): carbon isotopes and fluid inclusions'. Am. J. Sci., 280, 745-771.

Lamb, W. and Valley, J. 1985. C-H-O fluid calculations and granulite genesis'. This volume.

Maijer, C. (ed.) 1984. 'Excursion guide of the South Norway geological excursion'. NATO Adv. Stud. Inst., Moi near Stavanger, Jul. 16-30, 207 p. (in press, Norges geol. Unders.).

Moine, B., Roche, H. de la and Touret, J. 1972. 'Structures géochimiques et zonéographie métamorphique dans le Précambrien catazonal du Sud de la Norvege (région d'Arendal)'. Sci Terre, Nancy, 17 1-2, 131-164.

Newton, R.C. 1985. 'Temperature, pressure and metamorphic fluid regimes in the amphibolite-facies to granulite--facies transition zone'. This volume.

Newton, R.C., Smith, J.V. and Windley, B.F. 1980. 'Carbonic metamorphism, granulites and crustal growth'. Nature, 288, 45-50.

Otto, J.W. 1984. 'Melting relations in some carbonate - silicate systems: sources and products of CO_2-rich liquids'. Ph.D. Thesis, University of Chicago, 220 p.

Pecher, A. et Boullier, A.M. 1984. 'Evolution a pression et température élevées d'inclusions fluides dans un quartz synthétique'. Bull. Min., 107-2, 139-154.

Pineau, F. 1977. 'La géochimie isotopique du carbon profond'. These Etat, Université Paris, 7, 300 p.

Pineau, F., Javoy, M., Behar, F. and Touret, J. 1981. 'La

géochimie isotopique du facies granulite du Bamble (Norvege) et l'origine des fluides carbonés de la croûte profonde'. Bull. Mineral., 104, 630-641.
Ramberg, I. and Barth, T.F.W., 1966. 'Eocambrian volcanism in Southern Norway'. Norsk Geol. Tidsskr., 46-2, 219-237.
Roedder, E. 1984. 'Fluid inclusions'. Reviews in Mineral. Am Mineral. Soc., 12, 644 p.
Smalley, P.C., Field, D., Lamb, R.C. and Clough, P.W.L. 1983. 'Rare earth, Th - Hf - Ta and large-ion lithophile element variations in metabasites from the Proterozoic amphibolite - granulite transition zone at Arendal, South Norway'. Earth Plan. Sci. Lett., 63, 446-458.
Swanenberg, H. 1980. 'Fluid inclusions in high grade metamorphic rocks from S.W. Norway'. Geologica Ultraiectina, Univ. Utrecht, 25, 147 p.
Torske, T. 1977. 'The South Norway Precambrian region: a Proterozoic cordilleran type orogenic segment'. Norsk Geol. Tidsskr., 57, 97-120.
Touret, J. 1969. 'Le socle Précambrien de la Norvege méridionale'. These Etat, Univ. Nancy, 3 vol., 609 p.
Touret, J. 1971. 'Le facies granulite en Norvege méridional, I: les associations minérales, II: les inclusions fluides'. Lithos, 4, 239-249 et 423-436.
Touret, J. 1972. 'Le facies granulite en Norvege et les inclusions fluides: paragneiss et quartzites'. Sci. Terre, Nancy, 17 1-2, 179-193.
Touret, J. 1974. 'Facies granulite et fluides carboniques'. Cent. Soc. Geol. Belgique (vol. P. Michot, Liege), 267-287.
Touret, J. 1977. 'The significance of fluid inclusions in metamorphic rocks'. In Fraser, D. (ed.), Thermodynamics in Geology, D. Reidel Pub., NATO ASI, Ser., 203-227.
Touret, J. 1979. 'Les roches a tourmaline - cordiérite - disthene de Bjordammen (Norvege), sont elles liees a d'anciennes évaporites ?', Sci. Terre, Nancy, 23-2, 95-97.
Touret, J. 1981. 'Fluid inclusions in high grade metamorphic rocks'. In: Hollister, L.S. and Crawford, M.L. (eds), Short Course Mineral. Association Canada, 6, 182-208.
Touret, J. 1982. 'An empirical phase diagram for a part of the N_2 - CO_2 system at low temperature'. Chemical. Geol., 37, 49-58.
Touret, J. 1984. 'Fluid inclusions in rocks from the lower continental crust'. J. Geol. Soc. London, Spec. Pap. (Nature of the lower crust), in press.
Touret, J. and Bottinga, Y. 1979. 'Equation d'état pour le CO_2; applications aux inclusions carboniques'. Bull. Minéral., 102, 577-583.
Touret, J. and Dietvorst, P. 1983. 'Fluid inclusions in high grade anatectic metamorphites'. J. Geol. Soc. London, 140-4, 635-649.

Touret, J. and Olsen S.N. 1985. 'Fluid inclusions in migmatites'. In: Ashworth, J.R. (ed.). Migmatites, Blakie Pub., 265-286.

Verschure, R.H. 1985. 'Geochronological framework for the late-Proterozoic evolution of the Baltic shield in South Scandinavia'. This volume.

Verschure, R.H., Maijer, C., Andriessen, P.A.M., Boelrijk, N.A.M., Hebeda, E.H., Priem, H.N.A. and Verdurmen, E.A.Th. 1983. 'Dating explosive valcanism perforating the Precambrian basement in Southern Norway'. Norges geol. Unders., **380**, 35-49.

Vlaar, N.J. 1985. 'Precambrian geodynamical constraints'. This volume.

GEOCHEMICAL CONSTRAINTS ON THE EVOLUTION OF THE PROTEROZOIC CONTINENTAL CRUST IN SOUTHERN NORWAY (TELEMARK SECTOR)

P.C. Smalley[1] and D. Field[2]

[1] Institute for Energy Technology, P.O.Box 40, 2007 Kjeller, Norway
[2] Geology Dept., The University, Nottingham NG7 2RD, UK

ABSTRACT. The Proterozoic continental crust of the Telemark sector consists of supracrustal gneisses and several later plutonic suites intruded between 1.6 and 1.25 Ga ago. In chronological order, these are: metabasites and tonalites; granites; and charnockite-alkaline complexes. The metabasites and tonalites are calcic to calc-alkaline with MORB-like HFSE (P, Zr, Ti, HREE, Y) abundances, but they are enriched in LILE and deficient in Ta and Nb relative to other incompatible elements. The later plutonic rocks become progressively more alkaline, have higher Fe/Mg ratios, and are highly enriched in LILE <u>and</u> HFSE; Nb-deficiencies are less pronounced, but are accompanied by relative deficiencies in Sr, P and Ti.

The chemical evolution is similar to that observed in modern cordilleran orogens, in the transition from island and primitive continental arcs to mature continental and back arc-anorogenic environments, and is interpreted in terms of variation in sub-crustal source composition.

The earliest rocks were derived from previously depleted mantle which had experienced subduction-related enrichment in LILE via hydrous fluids originating in the subducted slab at the time of, or shortly before, magma generation. The later rocks have an increasing component derived from a within-plate mantle type, which had been modified by the introduction of non-subduction-related CO_2- rich LILE- and HFSE- bearing fluids or partial melts.

These geochemical signatures developed during accretion of the Proterozoic south Norwegian crust above a W-migrating, E-dipping subduction zone.

1. INTRODUCTION

Much progress has been made in the use of the chemical characteristics of recent volcanic rocks to fingerprint different tectonic environments, magma generation processes, and mantle reservoirs (1-6). However, <u>plutonic</u> rocks are volumetrically more important in crust formation (7), and they predominate in deeply eroded Proterozoic crustal segments.

Despite the intrinsically more complex nature of plutonic systems
(for example, samples may not represent liquid compositions), recent
work (8-10) has shown that chemical relationships do exist with contemporary overlying volcanic suites, and that their chemistries vary
coherently between different tectonic settings. On the basis of major
and trace element interrelationships, Brown et al. (9) distinguished
granitoids characteristic of four stages of magmatic arc maturity : (a)
island and primitive continental, (b) normal continental, (c) mature
continental and (d) back-arc or anorogenic. Chemical variations were
attributed partly to crustal contamination, but mainly to sub-crustal
magma sources, representing mixtures between depleted mantle, a
subduction-zone component, and within-plate mantle (4-6). Within a
single orogen (e.g. the Andes) there can be continuous variation from
intrusions of types (a) to (d) both laterally, with increasing distance
from the subduction zone, and vertically, i.e. with time, in a
particular area.

The present study investigates temporal chemical variation within
intrusive suites from a ca 100 km^2 area of the Proterozoic crust of
southern Norway. The objectives are : (a) to characterize the magma
sources for these rocks, (b) to determine the tectonic environment of
Proterozoic crustal growth in S. Norway and (c) to investigate whether
the same magma generation processes which produced chemical fingerprints
in Phanerozoic rocks were also operative in Proterozoic times.

2. SUMMARY GEOLOGICAL HISTORY

The study area is situated in central southern Norway, in the
Telemark sector of the Baltic Shield (Figure 1). This region consists of
high-grade matamorphic rocks which underwent a major gneiss-forming
event ca 1.55 Ga ago (11-15) . The predominant metamorphic grade is
amphibolite facies (orthopyroxene absent in metabasites; sillimanite
absent, muscovite present in metasediments), whereas the neighbouring
Bamble sector (Figure 1) is generally of higher grade, reflecting a net
downthrow to the NW on the Porsgrunn-Kristiansand shear zone (PKSZ,
Figure 1).

Previous work in this part of the Telemark sector (16) has tried to
explain the geology in terms of extensive supracrustal formations, said
to be correlatable with the oldest parts of the low-grade "Telemark
Supracrustals" further north (Figure 1). However, more recent fieldwork
in the Telemark sector (17) has revealed the presence of several important
intrusive suites, some of which occur within the so-called supracrustal "formations". Our detailed sequence of events can be summarized
as follows: (i) Deposition of supracrustals, mainly sediments (sandstone, arkose, greywacke) ± felsic volcanics. (ii) Intrusion of basic
dykes (metabasites) and grey tonalitic gneisses. Metabasites both
intrude and are intruded by tonalites; the intrusion of both suites
evidently overlapped in time. The metabasites are more common in Bamble
where they form a large swarm near Arendal, (18, 19). (iii) High grade
metamorphism and formation of gneissic-migmatitic structure, affecting
the supracrustals and earlier members of the tonalite and

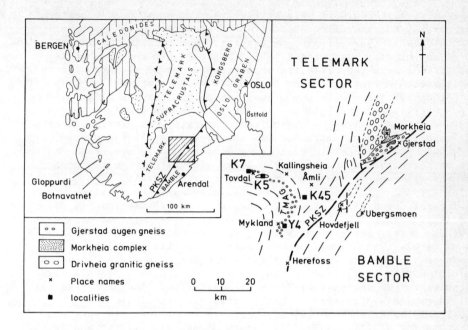

Figure 1. Situation of sampling localities and lithological units mentioned in the text. Lined ornament = undifferentiated gneisses, including supracrustal, tonalitic, and granitic lithologies. The elliptical body around locality K5 is a tonalitic gneiss. PKSZ = Porsgrunn-Kristiansand shear zone, TMAG = Tovdal-Mykland augen gneisses (small circles). Barbed and dashed lines are shear zones.

metabasite suites. (iv) Intrusion of pink granitic plutons, including large volumes of augen gneiss. These rocks are non-migmatitic and locally discordant to the earlier rocks and the regional gneissic foliation (i-iii above), although they may themselves be deformed. (v) Intrusion of the Gjerstad augen gneiss, a clearly discordant body of potassic augen granite-charnockite with rapakivi-type chemistry, and the Morkheia-Mykland alkaline monzonite-syenite complexes.

The relative ages of the above suites are well documented, but later low-grade events have extensively updated Rb-Sr whole-rock radiometric ages in many cases (11, 14, 15). Critical scrutiny of all age data is therefore necessary (20). Events i-v likely took place within the time period 1.6 - 1.25 Ga (11-15).

3. METABASIC SUITE

As in the adjacent Bamble sector, the metabasites in Telemark form subconcordant minor dykes and, more rarely, larger bodies, metamorphosed to amphibolite grade. These metabasites are earlier than, and quite distinct from, the so-called "hyperites" (coronitic (meta-) gabbros).

At Arendal (Figure 1), the metabasites have totally metamorphic textures and mineral assemblages, which petrographic and geochemical investigations have indicated are probably the result of intrusion and crystallization during the major high-grade event (amphibolite-granulite grade), in equilibrium with the prevailing conditions (18, 19).

By contrast, in the Telemark sector (and inland Bamble) the basic intrusions have not always equilibrated texturally and mineralogically to the conditions of metamorphism. In the centres of larger bodies, primary pyroxene and ophitic textures are occasionally preserved. Such bodies may have been either slightly later with respect to the high-grade metamorphism, or emplaced at higher crustal levels.

The Arendal and Telemark metabasites display a tholeiitic Fe-enrichment trend (18). Figure 2 shows the $C/N+K$ $(=CaO/(Na_2O+K_2O))$ - silica relationships for metabasites from both areas. In all of the samples $C > (N+K)$, but extrapolation of the trend yields an alkali-lime index of ca 61, transitional between calcic and calc-alkaline.

Figure 3 shows the field of Arendal metabasite analyses on a primordial mantle-normalized incompatible element plot, together with two representative analyses of Telemark metabasites. The main features exhibited by individual Arendal metabasites (19) are: (i) LILE (large-ion lithophile elements) are variable, but generally enriched relative to the HFSE (high field-strength elements); (ii) K, and usually Rb, are strongly enriched relative to the REE. All samples show low Th abundances relative to the other LILE; (iii) All samples have prominent Ta-troughs; (iv) The HFSE (Nd-Yb) have a convex-upward curved pattern, with slight peaks at Sr and Eu.

These features are characteristic of basaltic rocks formed in relatively primitive island arc environments, the result of partial melting of depleted mantle (MORB-source type) which had subsequently become enriched in LILE by the addition of subduction-related hydrous fluids and/or partial melts derived from subducted oceanic lithosphere (19).

Rb, Ba, K, Sr and, in most cases, Th are all enriched relative to MORB (Figure 3) - a common feature of basalts erupted at destructive plate margins (1, 2, 4-6). By contrast, HFSE values in the Arendal metabasites are similar to or, in the case of Ta, lower than in typical N-MORB, likely because Ta (and the other HFSE) are not so mobile in hydrous fluids, and are retained in the subducting slab or overlying mantle in new anhydrous high pressure minerals (garnet, omphacite), or in minor phases stabilized by the high $P(H_2O)-f(O_2)$ conditions in the subduction zone environment (e.g. sphene, ilmenite, zircon, apatite, rutile). The low Hf and Zr contents of some samples from very close to the Skagerrak coast (19) suggest that zircon may have been a residual phase during subduction-enrichment of their mantle source.

The Arendal metabasites and the broadly similar Telemark samples (Figure 3) thus fit a subduction-related genetic model. The relatively high P, Ti, Y and Zr contents of sample M7.2 (Figure 3) are possibly a result of a smaller degree of partial melting or more pronounced igneous fractionation. More importantly, both Telemark samples (especially

GEOCHEMICAL CONSTRAINTS ON THE PROTEROZOIC EVOLUTION OF S. NORWAY

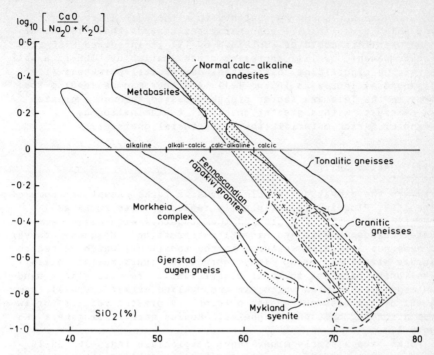

Figure 2. Log (C/(N+K))-silica diagram for south Norwegian plutonic suites. Trends for andesites and rapakivi granites from refs. 8 and 9. Alkali-lime index is silica % at which log (C/(N+K))=0.

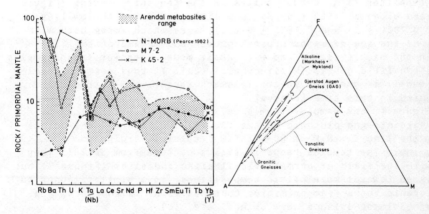

Figure 3 (left). Primordial mantle (2) normalized incompatible elements in Arendal metabasites (19), N-MORB (4), and metabasites from Morkheia (M7.2) and Kallingsheia (K45.2).
Figure 4 (right). $K_2O+Na_2O-FeO^*-MgO$ relationships in south Norwegian plutonic suites. T = tholeiitic field, C = calc-alkaline field.

K45.2) have higher K, Ba and Rb contents than the coastal rocks, suggesting an increase in LILE concentrations towards the interior of southern Norway. This could be explained by (i) an increased LILE-rich subduction component, or (ii) increased contamination by LILE-rich wall rocks. It may be significant that present day crustal thickness (i.e. level of exposure) increases inland (21), and thus magmas feeding the basic dykes in the Telemark sector probably passed through a greater thickness of crust, with a greater chance of contamination by pre-existing LILE-rich material (the supracrustal gneisses).

4. TONALITIC GNEISS SUITE

These are the oldest observable intrusive rocks except for some of the metabasites. Ploquin (16) originally regarded these rocks as volcaniclastic sediments, and grouped them together with his supracrustal "Banded Gneiss and Migmatites Formation". We have, however, observed numerous intrusive features in the tonalites, which we regard as a discrete structural group. They intrude the supracrustal gneiss ($\pm$ basic dyke) units parallel to their planar fabric, resulting in a banded injection migmatite complex which, in the Mykland area (Figure 1), now forms ca 40% of the basement gneiss outcrop. We present results for two small, but discrete, homogeneous gneissic bodies near Kallingsheia (K5, K45, Figure 1).

The rocks have a simple mineralogy; plagioclase (An25-30, 50-70 modal %), K-feldspar (5-10%), quartz (20-25%), hornblende, biotite, and accessory Fe-Ti oxide, sphene and apatite. Both bodies have suffered the major high-grade metamorphic event (event iii, see section 2).

Although updating of Rb-Sr whole-rock systems has occurred in the Telemark tonalites (15), K-poor suites in the Kongsberg sector (Figure 1) have been dated at 1580 $\pm$ 50 Ma, and 1520 $\pm$ 50 Ma, with low I(Sr) values of 0.7024 $\pm$ 14 and 0.7025 $\pm$ 6 (22), ages which agree both with estimates of the age of major crustal growth and metamorphism (1.53 - 1.6 Ga) in the Bamble sector and with Rb-Sr model ages for fresh samples of the K5 tonalitic gneiss (Kallingsheia) (23). It thus seems probable that the tonalitic gneisses in the Telemark sector were intruded and metamorphosed in this time interval.

In Figure 4, the samples form a calk-alkaline trend typical of orogenic volcanic and plutonic suites (8, 10). On Figure 2 the points intersect the line C = N+K at 60-65% SiO_2 (calcic, verging on calc-alkaline). The samples range in silica content from 58-70%. They overlap with the field of normal calc-alkaline andesites (Figure 2) but are displaced towards more calcic compositions, in a similar manner to calcic plutonic rocks from primitive continental or island arcs, e.g. New Britain-Solomon Islands, New Guinea, etc. (8).

Figure 5 includes the incompatible element data field for ten samples from the K5 tonalite and an individual analysis from the smaller K45 body. The patterns slope smoothly upwards towards the more incompatible elements (i.e. to the left) with a prominent trough at Nb, and minor troughs at P and Ti. The K5 tonalite has the steepest patterns, similar to the K45 sample with respect to the LILE (Rb to Sr), but with

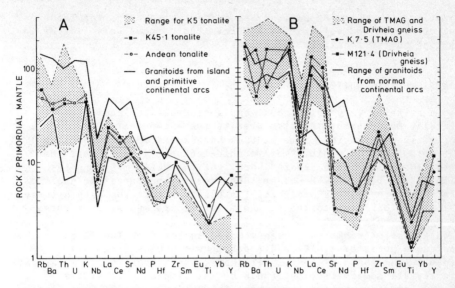

Figure 5. Comparison of incompatible element patterns for: A- tonalites from Kallingsheia (K5, K45.1) and the Andes (24), and granitoids from island and primitive continental arcs (9); B- south Norwegian granitic gneisses (locations on Figure 1) and granitoids from normal continental arcs (9).

more pronounced depletion of the less incompatible elements P to Y. The K45 pattern resembles those of typical Andean tonalites (Figure 5A), but the K5 tonalite with its low P, Zr, Ti and especially Y contents is more reminiscent of typical heavy REE-depleted Archaean tonalites (7), a similarity that is discussed later. Mostly, the Telemark tonalites lie within, or very close to, the field for Mesozoic and Cenozoic primitive continental and island arc granitoids (Figure 5A)(9). LILE and light REE enrichment is not so pronounced, and Nb-depletion more evident than in rocks from more evolved continental arcs (c.f. Figure 6).

The tonalite trace-element patterns, particularly for the K45 sample, are similar to those of the Arendal metabasites (Figure 3), and can be explained in a similar manner. All of the more incompatible elements (Rb-Sr on Figure 5) are enriched relative to MORB except Nb, which has similar or only slightly higher values. Contamination of a depleted mantle source region by hydrous fluids derived from subducting oceanic crust can explain the enrichment of the elements mobile in aqueous fluids, Rb, Ba, Th, K, Sr and possibly light REE, while retention of Nb in stable minor phases in the residue is reflected by the relative depletion of this element. The higher LILE contents of the tonalites relative to the metabasites reflects a larger contribution from subducted oceanic material. The possibility of modification due to contamination with pre-existing continental crust is unlikely, as the only possible contaminants at this early stage in the development of the continental crust were the basic dykes and supracrustals, the latter probably representing volcaniclastic or volcanic rocks very similar in composition to the tonalites themselves.

5. GRANITIC GNEISS SUITE

This suite is very voluminous, and includes large bodies of structurally homogeneous foliated granite and augen gneiss of monzogranitic composition. Some plutons have suffered the major (ca 1.55 Ga) metamorphism, and exhibit metamorphic-tectonic fabrics, but the majority, including those studied here, cut the early metamorphic-migmatitic structures in the older supracrustal and tonalitic gneisses, and are therefore post-metamorphic. The planar structures within these bodies, always parallel to the pluton margins, are magmatic features formed during emplacement.

Two groups of rocks have been studied in detail : The Tovdal-Mykland augen gneisses (TMAG) and the Drivheia gneiss (Figure 1). The TMAG form a long chain of individual sheet-like intrusions. They are not necessarily all comagmatic, but those bodies sampled have similar mineralogy, petrography and chemistry. The Drivheia gneiss is a single large augen gneiss body, elliptical in shape, ca 5 km wide and at least 30 km long.

Although mostly later than the tonalitic plutons, some of the granitic rocks were present at the time of metamorphism, ca 1.55 Ga ago. Similar K-rich granitic and charnockitic gneisses from Kongsberg and Bamble have yielded ages of ca 1.55 Ga (10, 22). Both the Drivheia gneiss and the TMAG from an outcrop in Mykland yielded young, reset Rb-Sr ages, whereas TMAG samples from Tovdal (Figure 1) yielded an isochron age of 1290 $\pm$ 50 Ma, but only after three scattered points had been deleted (17). The presently available evidence is equivocal, but the granitic gneisses may relate to intrusive episodes from ca 1.55 Ga to as late as ca 1.3 Ga.

The mineralogies are rather constant : quartz (20-30 modal %), K-feldspar (25-40%), plagioclase An8-15 (24-38 %), hornblende and biotite, with accessory allanite, apatite, zircon, opaques $\pm$ sphene. The K-feldspar augen megacrysts originated as igneous orthoclase, but are now largely inverted to microcline.

The granitic gneisses have markedly different major-element chemistry to the earlier tonalites. On Figure 2 these two suites plot in discrete fields : the granitic rocks are mostly richer in silica, and at equivalent silica contents are displaced towards lower C/(N+K) values. Consequently the granitic suite extrapolates to a calc-alkaline alkali-lime index, and plots almost entirely within the field of normal calc-alkaline andesites (Figure 2). The kink in the trend at the lowest C/(N+K) values towards lower silica contents is reflecting the presence of "cumulus" K-feldspar in the alkali-rich samples.

The granitic rocks have higher Fe/Mg ratios than the tonalites, and in Figure 4 they plot along the boundary between calc-alkaline and tholeiitic fractionation trends, with slightly higher Fe/Mg ratios than typical Phanerozoic calc-alkaline batholithic rocks (8-10).

Figure 5B reveals highly distinctive incompatible element patterns. Rb, Ba, Th, K, La, Ce, Zr and Y are all enriched relative to the tonalites (Figure 5A), but the granitic rocks retain a clear Nb-trough and also show pronounced Sr, P and Ti troughs, these elements frequently showing sub-MORB concentrations.

Probable effects of mineral fractionation during crystallization include exaggerated deficiencies in Sr, Ba (in sample M121.4 (Figure 5), Ti, Nb and P (plagioclase-hornblende- apatite $\pm$ K-feldspar $\pm$ Fe-Ti oxide). The effects of possible crystal fractionation are, however, minimized during comparison with the geochemical fields for different plutonic arcs compiled by Brown et al. (9), as these relate to samples within the same range of silica contents (65-75%).

The incompatible element patterns are distinct from those of island and primitive continental arc plutonics (c.f. Figure 5A), the high Rb, Ba, Th, K, La, and Ce contents closer to those found in normal continental arcs (Figure 5B). Nb contents are higher than in primitive arc rocks, but the retention of distinct Nb and Ti troughs is also a feature of normal continental arcs (Figure 5B). The strong depletion of Sr $\pm$ Ba and P is a feature seen in more mature plutonic arcs (c.f. Figure 6), but because these elements are strongly influenced by fractionation they cannot be regarded as diagnostic. Overall the suite is most similar to plutonic rocks intruded in "normal" Phanerozoic continental arcs, i.e. the main calc-alkaline batholith-forming suites of cordilleran continental margins. Minor differences can be attributed to magmatic fractionation. The higher Fe/Mg ratios of the Telemark granitic suite compared to most Phanerozoic calc-alkaline suites probably relates to a lower oxygen fugacity during fractionation, possibly emphasized by the deeper level of exposure in southern Norway.

The LILE-enrichment and _relative_ Nb deficiency can be explained by the presence of a subduction-related component in the magma source region, whereas the _absolute_ increase in Nb, Zr and Y abundances compared to the tonalites can be accounted for by a component from a within-plate mantle type (4-6, 9), which probably consists of a previously heterogeneous (variably depleted) mantle which was subsequently metasomatized by the introduction of CO_2-rich fluids, partial melts, or both (4, 25, 26), which were not directly related to the subduction process.

I(Sr) values for the granitic gneiss suites of Telemark, Kongsberg and Bamble fall in the range 0.7014-0.7045 (11, 17, 22). These low values preclude the involvement of significantly older continental crust _or_ enriched mantle with high Rb/Sr ratios. If subduction-related enrichment of the mantle did involve significant increases in Rb/Sr, this could not have occurred much before the time of magma generation.

As pre-existing rocks were either little older, or had very low Rb/Sr ratios (metabasites, tonalites), the presence of a crustal component cannot be formally precluded from the Sr isotopic evidence. However, it is unlikely that such large volumes of HFSE-rich granite could have been formed by partial melting of the supracrustal-metabasite-tonalite suites, which are poor in HFSE.

6. CHARNOCKITIC-ALKALINE COMPLEXES

This group was emplaced at a relatively late stage in the intrusive history. Volumetrically minor in extent, the distinctive lithologies fall into two categories : alkaline monzonite-syenite and rapakivi-type charnockite-granite. The former group is represented by the Morkheia and Mykland complexes, the latter by the Gjerstad augen gneiss (Figure 1) (14, 16, 17, 27).

The Morkheia complex outcrops at the SE edge of the Telemark sector. It forms a star-shaped intrusion of 15 km^2 consisting mainly of clinopyroxene ($\pm$ fayalite) bearing monzonite-quartz monzonite. It discordantly intrudes all gneiss-migmatite structures in the country rocks, and also minor amounts of gabbro and gabbroic anorthosite which may represent early cogenetic cumulates. Igneous layering in parts of the intrusion and a wide range of chemical compositions indicate that the monzonite magma fractionated extensively in situ (c.f. 27). The Mykland complex is much smaller, and consists of fayalite bearing syenite-quartz syenite.

The detailed geology and geochemistry of both complexes (17) will be presented by us elsewhere. Rb-Sr isotopic data suggest that the Morkheia complex was emplaced ca 1.25 Ga ago.

The Gjerstad augen gneiss (Figure 1), which was emplaced 1237 $\pm$ 53 Ma ago (14) intrudes the Morkheia monzonite, and is therefore younger. The augen gneiss is a large monzonitic to monzogranitic pluton which owes its structure to strong mylonitic deformation along the PKSZ (Figure 1) shortly after its emplacement. The "augen" consist of igneous orthoclase, or microcline, the product of inversion during deformation. In some places the augen gneiss is charnockitic, containing ortho- and clinopyroxenes as well as hornblende as primary igneous phases.

The monzonite-syenite complexes may correlate with the Gloppurdi and Botnavatnet layered syenites in SW Norway (Figure 1) (28), but no other similar suites have been described from the Telemark sector. The Gjerstad augen gneiss also appears to be unique in Telemark, although possibly equivalent charnockitic augen gneisses occur at Hovdefjell and Ubergsmoen in the Bamble sector, 30 km SW of Gjerstad (Figure 1).

Figure 4 illustrates that the monzonitic-syenitic rocks have very high Fe/Mg ratios. MgO falls below 0.01% in almost 1/3 of the analysed samples. This is partly a result of the in situ fractionation, but as even the least evolved monzonites have high Fe/Mg ratios (FeO /(FeO + MgO) > 0.91) it is evident that this must also have been a feature of the parental magma. The Gjerstad augen gneiss has slightly lower Fe/Mg ratios, but the values are still higher than for the earlier augen gneisses and tonalites, and the samples lie totally within the tholeiitic field (Figure 4).

The C/(N+K) silica-variations are highly distinctive (Figure 2). The basic to acid compositions of the Morkheia complex define a linear trend with negative slope, parallel to those obtained from the older granitic and tonalitic rocks but displaced towards more alkali-rich compositions at equivalent silica content. The alkali- lime index of 48 for the Morkheia complex is in the alkaline range. Extreme fractions are also peralkaline (17). The "flattening off" at the high-silica end of

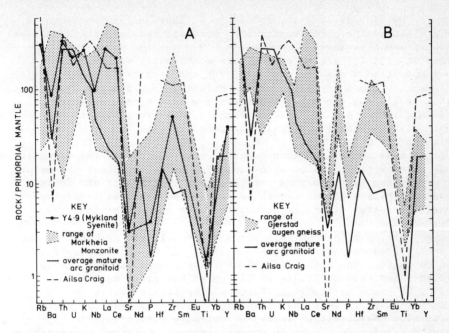

Figure 6. Comparison of incompatible element patterns for: A, the Morkheia and Mykland alkaline suites, B, the Gjerstad augen gneiss; with average mature arc granitoid and the Ailsa Craig alkaline anorogenic granitoid (9).

the Morkheia trend (Figure 2), shown to a greater extent by the Mykland syenite samples, could be a result of alkali-feldspar fractionation, but contamination by silica-rich country rocks is also a possible contributary factor.

The Gjerstad augen gneiss contains some alkali-rich samples, which overlap with the Morkheia and Mykland range (Figure 2). However, they do not form a well-defined trend, but a spread between the calc-alkaline granitic suite and the alkaline Morkheia suite, at 60-71% SiO_2. By extrapolation, the Gjerstad augen gneiss approximates to an alkali-calcic alkali-lime index. Alkali-calcic granitoids can occur in two environments: subduction-related in mature continental arcs, and tension-related in stabilized continental crust (8). The Proterozoic Fennoscandian rapakivi granites, of the latter type (8, 29), define a trend passing centrally through the Gjerstad augen gneiss field on Figure 2.

The mantle-normalized incompatible element ranges for the Morkheia monzonites and Gjerstad augen gneiss (Figure 6), are similar in shape, and the augen gneiss analyses lie almost entirely within the Morkheia range. The major features are : strong enrichment of Rb, Ba, Th, K, REE, Zr, Y; small or insignificant Nb troughs; very large Sr, P, Ti troughs. In situ fractional crystallization has caused decreases in Sr, P, Ti ± Ba (fractionation of plagioclase ± K-feldspar, apatite, mafics) and increases in REE, Zr, and K in the residual liquids.

For the Mykland syenite sample (Y4.9, - Figure 6A) the pattern is similar to that of the more fractionated Morkheia samples, but with even higher Rb, K and Y contents.

Brown et al's (9) estimate of mature continental arc average granite composition shows the high LILE, Nb and Y contents combined with Sr, P, Ti and Ba troughs typical of the Morkheia, Mykland and Gjerstad rocks (Figure 6), but the Norwegian rocks are more highly enriched in light and middle REE and Zr. In this respect they are closer to anorogenic alkaline rocks, typically enriched in REE, Zr, Y, Nb, but with relative depletion of Ba, Sr, P, Ti (9, 30, c.f. Ailsa Craig, Figure 6). A tension-related or anorogenic environment for the Telemark alkaline-charnockitic rocks is also supported by petrological and major element affinities with the alkaline complexes of Labrador and South Greenland (Morkheia-Mykland) and to rapakivi-type granites (Gjerstad augen gneiss), suites generally associated with such conditions (29, 30).

Accordingly, the incompatible elements show little evidence of the subduction-related component that was evident in the earlier tonalite and granite suites. The high LILE and HFSE abundances with little or no Nb deficiency indicate an origin involving within-plate mantle, enriched by CO_2-rich metasomatic fluids and/or partial melts. A significant feature is that the Morkheia, Mykland and Gjerstad suites are all partly "charnockitic" (anhydrous primary mineralogies) reflecting H_2O-deficient CO_2-rich parental magmas. Chemical modification of the mantle by CO_2-rich media and magma generation could thus have occurred contemporaneously as part of a single process (26).

Furthermore, the low I(Sr) values involved (0.703-0.705; 14, 23) indicate that significant enrichment of Rb relative to Sr in the mantle source could not have occurred much before the time of magma generation ca 1.25 Ga ago, as well as precluding significant contamination of the parental magmas by older crustal rocks with high Rb/Sr ratios.

7. MAGMA SOURCES AND TECTONIC ENVIRONMENT

The rocks have revealed regular chemical changes with time : increase in alkalinity, Fe/Mg ratio (Figures 2 and 4), and Rb,Ba,Th,K, Nb,La,Ce,Zr and Y concentrations ; greater deficiencies in Sr,P,Ti ± Ba and less prominent Nb deficiencies with respect to the other incompatible elements. There is generally an increase in incompatible element abundance levels and HFSE/LILE ratios. Differences in exposure level and degree of fractionation alone cannot explain these chemical changes. Although fractionation paths did change with time (e.g. Figure 4), studies of within-pluton fractionation (17) showed that the detailed fractionation trends were not always in the same sense as the temporal trends. Similarly, differing degrees of crustal contamination could not have generated the observed temporal changes without significantly affecting the I(Sr) ratios, or producing lower HFSE abundances and HFSE/LILE ratios than those observed.

The evidence presented shows that the more likely possibility is

that the differing chemistries are reflecting variations in parental
magma compositions, inherited from sub-crustal sources. The observed
differences cannot be explained solely by differing degrees of partial
melting of homogeneous mantle, as smaller degrees of partial melting,
giving rise to highest incompatible element abundances, would also yield
the lowest HFSE/LILE ratios, contrary to the observed trends. We deduce
that the different suites were derived from heterogeneous mantle
compositions.

Similar temporal chemical trends in Phanerozoic orogens (see
following section) have been interpreted as the result of mixing between
components with different source compositional end members:
heterogeneous depleted mantle, a hydrous subduction-related component
derived from the subducting slab, and within-plate mantle (6, 9). The
various magma compositions essentially reflect differing conditions
during magma generation and/or metasomatism of the upper mantle magma
source by fluids or partial melts of deeper origin; high $P(H_2O)$ and
$f(O_2)$ above subduction zones, CO_2-rich in within-plate environments.

The chemistries of the early basic rocks in Telemark, and Bamble
(19), suggest that subduction of oceanic crust and subduction-enrichment
of the overlying mantle were active processes ca 1.6 Ga ago. Previous
work has implied that such processes had already started by the early
Proterozoic (2), and thus may have been important in magma generation
and continental growth throughout much of geological time.

The tonalitic gneiss suite also fits an evolutionary model
involving subduction-enrichment of the mantle source, but the high LILE
contents indicate a smaller degree of partial melting and/or a greater
subduction-related component. The prominent Nb deficiency (Figure 5A)
reflects the stability of a Nb-bearing minor phase in the source of
either the magma or subduction-related metasomatic fluids. The
deficiency of Y in the K5 tonalite could be reflecting the presence of
residual garnet in the magma source, indicating a component of partial
melting of quartz-eclogitic subducting oceanic crust. Although such a
model might hold for magma generation under high geothermal gradients in
Archaean times, explaining the characteristic heavy REE-Y deficiency of
Archaean tonalites (7), thermal modelling of recent subduction zones
indicates that partial melting of the subducting slab would not be
expected under present-day geothermal gradients (31). Furthermore, some
of the Telemark tonalites (e.g. K45) have "normal" Y contents, similar
to those of present-day tonalites. A likely explanation is that some of
the metasomatic fluids derived from dehydration of subducted oceanic
crust were generated in equilibrium with garnet. This avoids the
necessity for a melt component from the slab to explain the depletion of
heavy REE and Y in some modern calc-alkaline volcanics (3).
Consequently, the low Y contents of the K5 tonalite can be explained
without having to invoke a higher geothermal gradient for the
Proterozoic.

A within-plate mantle reservoir existed under southern Norway ca
1.25 Ga ago, the source for alkaline magmatism. The granitic gneiss
suite, intruded in the time interval between the tonalites and alkaline
rocks, also have intermediate chemistry, with moderately high LILE and
HFSE abundances, but a prominent Nb-deficiency. This probably relates to

a mixture of components from both subduction-related and within-plate sources. The temporal chemical changes in magma chemistry thus relate to a gradual change from subduction-related to within-plate source materials, and reflects the decreasing influence of (i.e. the greater distance from) an active subduction zone, and thickening of the continental crust.

We conclude that the early basic rocks and tonalites are the equivalents of subduction-related intrusions from recent island and primitive continental magmatic arcs. The calc-alkaline granitic gneiss suite is more characteristic of normal continental arcs, including the voluminous calc-alkaline batholith-forming suites of cordilleran continental margins. The later charnockitic-alkaline magmatism is typical of a back arc-anorogenic environment.

8. TECTONIC SCENARIO

Our comparison of the intrusive rocks from the Proterozoic Telemark sector with Phanerozoic rocks has revealed similarities that can be interpreted in terms of tectonic setting.

Although detailed major and trace element data are sparse for the rest of the Proterozoic terrain of southern Norway-SW Sweden, it is clear that the lithological sequence (supracrustals, metabasites, tonalites, granites, charnockite-alkaline complexes) is broadly similar throughout this area (c.f. 22, 32, 33). One important difference, however, is the timing of each intrusive period across the orogenic belt. Whereas the oldest gneisses in Telemark, Kongsberg and Bamble are ca 1.55 Ga old, those further E in Østfold and SW Sweden (Figure 1) are 1.65-1.7 Ga old (15, 32, 33). The Svecofennian terrain in SE Sweden is yet older (1.7-1.9 Ga).

Ca 1.7 Ga ago calc-alkaline rocks were thus being intruded in SW Sweden, in a primitive-normal cordilleran magmatic arc on the SW margin of the Svecofennian continent. In Telemark there was as yet no continental crust - only clastic sediments derived from the growing continent to the E and deposited upon oceanic crust. By 1.55 Ga ago the axis of primitive-normal arc magmatism had migrated westwards to the Telemark-Kongsberg-Bamble area, whereas in SW Sweden potassic normal-mature arc granites were being emplaced. Continued westward migration of the subduction zone led to its decreasing influence on the chemistry of the magmas generated in Telemark. By 1.25 Ga ago there was little or no subduction-related igneous activity in southern Norway. The Proterozoic crust of SW Sweden and S Norway thus developed during the period 1.7 - 1.25 Ga during the westward migration of an east-dipping subduction zone at a convergent ocean-continent plate margin - probably a continuation of the lateral migration of the destructive plate margin at which the earlier (1.7-1.9 Ga) Svecofennian crust to the east was accreted (34, c.f. 35).

The succession of rock types, spatial and temporal geochemical evolution and scale of time involved in the development of southern Norway are all analogous to the evolution of modern continental cordillera, such as the Andes (9). Together with the geochemical evidence for the existence of similar mantle reservoir compositions underlying southern Norway in the mid-Proterozoic to those underlying the present day South American cordillera, and for similar mantle modification events, it can be concluded that the processes controlling magma source composition, magma generation and crustal accretion at cordilleran continental margins were essentially the same 1.5 Ga ago as they are today.

9. ACKNOWLEDGEMENTS

The analytical work was done at the University of Nottingham, while PCS was in receipt of an NERC Research Studentship. The participation of PCS in the NATO ASI was sponsored by the Institute for Energy Technology. J. Wilkinson and D. Jones provided skilled technical assistance, and T. Jacobsen typed the manuscript. G.C. Brown kindly provided a preprint of ref. 9.

REFERENCES

1 Wood, D.A., Joron, J.-L and Treuil, M. 1979, Earth Planet. Sci. Lett. **45**, pp 326-336.
2 Wood, D.A., Tarney, J. and Weaver, B.L. 1981, Tectonophysics **74**, pp 91-112.
3 Saunders, A.D., Tarney, J. and Weaver, S.D. 1980, Earth Planet. Sci. Lett. **46**, pp 344-360.
4 Pearce, J.A. 1982, in "Andesites" (R.S. Thorpe ed.), J.Wiley & Sons, pp 528-548.
5 Pearce, J.A. 1983, in "Continental Basalts and Mantle Xenoliths" (C.J. Hawkesworth & M.J. Norry eds.), Shiva, pp 230-249.
6 Thorpe, R.S., Francis, P.W. and O'Callaghan, L. 1984, Phil. Trans. R. Soc. Lond. **A310**, pp 675-692.
7 Weaver, B.L. and Tarney, J. 1982, in "Andesites" (R.S. Thorpe ed.), J. Wiley & Sons, pp 639-661.
8 Brown G.C. 1982, in "Andesites" (R.S. Thorpe ed.), Wiley & Sons, pp 437-461.
9 Brown, G.C., Thorpe, R.S. and Webb, P.G. 1984, J. geol.Soc. Lond. **141**, pp 413-426.
10 Thorpe, R.S. and Francis, P.W. 1979, in "Origin of granitic batholiths : geochemical evidence" (M.P. Atherton and J. Tarney eds.), Shiva, Kent, pp 65-75.
11 Field, D. and Råheim, A. 1979, Earth Planet. Sci. Lett. **45**, pp 32-44.
12 Field, D. and Råheim, A. 1981, Precambrian Res. **14**, pp 261-275.
13 Field, D. and Råheim, A. 1983, Precambrian Res. **22**, pp 157-161.
14 Smalley, P.C., Field, D. and Råheim, A. 1983, Isot. Geosci. **1**, pp 269-282.
15 Smalley, P.C., Field, D. and Råheim, A. 1984, Terra Cognita Special Issue, (ECOG VIII), pp 9 (abstr. 84).
16 Ploquin, A. 1980, Sci. de la Terre, Mem. **30**, 389pp.

17 Smalley, P.C. 1983, unpublished Ph.D. thesis, Nottingham Univ., 435 pp.
18 Clough, P.W.L. and Field, D. 1980, Contrib. Mineral. Petrol. 73, pp 277-286.
19 Smalley, P.C., Field, D., Lamb, R.C. and Clough, P.W.L. 1983, Earth Planet. Sci. Lett. 63, pp 446-458.
20 Smalley, P.C., Field, D. and Råheim, A. 1984, Terra Cognita Special Issue, (ECOG VIII), pp 9 (abstr. 85).
21 Calcagnile, G. 1982, Tectonophysics 90, pp 19-35.
22 Jacobsen, S.B. and Heier, K.S. 1978, Lithos 11, pp 257-276.
23 Smalley, P.C., Field, D. and Råheim, A. unpublished data.
24 Lopez-Escobar, L., Frey, F.A. and Oyarzùn, J. 1979, Contrib. Mineral. Petrol. 70, pp 437-450.
25 Wood, D.A. 1979, Geology 7, pp 499-503.
26 Menzies, M.A. and Wass, S.Y. 1983, Earth Planet. Sci. Lett. 65, pp 287-302.
27 Milne, K.P. and Starmer, I.C. 1982, Contrib. Mineral. Petrol. 79, pp 381-393.
28 Hermans, G.A.E.M., Tobi, A.C., Poorter, R.P.E. and Maijer, C. 1975, Nor. geol. Unders. 318, pp 51-74.
29 Emslie, R.F., 1978, Precambrian Res. 7, pp 61-98.
30 Sørensen, H. 1974, "The Alkaline Rocks", Wiley & Sons, 622 pp.
31 Anderson, R.N., De Long, S.E. and Schwartz, W.M. 1978, J. Geol. 86, pp 731-739.
32 Hansen, B.T. and Persson, P.O. 1982, Terra Cognita 2, pp 57.
33 Skjernaa, L. and Pedersen, S. 1982, Precambrian Res. 17, pp 215- 243.
34 Hietanen, A. 1975, J. Res. US Geol. Surv. 3, pp 631-645.
35 Berthelsen, A. 1980, 26th Int. Geol. Congr., Paris, Colloq. 108, pp 5-21.

GEOCHEMICAL EVOLUTION OF THE 1.6 - 1.5 Ga-OLD AMPHIBOLITE-GRANULITE FACIES TERRAIN, BAMBLE SECTOR, NORWAY: DISPELLING THE MYTH OF GRENVILLIAN HIGH-GRADE REWORKING

D. Field[1], P.C. Smalley[2], R.C. Lamb[3] and A. Råheim[2]

[1] Department of Geology, University of Nottingham, Nottingham NG7 2RD, UK
[2] Institute for Energy Technology, P.O. Box 40, 2007 Kjeller, Norway
[3] Charterhouse Petroleum plc, London EC1N 6SN, UK

ABSTRACT. The quartzo-feldspathic charnockitic orthogneisses in the coastal regions of the Bamble sector are highly fractionated in K, Rb and other LILE. This fractionation related to the imposition of the granulite facies assemblages, and thus to the single high grade metamorphic event that can be recognised in these rocks. Rb-Sr isotopic studies constrain this event to be pre-Grenvillian. It occurred some 1.54 Ga ago. The age is confirmed by newly presented data. In this part of the Province the Sveconorwegian cycle (1.2 - 0.9 Ga) is represented only by an intrusive event(s). Associated low-temperature hydrous fluids were introduced, which caused low-grade mineralogical alterations in the pre-existing gneisses and which disturbed, and sometimes reset, the Rb-Sr isotopic systems. There was no regional Grenvillian high-grade reworking in the Bamble sector.

1. INTRODUCTION

The unifying feature of the high-grade gneiss regions of the "Sveconorwegian Province" is that they have all yielded K-Ar mineral ages in the range 1.2 - 0.9 Ga (1-3). It was largely on the early assumptions that these data relate to a major orogenic event that the "Sveconorwegian Orogeny" or "Regeneration Period" (4) was born and correlated with the Grenville Orogeny in N. America (5,6).
Apparent confirmation came in 1969 from work in the Bamble sector, when the first detailed K/Ar study dated a thermal maximum at ≈ 1100 Ma (2), and was interpreted by O'Nions et al. (2) as relating to "the main (high-grade) metamorphic episode". When subsequent Rb-Sr data from within Bamble (7) and the related Kongsberg sector (8) seemed to confirm that the ≈ 1.1 Ga event involved resetting of the isotopic systems during high-grade metamorphism and genuine orogenic reworking, the die was cast and the concept of the "Sveconorwegian Orogeny" became generally accepted. Paradoxically it was these studies (7,8) which also provided the first radiometric evidence for an earlier, but poorly constrained (1.7 - 1.5 Ga), high-grade gneiss-forming event.

It was only as a result of later work on the coastal charnockitic rocks of Bamble (Figure 1) that it became clear that the resetting to Grenvillian dates in this part of the Bamble sector was temporally and spatially related to late, undeformed, post-tectonic granite sheets and pegmatite dikes, dated at 1063 ± 20 Ma (10-12). From this and subsequent work (9, 13-15) it became clear that not only the Arendal charnockites, but rocks elsewhere within Bamble, and indeed Telemark, had suffered only low-grade mineralogical alteration effects during the Grenvillian cycle, related to the post-orogenic instrusions and/or shear zones. Here, we summarise some of the key features of these earlier data and present new results which confirm that there was no regional Grenvillian orogenic reworking in Bamble - the orogenesis and associated high-grade metamorphism occurred much earlier, at ≈ 1.54 Ga ago.

2. GEOLOGICAL SETTING

Figure 1 summarises the geology and gives the locations of the outcrops within the Bamble sector for which we now have Rb-Sr isotope data. A significant feature of the terrain is the amphibolite-granulite transition zone along the Skagerrak coast, which can be described by a zonal scheme (Figure 1). Zone A is predominantly upper amphibolite facies (although a large area around locality D9, at Langsjøen, has recently been identified as belonging to the granulite facies (16)), and is separated from Zone B by a well-defined orthopyroxene isograd in metabasites (17-19). In Zones A and B the predominating acid-intermediate host gneisses have a broadly granitic mineralogy, but in Zones C and D they are charnockitic. The gneisses of Zone C contain orthopyroxene-hornblende-biotite-K-feldspar assemblages whilst the co-genetic suite of Zone D, which contains the highest grade rocks of the area, are virtually devoid of K-feldspar, hornblende and biotite - they are nearly anhydrous. Mineralogically, they fall in the range tonalite - trondhjemite (20). The C to D boundary does not represent a well-defined isograd, and the apparently sharp transition in mineralogy of the charnockitic gneisses across this boundary may, in part, result from displacement along a late brittle normal fault which coincides with the C-D line on Figure 1.

In common with many other granulite facies terrains, the rocks have undergone fractionation of the LILE (large-ion-lithophile elements), but there are three special characteristics which distinguish the Arendal rocks from those of other areas (17 - 21):
(a) the deficiencies of the LILE are restricted to Zone D, i.e. the fractionation has occurred between Zones C and D, entirely within the granulite facies and not at the amphibolite-granulite facies boundary,
(b) the deficiencies in K and Rb, in particular, are extreme but Ba, Zr, Cs and Th have also been affected,
(c) the deficiencies are shared by both the charnockitic gneisses and the metabasites.

Table I summarises the chemical data for the charnockitic gneisses

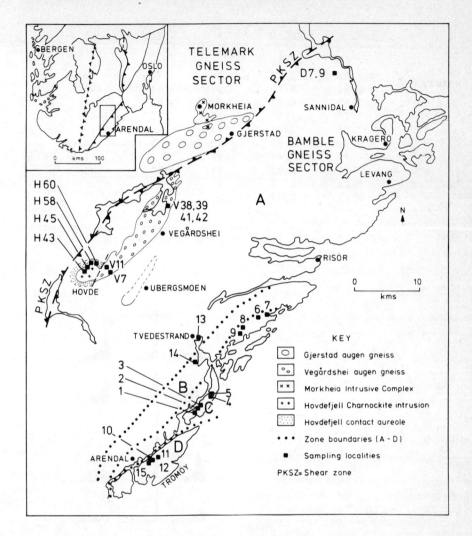

Figure 1. Zonal scheme across the amphibolite-granulite facies transition and sample locations for Rb-Sr geochronology.

from Zones C and D, and Figure 2 demonstrates the completeness of the chemical fractionation in this suite. The possibility that the gneisses in zones C and D represent two unrelated suites, granitic in zone C and tonalitic-trondhjemitic in zone D, as may be suggested by Figure 2, is rejected because of the many chemical and petrographic similarities in both zones. Particularly, the rocks from both zones have high Fe/Mg ratios. The zone D samples are thus quite different from other South

	ZONE C	ZONE D		ZONE C	ZONE D
SiO_2	67.48	68.35			
Al_2O_3	13.82	13.83	Rb	139	<9
TiO_2	0.84	0.53	Sr	163	149
Fe_2O_3	5.65	6.01	Ba	896	206
MgO	1.07	1.90	K/Rb	290	>1300
CaO	2.88	3.64	Ba/Rb	8.58	>105
Na_2O	3.21	4.67	Rb/Sr	1.15	<0.05
K_2O	4.25	0.47	Ba/Sr	6.10	1.6
MnO	0.07	0.09	Zr	478	133
P_2O_5	0.26	0.12			
H_2O+	0.49	0.23			

TABLE I. Mean values for the acid-intermediate charnockitic gneisses from zone C (n = 40) and zone D (n = 79). Data from (21). Major elements as wt% oxide; traces as ppm.

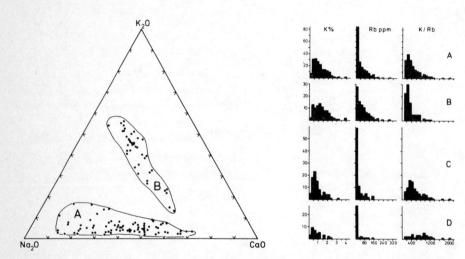

Figure 2. (left) K_2O-Na_2O-CaO diagram for the charnockitic gneisses from Zone D (field A) (20) and Zone C (field B).

Figure 3. (right) Histograms by zone for K, Rb and K/Rb in the metabasites (17).

Norwegian tonalitic gneiss suites (e.g. 22). Figure 3 shows how the metabasites have also been fractionated with respect to K and Rb.

The deficiencies of the LILE in the acid-intermediate gneisses of Zone D were originally interpreted in terms of a metamorphic dehydration mechanism (20), but modelling of the REE and LILE data has shown that the more likely explanation is that the rocks acquired their distinctive mineralogies and chemistries by an essentially primary fractionation, which involved the separation of cumulus (K, Rb-deficient) phases from magma which was emplaced directly under the granulite facies conditions, leaving a residual liquid from which the Zone C ('normal' K, Rb) charnockites crystallised. Studies of the Arendal metabasites (18, 19) have shown that their chemistries too can be largely explained by the effect of direct crystallization from a magma to their present "metamorphic" mineralogies. The single exception is the extreme depletion of Rb in zone D, which probably requires an additional metamorphic depletion. Whichever is the preferred explanation, the key point in the context of the following isotopic studies is that the chemical fractionation in Rb undoubtedly relates to the granulite facies event, whether reflecting primary crystallization of a magma under such conditions or modification during metamorphism, and that subsequently the charnockitic rocks suffered only relatively minor low-grade mineralogical alterations (sericitisation, chloritisation, serpentinisation) without the imposition of any new penetrative fabrics, low-grade or otherwise.

Recent structural models for the Bamble sector (23) have suggested that this area forms a major shear zone, the dominant constant NE-SW foliation being the result of large scale dextral transcurrent movement later than $\approx$ 1150 Ma, with large "augen"-like structures forming relics of an older structural trend. Our geochronological studies in the Arendal area are of rocks which contain the typical NE-SW foliation direction and, if this model is correct, should belong to rocks affected by the structural reworking rather than the older structural relics. Furthermore the high grade mineralogies are in all cases oriented parallel to this foliation, and there is no doubt that the high-grade metamorphism and major gneiss-forming and NE-SW foliation forming events were part of the same "orogeny".

There is thus absolutely no evidence to suggest that the coastal charnockites might represent an isolated massif which has somehow escaped later peripheral reworking: quite the reverse, in fact, for the NE-SW fabric relating to the high-grade event is contiguous from the area of our sampling into other parts of Bamble.

3. Rb-Sr GEOCHRONOLOGY

Analytical techniques are described in reference 10. The key data are summarised in Figure 4. Previously reported results are shown as fields which encompass the data points. Newly reported data are shown as individual points (localities 1, 15 and D9).

The most important results from the earlier investigations of the co-genetic coastal charnockites were that Series 12 (Zone D; K, Rb-deficient) and Series 3 (Zone C; 'normal' K, Rb) produced pre-Grenvillian ages of 1537 $\pm$ 118 Ma (I.R. = 0.70344 $\pm$ 26; MSWD = 12.13) and 1535 $\pm$ 123 Ma (I.R. = 0.70349 $\pm$ 0.00928; MSWD = 1.41), respectively (10). These were the only two outcrops to show no presently exposed evidence of the granite sheet/pegmatite dike intrusions, which elsewhere caused retrogresion of the adjacent charnockitic rocks, although granite sheets have been observed within a few hundred metres of these

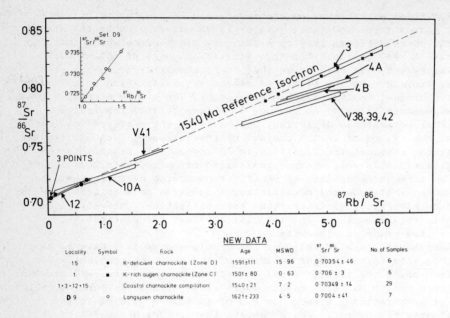

Figure 4. Summary of selected geochronological data for the coastal charnockites, the Vegårdshei gneisses and the Langsjøen charnockite. Analytical techniques - see reference 10. Rb, Sr concentrations: set 1 by XRF, set 15 by isotope dilution. λ Rb = 1.42 x 10^{-11} a^{-1}.

outcrops. The combined data for Series 3 + 12 yielded 1536 ± 26 Ma with I.R. = 0.70345 ± 14 (MSWD = 6.22) : they share a common age, and a common, low, initial ratio.

More recently acquired data at localities 1 and 15 (Figure 1) have confirmed these pre-Grenvillian ages: Locality 1 (Zone C, near locality 3) has yielded 1501 ± 80 Ma with I.R. = 0.70564 ± 556 (MSWD = 0.63) and locality 15 (Zone D, near locality 12) has yielded 1591 ± 112 Ma with I.R. = 0.70354 ± 46 (MSWD=15.96;Figure 3). Neither outcrop shows any obvious evidence of the granite sheets or pegmatite dikes. Thus there are now four outcrops which have yielded pre-Grenvillian ages, all within error of each other. Together, the combined data for localities 1, 3, 12 and 15 (n = 29) give an age of 1540 ± 21 Ma with an I.R. of 0.70349 ± 14 (MSWD = 7.2).

In both zones (C and D), some outcrops directly <u>adjacent</u> to those yielding the ≈ 1.54 Ga age have given quite different results (10-12). At locality 10 (near 12, 15; zone D) there is an undeformed granite sheet in outcrop and, at locality 4 (near 1,3; zone C), a pegmatite dike. In both cases there is retrogression adjacent to the intrusions. Even though samples were collected from outside the obviously affected regions (in which the originally green charnockitic rocks have developed a pink colouration) the rocks yielded best-fit estimates of 1065 ± 191 Ma (Series 10A; MSWD = 193.1) and 1075 ± 122 Ma (Series 4A; MSWD

= 1.10), both within error of the age of the granite sheet (Series 10B; 1063 ± 20 Ma; MSWD = 0.68) (10). The only difference between these samples and those at localities 1, 3, 12 and 15 (≈1.54 Ga) is that secondary, low-grade mineral alterations (e.g. chloritisation of pyroxene, sericitisation of feldspar) have affected the Series 10A and 4A suites, but not the others.

Rocks from the pink retrogressed zone at locality 4 (Series 4B) also define the younger age (1131 ± 135 Ma; MSWD = 0.55) (11,12), as do mineral isochrons for samples from Series 4A, 10A and 12 (10). Thus, although the whole-rocks at locality 12 retain the 1.54 Ga-old imprint, the minerals were reset to within error of the age for the granite sheet (10).

This, together with the high (> 2.5) MSWD values for sample sets 12, 15, and the compilation 1+3+12+15, indicates that even some of the whole-rock systems apparently retaining the ≈ 1540 Ma age did not totally escape the ≈ 1060 Ma event without some disturbance. There is, of course, the possibility that unexposed ≈1060 Ma granite sheets exist close to these outcrops, although their effect was not sufficient to "pinkify" the green gneisses.

The relatively small scale of our sampling (≈ 2 kg samples, each set from a few metres radius area within an outcrop) would, of course, lead to increased sensitivity to secondary events relative to larger samples collected over a wider area. However, our method of sampling reduces the possibility of obtaining a meaningless age intermediate between the primary and secondary events (11). Another advantage with the sampling method we have employed is that each set of samples from an outcrop may be viewed as a single large sample, hopefully representative of the outcrop sampled, i.e. a set of ten 2 kg samples collected over a 2 m radius area may approach to the mean composition not just of a 20 kg sample, but of the entire 2 m radius area. If, therefore, we treat each of sets 1, 3, 12 and 15 as single large samples, using their mean $^{87}Rb/^{86}Sr$ and $^{87}Sr/^{86}Sr$ ratios, a 4-point isochron can be calculated, giving 1534 ± 39 Ma with I.R. = 0.7035 ± 26. The age and I.R. overlap entirely with those given by the individual small samples, although slightly less precise. Important, however, is that the MSWD value is much lower (3.00). This MSWD value was calculated using very conservative estimates for the errors on each data point, equal to the means of the errors in $^{87}Rb/^{86}Sr$ and $^{87}Sr/^{86}Sr$ for the individual samples. The MSWD value of 3.00 is thus a maximum value, and little, if any, geological scatter is implicated. Possible disturbance during the ≈ 1060 Ma event at the scale of the 2 kg samples is thus absent at the scale of the outcrop.

Sample sets that were reset during the ≈ 1060 Ma event (e.g. 10A,4A,4B) now lie below the ≈ 1540 Ma isochron (Figure 4), likely having lost Sr relatively rich in ^{87}Sr (10). Regionally collected samples, however large, including rocks from these outcrops would thus not yield either the ≈ 1540 Ma or ≈ 1060 Ma ages.

There was (9-12) and, in the light of the new data there remains, only one possible interpretation of these results : that (a) the older (1540 ± 21 Ma) age relates to the imposition of the granulite facies assemblages and NE-SW structural trend, and thus to pre-Grenvillian orogenesis and (b) the younger (≈ 1060 Ma) ages were generated by resetting of the Rb-Sr systems during the passage of relatively low temperature fluids associated with the granitic sheets and pegmatite dikes and causing rehydration and (minor) retrogression of the original granulite facies mineralogies.

The two alternative models, (a) that the ≈ 1540 Ma age reflects the high grade metamorphism and the ≈ 1060 Ma age a major structural reworking (23), or (b) that the ≈ 1540 Ma age represents an intrusive event, followed by a ≈ 1060 Ma high grade metamorphism can be rejected on the following grounds.

Model (a) would require that the ≈ 1540 Ma ages came from old structural relics in a structurally reworked terrain, and that there be a correlation between the occurrence of the younger ages (≈ 1060 Ma) and progressive structural overprinting. This is not, in fact, the case. As described in section 2, none of the sample sets come from areas which may represent structural relics which survived the major deformational events. Furthermore, in some cases ≈ 1540 Ma and ≈ 1060 Ma ages have been obtained from gneisses within a single continuous roadcut. In such cases it is evident that the gneissic fabric is continuous throughout, with no structural overprinting of the rocks yielding the younger ages. The only differences are purely mineralogical, the younger ages coming from rehydrated rocks in the vicinity of post-tectonic granite sheets.

Model (b) is best tested by consideration of the Sr evolution diagram for the coastal charnockites (Figure 5). From the common starting point at 1540 ± 21 Ma, the full range of Sr evolution curves for the K, Rb-deficient rocks (localities 12, 15) is defined by samples 12.19 and 12.12 (Figure 5, field A). The full range for the "normal"-K, Rb rocks from Zone C (localities 1,3) is defined by samples 3.11 and 1.11 (Figure 5, field B). By ≈ 1060 Ma ago, the two groups of co-genetic rocks would have evolved quite different $^{87}Sr/^{86}Sr$ ratios, with no overlap (Figure 5). The relative positions of reset suites 10A (in field A) and 4A, 4B (in field B) on this diagram shows conclusively that they had quite different $^{87}Sr/^{86}Sr$ initial ratios at the time of resetting, consistent with a time-integrated fractionation of Rb relative to Sr of several hundred million years, coinciding with the Sr evolution paths for the suites retaining the older age. The chemical fractionation in Rb, related as it is to the only high-grade event - the granulite facies metamorphism - is constrained to have occurred during the 1.54 Ga-old event, and simply could not have occurred during the Sveconorwegian (1.2 - 0.9 Ga) cycle.

Neither can models (a) and (b) explain the 1063 ± 20 Ma age of the granite sheets : if ≈ 1060 Ma were the age of either (a) major structural reworking, or (b) high grade metamorphism, it is difficult to envisage how this could also be the age of completely undeformed and unmetamorphosed discordant granitic sheets.

Further support for the ≈ 1540 Ma age of the high grade metamorphism in the Bamble sector is provided by :

(i) An apparent age of 1509 ± 96 Ma (MSWD = 24.6) on the high-grade garnetiferous - quartz - biotite - plagioclase - sillimanite - K-feldspar - graphite - bearing metasediments at locality 14 (9).

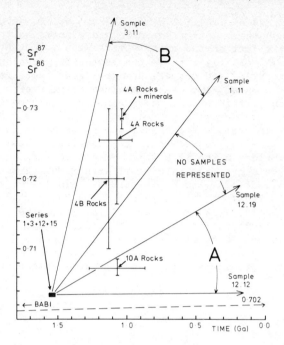

Figure 5. Sr evolution diagram for the coastal charnockites. Field A delimits the Sr evolution paths for Series 12 and 15 (zone D), and field B the paths for Series 1 and 3 (zone C).

(ii) An isochron age of 1397 ± 57 Ma (MSWD 1.12) on an anatectically-derived homophanous granite which represents the last recognisable episode related to the high P-T- event constrains the orogenic structures and metamorphism to be older (9).

(iii) The charnockite samples from Langsjøen (D9) have yielded an old, but poorly defined, apparent age of 1621 ± 233 Ma (I.R. = 0.7004 ± 41; MSWD = 4.49) (Figure 4). Despite the large error term, the age is seemingly constrained to be pre-1388 Ma. This is the first "early" age to be obtained from rocks outside the coastal region. Because these charnockitic rocks are unequivocally discordant towards earlier supracrustal and banded gneisses there is now known to be an even older basement, as yet undated. Apart from minor deformation associated with the Porsgrunn-Kristiansand shear zone, which likely explains the geological scatter of the Rb-Sr data, the charnockite clearly has not undergone any subsequent reworking which might possibly be related to the Sveconorwegian cycle.

(iv) The Ubergsmoen and Vegårdshei augen gneisses (Figure 1) have yielded respectively a U-Pb zircon age of ≈ 1275 Ma (24) and a Rb-Sr whole rock age of 1276 ± 105 Ma (V41, Figure 4), both indistinguishable from the 1237 ± 53 Ma age (15) of the nearby Gjerstad augen gneiss in Telemark (15) (Figure 1). The Hovde charnockite (Figure 1) has yielded a younger U-Pb zircon age of 1168 ± 2 Ma (A. Råheim & T.E. Krogh, unpubl. data). Although these intrusions have had a profound thermal-deformational contact effect on the country rocks the important feature is that they all clearly intrude the regionally developed gneiss-migmatite structures, and yet are unmigmatized themselves. The <u>regional</u> gneiss-forming event is thus constrained to be pre- ≈ 1250 Ma. All of these rocks, in certain places, have yielded ≈ 1060 Ma Rb-Sr whole rock dates, which can usually be related to low grade mineral alteration (13, 15). For example Vegårdshei gneiss sets V38, 39, 42 (Figure 1) from 10 m wide sub-zones within the same continuously exposed 400 m wide section as the 1276 ± 105 Ma-old set V41, but with clearly sericitized feldspar gave dates (all with MSWD < 2.5) of 1009 ± 81 Ma, 1103 ± 69 Ma and 1129 ± 89 Ma (Figure 4). Similarly, a previously unpublished Rb-Sr whole rock age for the Hovde charnockite (H45, Figure 1) yielded 1058 ± 80 Ma (I.R. = 0.709 ± 1, MSWD = 1.63, n= 10), significantly (2σ) younger than the U-Pb zircon age. In these rocks the resetting at ≈ 1060 Ma also occurred at low grade (greenschist or lower) and was related to the passage of hydrous fluids (related to the Arendal granite sheets ?) particularly along shear zones (13).

4. CONCLUSIONS

To paraphrase Bell (25), who was pleading for the "real" Grenville Orogeny in North America to "please stand up", there are two major questions to which we must address ourselves - what do we mean by the term Sveconorwegian (Grenville) Orogeny and how many events have affected the southern Scandinavian Province ? (26). If one agrees with Bell (25) that the term "orogeny" in its present usage means something more than simple uplift (which migh explain the argon dates), and encompasses the more geologically significant processes of regional high-grade metamorphism, deformation, and intrusive activity, then the results from the Bamble sector are unequivocal - the "Orogeny" occurred some 1.54 Ga ago. This was the time of gross crustal accretion, granulite facies metamorphism in the Arendal region, isoclinal folding and formation of a regional gneissic structure with a NE-SW trend. Later events were certainly not "orogenic" in the same sense as the 1.54 Ga event, but included the intrusion of several large augen gneiss bodies (≈ 1250 Ma), with, in places, the development of pyroxenic assemblages in wide contact aureoles. Penecontemporaneous ductile deformation was restricted to the intrusions and to relatively narrow shear zones. The majority of this activity was pre 1.2 Ga and thus, strictly, also pre-Sveconorwegian (1.2 - 0.9 Ga).

The typical "Sveconorwegian" radiometric ages of 1100 to 1050 Ma in Bamble, often interpreted as dating a metamorphic peak, relate neither to major deformation nor high-grade metamorphism, but rather to a period of retrogression and rehydration effected by relatively low-T fluids associated with discordant minor granite sheets and pegmatites.

5. ACKNOWLEDGEMENTS

This work was undertaken whilst PCS and RCL were in receipt of Research Studentships from NERC and the University of Nottingham respectively. DF and AR were in receipt of NATO Grant No. 1391 for their early studies and DF's work was supported latterly by NERC Grant No. GR/3773. The isotopic determinations were undertaken at the Geologisk Museum, Oslo, and we thank W.L. Griffin for the help and advice he gave RCL and PCS. We particularly thank T. Enger (Oslo), J. Eyett, D. Jones and J. Wilkinson (Nottingham) for their skilled technical assistance. T. Jacobsen typed the manuscript.

REFERENCES

1. Kratz, K.O., Gerling, E.K. and Loback-Zhuchencko, S. 1968, Can. J. Earth Sci. 5, pp 657-660.
2. O'Nions, R.K., Morton, R.D. and Baadsgaard, H. 1969, Nor. Geol. Tidsskr. 49, pp 171-190.
3. Lundqvist, T. 1979, Sver. Geol. Unders. 758, pp 1-87.
4. Magnusson, N.H. 1969, Geol. Fören. Stockholm. Forh. 82, pp 407-432.
5. Barth, T.F.W. and Dons. J.A. 1969, In "Geology of Norway" (O. Holtedahl, Ed.), Nor. Geol. Unders. 208, pp 6-67.
6. Wynne-Edwards, H.R. and Hasan, Z. 1970, Am. J. Sci. 268, pp 289-308.
7. O'Nions, R.K. and Baardsgaard, H. 1971, Contrib. Mineral. Petrol. 34, pp 1-21.
8. O'Nions, R.K. and Heier, K.S. 1972, Nor. Geol. Tidsskr. 52, pp 143-150.
9. Field, D. and Råheim, A. 1981, Precambrian Res. 14, pp 261-275.
10. Field, D. and Råheim, A. 1979, Earth Planet. Sci. Lett. 45, pp 32-44.
11. Field, D. and Råheim, A. 1979, Nature 282, pp 497-498.
12. Field, D. and Råheim, A. 1980, Lithos 13, pp 295-304.
13. Field, D., Lamb, R.C. and Råheim, A. 1981, Terra Cognita 2 (ECOG VII), p 55 (abstr.).
14. Smalley, P.C., Field, D. and Råheim, A. 1984, Terra Cognita Special Issue (ECOG VIII), p 9 (abstr. 85).
15. Smalley, P.C., Field, D. and Råheim, A. 1983, Isotope Geosci. 1, pp 269-282.
16. Milne, K.P. 1981, unpublished PhD thesis, London University.
17. Field, D. and Clough, P.W.L. 1976, Jour. Geol. Soc. Lon. 132, pp 277-288.
18. Clough, P.W.L. and Field, D. 1980, Contrib. Mineral. Petrol. 73, pp 277-286.
19. Smalley, P.C., Field, D. Lamb, R.C. and Clough, P.W.L. 1983, Earth Planet. Sci. Lett. 63, pp 446-458.
20. Cooper, D.C. and Field, D. 1977, Earth Planet. Sci. Lett. 35, pp 105-115.

21. Field, D., Drury, S.A. and Cooper, D.C. 1980, Lithos 13, pp 281-289.
22. Smalley, P.C. and Field, D., this volume.
23. Falkum, T. and Petersen, J.S. 1980, Geol. Rundschau 69, pp 622-647.
24. Chessex, R., cited in Maijer, C. 1984, Excursion Guide for the south Norway Geological Excursion, N.A.T.O. A.S.I., Moi, Norway.
25. Bell, K. 1981, Nature 290, pp 89-90.
26. Field, D., and Råheim, A. 1983, Precambrian Res. 22, pp 157-161.

A PRELIMINARY STUDY OF REE ELEMENTS AND FLUID INCLUSIONS IN THE HOMME GRANITE, FLEKKEFJORD, SOUTH NORWAY

T. Falkum[1], J. Konnerup-Madsen[2] and J. Rose-Hansen[2]

(1) Geologisk Institut, Aarhus Universitet, 8000 Aarhus C, Denmark
(2) Institut for Petrologi, Øster Voldgade 10, 1350 Copenhagen, Denmark

ABSTRACT. The results of a preliminary study of REE evolution and fluid inclusions in the Homme biotite granite are presented. Both sets of data are tentatively related to one of two mechanisms: (1) assimilation of felsic granulite facies gneisses, and/or (2) formation of the Homme granite by fractionation from a charnockitic source. The REE pattern does not appear to be controlled by mineral fractionation.

INTRODUCTION

A model for the emplacement of the 990 Ma Homme biotite granite, based on field work, major elements and radioelements, has been presented by Falkum (1976a,1976b) and Falkum & Rose-Hansen (1978). According to these studies the Homme granite was emplaced by upward movement along easterly dipping foliation planes in the country rock gneisses, and subsequently spreading out towards the south, tilting the country rocks. The earliest parts of the granite occur along the western and northeastern contacts while the most differentiated parts are towards the southeast (Fig. 1).

REE STUDY

The Homme granite is high in radioelements compared to the surrounding country rock gneisses and other intrusive rocks in the southern part of Norway. High contents of radioelements correlate with high contents of K and Si in the granite.
The REE in the Homme granite show a general decrease in total REE and a decreasing Eu-anomaly with presumed increasing differentiation of the granite (Fig. 2). Samples which, on the basis of their major element composition, are considered to contain high contents of assimilated country rocks have REE patterns similar to the surrounding granitic gneisses and the Kvinesdal granitic gneisses. Granitic layers in the migmatitic banded gneisses also have weak negative or weak positive Eu-anomalies.

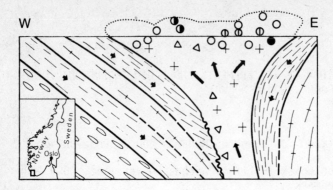

Figure 1. Schematic cross-section showing the present mushroom-shaped form of the Homme granite. Arrows indicate flow directions, triangles the presence of country rock xenoliths in the granite. Locations of samples indicated by circles. The samples used in the fluid inclusion study have been distinguished according to types of fluid inclusions present in addition to aqueous inclusions: ◐ only aqueous inclusions, ◐ plus mixed type aqueous-CO_2 inclusions, ● plus CO_2 inclusions.

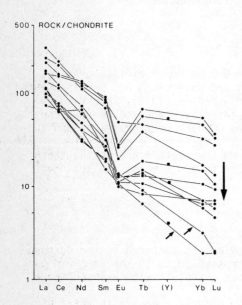

Figure 2. Chondrite-normalized REE patterns in the Homme granite. Vertical arrow indicates general evolution of presumed differentiation. Samples marked by ↗ depleted in LREE either due to assimilation of granitic country rock or to fractionation of the Homme granite from a charnockitic source. Analysis of REE performed by means of Instrumental Neutron Activation except (Y), which was analyzed by XRF.

A depletion in LREE similar to that observed in the Homme granite has been described by Miller & Mittlefeldt (1982) from plutons of the Old Woman Piute Range, where the depletion of LREE was considered to be due to uptake of LREE by allanite and monazite. A similar mechanism was tested for the Homme granite. Microprobe analyses of Ce, La and Nd in zircon, biotite and magnetite from both the Homme granite and the surrounding gneisses revealed levels below detection limit (0.2 wt %) and only titanite, apatite and allanite (all of which are present in only very small quantities) in the Homme granite have higher values (titanite: 0.5 wt % Ce, 0.8 wt % Nd - apatite: 0.2 wt % Ce, 0.4 wt % Nd). Monazite was looked for both microscopically and by microprobe but was not observed. Therefore a similar mechanism to that proposed by Miller & Mittlefeldt (1982) cannot be invoked to explain the REE pattern in the Homme granite.

FLUID INCLUSION STUDY

Examination of fluid inclusions in igneous matrix quartz from the Homme granite revealed the presence of three main types of inclusions: (1) essentially pure CO_2 inclusions, (2) moderately saline aqueous inclusions, and (3) mixed type aqueous-CO_2 inclusions. Most of the inclusions, irrespective of type, occur along more or less well healed fractures in the quartz grains and are considered to reflect fluids circulating through these rocks at subsolidus stages during uplift and cooling. A minor number of the CO_2 inclusions, however, occurs isolated and may represent earlier entrapped fluids. The results of microthermometry on the three types of inclusions are given in Fig. 3.

The CO_2 inclusions have temperatures of final melting of CO_2-solid at -56.6 °C (+/- 0.2 °C), indicating the composition to be very pure CO_2. From the temperatures of observed CO_2-homogenization, densities of about 0.9 g/cm³ can be inferred. Such densities are similar to those previously obtained on early CO_2 inclusions in charnockites from the Farsund igneous complex (Konnerup-Madsen, 1979). If a near-solidus origin for these inclusions is assumed conditions of entrapment at about 4 kb and 600 °C are indicated.

Aqueous inclusions are by far the most abundant type in all examined samples and have salinities between 5 and 21 equivalent wt % NaCl. Temperatures of first observable melting significantly below the eutectic in the H_2O-NaCl system, however, indicate the presence of other cations such as K^+, Ca^{2+} and/or Mg^{2+}. Homogenization temperatures vary between +150 °C and +315 °C. Although homogenizations into both liquid and vapour were observed, indications suggestive of boiling of the aqueous fluid are scarce. Densities of the aqueous fluids corresponding to the observed salinities and homogenization temperatures vary from 1.1 to 0.8 g/cm³.

The mixed type aqueous-CO_2 inclusions invariably occur in complex intersections of fracture planes and most probably reflect mixing of earlier CO_2 fluids and late aqueous fluids. The mixed type inclusions show a wide range in CO_2/H_2O volume ratios. All show homogenization of the CO_2 liquid and vapour phase around +28 °C, with homogenization into liquid. The salinity of the aqueous liquid, deduced from invariant melting temperatures of the clathrate hydrate, is about 11.3 equivalent wt % NaCl and hence within the range of salinities obtained on the aqueous fluid inclusions.

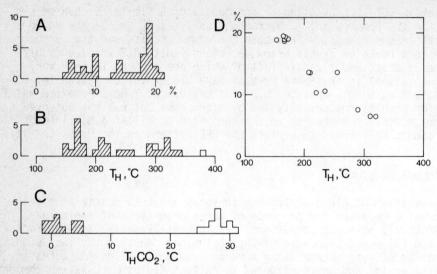

Figure 3. Microthermometry results on fluid inclusions in igneous matrix quartz from the Homme granite. (A) Salinity of aqueous inclusions (given as equivalent wt % NaCl). (B) Temperatures of homogenization of aqueous inclusions. Ruled and blank areas indicate homogenization into the liquid and vapour phase, respectively. (C) Temperatures of homogenization of the CO_2 liquid and vapour phases in CO_2-bearing inclusions. Ruled and blank areas indicate temperatures of homogenization for CO_2 and aqueous-CO_2 inclusions, respectively. All homogenizations occurred into the liquid phase. (D) Salinity versus liquid homogenization temperature for aqueous inclusions.

CONCLUSIONS

The results of this study indicate that the observed REE pattern in the Homme biotite granite may possibly be related to one of two mechanisms, (1) assimilation of felsic granulite facies gneisses, and/or (2) formation of the Homme granite by fractionation from a charnockitic source similar to that of the Kleivan granite (Pedersen, 1980). The presence of high-density CO_2 inclusions in the igneous matrix quartz is in accordance with both hypotheses (e.g. Touret, 1974; Konnerup-Madsen, 1977). More detailed work would, however, be needed for a further evaluation of the actual influence of the two mechanisms.

ACKNOWLEDGEMENTS

The authors are grateful to Henning Sørensen for his stimulating interest during this study. R. Gwozdz performed the NAA, J. Rønsbo the microprobe, and J. Bailey the XRF analyses. The financial support provided by a Niels Bohr Fellowship (to JK-M) during completion of this work is gratefully acknowledged.

REFERENCES

Falkum, T., 1976a. Some aspects of the geochemistry and petrology of the Precambrian Homme granite in the Flekkefjord area, southern Norway. Geol. Fören. Stockh. Förh., 98, 133-144.

Falkum, T., 1976b. The structural geology of the Precambrian Homme granite and the enveloping banded gneisses in the Flekkefjord area, southern Norway. Norges Geol. Unders, 323, 79-101.

Falkum, T., and Rose-Hansen, J., 1978. The application of radioelement studies in solving petrological problems of the Precambrian Homme granite in the Flekkefjord area, south Norway. Chem. Geol., 23, 73-86.

Konnerup-Madsen, J., 1977. Composition and microthermometry of fluid inclusions in the Kleivan granite, south Norway. Am. J. Sci., 277, 673-696.

Konnerup-Madsen, J., 1979. Fluid inclusions in quartz from deep-seated granitic intrusions, south Norway. Lithos, 12, 13-23.

Miller, C.F., and Mittlefeldt, D.W., 1982. Depletion of light rare-earth elements in felsic magmas. Geology, 10/3, 129-133.

Petersen, J.S., 1980. Rare-earth element fractionation and petrogenetic modelling in charnockitic rocks, southwest Norway. Contrib. Mineral. Petrol., 73, 161-172.

Touret, J., 1974. Facies granulite et fluides carboniques. Ann. Soc. Geol. Belg. Bull., vol. P. Michot, 267-287.

SUBJECT INDEX

A

accretionary wedges, 16
Adirondacks (see also anorthosite massifs), 28, 29, 77,78,
  81, 85-87, 95-98, 123, 127, 175-215
    geological evolution, 175-215
    isotopic age, 185, 186, 188
    lithology, 175-185
    metamorphism, 196-201, 217-236
    mineral deposits, 186-188
    seismic reflections, 28
    structural geology, 188-196, 233
akermanite, 217-236
Al-silicates, 79, 84, 503
albitites, 286
alkali-lime index, 554, 555
alkaline igneous rocks, 296-300, 348-351, 383, 560
    monzonite-syenite, 560
allochthon, 163-174, 336
Åmål formation, 387
    mega-units, 252
    supracrustals, 347, 413
amphibole, 107
    brown, in granulite facies, 480
    stability, 93, 94, 96
amphibolite facies, 75-104
    Bohuslän, 362
    Grenville Province, 138-140, 144, 145, 155, 167, 196-201
    Kongsberg-Bamble, 259-290, 424, 500, 567-578
    Rogaland-Vest Agder, 424, 478, 500
    West Uusima, 369-380
anatexis, 10, 12, 16, 278, 283, 460
andalusite, 478, 503
Andes, 552
anorogenic granite (syngenetic with anorthosite), 40
anorthosite, 4, 19, 39-60, 137, 140, 142, 155, 217-236, 452, 560
    archaean, 43
    formation of, 16, 50
    shallow intrusion, 217-236
    suite, 454, 464, 465, 467
anorthosite massifs (Proterozoic), 39-74
    Aana Sira, 48, 417, 450, 453, 454, 463, 508
    Adirondack (Marcy), 46, 49, 62-64, 68, 85-88, 175-216, 217-236
    age, 40
    associated rocks, 42, 54, 56, 61-63, 68
    depth of intrusion, 42, 51
    Egersund-Ogna, 48, 56, 417, 449-476, 478
    geochemistry, 44-50, 449-476
    geotectonic settings, 43, 44, 62, 63, 233

Haaland-Helleren, 417, 419, 450, 453, 463, 466
Harp Lake, 45, 46, 64, 68
Hidra (leuconoritic), 49, 417, 421, 459
isotopes, 61-74
Kiglapait, 62, 64
Laramie, 26-28, 34, 49
Mealy Mountains, 45, 46, 49, 64, 68
mineralogy, 44-47, 54
Nain Complex, 61-69, 88-90, 506
North-Haaland Complex, 417
parent magmas and their generation, 42, 43, 50-54
polybaric crystallization, 42, 51-53, 217-236
Rogaland, 49, 417, 449-476
role of fluids, 217-236
Sr isotopes, 42, 47-50
trace elements, 42, 47-49, 56
anthophyllite , see orthoamphibole
apatite, 54, 454, 455
    fission-track ages, 391
Aphebian, 154
^{40}Ar-^{39}Ar, 153, 156
arc maturity, 552
Archaean, 4, 10, 138, 142, 143, 237-246, 381
    anorthosite, 43
    continental lithosphere, 10
    craton of South Greenland, 237, 238, 240, 241, 336, 337
    early rocks in Greenland, 237, 242
    high-grade terrains, 4
    plate tectonics, 6
    'proto-continent', 9, 253, 256
    thermal dynamical state, 9
    tonalite, 9, 557, 563
    upper mantle, 9
Arendal, see Bamble Sector
Arendal metabasite, 554
arendalite, 540
assemblage
    amphibole - garnet, 495
    amphibole - quartz, 477, 480-482
    biotite - quartz, 478
    clinopyroxene - garnet - wollastonite, 222
    cordierite - garnet, 372, 482
    cordierite - K-feldspar, 372
    cordierite - spinel/magnetite, 477, 482, 483
    garnet - biotite - quartz - sillimanite, 477, 484-487
    garnet - sillimanite, 477, 482-484
    muscovite - quartz, 372
    muscovite - kyanite - biotite - garnet, 167
    orthopyroxene - cordierite - spinel/magnetite, 477, 488-490
    orthopyroxene - sapphirine, 156, 489, 495
    osumilite - orthopyroxene - spinel/magnetite, 477, 484-487

osumilite - sillimanite - spinel, 485
plagioclase - (opx - spinel/magnetite), 491-493
quartz - spinel/magnetite, 477, 482, 484
sillimanite - biotite - quartz, 482
sillimanite - biotite - K-feldspar, 372
Atikonak Lake area, 168
augen gneisses
    Feda, 317, 417, 427
    Gjerstad, 273, 282, 553, 561, 575
    Hovdefjell, 273, 560, 569, 575
    Sirdal, 477, 479
    Tovdal-Mykland, 551, 557, 558, 561
    Ubergsmoen, 268, 273, 282, 424, 560, 575
    Vegårdshei, 575
autochthon, 163-174

B

back-veining, 283
Baltic Shield, 247-258, 333, 336, 337, 381, 383, 385
Bamble
    Sector, 76, 78, 97, 259-290, 386, 388, 396, 424, 499-516, 517-555, 552, 553, 567
    shear zone, 310, 314, 318
Bandak Group, see Telemark supracrustals
banded gneisses, 311, 312, 315
"   migmatites, 311, 312, 315, 477
basalt-eclogite transition, 6, 8
basaltic crust, 3-20
    composition, 466, 467
    generation of, 8
basic intrusions, 264, 266, 272-274, 279-283
binary vapor line, 227
biotite, stability, 93, 94, 96
Bjerkreim-Sokndal Lopolith, 48, 390, 417, 419, 421, 450, 458, 478, 479, 488, 492, 495
Blomskog belt, 413
Bohuslän, Proterozoic evolution, 345-357, 359-367
Borden Basin (Baffin Island), 336, 339
Botnavatn Massif, 390, 417, 478
Bouguer anomalies, see gravity anomalies
brines (lithological control, origin), 526
brittle deformation, 143

C

calc-alkaline suite, 155, 350
calc-silicate rocks, see skarns
calcium solubility, 110

Caledonian
    Front, 479, 482, 488, 492, 495
    metamorphism, 384, 492, 502
Canadian Shield, 247-258, 337
carbon isotopes, 538, 542
carbonates, 517, 524, 542
carbonatite, 517, 542, 543
Cascade Slide Xenolith, 217-236
Central Gneiss Belt, 135, 138, 141, 142, 144, 160
Central Highlands (Adirondacks), 175-216
Central Metasedimentary Belt, 135, 138, 140, 141, 144, 203
CH_4, 517-550
charnockite, 155, 267-273, 385, 478-480, 508, 568
    Farsund, 417, 421, 450, 459, 460
    Hovde, 575, 576
charnockitic-alkaline complex, 560
charnockitic augen gneiss, 273
    "   dykes, 459
charnockitic-enderbitic rocks, Bamble Sector, 268, 278, 279,
    281, 287, 524-527, 540, 543, 544, 568, 575
chemical resetting, 500, 506, 507
chlorine, 105-107, 112, 115
chloritisation, 573
Churchill Province, 163
clinopyroxene, 222
CO_2, 517-550
    density, 375
    fugacity, 199, 200
    -H_2O-ratio, 105-118
    inclusions, see fluid inclusions
    -streaming, 97, 119-132, 217-236, 370, 376, 377
COCORP, see seismic reflections
composite isograd, 502
contact metamorphism, 217-236, 390
contamination, 465, 466
    with old retarded lead, 237-246
continental
    accretion, 15
    arcs,
    collision, 17, 329, 544
    rift, 292
cooling of the crust, 386
copper, 328
cordierite, 95, 155, 156, 196, 197, 264, 269, 482, 526, 529,
    531, 537
cordilleran environment, 306, 551-566
corona, 222, 269, 283
coronitic metagabbro, see hyperite
correlation Labrador-Sweden, 247-258
Cr content in opx, 465

crustal blocks, 386
" contamination, 43, 47, 50, 53, 61-74
" reflection profiling, see seismic reflections
" stretching, 10
" thickening, 4, 12, 16, 144, 145
" thickness, 201
" thinning, 10, 16
cryptic layering, 454
crystal mush, 417, 453, 462, 466
cumulates, 454, 458, 467
cuspidine, 224

D

Dal group, 346, 395, 413
Dala
    granites, 394
    Province, 335
    supracrustals, 337
    volcanics, 334-336
Dalsland Boundary Thrust, 386, 413
decarbonation, 217-236
deformation phases, 352, 353
    Grenville Province, 163-174, 190-203
    Southern Norway, 277-285, 315-318, 415-417
    West Uusimaa, 371, 373
depleted mantle, 552, 557, 563
    peridotite, 6, 8, 13
depletion of Rb, 571
destructive plate margin, 292, 296, 305
Diana complex, 193, 198, 223
diapir, 183, 199, 453, 462-464, 470
dike swarms, 16
dolerite dykes, 241
    Breven-Hällefors type, 385
domical uplift, 234
double crustal thickness, 201, 544
Drivheia gneiss, 558
ductile deformation, 188, 194, 201, 203
" shearing, 133-150, 279, 283

E

Eia-Rekefjord jotunitic intrusion, 417, 421
enderbitic gneisses, see also charnockitic-enderbitic
  rocks, 271
epeirogenesis, see uplift
epidote-amphibolite facies, see greenschist facies
equilibrium domains, 499, 509
Europium-anomaly, see REE

Evje area, 415
    -Iveland intrusions, 283

F

fahlbånds, 264
fault (-zone)
    Alpine Fault New Zealand, 339
    Carthage-Colton Mylonite Zone, 176-179, 190-194, 196-198, 202, 203
    Central Värmland Mylonite Zone, 346
    Chibougimou-Gatinau Line, 193
    Great Breccia, see friction breccia
    Kongsberg-Bamble (several), 259-290
    Mandal-Ustaoset Line, 323-332
    Oslofjord fault, 335
    Porsgrunn-Kristiansand fault, shear zone (= Kristiansand-Bang Shear Zone), see friction breccia
Faurefjell metasediments, 336, 477, 506, 510
Fe-Ti-oxide, 126
    ore bodies, 454
ferrodiorite, 42, 46, 47, 54
Fen-complex, 542, 543
fibrolite in garnet, 483
filter-press mechanism, 453, 464, 466
Fjordzone, deformation, 285
Flinton Group, 134, 136
flooding of CO_2, see CO_2-streaming
fluid-absent metamorphism, 119-132, 219, 225, 227-229
fluid inclusions, 84, 91, 94, 95, 106, 121, 129, 279, 374, 376, 377, 517-550, 581
fluids, see also fluid inclusions
    acidity, 111, 114
    aqueous, see H_2O
    in deep shear zones, 105-118
    infiltration, 217-236
    in metamorphism, 75-104, 119-132, 196, 199, 517-550
fold belts, see orogeny
    generation of, 5
fold interference patterns, 269
Folded Basic Intrusions, 478
foliated anorthosite, 452, 453
    "   inclusion, 452, 454
    "   leuconorite, 453
foreland, 166
fractionation
    polybaric, 42, 51, 53
    K + Rb, 570
    LILE, 568

SUBJECT INDEX

friction breccia, 260-267, 283-286, 386, 552, 553
Front, see orogeny
Frontenac Arch, 176, 177

G

gabbro+diorite-tonalite bodies, 265, 279
gabbroid coronites, 269
garnet, 188, 222
    decomposition, 488-490
Garsaknatt Outlier, 417, 421
geobarometry, 75-104, 199, 219-221, 372-376
    anorthite-enstatite-garnet-quartz, 82, 85, 87, 91, 92
    anorthite-fayalite-garnet, 83, 87
    anorthite-ferrosilite-garnet-quartz, 83, 87
    "charnockite", 82, 83
    CO_2 equation, 84
    cordierite-garnet-sillimanite-quartz, 83, 91, 92
    fluid inclusions, 94, 95, 527, 528, 534
    garnet-cordierite, 88
    garnet-cordierite-opx-quartz, 89
    garnet-cpx-plagioclase-quartz, 91
    garnet + hypersthene, 374
    garnet-sillimanite-plagioclase-quartz, 89, 91, 92
    Ghent barometer, 82
    olivine-orthopyroxene-quartz, 82, 88
    regional results, see geothermometry
geochronology, 136, 140, 143, 151-162, 165, 185, 186, 189, 292, 301-306, 353, 381-448, 567-578
geodynamical constraints, 3-20
geothermal gradients, 4, 112-114, 377, 563
geothermometry, 75-104
    Adirondacks, 85-88, 177, 198, 219-221
    Bamble and Rogaland, 499-516, 517-550
    cordierite-garnet, 88
    cordierite-garnet-sill-quartz, 83, 91, 92
    fluid inclusions, 527, 528, 534
    garnet-biotite, 78, 80, 374-376
    garnet-pyroxene, 78, 80, 90
    ilmenite-ulvøspinel, 499, 509, 511, 513
    Karnataka, South India, 90, 91
    magnetite-ilmenite, 79, 81, 86
    methods, 78
    Nain Complex, 88, 89
    petrogenetic grid constraints, 79
    plagioclase-amphibole, 374-376
    pyroxene-amphibole, 374
    Rogaland-Agder, 499-515
    two feldspars, 79, 81, 86
    two pyroxenes, 78, 81, 90, 374-376, 509, 510, 513, 534

West Uusimaa, 372, 374-378
Williyma Complex, N.S. Wales Australia,
Gjerstad-Morkeheia Complex, see Morkheia Complex
Gloppurdi pyroxene monzonite massif, 390, 417, 478, 480, 482
gneiss
    hornblende-granitic, 184
    mangeritic and quartz-syenitic, 184
Göta älv shear zone, 346
granite, 558
    Åmål, 384
    anorogenic, 40, 281
    Bohus-Iddefjord-Flå, 349, 350, 413
    continental rift setting, 41
    Dala, 394
    diapirism, 10, 13
    Grimstad, 273
    Hedal, 285
    Herefoss, 268, 273
    Holum, 313, 317
    Homme, 417, 579
    Killarney, 143
    Kleivan, 450, 458, 459, 461
    Kongsbergian, 272, 277-279
    Lane-type, 385
    late-tectonic, 281, 313, 396
    Levang, 272, 281
    post-tectonic, 285, 313, 324, 396
    Rapakivi, 378, 560, 561
    Sjelset pyroxene-, 478, 479, 482, 508
    Småland-Värmland, see under Småland
    Telemark, 265, 272, 283, 285
    Trans-Labrador batholith, 155, 157, 166, 248-250
    Trysil, 394
granitic gneisses, 311, 312, 315
granulated, 451, 452
granulite facies, 10, 75-104, 106, 109, 111, 119
    Adirondacks, 196-201, 217-236
    Grenville Province, 139, 140, 144, 145, 155, 196-201
    Kongsberg Bamble, 259-290, 391, 499-549, 567-578
    low-pressure, 369-380, 477-498, 499-515
    Rogaland-Vest Agder, 386, 424, 477-497, 499-515
    tectonic mechanisms, 97
    transition to amphibolite facies, 75-104, 369-380, 478, 499-515, 517-550
granulites, see granulite facies
graphite, 119-132, 264, 269, 529, 543
Gräsmark Formation, 335, 336
gravity anomalies, 178, 183, 184, 249, 251, 255, 378
Greenland, 105-118, 237-246, 248, 249, 253, 337
greenschist facies, 155, 167, 271, 285, 293-296, 301-305, 391, 416, 421
greenstone belts, Archaean, 4, 5

Grenville Front, 63-69, 133-150, 203, 319, 384, 386
" Low, 171
Grenville Province, 39-60, 61-74, 151-216, 218, 247-257,
  397, 411, 461
    tectonic framework, 133-149, 163-174, 190-203
Grenville supergroup, 133-150
grossular, 225
Groswater Bay Terrane, 164, 248, 251, 252

H

H_2O, see also fluids, 396, 517-550, 557
    fugacity, 76, 77, 93, 96, 188, 199, 200
Hakhovik Province, 248, 249, 253
halite, 517-550
Hardanger sequences, 337
harkerite, 224
Hastings Lowlands, 134, 136, 140
heat conduction, 11
heat flow, 4, 370
  " from the mantle, 3-20
homogenization temperature, 528-531, 533, 535
hornblende, see amphibole
    porhyroblast, 294, 295, 301
'hot spot', 18, 378
Hovdefjell, 273
Hudsonian Orogeny, 157, 237
Huronian, 135, 138, 142, 143
hybrid rocks, see crustal contamination
hyperfeldspathic liquids, 469
hyperite, 251, 252, 269, 272, 283, 439, 543
    dolerites, 385
hypersthene, see orthopyroxene
    line, 478-480, 482, 492
    porphyroblasts, 373

I

Iapetus Ocean, 255, 256
immiscibility CO_2-H_2O, 522, 536, 543
inclusions, 453, 460, 464
incompatible element plots, see spidergrams
infiltration of CO_2, see CO_2-streaming
intracontinental tectonics, 3
isobaric cooling, 219, 512, 517, 534, 535, 537, 539
  " heating, 376
isochores, 377, 527, 528, 533-538
isochrons, see geochronology, Rb-Sr, Sm-Nd

isograds, 172, 274, 287, 477-498
    (amphibole + quartz)-out, 481, 482
    biotite-in, 502
    Caledonian green biotite, 479, 492
    cordierite-in, 197, 500, 501
    garnet decomposition, 477, 479, 488-490
    muscovite-out, 197, 500
    orthopyroxene-in, 197, 271-274, 372, 373, 478-480, 482,
        492, 500, 502, 513, 519, 524, 525, 568
    orthopyroxene-in, mafic, 478, 519, 524, 525, 568
    osumilite-in, 479, 482, 487, 489, 490, 500, 502, 506,
        507, 511, 513
    pigeonite-in, 479, 492, 494, 500, 502, 511
isotherms, 177, 197, 199, 221
isotopes, see also at elements, 105-118, 153, 467-469
    unequilibrated, 398
isotopic ages, see geochronology
    "   enriched mantle, 66, 68

J

jotunite, 42, 46, 47, 54, 454, 456, 470

K

K_D spinel-cordierite, 507, 509
Kappebo formation, 413
Karelian belt, 370
Ketilidian mobile belt, 238-240
kinematic indicators, 139, 140, 142, 145
Kisko shear zone, 372
klippe, 156
Knob Lake Group, 166
Kola nucleus, 336, 337
komatiites, 3-20
Kongsberg-Bamble mobile belt, -Sector, see also Bamble Sector,
    259-290, 314, 323
Kongsberg Sector, 259-290, 567
Kornepurine, 503
kyanite, 167, 221, 251, 285, 503, 537, 538

L

Labrador, 137, 144, 151-161, 163-174, 466
Lac Joseph allochthon, 159, 167
Lake George group, 176, 178, 180, 182-185, 193
Lake Melville terrane, 248, 252
Large Ion Lithophile, see LILE
Laurentian Shield, 336

layered basic intrusions, 46, 47
leuconoritic anatexis, 453, 464
Levang granitic dome, 268, 281, 287
lherzolite, 8, 13
LILE, 96, 271, 296, 524, 539, 554-556, 568
linear
    fabric, 193
    tectonic zones, 386
lineation, 138, 139, 142
lithospheric buoyancy, 7, 13
    "   doubling, 5, 14-19
    "   shifting, 18
lithostratigraphy Grenville Province, 134
lithotectonic blocks, 154
lithotectonic domains, 133-150
low-angle ductile fault, 192
" -grade, hydrous alteration, 396
" -pressure metamorphism, 477-498
" -pressure thermal dome, 370, 377
" water pressures, 372, 374
lower crust, 463, 468
Lyngdal granodiorite, 417, 421, 450, 461

M

mafic intrusions, 378
magmatic differentiation, 525, 539
magnetic anomaly, 183, 202, 326-328
magnetite-ilmenite deposits, 186
manganese distribution, 511
mangerite, 184, 454, 455, 457, 465
mantle convection, 4
mantled dome, 453, 462
marble, 133-135, 137, 141, 142, 270, 274, 477
meionite, 372
mesoperthite, 455
meta-acidic volcanics, 526
meta-anorthosite, 183
metabasalts, 291-308, 438, 477, 478
metabasite, 269, 272, 279, 291-307, 535, 540, 553
    dyke, 442
metadolomite, 477
metagraywackes,
metamorphic envelope, 415
    "   pressure, see geobarometry
    "   stages, 415, 422, 499-516
    "   temperatures, see geothermometry
    "   volatilization, 217-236
metamorphic zones, zoning (see also assemblage, isograd)
    Bamble Sector, 271-274, 499-516, 517-550, 567-578

Labrador, 163-174
Rogaland-Vest Agder, 477-498, 499-516
West Uusima, 369-380
metamorphism, see amphibolite facies, granulite facies, greenschist facies
fluid-absent, 217-236
fluid regimes in metamorphism, 93-95, 111-116, 119-131, 199, 200, 224-229, 374-377, 517-550
HT/LP granulite facies, 369-380, 477-498, 499-516
metapelites, see also supracrustals, metasediments, 163-174, 372-376, 416, 477-515
metaquartzite, 442
metarhyolite, 291-308
metasediments, see also supracrustals, metapelites, 155, 477, 517, 526, 535, 574
metasomatism, 222, 563
metastability, 531, 534
metavolcanics, 438, 439
meteoric water, 233
Mg/Fe ratio, see also geothermometry, 508, 560
migmatite,
    banded, 477, 478
    garnetiferous, 336
mineral deposits, see ore deposits
"   homogeneity, 110
"   textures, 499, 503
mobile belt, see orogeny
Modum Complex, 260, 263, 264, 269, 277, 281, 282
Moho, 34, 201, 463, 544
molybdenum, 328, 329
monazite ages, 416, 419
monticellite, 217-236
MORB, 184, 296, 299, 300
Morkheia Complex, 273, 275, 281, 282, 560
μ-values, 237-246
multi-stage leadloss, 399
Mykland complex, 560
mylonite zone, see fault (-zone)
mylonites, 137, 139, 141-143, 267, 277, 279, 282, 372

N

Nagssugtoqidian (Huronian) Orogeny, 237-239, 242, 243
Nain Province, see anorthosite massifs, Nain Complex
natural spiking, 398
Nb-depletion, 557
Nd, see Sm-Nd
negative buoyancy, 14
Neohelikian, 155

netveining, 362
nickel, 328
Nicolaysen diagram, 304, 397
nitrogen, 105-118, 517-550
nodules, 268, 269
Nordre Stromfjord Shear Zone, 109, 112
North-Haaland anorthosito-leuconoritic complex, 417, 419
Northwest Lowlands (Adirondacks), 175-216
N.Y.-Alabama lineament, 202

O

oceanic lithosphere, 4, 15
    subducted, 554
oceanic spreading ridge, 6, 8, 13
olivine, 39-60, 454, 455, 462
    -metagabbro, 185
ophiolite thrust sheets, 18
ore deposits (Adirondacks), 186, 187
Oregon dome, 189, 192
Orijärvi shear zone, 372
Orogeny, 3-20
    Alpine, 17
    Caledonian, 381, 383, 386, 416
    Dalslandian, 383
    Elzevirian, 136, 137
    Gothian, 159, 318, 382, 383, 413
    Grenvillian, 136-138, 143-145, 151-161, 163, 383
    Hudsonian, 166, 172, 237
    Kongsbergian, 160, 260, 286, 318
    Labradorian, 151-161, 164
    Mazatzal, 159
    "Ottawan", 136
    Penokean, 159
    "Proterozoic", 17
    Pyrenean type, 18
    Svecokarelian, 314, 318, 320, 370, 382
    Sveconorwegian, most papers
orthoamphibole-cordierite rocks, 264, 270
orthocumulate, 459
orthogneiss, 155
orthopyroxene, 39-60, 119-132, 155, 156, 270-273, 372, 373
    Al-rich megacrysts, 42, 47, 50, 417, 425, 451, 461, 464
Oslo (-Skagerrak) Graben, 285, 310, 317, 383, 399
Oslofjord shear zone, fault, 262, 335, 339
Ostfold-Marstrand Belt, 359-368

osumilite, 390, 482, 484-487, 503
    properties, 484
Oswegatchie, 176, 180, 184
oxygen fugacity, 76, 95, 126, 543, 559
    isotopes, 184, 187, 200, 221-236, 456

P

P-T boxes, 377, 512, 533, 535, 538
Paleohelikian, 154
paleomagnetism, 315, 336, 385
palingenesis, 417
paragneiss, see metasediments
parautochthon, 163-174
parental magma of anorthosite, 16, 50, 460-470
Parry Sound domain, 139, 141, 144
partial melting, 466, 469, 470
    water pressures, 372, 374
partially reset ages, 398
Pb-isotopes, 237-246, 456
Pb-Pb isochrons, 240, 243, 464
pegmatites, 188, 273, 285, 572, 576
peralkaline, 560
peridotite
    latent heat of melting, 5
    as source for anorthosite, 40
pillow lavas, 348
Piseco Group, 176-194
plagioclase, 39-60, 454, 455, 462
    , exsolved lamellae in pyroxene, 64
    megacrysts, 417, 451
    Schwantke's molecule, 492
planar fabric, 193
plate boundaries (margins), 145, 237, 239, 240, 243, 292,
  296, 306, 544
   "   collision, 311
   "   driving forces, 6, 15
   "   interaction, convergent, 201
   "   production, 14
   "   -tectonics, 3-20, 286, 287, 317-320, 333-344, 425,
        426, 551-566
plutons, see also granite, 136, 309-322
    Grenville, 138-140, 143-145
    late- and post-orogenic granite, see granite, post-tectonic
    South Norway, 265, 272, 277-285, 417, 419
Polar wandering paths, 249
polymetamorphism, 217-236, 399, 499, 503
Post-tectonic granites, see granites
Pregothian
    mega-units, 248, 251, 252
    sequence, 413

SUBJECT INDEX

Pré-Grenvillian orogenic belts, 253
pressure, 220, 221, 462
Presvecokarelian continental nucleus, 383
Presveconorwegian intrusives, 359-368
primordial mantle, 555
Proterozoic plate interaction, 14
     "   supercontinent, 277, 337
Protogine zone, 346, 386
pumpellyite-prehnite facies, 416, 421
pyribolites, 269, 271, 317
pyroxene, see also clinopyroxene, orthopyroxene, 62-64, 68
    -megacrysts, 16
    -syenite suites, see Gloppurdi, Botnavatn Massif

Q

quartz-diopside gneiss, 477
   "   -mangerite, 454
   "   -sillimanite 'nodules', 264
quartzite "Complexes", 260, 287

R

radiogenic heat production, 14
radiometric anomalies (U, Th), 327, 328
Raman microspectrometry, 518, 529, 542
rare earth elements, see REE
Rb, 61-69, 292, 301-304, 399
Rb-Sr, 64, 108-110, 115, 153, 156, 301-306, 351, 364, 381-433,
   456, 459, 559, 562, 571-576
reactivation, 464
reclined folds, 315
Redlich Kwong equation, 94, 120
REE, 47, 54, 456, 460, 465, 466, 470
     Homme granite, 579
     in shear zones, 107-109
regeneration, 260
regional contact metamorphism, 390
remobilization, 394
resetting
     of isochrons, 567
     of U-Pb zircon systems, 391
     of whole rocks systems, 394, 397
'retarded' lead, 239
retrogression, 571, 572, 574, 576
     of granulites, 105-117
ribbon lineations, 190-194, 202, 203

rift formation, 18
" zone, 88, 98, 99
Rinkian mobile belt, 237-239
Rjukan Group, see Telemark supracrustals
Rogaland-Vest Agder, 310-315, 388, 390, 391, 415, 517,531, 534, 538, 539
     intrusive complex, 310, 314, 417
     geothermometry and -barometry, 499-515
     metamorphic zoning, 477-498
     stages of metamorphism, 416, 420, 499-516
     supracrustals, 416, 477-498, 499-516
rotation of Baltic Shield, 247-258, 286, 311, 315, 319, 333-344
     of Bamble segment, 286

S

Salinian block (California), 338
salinity, 524, 527
sample collecting for geochemical study, 397, 573
sapphirine, 155, 156, 503
saussurite, 398
scapolite, 96, 107, 115, 285, 286
scapolitisation, 286
seismic activity, 195
seismic reflections, (COCORP), 21-38
    Adirondack, 28, 29
    arcuate and criss-crossing, 22, 25, 31-33
    Canada, 33
    in crystalline rocks, 22, 23
    of intrusions, 33, 34
    Laramie anorthosite complex, 26, 34
    Michigan and Kansas, 32, 33
    Minnesota, 23-25
    multicyclic, 22-25
    mylonite zones, 23, 24, 26, 28, 33
    Oklahoma and Texas, 30, 31
    out-of-the-plane events, 22
    side-arriving seismic events, 33
    suture zone, 23
    transparant zones, 33
    wide angle reflections, 28, 33, 34
    Wyoming, 25, 26, 29
Seljord Group, see Telemark supracrustals
sericitisation, 573
Setesdal gneiss, 312
Sirdal-Ørsdal area, 416, 477-498, 499-516
Shabogamo Intrusive Suite, 167
shear zones, 105-117, 372, 373, 378, 568, 571, 576
    retrogression along -, 105-117
sheath folds, 193, 194, 203
silica, 53

sillimanite, 155, 156, 264, 269, 277, 478, 482, 484, 487, 503
    Fe^{3+} in-, 484, 487
skarns, 65, 188, 200, 217-236, 270, 477, 517, 527
Sm-Nd isotopes, 61-74, 220, 292, 301-306, 435-448, 470
Småland-Värmland (granitoid) belt, 159, 248-250, 334, 383, 384, 385, 394, 395, 399
Sokna, 262
solid inclusions, 520, 522, 524, 528, 540, 541, 543
solubility alkali metal ions, 111
South-Haaland basement complex, 417
South-western Gneiss Region, 383-387
Southern India (Karnataka), granulite-facies transition zone, 77, 78, 90-92, 95, 97
Southern Norway (see also Rogaland-Vest Agder, Kongsberg, Bamble, Telemark)
    ages, 309-322, 381-410, 411-434, 435-448
    geotectonic setting, plate-tectonic evolution, 306, 309-322, 333-344, 411-434
    lithostructural units, 259-290, 309-322, 323-332, 411-434
Southern Province, 135, 138
Sparagmite Basin, 339
Spidergrams, 291, 296
spinel, 52, 482, 499-516
Sr, see Rb-Sr, trace elements
Sr content in plagioclase, 47, 49, 56, 451
Stora-Le-Marstrand supracrustal formation, 335, 347, 384, 413
strain markers, 139, 140
stromatolite, 176, 180
strontium isotopes, 42, 47-50, 456
    isotopic homogenization, 397
subcrustal accretion, 4
subduction-related enrichment, 554
    "   -related hydrous fluids, 554
    "   zone, 315, 334
subsidence, 16
sulfur, 105-118
sulphidic layers, 277
superior Province, 135, 138, 143, 163
supracrustals
    Åmål-, 384
    Bamble, 259-290, 499-516, 517-550
    Dal Group, 385
    Grenville, 133-149, 154-156, 163-174
    Jotnian, 385, 395
    Rogaland-Agder, 311, 477-515
    Telemark, see Telemark Supracrustals
suture, 145
    cryptic, 145
    Småland,
Svecofennian
    belt, 370

craton, see Baltic Shield
platform, 413
Sveconorwegian,
Front, 259-290, 319
metamorphism, orogeny etc., discussed in most papers
symplectite
biotite - quartz, 480
cordierite - K-feldspar - quartz, 484, 487
cordierite-quartz-kalifeldspar-orthopyroxene, 506
cordierite-spinel, 505
orthopyroxene - cordierite - spinel/magnetite, 488
orthopyroxene-spinel-cordierite, 505
plagioclase - orthopyroxene - spinel/magnetite, 492, 493
synemplacement deformation, 461
system
C-O-H, 111, 119-131, 224-229
C-O-H-N-S, 112, 115
$CaMgSi_2O_6-Mg_2Si_2O_6$, 81
$CaO-MgO-Al_2O_3-SiO_2$, 51-53
$K_2O-MgO-Al_2O_3-SiO_2-H_2O$, 94
$MgSiO_3-FeSiO_3-CaSiO_3$, 81

T

Ta-troughs, 554
talc-tremolite deposits, 188
Telemark basement gneiss, 259-262, 274, 283, 391
" metabasite, 291-307, 551-566
" -Rogaland-Agder block, 340
" Sector, Telemark Sector, 259-265, 272-276, 283-285,
415, 526, 543, 552, 575
Telemark Supracrustals, 291-307, 315-318, 323, 325, 326, 329,
334, 387, 388, 394, 398, 438, 551-566
age, 291-307, 435-448
geochemistry, 296-300, 551-566
lithostratigraphy, 292-294
geotectonic evolution, plate tectonics, 291-307, 315-318,
334, 551-566
temperature, see geothermometry
thermal dome, 369-380
" perturbation, 12
thermometry, see geothermometry
thrust, 137, 139, 142-144
" faults, 163
" nappe, 156
tonalite, 348, 361, 556, 557, 568
" islands, 9
trace element, see also Rb-Sr, Sm-Nd, 61-74, 296-300, 312, 313,
397, 449-476, 551-566
Trans-Labrador batholith, see granite
Tromöy, 270, 271, 524-527, 540, 543, 544, 569, 575

SUBJECT INDEX

Trysil area, 384
Tuddal formation, see Telemark Supracrustals
two-component bulk mixing, 443

U

U-Pb, see zircon age
Ubergsmoen, see augengneisses
underplating of hot mantle blobs, 12
Ungava, 166
uplift, 98, 99, 283, 373, 376, 383, 386, 517, 527, 534, 535, 539

V

Varberg Region, 385
Vegårshei, 273, 281, 572, 576
Vemork formation, 438
Vest-Agder district, see Rogaland-Vest Agder
Vinor intrusions, 264, 283

W

Wabush-Labrador City area, 167
water fugacity, see H_2O
West Uusimaa complex, 369-380
wilkeite, 224
Willsboro skarn belt, 217-236
Willyama Complex, South Australia, 92, 93
Wilson cycle, 17
Wilson Lake allochthon, 168
wollastonite, 187, 188, 200, 217-236, 372
    -clinopyroxene-garnet skarns, 222

Z

zircon
    age, 153, 353, 391, 419, 421, 423, 454, 455
    concordia chords, 416
    U-Pb discordias, 390